Math in Our World: A Quantitative Reasoning Approach

2026 Release

Dave Sobecki

Brian Mercer
Professor of Mathematics
Parkland College

Mc Graw Hill

MATH IN OUR WORLD: A QUANTITATIVE REASONING APPROACH, 2026 RELEASE

Some ancillaries, including electronic and print components, may not be available to customers outside the United States.

This book is printed on acid-free paper.

2 3 4 5 6 LBC 30 29 28 27 26

ISBN 978-1-266-21451-6 (loose-leaf)
MHID 1-266-21451-8 (loose-leaf)
ISBN 978-1-264-74413-8 (instructor)
MHID 1-264-74413-7 (instructor)

Portfolio Manager: *Lauren Whalen*
Product Developer: *Katie Peterson*
Marketing Manager: *Candice Madden*
Content Project Managers: *Melissa Leick*
Manufacturing Project Manager: *Sandy Ludovissy*
Content Licensing Specialist: *Erika Anderson*
Cover Image: *denio109/Shutterstock*
Compositor: *Straive*

All credits appearing on page or at the end of the book are considered to be an extension of the copyright page.

Library of Congress Cataloging-in-Publication Data

Names: Sobecki, Dave author | Mercer, Brian A. author
Title: Math in our world : a quantitative reasoning approach / Dave Sobecki, Brian Mercer, Professor of Mathematics Parkland College.
Description: 2026 release. | New York, NY : McGraw Hill, [2026] | Includes index.
Identifiers: LCCN 2025032798 | ISBN 9781266214516 paperback
Subjects: LCSH: Mathematics--Textbooks | Mathematics--Study and teaching (Higher) | LCGFT: Textbooks
Classification: LCC QA39.3 .B5974 2026
LC record available at https://lccn.loc.gov/2025032798

The Internet addresses listed in the text were accurate at the time of publication. The inclusion of a website does not indicate an endorsement by the authors or McGraw Hill LLC, and McGraw Hill LLC does not guarantee the accuracy of the information presented at these sites.

mheducation.com/highered

About the Authors

Dave Sobecki

Dave was born and raised in Cleveland, and in spite of starting college with a major in creative writing, he somehow ended up with a Doctorate degree in math. Go figure. Dave spent two years at Franklin and Marshall College in Lancaster, Pennsylvania, followed by 22 years at the Hamilton campus of Miami University in southwest Ohio. He recently relocated to North Port, Florida. Dave has won a number of teaching awards in his career, and has written or co-authored either nine or nineteen books, depending on how you count them. He's also working on his first novel. When not teaching or writing, Dave's passions include Ohio State football, Cleveland Indians baseball, heavy metal music, travel, golf, collecting fine art, and most importantly spending time with his wife Cat and over-attached retrievers, Macleod and Tessa.

Courtesy of Dave Sobecki

Brian Mercer

Brian is a tenured professor at Parkland College in Champaign, Illinois, where he has taught developmental and transfer math courses for 27 years. He began writing in 1999 and has currently co-authored eight textbooks, with others in the planning stages. Outside of the classroom and away from the computer, Brian is kept educated, entertained, and ever-busy by his wonderful wife Nikki, their two children, Charlotte (18) and Jake (17), and dog Molly. He is an avid St. Louis Cardinals fan and enjoys playing softball and golf in the summertime with colleagues and friends.

Courtesy of Brian Mercer

Table of Contents

View Stock/age fotostock

Unit 3: PLACE YOUR BETS

James W. Hall

Unit 4: STATISTICALLY SPEAKING

SofikoS/Shutterstock

Unit 5: BUILDING MODELS

2026 RELEASE

MATH IN OUR WORLD

A Quantitative Reasoning Approach

McGraw Hill

SOBECKI | MERCER

John Lund/Blend Images

Colin Anderson/Blend Images

Unit 7: UNCOMMON SENSE

clintscholz/iStock/Getty Images

Unit 8: HOW DO YOU MEASURE UP?

Units 9, 10, and 11 are available online. Though not included in this desk copy, they can be added to custom versions of the text built through Create or accessed in the Instructor Resources area of ALEKS.

Peter Joneleit/Zuma Wire Service/ZUMA Press/Alamy Stock Photo

Unit 9: UP FOR A VOTE

Unit 10: WARNING: GRAPHIC CONTENT

John Lund/Tiffany Schoepp/Blend Images

mirtmirt/Shutterstock

Unit 11: DISCOVERING NEW NUMBERS

A Letter to Students

This book has the term "Quantitative Reasoning" in the title, so we probably ought to begin by describing what that means. But nothing is for free in this course, so before we do that, we're going to ask you to think about an answer to this question: Why are you taking this class? For that matter, why are you in college at all? It would be great if "To get an education" was your answer, or at the very least crossed your mind. It might be even better if you thought "To get smarter." That should be the goal of everyone that's going to school, really.

So can you actually get smarter? OF COURSE YOU CAN. You almost certainly know that you can get stronger and more fit than you currently are by exercising your body. Then why in the world wouldn't you think you can get smarter by exercising your brain? You can, but not if you occupy a chair and expect your professor to just magically fill your head with math. That's not mental exercise. True mental exercise requires challenging yourself to think, to solve problems, to make connections, and to explain your understanding of concepts. Which brings us back to our part of the bargain, a description of what quantitative reasoning means.

At its core, this course is about exercising your brain by challenging you to really LEARN, not just to memorize formulas or mimic step-by-step procedures. When you hear the word "reasoning," it probably brings to mind something like "figuring stuff out." By including the term "Quantitative Reasoning" in the title of this book, we're simply pointing out the fact that we're not going to just ask you to do a bunch of calculations, simplify a bunch of expressions, or solve a bunch of equations. Instead, the focus is going to be practicing your reasoning skills by studying situations where mathematical thinking can help you be smarter and more successful, both as a student and as a citizen of 21st-century planet earth.

Certainly being comfortable with numbers is important in a monetary society. But it goes beyond that: Someone once told us that nobody is going to pay people to solve problems that have already been solved, and that kind of became a framework for this course. We gathered as much information as we could about what employers want college graduates to be able to do these days, and three things consistently jumped out at us: being able to work effectively in groups, being able to solve problems and figure stuff out without always being told what to do, and being able to use technology well. It's those three things that became the key principles of this book.

The Approach: What Truly Makes This Course Different

The thing that makes this a course in quantitative reasoning, rather than a survey of mathematical topics, is the way the material is presented, and more importantly, how we expect you to interact with the material. If you're dead-set on sitting in the back of the room and being talked at for an hour, you'll find out pretty quickly that you're in the wrong place: Active participation is required. You won't find large blocks of text and 8 to 10 solved examples per section in this book, like you're probably used to seeing in other math textbooks. What you will find is a series of activities designed to encourage (and by "encourage" we mean "force") you to take responsibility for your own learning. Discovery learning and productive struggle are the key phrases that describe our philosophy.

Discovery learning is the process of learning new ideas on your own, rather than having them presented to you. That doesn't mean that you're expected to "teach this course to yourself." Your professor is by far the most valuable resource you'll have on this journey. But his or her job is to help you to learn, not to show you exactly what to do. In a very real sense, the goal of this course is to help you become mathematically literate. We can define mathematical literacy as the quantitative and reasoning skills anyone should have to be successful in society. Think about literacy in the reading sense. You can't teach someone to read by just reading to them: At some point you have to guide them through reading on their own, and that's the role of your professor in this course.

Productive struggle is described really well by the exercise analogy from earlier: The best way to really learn something, and to make your brain stronger in the process, is to have to work at it. If you get every question in this book right on the first try, then we've failed as writers, to be honest. Life can be a hard place, especially when you're responsible for your own care and feeding, and we can promise you that you will get NOWHERE in this world if you give up every time you encounter some resistance. In our view, the single most valuable skill you'll practice in this course is GRIT: the ability to overcome obstacles, and keep trying until you reach a goal.

Because working effectively in groups is so important, the book has been designed as a series of in-class activities for you to work through in a group setting. There are two things you need to do in order to maximize your chance for success in this course: Be willing to put in a lot of work, and be willing to learn with and learn from your group mates, as well as to help teach them from time to time. If we were pure evil and decided to kidnap you and dump you out in the wilderness somewhere, you surely would appreciate not being alone, right? As you wander the path toward an adult education, having peers there with you should be comforting, and should be a resource that you take advantage of whenever possible.

Once class is over, there will be a lot of work you're expected to do on your own, in a variety of different settings. If you turn ahead to one of the pages in the book marked "Portfolio," you can see the different categories of homework, and no doubt your professor will talk about them on the first day of class. If you keep thinking back to the exercise analogy, you'll have a much easier time getting motivated to keep up with the homework. Why are you in this class? To get smarter. That will happen if you work hard toward the goal, and are never, ever afraid to think on your own. You were smart enough to make the decision to further your education in college, so you're smart enough to think, learn, and succeed in quantitative reasoning. Set a goal, work toward it, and believe in yourself. And don't be afraid to drop us a line if you have questions about the book, suggestions for improvement, or just want to share your thoughts on how the course is going.

davesobecki@gmail.com
bmercer51@gmail.com

The Content

The backbone of any well-designed math course is content, so let's talk about the content that we've chosen as the vehicle for becoming quantitatively literate.

Unit 1: Mathematical Reasoning and Problem Solving

In our view, the ideal way to begin this course is to explicitly talk about the different kinds of reasoning that humans do, and to define and practice the types of problem-solving skills that will form the backbone of the course. We also encourage students to practice graph interpretation and estimation skills that should be (but are not) possessed by everyone in a technological and monetary-based society.

Unit 2: Consumer Math

Did we mention that we live in a monetary society? Consumer math is kind of the low-hanging fruit of quantitative reasoning content: Everyone who's ever spent their own money on anything has a natural understanding of how financial literacy is important. The focus of this unit isn't "here's how to do financial calculations"; it's "let's understand the concepts and situations that lead to important financial considerations." While we absolutely will study formulas that calculate interest, loan payments, and the like, we certainly will not just provide the formulas, ask students to plug a bunch of numbers in, and call that education. Instead, students are asked to think like consumers, and develop a true understanding of the math that serves us well in navigating the rocky financial waters of modern society.

Unit 3: Probability

Almost everything that we do is governed at least in part by chance and probability, so a book devoted to quantitative reasoning should go beyond simply calculating probabilities and odds: The focus should be on being able to interpret the story told by those calculations. We certainly feel that performing probability calculations is important, as long as the focus is on understanding rather than formulas. But we also believe that it's important to move beyond simple probability calculations, incorporating technology to allow us to study complex outcomes in a complex world.

Unit 4: Statistics

The logical follow-up to a study of probability is continuing the mindset by developing key ideas in the study of basic statistics. The focus is different methods of gathering, organizing, and interpreting data. Anyone can find the average (whatever that means) of a list of numbers, but that's a hollow exercise without the ability to interpret the story told by that statistic. While we don't feel that quantitative reasoning necessarily requires high-level statistical analysis, we do feel that our society will be much the better if everyone has experience with sifting through the mountain of information they're exposed to every day, and separating the gold nuggets from the cow pies, if you will. So a major focus of, and the culmination of the unit, is recognizing ways that statistics can be manipulated to tell a story that has at best a passing resemblance to the truth. While politicians and advertisers may not appreciate these efforts, we think that students interested in getting an education will.

Unit 5: Math Modeling

It's the sad truth that most college math students honestly believe that algebra was developed as a completely abstract field, with applications following along at some later point. We know, of course, that this is completely backward, but traditional curriculum design tends to foster this all-too-common misconception. In Unit 5, we cover important algebra topics from an applied approach, hopefully allowing students to finally see how the algebra that they've learned in the past is useful and relevant.

Unit 6: Set Theory

Humans have a natural tendency to categorize things in groups, from numbers to people. In a very real sense, the fundamentals of set theory are central to pretty much everything we do in math. So we study set theory both as an accessible way to apply mathematical thinking to our world and as a vehicle for thinking about the type of organized thought that's useful in any math-related field.

Unit 7: Logic

Digging further into the fundamentals of mathematical and scientific thought, we cover formal logic in a way that helps students to always remember that the *p*s and *q*s they're working with aren't letters, but statements about the world in which they live. We feel like the way to keep students engaged in the study of truth values for statements and validity of arguments is to ask them to build the rules for truth values through examples, rather than providing and illustrating the rules. All of us hope that through math, students learn to think logically and analytically: An explicit study of logical thought helps to foster that type of thinking.

Unit 8: Measurement

Measuring sizes is not only one of the oldest applications of math; it's also one of the most common even today. The study of mathematics should, of course, be part of any college program for a variety of reasons; in terms of direct application to many fields of study, working with measurement and units ranks highly. Ask any instructor of general sciences how much more they could accomplish if they didn't have to spend weeks covering key ideas in measurement and unit conversion, and you'll gain an appreciation for how useful this topic can be for students.

Unit 9: Voting Methods

Many students think that voting is simply casting a ballot and adding up the votes to see who got the most. The study of different voting methods provides an accessible example of mathematical thinking in society. It also opens students' eyes to the wide variety of places where voting is used, and it helps them to understand the complexities inherent in the topic.

Unit 10: Graph Theory

Think you know what a graph is? If you're thinking in the algebraic sense, you might be surprised to find out that there are completely different types of graphs that have nothing to do with x and y. You might be even more surprised to find out the wealth of interesting problems that can be studied using the theory of graphs. And break out those colored markers! Graph coloring is a thing.

Unit 11: Numeration Systems

Think you know what a number is? Is it different from a numeral? Are there different ways to represent numbers? This unit goes to the very foundation of everything that math was built upon: numbers. Your mind just might get blown when you study systems that use completely different symbols to represent numbers, and systems that use our familiar symbols in completely different ways.

One of the best things about a course of this nature is that the material is not completely linear. It's perfectly possible for instructors to choose the content that's most relevant for his or her students, and reordering of some of the topics is reasonable as well. We've devised an ordering that works well for us, but colleges or instructors can certainly choose a topic list and order that best meets the needs of their student populations.

What's New in This Release?

Writing the first edition of a math text, especially in an evolving area like quantitative reasoning (QR), is part art and part science. You use your training and experience as an instructor to decide on the approach and the most appropriate topics. You travel a lot and talk to anyone who doesn't run away when they see you coming to gather more professional opinions. You count on your crack publisher's team to conduct surveys and focus groups. Then you put it all together and make some educated guesses, hoping that the result hits the mark.

We feel like we did a pretty good job of that in the first edition of this book. Then we set out to do it all again, gathering feedback from a multitude of sources. In the second release, we addressed the content, expanding some topics and adding some completely new units. We also added the Solved Examples that were requested by many students and instructors.

In this release, the focus was on currency. Not the kind you spend, although of course we cover consumer math fully. In a continued and determined effort to keep the focus of this course on context and application rather than abstract math, much of what we do is based on real-life data and events. As such, it's important to us to keep this context updated so that it retains its relevance to any generation of students. Our general rule of thumb is anything that's more than say 8 years past is ready to be updated or replaced.

In addition, we're constantly receiving feedback from students and instructors (and keep it coming!), so that any errors (rare though they may be) are corrected, and any passages or topics that can use some clarification are updated as well.

We hope that you find this release to be an upgrade, and we'd love to hear your thoughts on how we did.

Acknowledgments

So many people were involved in bringing this project to life and improving the second edition that it's tempting to just thank the whole world and move on, but some folks deserve special recognition. We sincerely hope we didn't omit anyone in this category. If we did, blame it on Dave, and he owes you some beverages of your choice.

We've spent something like 5 or 6 years traveling all over the country visiting conferences and campuses and conducting focus groups and symposia. We thank everyone who attended and shared their ideas on improving college math for non-STEM students. You are the main reason this project exists.

Elizabeth Betzel
Columbus State Community College

Colleen Bye
Utah Valley University

Erin Cooke
Perimeter College at Georgia State University

Steven Cosares
LaGuardia Community College

Jodi Cotten-Konsur
Westchester Community College

Lisa Creighton
Central Piedmont Community College

Tim Delworth
Purdue University, West Lafayette

Raina J. Eckhardt
Manchester Community College

Jennifer Fitzgerald
Northern Essex Community College

Debra Hall
James Madison University

Julie Hartzler
Des Moines Area Community College

Lori Heymans
Northern Essex Community College

Heather Howington
University of North Georgia

Gretta Johnson
Amarillo College

Rachel King
Northern Essex Community College

Kurt Klopstein
Bakersfield College

Bernadette Kocyba
J. Sargeant Reynolds Community College

Chris Mansfield
Durham Technical Community College

Susan McLoughlin
Union County College

Linda Murphy
Northern Essex Community College

Jonathan Oaks
Macomb Community College

Rebecca Rose
Northern Essex Community College

Robin Rufatto
Ball State University

Mike Sieve
Ridgewater College

Jim Sullivan
Northern Essex Community College

Prudence York-Hammons
Temple College

Our deepest appreciation goes to all of our friends in North Carolina who inspired and supported this project. In particular, Paula Savich from Mayland Community College was the first person who suggested a version of *Math in Our World* in a group project–based format. Chris Mansfield from Durham Tech went above and beyond, working with us on developing a content plan. Several instructors provided feedback on the various drafts of the manuscript.

Kari Arnoldsen
Snow College

Kyle Carter
The University of West Georgia

Maria DeLucia
Middlesex County College

Holly Dickin
Ball State University

Crystal Lorch
Ball State University

LeAnn Neel-Romine
Ball State University

Eric Neumann
Community College of Philadelphia
Kathy Pinzon
Georgia Gwinnett College
Linda K. Schmidt
Spartanburg Community College
Jennifer Stanley
West Georgia State
Donna Toll
Ball State University
Erin Wilding-Martin
Parkland College
Janet Yi
Ball State University

We were very lucky to put together a board of advisors, listed below. The board's thoughtful discussions and detailed content reviews helped to shape our somewhat rough-hewn early attempts into the shiny product you're looking at now.

Nikki Armstrong
Southwestern Oregon Community College
Jordan Bertke
Central Piedmont Community College
Cindy Box
Perimeter College at Georgia State University
Erin Cooke
Perimeter College at Georgia State University
Wendy Davidson
Perimeter College at Georgia State University
Debbie Hanus
Brookhaven College
Barbie Hoag
Oakland Community College
Julie Gunkelman
Oakland Community College
Kelly Kohlmetz
University of Wisconsin–Milwaukee
Vikki Maurer
Linn-Benton Community College
Kevin McLeod
University of Wisconsin–Milwaukee
Lori Murphy
Rockingham Community College
Dawn Peterson
Itawamba Community College
Erin Wilding-Martin
Parkland College

We also had the rare opportunity to have the following courageous instructors class testing the book as we wrote it: Deb Hanus, Nikki Armstrong, Barbie Hoag, and Marcus Szwankowski (Front Range Community College).

At McGraw-Hill Higher Education, we thank: Vice President, Portfolio Tim Vertovec and Portfolio Director, Mathematics and Chemistry Ian Townsend for their continued support of our unconventional approach to . . . pretty much everything; Portfolio Manager Lauren Whalen for expertly leading the non-STEM team; Product Developer Katie Peterson, who provides so much more than what her job description calls for; Marketing Manager Candice Madden, who spreads the word far and wide; accuracy checker Jennifer Blue and copy editor Kevin Campbell, who never make fun of our silly mistakes; and Lead Core Content Project Manager Melissa Leick who kept it all together.

Supplements

ALEKS

ALEKS is an online learning and assessment platform for Math and Chemistry courses. Powered by AI, instructors choose the content, while ALEKS leverages over 25 years of real student data to measure where students have gaps and place them into personalized learning paths. Instructors have the flexibility to assign homework, share curated content, track progress, and provide support anytime, anywhere. ALEKS helps students stay engaged, build knowledge, and achieve real progress.

Your Book, Your Way

McGraw Hill's Content Collections Powered by Create is a self-service website that enables instructors to create custom course materials—print and eBooks—by drawing upon McGraw Hill's comprehensive, cross-disciplinary content. Choose what you want from our high-quality textbooks, digital products, articles, cases, and more. Combine it with your own content quickly and easily, and tap into other rights-secured, third-party content such as cases, articles, readings, cartoons, and labs. Content can be arranged in a way that makes the most sense for your course, and you can select your own cover and include the course name and school information as well. Choose the best format for your course: color print, black-and-white print, or eBook. The eBook can be included in your McGraw Hill Connect® course and is available on the free ReadAnywhere® app for smartphone or tablet access as well. When you are finished customizing, you will receive a free digital copy to review in just minutes! Visit McGraw Hill Create—mcgrawhillcreate.com—today and begin building!

Evergreen™

Content and technology are ever-changing, and we know how important it is for instructors to keep courses up to date. That's why *Math in Our World: A Quantitative Reasoning Approach* uses an Evergreen delivery model—providing the latest content, tools, and accessibility updates hassle-free.

With Evergreen, you can engage students and freshen up assignments with up-to-date coverage of select topics and new questions—all without switching editions or building a new course.

Instructor's Testing and Resource Online

This computerized test bank, available online to adopting instructors, utilizes TestGen® cross-platform test generation software to quickly and easily create customized exams. Using hundreds of test items taken directly from the text, TestGen allows rapid test creation and flexibility for instructors to create their own questions from scratch with the ability to randomize number values. Powerful search and sort functions help quickly locate questions and arrange them in any order, and built-in mathematical templates let instructors insert stylized text, symbols, graphics, and equations directly into questions without need for a separate equation editor.

Videos

To ensure instructors have the support needed to help students learn concepts, a Solution Video is available for every exercise found in the text, including Prep Skills exercises. The intent in providing a full suite of videos is to support students as they discover concepts on their own.

In addition to these Solution Videos, DYGI (Did You Get It?) videos help students check their conceptual understanding, while Tech Hints videos provide extra help with technology, such as Excel. This video suite was expanded for the second edition to provide instructors more resources to deliver a discovery-based learning course in an online or hybrid environment.

Solved Examples

Many of the more challenging topics include a Solved Example so that students have additional structure and support as they are learning concepts on their own. Intended to help students with concepts that they might have a hard time discovering independently, these examples also serve as a helpful resource for students in an online or hybrid environment. Each of the solved examples is also accompanied by a video lecture that walks through the problem and solution, which can save class time as well as provide an extra out-of-class resource for students.

Corequisite Workbook

A corequisite workbook accompanying this text provides complete coverage of the prerequisite skills necessary for each unit and lesson. Individual sections of the workbook are available in the online resources area and can serve as worksheets for students needing review in specific skill areas.

Additional Resources

PowerPoint decks for students and instructors are available for each unit of the text. These and other resources can also be found in the online resources area.

McGraw Hill ALEKS®

Students start your course with varying levels of preparedness. ALEKS is a course assistant that helps you meet each student where they are and provides the necessary building blocks to get them where they need to go. You determine the assignments and the content, and ALEKS will deliver customized practice until your students truly get it.

Experience the ALEKS Difference

Easily Identify Knowledge Gaps

Gain greater visibility into student performance so you know immediately if your lessons clicked.

- **ALEKS's Initial Knowledge Check** helps accurately evaluate student levels and gaps on day one, so you know precisely where students are at and where they need to go when they start your course.
- **You know when students are at risk of falling behind** through ALEKS Insights so you can remediate—be it through prep modules, practice questions, or written explanations of video tutorials.
- **Students always know where they are,** know how they are doing, and can track their own progress easily.

Gain More Flexibility and Engagement

Teach your course your way, with best-in-class content and tools to immerse students and keep them on track.

- **ALEKS gives you flexibility** to assign homework, share a vast library of curated content including videos, review progress, and provide student support, anytime anywhere.
- **You save time** otherwise spent performing tedious tasks while having more control over and impact on your students' learning process.
- **Students gain a deeper level of understanding** through interactive and hands-on assignments that go beyond multiple-choice questions.

Already benefitting from ALEKS?

Check out what's new! *mcgrawhill.info/aleks-new-releases*

Narrow the Equity Gap

Efficiently and effectively create individual pathways for students without leaving anyone behind.

- **ALEKS creates an equitable experience for all students,** making sure they get the support they need to successfully finish the courses they start.
- **You help reduce attrition,** falling enrollment, and further widening of the learning gap.
- **Student success rates improve**—not just better grades, but better learning.

Count on Hands-on Support

A dedicated Implementation Manager will work with you to build your course exactly the way you want it and your students need it.

- **An ALEKS Implementation Manager** is with you every step of the way through the life of your course.
- **You never have to figure it out on your own** or be your student's customer service. We believe in a consultative approach and take care of all of that for you, so you can focus on your class.
- **Your students benefit** from more meaningful moments with you, while ALEKS—directed by you—does the rest.

Unit 1
Everyone Has Problems

Bilanol/Shutterstock

Outline

Math in the Disinformation Age

While we surely hope this never happens, imagine for a moment that you've developed a potentially life-threatening illness. Would you consult:

a. A physician who specializes in the field
b. A politician
c. A cable talk show host
d. A self-proclaimed expert on social medial

Of course, you would go with choice (a). It seems obvious that in important matters, a reasonable, thoughtful person would consult with a trained expert. But is it really that obvious? Consider one of the most important issues facing the world today, climate change. Among trained, credentialed researchers, there is universal agreement that the planet is in fact warming. And, depending on the study, between 90 and 97% of researchers agree that human activity is at least partly responsible. And yet over 30% of Americans refuse to believe that climate change is happening at all. How is this disconnect possible?

The answers to that question could fill an entire semester's course on politics and human psychology, but surely one of the reasons is that in today's society we are bombarded with false and misleading information, and it takes a sophisticated, discriminating mind to navigate the treacherous waters of the disinformation age that we find ourselves in.

This is a skill like anything else—it needs to be learned, practiced, and improved. This first unit is based on **reasoning,** which we could define as using your brain to think logically and effectively before making decisions or drawing conclusions. A big part of reasoning well is understanding the difference between opinions, claims, and evidence. Modern technology gives us unprecedented access to information compiled by professionals in almost any area you can think of. Yet millions of people choose to rely on politicians, talk show hosts, and nameless people on social media for important information.

Here's a scary fact: when asked how much confidence they have in scientists, over 27% of Americans said very little or none at all. Think for a moment about all of the huge advances that have been made in computing technology, medicine, transportation, communications devices, and many other areas in just the last 30 years. Who do we have to thank for all of that? Cable commentators? People trying to get elected? Of course not.

The thing that distinguishes science from every other field is that it requires EVIDENCE to make predictions and draw conclusions—and you should too! If you take the word of every salesperson you ever come in contact with, you will end up broke and the unhappy owner of a lot of junk. And if you take the word of every random person with an agenda, you will end up badly misinformed, and unable to make important decisions effectively.

In Lesson 1 of this unit, we'll study different types of reasoning. By the time you've finished the unit, you should be able to evaluate the claims and conclusions in the following scenarios and classify each as deductive reasoning, strong inductive reasoning, weak inductive reasoning, or nonsense. Then talk about how you might go about investigating the claims.

1. The four streets closest to my hotel are called Kahana Ln., Kailani Rd., Kahua Trail, and Kukui St. I guess most of the street names in Maui begin with the letter K.
2. DNA analysis has shown that most living humans have a small percentage of Neanderthal DNA. The conclusion is that humans and Neanderthals must have interbred at some point.
3. While I was waiting in line at the grocery store, I read on the cover of the National Enquirer that a famous senate candidate advocated for euthanizing unwanted newborns.
4. A majority of voters in South Carolina have voted for the Republican presidential candidate in every election since 1980. The Democratic candidate would be wise to not waste resources campaigning in that state for the next election.

For answers, see Math in the Disinformation Age on page 55.

Lesson 1-1 Prep Skills

SKILL 1: RECOGNIZE PATTERNS IN NUMBERS

Recognizing a pattern in a list of numbers can allow us to predict the next few numbers in the list. One common pattern that can be easy to recognize is when every number in the list is obtained by adding, subtracting, multiplying, or dividing a constant by the previous number.

- 2, 5, 8, 11, 14 — Adding 3 to each number will lead to the next number.
- 5, 1, −3, −7, −11 — Subtracting 4 from each number will lead to the next number.
- 2, 20, 200, 2,000 — Multiplying each number by 10 will lead to the next number.
- 20, 10, 5, 2.5, 1.25 — Dividing each number by 2 will lead to the next number.

SKILL 2: IDENTIFY DIVISIBILITY OF NUMBERS

If a number is divisible by another number, it means one number divides evenly into the other number. (That is, if you divide the bigger number by the smaller, there's no remainder.) 6 is divisible by 3 because 3 divides into 6 exactly 2 times with no remainder. This can be written as 6 is divisible by 3 because $3 \times 2 = 6$.

There are some interesting tests for divisibility for larger numbers, but divisibility tests for small numbers basically involve remembering our multiplication tables.

- Is 18 divisible by 2? Yes, because $2 \times 9 = 18$. — Note that 18 is an even number and that all even numbers are divisible by 2.
- Is 18 divisible by 3? Yes, because $3 \times 6 = 18$. — Note that 18 is an even multiple of 3. If you start counting by 3, you would land exactly on 18. (3, 6, 9, 12, 15, **18**)
- Is 18 divisible by 5? No. If you divide 18 by 5, the result isn't a whole number. — Note that 18 is not an even multiple of 5. If you start counting by 5, you would not land exactly on 18. (5, 10, 15, 20)

SKILL 3: SIMPLIFY ALGEBRAIC EXPRESSIONS

Two of the fundamental tools used in simplifying algebraic expressions are the distributive property and combining like terms. Assessing and keeping track of the number of terms is critical. Terms are constants, variables, or expressions that are added or subtracted.

- $6x + 5 - 2x = 4x + 5$ — Combine like terms.
- $-3(2c + 7) = -6c - 21$ — Use the distributive property.
- $7y - 5(3y - 8) = 7y - 15y + 40$ — Distribute the −5.
  $= -8y + 40$ — Combine like terms.

PREP SKILLS QUESTIONS

1. Fill in the blanks with the most likely next two numbers.

   7, 11, 15, 19, 23, ______, ______

2. Fill in the blanks with the most likely next two numbers.

   1, 2, 4, 8, 16, ______, ______

3. Which of the following numbers are divisible by 3?

   12, 14, 19, 21, 36, 40

4. Which of the following numbers are divisible by 5?

   25, 36, 55, 70, 100, 111

5. Simplify the expression: $2(3a + 7) - 4a + 2$

6. Simplify the expression: $5y + 6 - 7(2y - 4)$

Lesson 1-1 Be Reasonable

LEARNING OBJECTIVES

- ☐ 1. Explain the difference between inductive and deductive reasoning.
- ☐ 2. Use inductive reasoning to make conjectures.
- ☐ 3. Use deductive reasoning to prove or disprove a conjecture.

foodfolio/Alamy Stock Photo

We live in an epoch where rational reasoning associated with evidence isn't universally accepted and is, in fact, in jeopardy. That worries me a lot.
—Rainer Weiss

A big part of being an adult is making decisions on your own—every day is full of them, from the simple, like what to eat for breakfast, to the critical, like a choice of major or career. If you make every decision based on a coin flip, you won't get very far in life. Instead, it's important to be able to analyze a situation based on logical thinking. What are the possible outcomes of making that decision? How likely is it that each choice will have positive or negative consequences? We call the process of logical thinking **reasoning.** It doesn't take a lot of imagination to understand how important reasoning is in everyone's life. So in this lesson, we'll study the process of reasoning, which I think we can all agree is a perfectly reasonable thing to do.

0. Write about an important decision you've made where you had to use reasoning.

1-1 Class

Consider this scenario: A friend invites you to check out their new apartment a few blocks away from where you live. It's a nice day, so you decide to walk over. On the way, you notice a dog loose in someone's yard, but it seems like a nice doggie so you don't think much of it—until the dog runs over and bites you on the leg. Owwww!

1. Discuss whether or not you'd be likely to take the same path the next time you visit your friend. Include both an argument for why you would take the same route and an argument for why you would not.

Dave Sobecki

2. Suppose that you decide to take the same path the next day. When passing by the same dog at the same house, he again runs over and bites you on the leg. Discuss how this might affect your thoughts on this particular route to your friend's apartment.

Assuming that you would find a different way to get to your friend's place next time, you have used what we call *inductive reasoning*.

Inductive reasoning is the process of reasoning that arrives at a general conclusion based on the observation of specific examples.

In this case, you are using two specific events (the dog bites) to draw a general conclusion (that walking along a different route would be a good idea). That's what makes this way of thinking inductive reasoning.

Now let's think about another scenario. You're still walking on the shortest path to your friend's new apartment, but on one street, you notice that there's a big hole in the sidewalk where construction is being done, with no way to get around it.

3. Discuss whether or not you'd be likely to take the same path the next day. Would you have to fall into the hole to decide?

A_Lesik/Shutterstock

There's a clear difference between these two scenarios: In the second, you wouldn't need to fall into the hole twice to decide that a different route would be a good idea. At least I hope you wouldn't. Instead, you would deduce from a known principle (I like to call it "gravity") that walking along this sidewalk wouldn't work. The use of the word "deduce" was important there: We call this type of reasoning *deductive reasoning*.

Deductive reasoning is the process of reasoning that arrives at a conclusion based on previously accepted general statements. It doesn't rely on specific examples.

Before we start using these two methods of reasoning to draw conclusions, let's make sure you understand the difference between the two. In Questions 4–8, decide whether inductive or deductive reasoning was used to draw the conclusion, and explain your choice in your own words.

4. The last six times we played our archrival in football, we won, so I know we're going to win on Saturday.

5. There is no mail delivery on holidays. Tomorrow is Labor Day, so I know my student loan check won't show up.

6. The syllabus states that any final average between 80 and 90% will result in a B. I got 80 and 82 on my first two tests, and if I get 78% on my final, my overall average will be 80.1%. That means I'll get a B.

7. Everyone I know in my sorority got at least a 2.5 GPA last semester, so I'm sure I'll get at least a 2.5 this semester.

Sidelight The Real Danger of Fake News

On December 4, 2016, a 28-year-old man from North Carolina walked into a pizza parlor in Washington, DC, and fired three shots from an assault rifle. Miraculously, nobody was injured. Why would he do this? Because he wholeheartedly believed an absurd online conspiracy theory that a former President of the United States and his wife were involved in a child sex trafficking ring that was being run out of the basement of that establishment (which, by the way, lacks a basement). By terrorizing a room full of people trying to enjoy a family meal, he really believed that he was saving abused children. The restaurant involved had been fielding harassing calls, posts, and death threats for months leading up to this near-tragedy.

This story is a real-life study in one of the biggest roadblocks to using reasoning: **confirmation bias.** This is the tendency to seek out information that confirms our preexisting beliefs, and to interpret new information in a way that supports them. An unbiased and well-reasoning person would easily recognize how ridiculous the story was. But many people were so strongly opposed to the politicians who were alleged to be involved that their brains just shut reason out, and they believed something that confirmed their views no matter how improbable it sounded. While fortunately no actual injuries occurred, one man ended up being sentenced to four years in prison because hundreds of thousands of people were willing to spread a farfetched story made up by some anonymous troll sitting at a keyboard.

8. All birds can fly. An ostrich is a bird. Therefore, ostriches can fly.

Did You Get It

Try this problem to see if you understand the concepts we just studied. The answer can be found at the end of the portfolio section.

1. Decide whether inductive or deductive reasoning was used to draw each conclusion.
   a. I've never met a golden retriever with a nasty disposition. I bet there aren't any.
   b. An apple a day keeps the doctor away. I eat an apple every day, so Dr. Phil has never come to my house.
   c. Today is Friday, so it will be Monday in three days.

1-1 Group

1. Numerous studies have shown that one of the best ways to do better in college classes is to study in pairs or groups. To help you get started, if you feel comfortable sharing contact information, exchange the information in the table below. The group you're in now will be your small group for the first unit of this course. When you get used to meeting in class, you'll likely find that meeting outside of class to study and work on homework is a good idea as well, so include some study times that would be convenient for you to meet.

Name	Phone Number	Email	Available times

In inductive reasoning, we're drawing conclusions based on observing specific examples or occurrences. An important aspect of being able to do this effectively is recognizing patterns. So we'll start there in our study of using inductive reasoning. Consider this list of numbers:

3, 6, 9, 12, 15

If you were asked to make an educated guess as to what number might come next on the list, a good place to start would be to recognize that after the first number, all others come from adding 3 to the previous number on the list. In that case, the most logical guess for the next number comes from adding 3 to 15, of course getting 18.

Math Note

The word **conjecture** is basically a synonym for "educated guess." So we would say that we *conjectured* that the next number on the list 3, 6, 9, 12, 15 would be 18.

2. Use inductive reasoning to make a conjecture about the next number on this list:

1, 2, 4, 5, 7, 8, 10, 11, 13

Explain how you decided on your answer.

3. Make a reasonable conjecture for the next figure in this sequence, and briefly describe your reasoning.

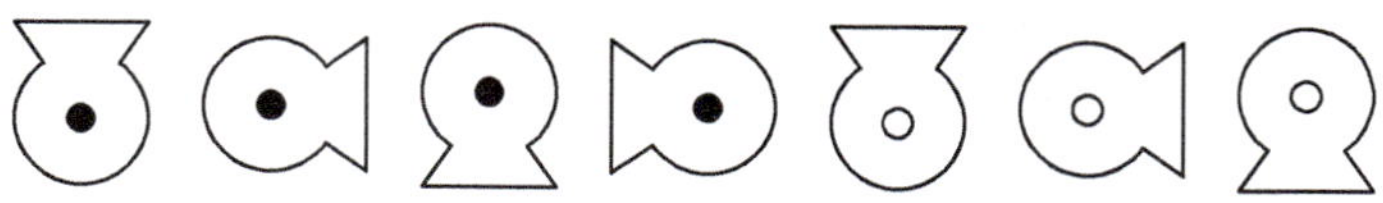

4. Think of a way that you have used inductive reasoning recently to draw a conclusion of some sort, and describe that thought process. Each member of the group should come up with their own example.

Did You Get It

2. Make a reasonable conjecture for the next two items on each list.
   a. 0 −2 4 −6 8 −10
   b. 26 y 24 w 22 u 20 s
   c.

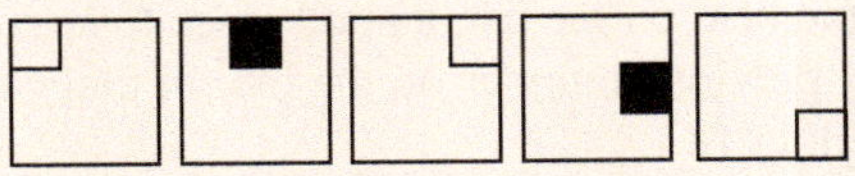

You could make a very real case that inductive reasoning is the essence of where scientific advances come from. You make some observations, then use inductive reasoning to draw a conclusion, and test to see if it seems to hold true. Let's see how this works.

5. If you add two odd numbers together, what will the result look like? Use inductive reasoning to make a conjecture.

6. How sure are you that your conjecture is true? Discuss.

Here's the problem with your conjecture, and it's HUGELY important in mathematical reasoning: You're guessing that the sum of EVERY pair of odd numbers has a certain property. ***But there's no way that you can actually test every pair of numbers because (of course) there are infinitely many to test.*** And that's the big issue with inductive reasoning: It's an incredibly useful tool in decision making, but in most cases you can't verify a conclusion for every possible case. So you can't be 100% certain that your conclusion is valid. We'll come back to your conjecture later in the lesson. For now, we'll look at a different problem.

Suppose that your history professor gives a surprise quiz every Friday for the first 4 weeks of class. At that point, you might use inductive reasoning to conjecture that you'll have a quiz every Friday. But you can't look into the future and know for sure, so it's entirely possible that your conjecture is not true. In fact, the first time you DON'T have a quiz on Friday, your conjecture has proven to be false.

This is a really useful observation: While it can be difficult to prove that a conjecture is true, it's much simpler to prove that one is false. All you need to do is find one specific example that contradicts the conjecture (like the first Friday without a quiz). This is known as a *counterexample.*

A **counterexample** is a single specific example that violates a conjecture. Even one counterexample is enough to prove that a conjecture is false.

7. This conjecture is false (Trust me, I'm a doctor.): A whole number greater than 10 can be divided by 3 with no remainder if the last two digits can be divided by 3 with no remainder. Find a counterexample that proves it's not true. (Hint: Pick some random numbers whose last two digits can be divided evenly by 3.)

A chord is a line connecting two points on a circle. If you draw a chord connecting two points, it divides the circle into two sections. With only one point, there's no chord to draw, and the circle has just one section:

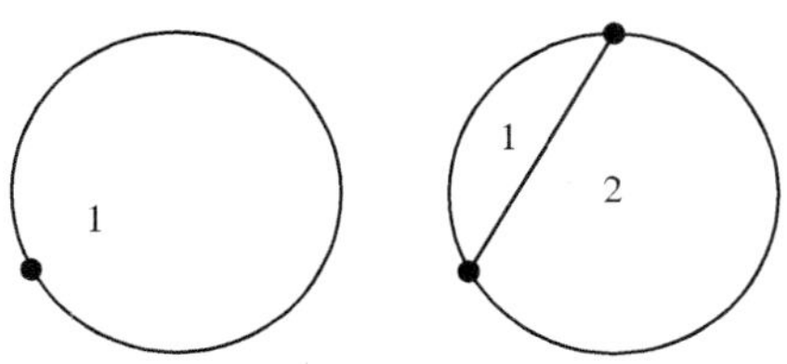

8. Using the diagrams below, count the number of sections a circle is divided into by chords connecting 3, 4, and 5 points. Numbering the sections as you count them is a big help.

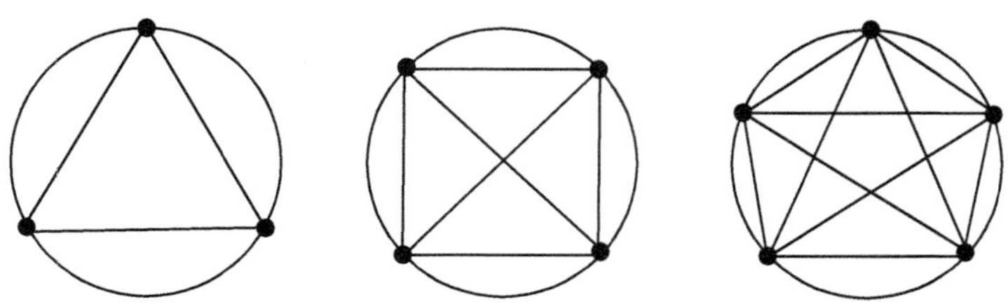

Number of Points	1	2	3	4	5
Number of Sections	1	2	____	____	____

9. Use your results from Question 8 to make a conjecture as to the number of sections with 6 points. Explain your reasoning.

10. Use the diagram to check your conjecture. What does that tell you about conjectures made using inductive reasoning?

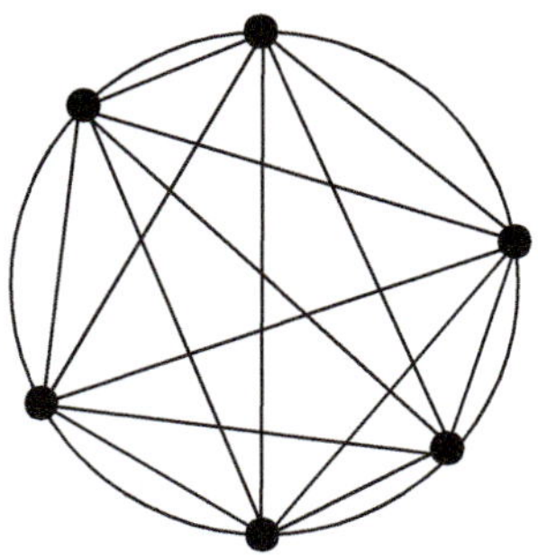

Question 10 illustrates the essence of math and science we mentioned earlier: We have an idea that something might be true, and we use inductive reasoning to test it. But the result shows why inductive reasoning can't be used to prove results: What appears to be true after looking at several examples can still turn out to be false. We'll explore this idea more in the Technology part of tonight's homework. Doing so will require using formulas to perform calculations in Excel; if you're not familiar with how to do this, these directions will help, and the video in class resources will help even more.

Using Technology: Evaluating a Formula with Multiple Variables in Excel

To enter a formula in Excel:

1. Set up a separate column for each quantity involved in the calculation.
2. Identify the column where you want to store the result of the calculation.
3. In the first cell of that column, type = to indicate that you're entering a formula.
4. Type the calculation that you want completed after the = sign. When you want to include the number in a particular cell as part of the calculation, you can either type the column and row of that cell (A3, for example), or you can use the cursor to click on the cell and it will be entered into your formula automatically.
5. When you've completed the formula, press ENTER.

See the Using Tech Video on Evaluating a Formula with Multiple Variables in Excel.

Did You Get It

3. Find a counterexample to prove that each conjecture is false.
   a. No month has more than eight letters in it.
   b. The sum of any two positive numbers with two digits also has two digits.

11. When you prove that a conjecture is false using a counterexample, are you using inductive or deductive reasoning? Discuss.

So inductive reasoning can be used to give you a really good idea that something may be true, but most often it cannot be used to *prove* that it's true. Our next step will be to look at how inductive and deductive reasoning can work together to both make a conjecture, and then prove it.

12. Pick a random number, and add 50 to it. Then multiply by 2 and subtract the original number. What is the result?

13. Repeat Question 12 for several other beginning numbers until you're able to use inductive reasoning to make a conjecture as to what the result will always look like.

Now let's use deductive reasoning to prove your conjecture. So far, you've used *specific* numbers in developing your conjecture: That's what makes what you've done so far inductive reasoning. In order to turn the same idea into deductive reasoning, you need to do these calculations on a number that's nonspecific. Sounds impossible, right? Not so fast, my friend. Rather than use another specific number, we'll use a symbol (a letter in this case) to stand in for ALL possible numbers. If that sounds like a variable to you, you're absolutely right! Good job. The letter we choose will represent a quantity (the number we pick) that can VARY. We want our symbol to stand for any old number, so let's use the letter a.

14. Add 50 to our variable number. The result will be an **algebraic expression,** which is a combination of numbers and variables using operations like addition, subtraction, multiplication, division, powers, and roots.

15. Next, multiply your answer to Question 14 by 2. Please don't forget that the distributive property exists.

16. Finally, subtract the original number from your answer to Question 15. What is the result?

1-1 Class (Again)

1. Explain why Questions 14–16 in the Group portion prove the conjecture, while Questions 12 and 13 just indicated that it was likely to be true.

Did You Get It

4. Think of any number. Multiply that number by 3, then add 30, and divide the result by 3. Next subtract the original number. What is the result?
   a. Use inductive reasoning to make a conjecture for the answer.
   b. Use deductive reasoning to prove your conjecture.

Now let's get back to your earlier conjecture about adding odd numbers. You DID conjecture that the result is always even, right? Good job. Go buy yourself a muffin or something. Next, we're going to think about how we would go about proving this conjecture.

Solved Example

Prove that the sum of any two odd numbers is always even.

SOLUTION

The most important thing to understand here is what you CAN'T do to prove a conjecture like this: Just pick a bunch of *specific pairs* of odd numbers and add them, then check to see if the result is even. No matter how many pairs you choose, even if the results keeps coming up even, this can NEVER prove that it will ALWAYS happen. For that, we'll need to add together two arbitrary odd numbers.

In order to do that effectively, we need to think about what exactly makes a number even or odd. A number is even exactly when it can be written as 2 times some other integer. So we can write an arbitrary even number by writing it as $2n$, where n is some integer. Next, think about the fact that every odd number is exactly one less than an even number. One is one less than 2, 13 is one less than 14, 157 is one less than 158, and so on. Since an arbitrary even number looks like $2n$, we can write an arbitrary odd number as one less than that, which of course is $2n - 1$. Now we're ready to prove our conjecture.

Two arbitrary odd numbers can be written as $2n - 1$ and $2m - 1$, where n and m are integers. Let's add them:

$$(2n - 1) + (2m - 1)$$

Now we need to do some algebra.

$(2n - 1) + (2m - 1) = 2n + 2m - 2$	Combine the like terms −1 and −1
$= 2(n + m - 1)$	Factor 2 out of each term
$= 2(\text{Some integer})$	Since n and m are integers, so is $n + m - 1$

Did You Get It

5. Prove that the sum of two even numbers is always even.

2. Explain why what we did in the solved example proves that the sum of two odd numbers is ALWAYS even.

3. The owner of a growing business has noticed that as she hires more employees, the company's weekly profit grows. She hires a consultant to run some numbers, and he informs her that the formula below is a good model of the profit in dollars she can expect to make each week. The variable x represents the number of employees. If she started with six employees, use inductive reasoning to investigate whether or not the weekly profit will keep going up if she keeps hiring more employees. Write your conclusion in the form of a sentence.

$-10x^2 + 480x + 308$

Within the study of reasoning, we can actually divide inductive reasoning into two subcategories: **weak inductive reasoning** and **strong inductive reasoning.** Weak inductive reasoning comes from using just a small handful of observations to draw a conclusion, while strong inductive reasoning is based on many repeated observations. As you may have guessed from the names, strong inductive reasoning is much more useful and reliable when trying to draw general conclusions from observed data.

For example, if someone says "It's been really cold this winter, so global warming isn't happening," that's the weakest of weak inductive arguments: a conclusion about long-term climate patterns is being drawn based on just one season. On the other hand, the statement "COVID-19 vaccines were found to have a 95.1% effectiveness when tested on 21,720 subjects" represents a very strong inductive argument because so much data was gathered, the majority of which indicated a positive result.

4. Classify each of the following as a strong or weak inductive argument.

a. All three of my roommates plan to vote for the Democratic candidate this November, so that candidate must have overwhelming support on our campus.

b. In September of 2024, the polling group 538 estimated that just 20.3% of Americans approve of the job being done by Congress after averaging 8 polls done by information-gathering organizations.

c. Two of the last three years have been the hottest on record, so obviously global warming is real.

d. A computer manufacturer randomly tested 937 outgoing laptops and found that just four were defective, so they concluded that the manufacturing process was sound.

1-1 Portfolio

Name ________________________________

Check each box when you've completed the task. Remember that your instructor will want you to turn in the portfolio pages you create.

Technology

1. ☐ Spreadsheets can be a hugely useful tool for using inductive reasoning since you can set up a spreadsheet to test the results of calculations for many different values quickly. In this Excel activity, we'll use a spreadsheet to help test a conjecture: Pick a random number, and add 50 to it. Then multiply the result by 2 and subtract the original number. Will the result always be 100 more than the original number?

ALEKS Practice

1. ☐ Include any written work from the ALEKS assignment along with any notes or questions about this lesson's content.

Applications

1. ☐ Complete the applications problems.

Reflections

Type a short answer to each question.

1. ☐ Describe the difference between inductive and deductive reasoning in your own words.
2. ☐ Which type of reasoning (inductive/deductive) do you prefer, and why?
3. ☐ Why can inductive reasoning never be used to PROVE a conjecture?
4. ☐ Name one thing you learned or discovered in this lesson that you found particularly interesting.
5. ☐ What questions do you have about this lesson?

Looking Ahead

1. ☐ Complete the prep skills for Lesson 1-2.
2. ☐ Read the opening paragraph in Lesson 1-2 carefully and answer Question 0 in preparation for that lesson.

Answers to "Did You Get It?"

1. a. Inductive **b.** Deductive **c.** Deductive **2. a.** 12 −14 **b.** 18 q **c.** 
3. a. September has nine letters. **b.** If you pick two numbers with a sum more than 100, the sum has more than two digits. **4. a.** The result is always 10.

b. $3a \Rightarrow 3a + 30 \Rightarrow \frac{3a + 30}{3} \Rightarrow \frac{3a + 30}{3} - a = a + 10 - a = 10$

5. See video in the online class resources for the proof, which looks a lot like the solution to the Solved Example.

Answers to Prep Skills Questions

1. 27, 31
2. 32, 64
3. 12, 21, 36
4. 25, 55, 70, 100
5. $2a + 16$
6. $-9y + 34$

1-1 Applications

Name ____________________

In Questions 1–4, decide whether inductive or deductive reasoning was used to draw a conclusion. Briefly explain your choice. If inductive reasoning was used, rate it as strong or weak. An actual sentence or two would be nice.

1. The last four congressional representatives from this district were all Republicans. I don't know why the Democratic candidate is even bothering to run this year.

2. Working as a nurse in a hospital requires at least a 2-year degree in this state, so when I was in the emergency room last week I asked the nurse where he went to college.

3. Experts say that opening email attachments that come from unknown senders is the easiest way to get a virus on your computer. Shauna constantly opens attachments from people she doesn't know, so she'll probably end up with a virus on her system.

4. My instructor has given us a "surprise" quiz every Friday so far this semester, so I'm going to be studying for a quiz Thursday evening.

1-1 Applications

Name ______________________________

For each conjecture in Questions 5–7, if it's false, provide a counterexample. If it's true, use deductive reasoning to prove it.

5. The square of every real number is greater than the number itself.

6. Pick two even numbers and subtract the larger number minus the smaller number. The result is even as well.

7. When an even number is added to the product of two odd numbers, the result will be even.

1-1 Applications

Name ______________________________

8. Suppose that you drive a certain distance at 20 miles per hour, then turn around and drive back the same distance at 60 miles per hour. Choose at least four different distances and find the average speed for the whole trip. Then use inductive reasoning to make a conjecture as to what the average speed is in general. (Hint: You'll need the formula distance = speed times time repeatedly!)

9. Use algebra to prove your conjecture from Question 8. (Hint: Use the exact same steps you did for each calculation in Question 8, but use a variable for the distance rather than a distance you chose.)

Lesson 1-2 Prep Skills

SKILL 1: RECOGNIZE PLACE VALUES

The place value of a digit in a number depends on where that digit is located in the number. The chart below should help.

Place Value																	
Millions			,	**Thousands**			,	**Ones**			.	**Decimals**					
hundred millions	ten millions	millions		hundred thousands	ten thousands	thousands		hundreds	tens	ones		tenths	hundredths	thousandths	ten thousandths	hundred thousandths	millionths

The number 342.7 has four digits.

The 3 is in the hundreds place. — Notice that the 3 is three places to the left of the decimal.

The 4 is in the tens place. — Notice that the 4 is two places to the left of the decimal.

The 2 is in the ones place. — Notice that the 2 is one place to the left of the decimal.

The 7 is in the tenths place. — Notice that the 7 is one place to the right of the decimal.

SKILL 2: ROUND NUMBERS

Rounding numbers

Rounding numbers involves a fairly simple procedure.

Step 1. Locate the digit in the desired place.

Step 2. Look one place to the right.

Step 3. If the digit to the right of the desired place is 4 or lower, the digit in the desired place remains unchanged, rounding down. If the digit to the right is 5 or higher (5, 6, 7, 8, or 9), the digit in the desired place increases by one, rounding up.

To round 34.726 to the nearest hundredth:

Step 1. The digit in the hundredths place is 2. — Notice that the 2 is two places to the right of the decimal.

Step 2. The digit to the right of the 2 is 6.

Step 3. Since 6 is greater than 5, the digit in the desired place should be increased by one (rounded up). — We round up because our procedure helps us see that 34.726 is closer to 34.73 than it is to 34.72.

So 34.726 rounded to the nearest hundredth is 34.73.

SKILL 3: PERFORM MENTAL ARITHMETIC

Performing mental arithmetic is just that. Mental. It's a useful skill that you have to PRACTICE. If you insist on always using a calculator, even for relatively simple operations, there's a pretty good chance that you'll never feel very good about your math skills. You may be surprised how practicing some mental arithmetic makes you feel much more comfortable with math in general. Use a calculator as a tool, not a crutch.

PREP SKILLS QUESTIONS

1. For the number 87.23, list the digit in each place value.

   a. Tens place b. Ones place c. Tenths place d. Hundredths place

2. Round each number to the designated place value.

   a. 8,431 to the nearest ten b. 11.46 to the nearest tenth c. $84.76 to the nearest dollar

3. Perform each calculation without using a calculator or computer.

   a. 12×8 b. 300×6 c. 25×8 d. $120 + 80$ e. $95 + 15$ f. $1{,}400 + 800$

Lesson 1-2 More or Less

LEARNING OBJECTIVES

☐ 1. Use rounding and mental arithmetic to estimate the answers to applied problems.

☐ 2. Obtain and interpret information from bar graphs, pie charts, and time series graphs.

YinYang/Getty Images

The key to good decision making is evaluating the available information—the data—and combining it with your own estimates of pluses and minuses.
—Emily Oster

Everyone likes buying items on sale, so we should all be familiar with the idea of finding a rough approximation for a sale price. If you're looking at a pair of shoes that normally sells for $70 and the store has a 40% off sale, you might figure that the shoes are a little more than half price, which would be $35, so they're probably around $40. We will call the process of finding an approximate answer to a math problem **estimation.** Estimation comes in handy in a wide variety of settings. When the auto repair shop technicians look over your car to see what's wrong, they can't know for sure what the exact cost will be until they've made the repairs, so they will give you an estimate. When you go to the grocery store and have only $20 to spend, you'll probably keep a rough estimate of the total as you add items to the cart. (Imagine buying a week's worth of groceries and keeping track of every price to the penny on your cell phone. Who has time for that?) If you plan on buying carpet for a room, you'd most likely measure the square footage and then estimate the total cost as you looked at different styles of carpet. You could find the exact cost if you really needed to, but often an estimate is good enough for you to make a sound buying decision.

0. Write about a time you've found estimation to be useful in your life.

1-2 Class

Rounding numbers is going to play a very significant role in our study of estimation, so we'll begin with a quick review of the rules for rounding. But of course, we'll do so in the context of an applied problem. (As you may have already noticed, that's kind of the point of this course.)

Solved Example 1

Let's say you're planning a group outing, with lunch included. The lunch is supposed to cost $3.95 per person, and 24 people sign up. Would $94.80 be a reasonable bill?

SOLUTION

The most precise approach would be to multiply $3.95 by 24 and see what the bill would be. But we're not asked if that amount is EXACTLY right—we're asked if it's reasonable. So a quick estimate would get the job done, and would be simpler than the actual multiplication, which pretty clearly would require a calculator.

Here's our strategy: First, we'll notice that $3.95 is pretty close to $4, and that 24 people is pretty close to 25 people. Why is that helpful? Because $4 times 25 is $100, and that doesn't require a calculator. Finally, we rounded the price up by a bit, and the number of customers up by a bit, so that $100 estimate has to be a bit on the high side. Bingo! A bill of $94.80 is just about what we'd expect.

Solved Example 1 does a nice job of illustrating a good, reliable procedure for doing estimation with numerical calculations:

1. Round the numbers being used to numbers that make the calculation simple.
2. Perform the operation or operations involved.

In order to practice rounding rules, we'll start with a very quick review of percents. This topic is covered in detail in Unit 2, so we'll just review the basics here. Percents are used to describe certain portions of a whole. As you know, 50% of some amount is half of it. We can use this idea and estimation to get a rough idea of certain percentages without needing a calculator.

1. There are about 212 million people in the U.S. over the age of 24. About half of them are female. Estimate the number of females over the age of 24.

2. Of those females, the U.S. Census Bureau reports that about 33% have a bachelor's degree or higher. Use this to estimate the number of females over age 24 who have a bachelor's degree or higher.

3. For males in that age group, about 12% have a graduate degree. About how many men is that?

We often use the word "nearest" when talking about rounding numbers. If we're dealing with currency, we typically will want to round either to the ones place (which is dollars), or the hundredths place (which is cents). In that case, we'll use the terminology in Solved Example 2.

Solved Example 2

The purchasing manager for a resort hotel places an order for 1,155 new linen napkins, each of which costs $2.48.

a. Find the exact cost.

b. The hotel accountant needs the amount rounded to the nearest dollar. What is it?

c. The general manager, more of a "big picture" boss, wants to know the cost to the nearest hundred dollars. What should our friend the purchasing manager report?

SOLUTION

a. The exact cost is just 1,155 times $2.48, which is $2,864.40.

b. The number of cents is less than 50, so we'd round down to $2,864.

c. The digit in the tens place is 6, so we round the 8 in the hundreds place up to 9. The price to the nearest hundred dollars is $2,900.

4. A one-year investment in a certificate of deposit (CD) draws 2.1% interest, which mathematically means that to find the new value after one year, you multiply the original investment by 1.021. If the beginning balance of an investment is $8,412.35, find the value at the end of one year, rounded to the nearest cent. Unless you're a savant of some sort, you'll need a calculator to get started.

One of the key concepts involved in rounding is *accuracy*. For our purposes, we'll define accuracy as how far off from the exact value of some quantity an estimate is. For the calculation in Question 4, you could round the initial investment of $8,412.35 up to $10,000, and that would make the computation considerably easier. But because $10,000 is quite a bit different from the actual amount, your estimated value would be very different from the actual value. In that case, we'd say that your answer was inaccurate.

In short, when using rounding and estimation, it's always important to think about the level of accuracy as well as the difficulty level of getting an estimate.

5. The owner of an apartment complex needs to buy six new refrigerators. She's considering choosing a model that costs $579.95 each. By rounding the cost to the nearest hundred dollars, estimate the total cost.

6. Instead of rounding to the nearest hundred dollars, round to the nearest ten dollars and estimate the cost. Then discuss the pros and cons of rounding to the nearest hundred as opposed to the nearest ten.

Did You Get It

Try this problem to see if you understand the concepts we just studied. The answer can be found at the end of the portfolio section.

1. When shopping for his books this semester, Arlen has $400 in financial aid money allocated. He needs two paper books at $147 each, an e-book costing $79, and an online homework system access code for $82. Rounding each value to the nearest ten dollars, estimate if Arlen's aid will cover his expenses.

Questions 5 and 6 bring up an interesting point: How do you know what digit to round to when estimating? The correct answer is "five." Just kidding—there IS no correct answer. It depends on the individual numbers. Deciding on how much to round is really a trade-off: ease of calculation versus accuracy. In most cases you'll get a more accurate result if you round less, but the calculation will be a little harder. Since there's no exact rule, it's important to evaluate the situation and use good old-fashioned common sense. And remember, when you are estimating, there is no one correct answer.

7. Suppose that you're considering a new cell phone plan where you have to pay $179 upfront for the latest phone, but the monthly charge of $39.99 includes unlimited data, talk, and text. A different plan gives you the phone you want for free (aside from sales tax of $12.53, which you are responsible for), and the unlimited data, talk, and text plan is $61.20 per month. Use the type of estimation we've practiced to decide which is the better choice. Are there factors other than total cost you might consider? Explain your calculations and your decision.

moodboard/SuperStock

1-2 Group

In this age of information, it's very common for interesting or useful information to be displayed in graphical form, rather than just listing out numbers. Why? Because this almost always makes it easier to get a quick overall view of what that data are telling us. Here's an example: The table below shows the percentage of total power generated that came from renewable sources for the countries with the six largest economies in the world as of 2023.

Country	% of Power from Renewable Sources
United States	22.7%
China	30.7%
Japan	23.8%
Germany	52.4%
United Kingdom	46.0%
India	19.5%

There is absolutely nothing wrong with this table. It's positively delightful. And if you want to read it line by line and concentrate on the percentages, you can get a very accurate comparison of how well these nations are doing in terms of transitioning to a green economy. But let's compare that to a **bar graph** representing the same data.

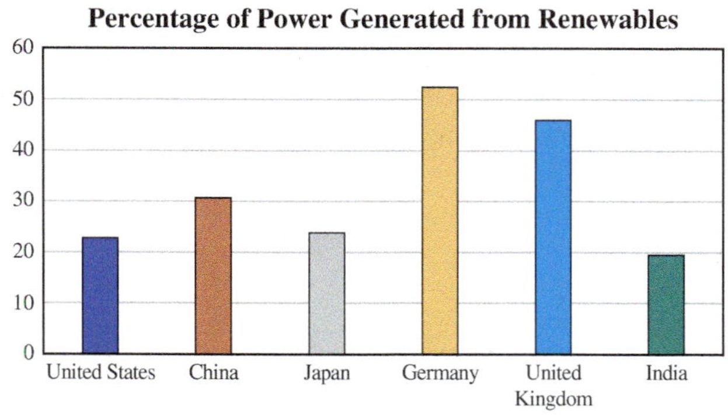

Source: Statistical Review of World Energy

Bar graphs are used to compare amounts or percentages using either vertical or horizontal bars of various lengths; the lengths correspond to the amounts or percentages, with longer bars representing larger amounts.

The graph makes it pretty easy to get an overall picture of what the data in the table indicate. First, we can easily see that Germany gets the highest percentage of its energy from renewables, and India gets the lowest percentage. We can also see that the percentages for China, Germany, and the U.K. are reasonably similar, as are the percentages for the U.S., Japan, and India.

So what does that have to do with estimation? In looking at the graph, we're visually estimating the percentages based on the heights of the bars, rather than getting an exact percentage like we did from the table.

1. Without looking at the table, use the bar graph above to estimate the percentage of electricity generated by renewable sources for each country shown. Then compare your estimates to the exact percentages shown in the table. You should give some thought to how accurately you think you can reasonably round the percentages based on the graph.

Math Note

Sometimes a graph will contain labels that provide the exact values of the data illustrated. In this case, we could have put the exact percentages at the top of each bar.

2. The same information is shown on the next bar graph. But we've altered the scale on the vertical axis. Describe several ways that this can change your perception of the differences in percentages for these countries.

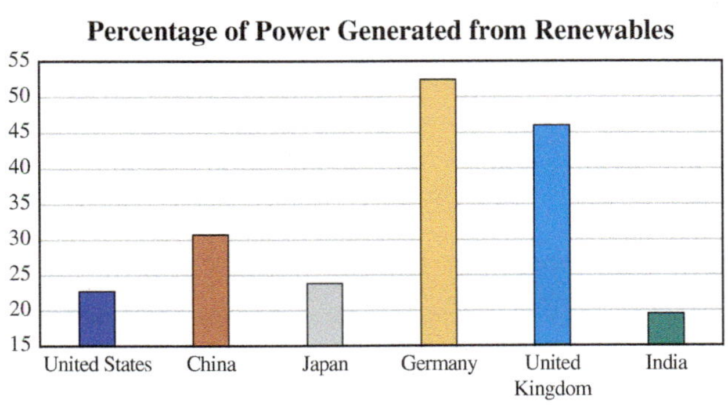

Source: Statistical Review of World Energy

Did You Get It

2. Use the second bar graph describing renewable energy production to estimate the percentages for the three top producers. Then describe your confidence in how accurate your estimates are compared to the ones you made from the first bar graph.

A **pie chart,** also called a **circle graph,** is constructed by drawing a circle and dividing it into parts called sectors, according to the size of the percentage of each portion in relation to the whole. The next series of questions will be based on the pie chart below, which illustrates the breakdown of different types of fatal workplace injuries in 2022.

3. What percentage of workplace fatalities were caused by transportation incidents?

4. What percentage of workplace fatalities were not due to exposure to substances or contact with objects and equipment?

Fatal Occupational Injuries

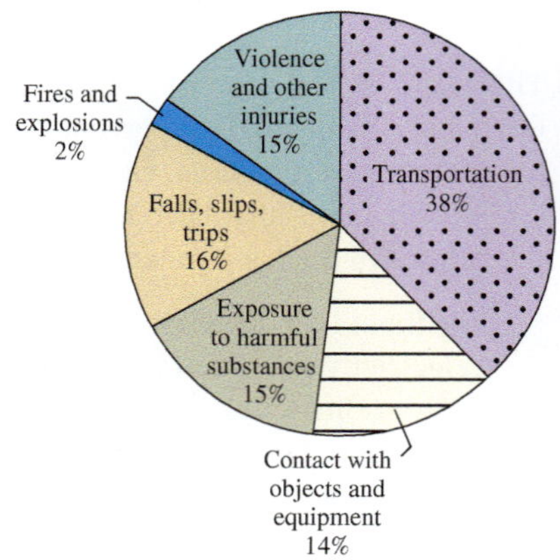

Source: U.S. Bureau of Labor Statistics

5. There were 5,464 workplace fatalities in that year. Use rounding and estimation to estimate how many were due to either falls or contact with objects and equipment. Explain how you got your estimate.

6. Use a calculator to find the exact number of deaths caused by either falls or contact with objects and equipment. Discuss the accuracy of your estimate, and why it was either too low or too high.

Did You Get It

3. Were there more fatal accidents resulting from transportation, or from the next three highest causes? What was the difference in percentages for these causes?

A **time series graph** or **line graph** shows how the value of some variable quantity changes over a specific time period. These are the types of graphs you're most likely to remember working with in algebra classes. And maybe you always wondered, "Why does anyone care about drawing graphs in algebra?" The answer to that question is exactly the point of studying different types of graphs. Graphs aren't just bad art; they're used to conveniently display useful numerical information.

The last handful of questions in this portion of the lesson are based on the time series graph below, which illustrates the average selling price on the international market of a barrel of crude oil for the years from 2010 to 2024.

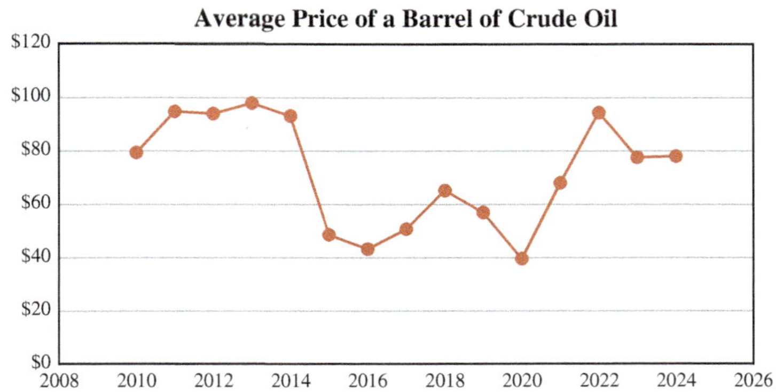

7. Estimate the average price of crude oil in 2011 from the graph. Don't forget to consider how "rounded" you think your answer should be.

8. In which year between 2010 and 2024 did the average price reach its highest point? Estimate the highest average price, and write a brief explanation of how you decided on your answer.

9. What was the biggest year-to-year decrease in oil prices between 2010 and 2024? Was it more or less than the biggest year-to-year increase? Explain your answers.

10. What does the graph tell us about what the average price will be for 2025? Talk about this with your group, and give it some serious thought.

Now that we've studied three different types of graphs, we'll wrap up the lesson by thinking about the types of situations for which each type would be most appropriate. In Questions 11–14, some information is discussed. Write which type of graph—bar graph, pie chart, or time series graph—that you think would be most appropriate to display that information. There's not necessarily a right answer, so make sure that you explain why you made your choices.

11. A CNN/*USA Today* poll is conducted to decide which of 5 Democratic presidential candidates voters plan to vote for in their state's upcoming primary.

12. CNBC reports on the amount of money that had been raised by each of those candidates up to that point in the election.

13. A CBS Evening News profile on one of the top candidates describes how their polling numbers have changed during the primary season.

14. The supporters of one controversial candidate are broken down demographically by race, income level, and education level.

1-2 Portfolio

Name ______________________________

Check each box when you've completed the task. Remember that your instructor will want you to turn in the portfolio pages you create.

ALEKS Practice

1. ☐ Include any written work from the ALEKS assignment along with any notes or questions about this lesson's content.

Applications

1. ☐ Complete the applications problems.

Reflections

Type a short answer to each question.

1. ☐ What did you learn in this lesson that you think will be helpful in your daily life? Be specific, and provide one or two examples of where you think you'll use your newly acquired mathematical skills.
2. ☐ Look back at the very beginning of this lesson, and discuss some pros and cons of illustrating the renewable power data graphically, as compared to in table form.
3. ☐ In what types of situations is estimation helpful, and when might estimation actually be harmful?
4. ☐ Name one thing you learned or discovered in this lesson that you found particularly interesting.
5. ☐ What questions do you have about this lesson?

Looking Ahead

1. ☐ Complete the prep skills for Lesson 1-3.
2. ☐ Read the opening paragraph in Lesson 1-3 carefully and answer Question 0 in preparation for that lesson.

Answers to "Did You Get It?"

1. Arlen will need roughly \$460, so he's going to come up short.
2. Germany 52.5%, UK 46%, China 31%. The finer scale allows us to make a more accurate estimate. **3.** The next three highest causes total 46%, which is 8% more than transportation.

Answers to Prep Skills Questions

1. a. 8 **b.** 7 **c.** 2 **d.** 3 **2. a.** 8,430 **b.** 11.5 **c.** \$85
3. a. 96 **b.** 1,800 **c.** 200 **d.** 200 **e.** 110 **f.** 2,200

1-2 Applications

Name ______________________________

1. In 12 months, you're planning a trip to either Miami or San Francisco, whichever is a more likely vacation destination for you. Use the Internet to estimate the cost of transportation and the cost of a hotel. Then estimate the total amount of money you'd need for a 5-day trip including food, beverages, and entertainment. Describe all details, including what level you're rounding to.

Use the information shown in the graph for Questions 2 and 3. The graph represents a survey of 1,385 office workers and shows the percent of people who indicated what time of day is most productive for them.

When Office Workers Feel That They're Most Productive at Work

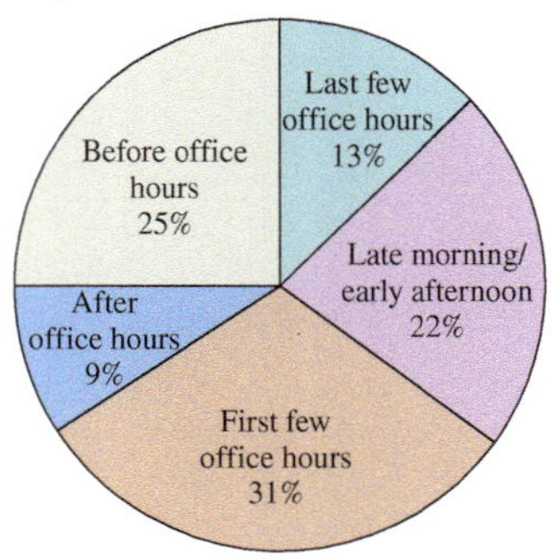

Source: USA Today

2. Estimate the number of people who feel they are most productive outside normal office hours. Show details, and discuss if you think you made an overestimate or underestimate.

1-2 Applications

Name ____________________

3. How many more people feel they're most productive in the first few office hours compared to those that feel they're most productive in the last few office hours? (Exact answer this time.)

The two bar graphs on this page describe Internet access by continent. The first shows the NUMBER of people that have Internet access in the place where they live, in millions. The second shows Internet penetration, which is the PERCENTAGE of people on each continent that have Internet access where they live. Use the graphs to answer Questions 4–7.

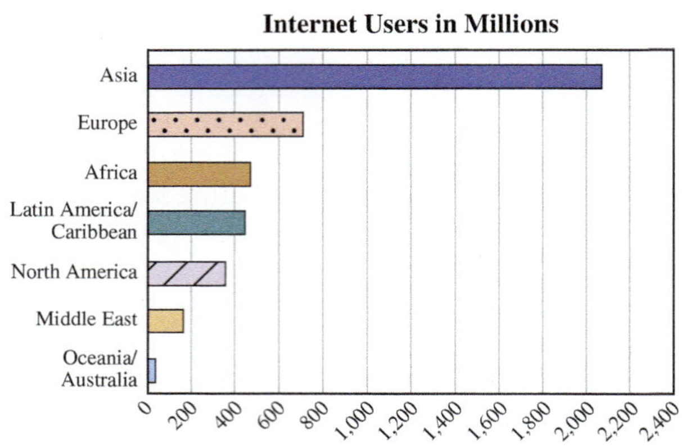

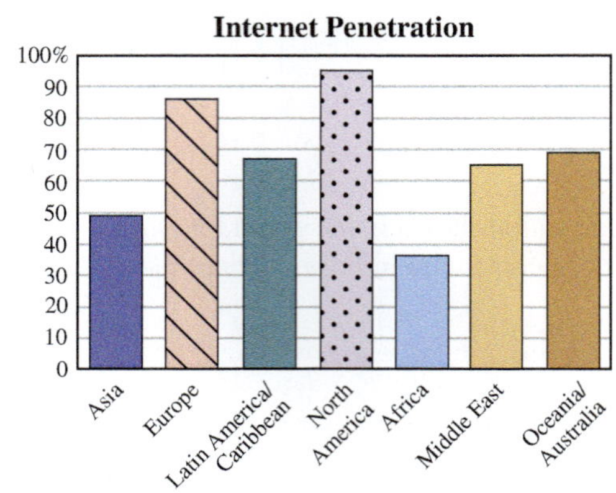

Source: Internet World Stats–www.internetworldstats.com/stats.htm

4. Which continent has the most people with Internet access at home? The least?

5. If you picked a person at random from each continent, on which continent would it be most likely that the person has Internet access at home? Least likely?

1-2 Applications

Name ______________________________

6. What do you think accounts for the fact that North America has the longest bar on the second graph, but is middle of the pack on the first one?

7. Does Asia have more Internet users than the rest of the world combined? Justify your answer.

Our final graph displays growth in Internet penetration from 2005 to 2023 in Africa compared to growth in Europe. Use this graph to answer the remaining questions.

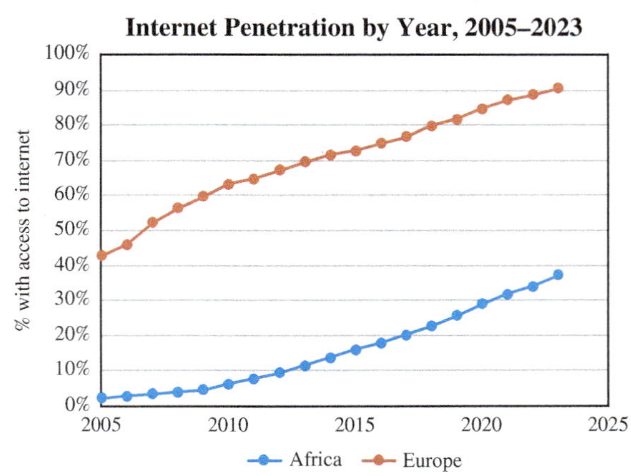

Source: International Telecommunication Union

1-2 Applications

Name ________________________________

8. By how much did the percentage change from 2005 to 2023 in Europe? What about Africa?

9. Is the gap between Europe and Africa getting bigger or smaller? Discuss.

10. Which had a more significant increase between 2005 and 2023, Africa or Europe? Explain how you decided.

Lesson 1-3 Prep Skills

SKILL 1: GET INFORMATION FROM A DIAGRAM

Getting information is mostly about attention to detail. Don't make assumptions: Look over the entire diagram carefully to make sure you understand what all of the labeling represents. Pay special attention to units on measurements—don't assume that every measurement in a diagram is going to be given in the same units. If one side of a rectangle is labeled 12 ft and a side that's clearly shorter is labeled with a length of 48, you should think about the fact that the shorter side can't be 48 feet, making you more likely to notice that it's actually 48 inches.

PREP SKILLS QUESTIONS

1. Find the perimeter of the rectangle.

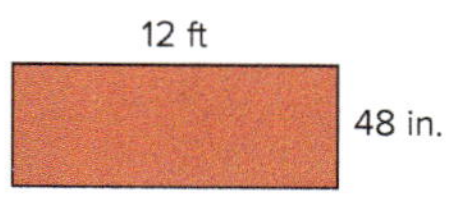

2. The perimeter of the figure is 59″. What is the length of the missing side?

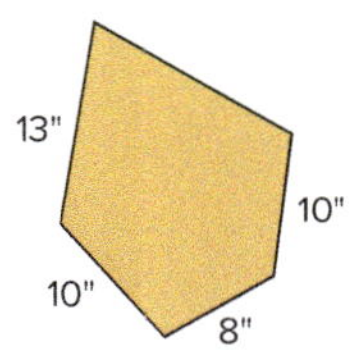

3. The total width of the room diagrammed below is 48 ft, and the spaces between the tables are all the same. How far apart are the tables?

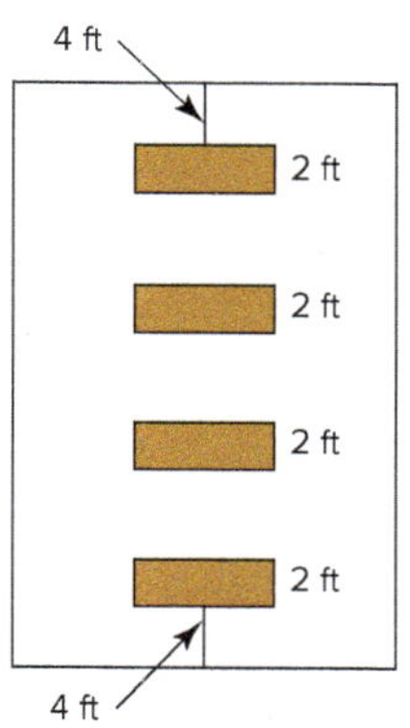

Lesson 1-3 You Got a Problem?

LEARNING OBJECTIVES

- ☐ 1. Identify the four steps in Polya's problem-solving procedure.
- ☐ 2. Apply Polya's procedure to solving problems.
- ☐ 3. Solve problems using different strategies: trial and error, drawing a diagram, using algebra, and comparing different outcomes.

Juanmonino/iStock/Getty Images

If you only have a hammer, you tend to see every problem as a nail.
—Abraham Maslow

Here's an idea that can help you understand your math classes better: Once in a while, think about what some of the math words you take for granted actually mean in English. For example, have you ever thought about why we call math problems "problems"? In real life, when a problem confronts you, you probably think about different strategies you might use to overcome an obstacle and then decide on the best course of action. So why not use that same approach to solving math problems? This is probably the single biggest reason why taking math classes is useful to anyone: Math is all about learning and practicing problem-solving strategies. While we'll study a general framework for problem solving, as the quote above suggests, thinking too rigidly can be a big detriment to solving problems effectively. The more strategies you develop, the more types of problems you'll be able to solve—not just in this course, but in the big, wide world outside the classroom door.

0. What does "problem solving" mean to you in math classes?

1-3 Class

To begin the lesson, we're going to dive right in and try to solve a problem, with a little bit of guidance. Then we'll pause for a bit and summarize some important ideas that can be learned from our first problem.

Problem 1: While working late on an important case, a group of lawyers sends an intern out to buy takeout. A nearby food truck has lobster rolls for $14 and crab cakes for $8, and the intern was given $95. If the plan is to buy one item for each of eight people while spending as much of the $95 as possible, how many of each item can they buy?

1. The first thing we should do is make sure that we understand the problem. This entails both clearly identifying what we're being asked to accomplish, and what information we're provided that might be useful in accomplishing that goal. Write down the important information provided in the text of the problem, and write a sentence clearly identifying what you're being asked to do.

2. Remember that when solving problems in your life, you don't just do a bunch of random stuff. Instead, you think about the situation and try to devise a plan of action. In this case, it might be helpful to try out some potential solutions and see if they fit the given conditions. We'll figure out the cost of different combinations and see if we can find the one that fits the criteria best. First, find the total cost if the intern goes with one lobster roll and seven crab cakes. Is this the best choice? Why not?

3. Now we've decided on a potential approach. The next logical step to take is applying our strategy to try and solve the problem. In this case, fill in the table to work out the cost of several different combinations of items. We started the table for you, but you'll need to decide on further combinations to try.

# of Lobster Rolls	# of Crab Cakes	Total Cost
1	7	
2	6	

# of Lobster Rolls	# of Crab Cakes	Total Cost

4. Which is the best combination? Why?

5. When solving any type of problem, your last step should be a reality check. Does your answer make sense? Does it solve the problem and fit the parameters that were established? Discuss your solution, focusing on realism in the situation, fitting what you were asked to do, and matching the provided information.

Did You Get It

Try this problem to see if you understand the concepts we just studied. The answer can be found at the end of the portfolio section.

1. When shopping for a campus event, two representatives of a student organization are given a budget of $120 to buy bags of candy ($9 each) and packages of cupcakes ($12 each). Based on the expected attendance, they'd like to have a total of 10 or 11 items. How many of each should they buy? (An interesting, absolutely true note about organizational budgeting: When you're given a certain amount to spend, if you don't spend it all, next time you need money for something, you'll get less than you're asking for.)

One thing we can get from the experience of solving Problem 1 is that the process of **trial and error**—trying different potential solutions and adjusting until you find an acceptable solution—can be a really useful strategy. But there's actually something much deeper embedded in the first five questions of this lesson: a general problem-solving strategy that can be applied to almost any situation.

A Hungarian mathematician named George Polya did years of research on the nature of problem solving in the first half of the 20th century. His biggest contribution to the field was an attempt to identify a series of steps that were fundamental to problem-solving strategies used by great thinkers throughout human history. Polya's strategy isn't necessarily earth-shattering: Its brilliance lies in its simplicity. It provides four basic steps that can be used as a framework for problem solving in any area, from math to home improvements.

Polya's Four-Step Problem-Solving Procedure

Step 1. *Understand the problem.* The best way to start any problem is to write down information that's provided as you come to it. Especially with longer word problems, if you read the whole thing all at once and don't DO anything, it's easy to get overwhelmed. If you read the problem slowly and carefully, writing down information as it's provided, you'll always at least have a start on the problem. Another great idea: Carefully identify and write down what it is they're asking you to find. This almost always helps you to devise a strategy.

Step 2. *Devise a plan to solve the problem.* This is where problem solving is at least as much art as science—there are many, many ways to solve problems. Some common strategies: making a list of possible outcomes; drawing a diagram; trial and error; finding a similar problem that you already know how to solve; and using arithmetic, algebra, or geometry.

Step 3. *Carry out your plan to solve the problem.* After you've made a plan, try it out. If it doesn't work, try a different strategy! There are many different ways to attack problems. Be persistent!

Step 4. *Check your answer.* It's always a good idea to think about whether or not your answer is reasonable, and in many cases you'll be able to use math to check your answer and see if it's exactly correct. If not, don't forget what we learned about estimation in Lesson 1-2—that can be a big help in deciding if an answer is reasonable.

The goal of this lesson is to use this framework to solve a variety of different problems, using different general strategies. The first of those was trial and error. In our second problem, we'll see the value of drawing a diagram.

Solved Example

A private college is given a grant from a wealthy alumnus to build a sculpture garden, where statues of the first eight presidents of the college will be installed in a straight line from earliest to most recent. Each statue will be 8 feet tall and 2 feet wide, and the sculptor decides on a consistent distance of 12 feet between statues to reflect the 12 decades since the school was founded. How much space will be needed from the beginning of the first statue to the end of the last one?

SOLUTION

Let's use Polya's method and see if we can come up with a reasonable approach.

Step 1: Understand the problem. At this stage, it's ALWAYS a good idea to clearly identify exactly what we're being asked to find. In this case, it's the total amount of space needed. We should also identify the useful information we've been given: There are eight statues, each of which are 2 feet wide. They will be installed in a straight line with 12 feet in between any two consecutive statues.

Notice that there's also an extraneous piece of information: We're told that the statues are 8 feet tall, but that has no bearing on the space between them. The world doesn't always give you exactly what you need and nothing more when solving a problem! It's a very valuable skill to be able to pick out what is and isn't useful.

Step 2: Devise a plan. Here's a good guideline to keep in mind: If you can draw a diagram, you probably should. That sounds like an excellent approach here. A diagram with the statues and all the relevant measurements will most likely allow us to calculate the total space between the first and last statues.

Step 3: Carry out the plan. Here's our diagram, with the 2-foot width of the statues and the 12-foot distance between them labeled:

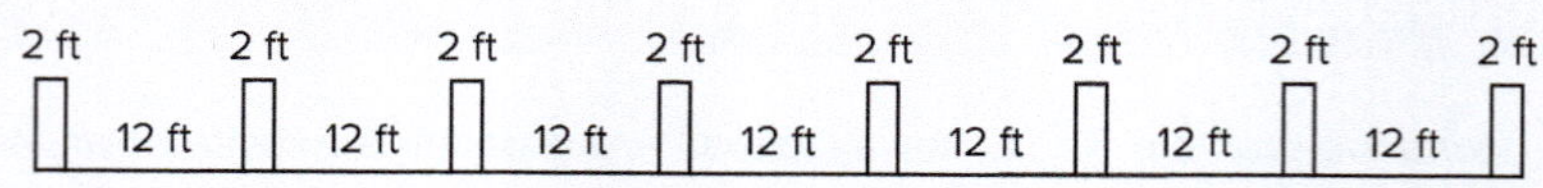

Now it's pretty easy to calculate the whole length. You could just add all the lengths one at a time, but you're smarter than that! There are eight widths of 2 feet, and seven 12-foot spaces in between, so the total amount of space needed is $8 \times 2 + 7 \times 12 = 16 + 84 = 100$ feet.

Step 4: Check your answer. Sometimes, an answer just kind of checks itself. Since we have the diagram, we can see that it matches all of the conditions laid out in the problem: eight statues that are each 2 feet wide, and 12 feet between consecutive statues. Exactly what we wanted.

Did You Get It

2. An architect is designing a rectangular sunken bathtub with dimensions 5 feet by 8 feet. The design calls for 2-inch rectangular tiles installed facing the ceiling around the outside of the tub as a border. How many tiles will be needed?

1-3 Group

Now that we've identified and practiced the four steps in Polya's problem-solving procedure, your group will solve a few other types of problems, keeping Polya's framework in mind. The first is another example of a problem for which a diagram is a big help.

Problem 2: A campus group is setting up a rectangular area for a tailgate bash. Since adult beverages will be served, city regulations require the area to be enclosed and access-controlled. They have 100 feet between two roads to use as width and 440 feet of fence to use. What length will use up the total amount of fence and therefore enclose the biggest space?

1. *Understand the problem.* Write down the important information provided in the text of the problem, and write a sentence clearly identifying what you're being asked to accomplish.

Nice One Productions/Corbis

2. *Devise a plan to solve the problem.* Why is a diagram likely to help? What should your diagram contain?

3. *Carry out the plan.*
   a. Draw a diagram of the tailgate area, and label the lengths of the sides that you know.

   b. How much fence is left for the two sides you don't know the length of?

c. What should the length of each side be? Justify your answer.

4. *Check your answer.* Is your answer physically reasonable? Does it match the conditions of the problem? Does it answer the question? Discuss.

Did You Get It

3. A poster is being designed to advertise an upcoming art show. It will be rectangular, and the printer requires a 1-inch margin on all four sides. To keep from paying an oversize fee, the promoter of the show wants the overall height of the poster to be 24 inches. The designer feels like the printed area needs to have a perimeter of about 75 inches. How wide should the poster be to make it as big as possible and meet these conditions?

Many applied problems involve making a choice between two or more options. In that case, you could flip a coin, but it doesn't seem likely you'll get very far in life using that approach. A better idea would be to use reasoning skills to make an informed decision. In some cases, that requires solving math problems, like the next one.

Problem 3: So you've graduated from college and you're ready for that first real job. In fact, you have two offers. Dude! Sweet. One pays an hourly wage of $20.75 per hour, with a 40-hour work week. You work for 50 weeks and get 2 weeks' paid vacation. The second offer is a salaried position, offering $41,000 per year, with an opportunity for a $5,000 performance bonus. Which job will pay more? Which sounds like the better opportunity financially?

5. *Understand the problem.* Write down the important information provided in the text of the problem, and write a sentence clearly identifying what you're being asked to accomplish.

6. *Devise a plan to solve the problem.* What would you need to know about the first job in order to compare the amount of money you'd make to the second one? How can you find that out?

7. *Carry out the plan.*

a. How much would the hourly job pay for each 40-hour week? Show your work.

b. How much does the hourly job pay per year?

c. Which job pays more, hourly or salaried? Which is the better opportunity? Explain.

8. *Check your answer.* Here's a clever approach: Do the opposite of the calculations you did in Question 7 to find what the hourly rate would be for the salaried job. Does this confirm your answer?

Did You Get It

4. One investment requires an initial deposit of $8,000 and generates interest at the rate of $1.35 per day. The other requires an initial investment of $12,000 and generates interest at the rate of $10.40 per week. Which will earn more interest in a year? Which is the more attractive investment?

Our last problem is intentionally left more open-ended. Rather than providing the specific steps, we're going to ask you to discuss it among your group members to develop and execute a strategy. Show all work, and it's probably a good idea to write some sentences describing your approach as well. Make sure that you LIST each step in the four-step method, then follow that up with what you did in your solution to use that particular step.

Problem 4: The grade in Marlene's history class will be determined completely by three tests, each worth 100 points. She scored 78 and 84 on the first two tests, but still hopes to get an A, which would require an average of 92. What's the minimum score she can get on the third test?

To close our lesson (and unit) on problem solving, we've compiled a list of strategies that can come in handy in ANY class, or any situation, where you need to solve a problem.

Strategies for Understanding Problems

One reason that problem solving is challenging is that every problem can be a little different, so you can't just memorize a procedure and mimic it over and over. You actually have to think on your own! But here are some suggestions for helping you to understand a problem and devise a strategy for solving it.

- If the problem describes something that can be diagrammed, a drawing almost always helps.
- If you find that you tend to read through problems and feel like you have no idea where to start, you should consider NOT reading the entire problem at first. This can sometimes lead to feeling overwhelmed or intimidated. Instead . . .
- Write down all the numeric information in the problem as you read it. This often helps to organize your thoughts. But maybe more importantly, it's a guaranteed way to get a START on every problem, so you avoid that moment of panic when you feel like you have no clue what to do.
- Sometimes making a chart to organize information is helpful, especially when you're trying to use trial and error.
- Don't expect that you'll know exactly how to solve a problem after reading it once! Most often you'll need to read through a problem several times, and even then you may need to just try some approaches before finding one that works. Don't be afraid to try! Nobody ever became a good problem solver by being afraid to make a mistake.

1-3 Portfolio

Name ______________________________

Check each box when you've completed the task. Remember that your instructor will want you to turn in the portfolio pages you create.

Technology

1. ☐ One downside of solving problems using trial and error is that you often have to do many repeated calculations. If only there were a tool that we could use to quickly do repeated calculations . . . see where this is headed? A little bit of spreadsheet setup time can really pay off. In Question 3 of the Class portion of this lesson, you created a table while solving a problem using trial and error. In this Excel activity, we'll tackle this same problem.

ALEKS Practice

1. ☐ Include any written work from the ALEKS assignment along with any notes or questions about this lesson's content.

Applications

1. ☐ Complete the applications problems.

Reflections

Type a short answer to each question.

1. ☐ Do you feel better about your ability to solve word problems after learning and practicing Polya's procedure? Be specific: Provide a specific example of why you do or do not feel better about your ability.
2. ☐ Of the problem-solving strategies we studied, which do you feel like you're likely to use most often in your life?
3. ☐ Name one thing you learned or discovered in this lesson that you found particularly interesting.
4. ☐ What questions do you have about this lesson?

Looking Ahead

1. ☐ Complete the prep skills for Lesson 2-1.
2. ☐ Read the opening paragraph in Lesson 2-1 carefully and answer Question 0 in preparation for that lesson.

Answers to "Did You Get It?"

1. The combination of four bags of candy and seven packages of cupcakes works out to be exactly \$120.
2. 160 tiles (did you forget the corners?) 3. The width should be 17.5 inches.
4. First investment: \$492.75; second investment: \$540.80. Which is better depends on how you feel about the larger initial investment.

Answers to Prep Skills Questions

1. 32 ft
2. 18 in.
3. 10 ft 8 in.

1-3 Applications

Name ______________________________

1. A physical therapy facility is building a new pool that is 60 feet long and 5 feet deep. They have ordered enough material for a 220-foot-long border around the edge. How wide should the pool be to ensure that the entire amount of border material is used?

2. A physical education teacher plans to divide the seventh graders at Wilson Middle School into teams of equal size for a year-ending mock Olympic event. The teacher wants each team to have between 5 and 9 students, and has a strong preference to have the same number of students on each team. The seventh grade at Wilson consists of three classes: one with 24 students, one with 26, and one with 21. How many students should be on each team?

1-3 Applications

Name ______________________________

3. A condo on the water in Myrtle Beach can be rented for $295 per day, with a nonrefundable application fee of $50, and a daily parking fee of $11. Another one down the beach can be rented for $2,100 per week, plus a $600 refundable security deposit and a $4 per day Internet fee. Which condo costs less for a week's stay?

4. Suppose you're taking a pass-fail math class and need to average 70% to get a passing grade. The grade is determined by five 50-point tests, and you're doing really well: On the first four tests, you scored 48, 45, 45, and 47. High five! The week of the final test, you have a really important chemistry test that you want to focus on, but you don't want to risk failing math. Find the range of scores you can get on test five that will result in failing the class.

Math in the Disinformation Age Revisited

Bilanol/Shutterstock

1. Weak inductive reasoning. There are probably hundreds of streets on Maui. It could just be a coincidence that the four nearest have names that begin with K, or it could be that the local neighborhood identifies itself by naming all streets with such names. If you're in Maui, which would be super cool, you could drive around and look at some street signs. If not, you could investigate this by doing an Internet search for street names in Maui, where you would quickly discover that there are names beginning with other letters.
2. Deductive reasoning. If human beings have Neanderthal DNA, it had to get there somehow, and the known laws of genetics say this could only happen through interbreeding. An Internet search for "Neanderthal DNA" will yield a wealth of information on this fascinating subject.
3. This is nonsense. The "newspapers" you see in supermarket checkout lanes are not considered serious journalism by any media experts. It would take just a couple of minutes online to find that such a claim was not true.
4. Strong inductive reasoning. There is no rule that says a Democratic candidate cannot win in South Carolina, but a history of over 12 consecutive elections voting for the Republican candidate is pretty strong evidence that the trend will continue. There are numerous websites that track the voting patterns of states, counties, and cities. A good start toward checking this claim would be a search for "South Carolina presidential elections."

Unit 2
Managing Your Money

Jasper White/Image Source

Outline

Math in Paying for an Education

"If you think education is expensive, try the cost of ignorance." This famous quote comes from Dr. Derek Bok, the former president of Harvard University. (I like it so much that it's the only quote used twice in this book.) There are many nuanced interpretations of exactly what Dr. Bok was trying to say, but if you think of it strictly in terms of financial costs, it says that your earning power increases dramatically as you attain more education. This isn't a cliché or a myth, and it's not an opinion to be debated on cable news channels. The table below is based on cold, hard facts.

Education Level	Median Salary in 2022
Less than a high school diploma	$ 35,464
High school diploma	$ 44,356
Some college, no degree	$ 48,620
Associate's degree	$ 52,260
Bachelor's degree	$ 74,464
Master's degree	$ 86,372
Professional or doctorate degree	$108,160

Source: Bureau of Labor Statistics

The downside, of course, is that college is expensive, and costs continue to rise at a rate far greater than inflation. According to USinflationcalculator.com, between 2000 and 2024, the price of average consumer goods in the U.S. went up by 82.7%. Ouch. But during the same period, college costs went up by 184.6%! The truth is that a small percentage of families have the means to pay all of the costs associated with college, meaning that student loans are a reality for an increasing segment of the population. Even the most optimistic students worry about whether or not they'll be able to afford college, and how much debt they'll run up while earning a degree.

This might lead you to a larger realization: Long ago, survival depended on overcoming physical challenges that modern humans can only imagine, but today, survival is more about navigating the waters of the modern financial system. There are over 300 million people in the U.S., and at any given time, it seems like half of them are trying to figure out a way to separate you from your hard-earned money. That's why this chapter is all about different aspects of our financial system that are particularly important to the average adult. Success in this chapter will help you to become a more well-informed consumer, which makes it less likely that those 150 million people will succeed in getting your cash.

One of the topics we'll study in this unit is student loans, and a goal will be to decide if a college education really pays off in the long run. Recently, the National Center for Educational Statistics reported that a typical college budget for an in-state public 4-year college for the 2022–2023 academic year averaged $27,150. (This takes into account tuition, fees, room and board, and other living expenses.) Let's study the ramifications of this intimidating statistic.

1. How much would it cost to earn a four-year degree with this budget?
2. If you are able to obtain grants and scholarships to pay for 25% of these costs, and can also pay $5,000 per year out of pocket, how much money would you have to borrow to cover the rest?
3. If you obtain a federal student loan with an interest rate of 6.8% and a term of 10 years for the amount in Question 2, what will your monthly payment be upon graduation if you pay interest on the loan while still in school? How much total will you pay in interest?
4. Repeat Question 3 if you capitalize the interest on your loan.
5. Based on the median salaries in the table, how long would you have to work after obtaining a bachelor's degree to make back the total amount spent on student loan payments and the amount paid out of pocket? (Use the loan with interest capitalized, and don't forget to factor in how much you would have been able to make without going to college at all!)

For answers, see Math in Paying for an Education Revisited on page 197.

Lesson 2-1 Prep Skills

SKILL 1: REDUCE FRACTIONS

A fraction is in **lowest terms** or is **reduced** if there are no factors (other than 1) common to the numerator and denominator.

- $\frac{4}{6}$ is not in lowest terms because 2 is a factor of both 4 and 6.
- $\frac{6}{25}$ is in lowest terms because the factors of 6 are 1, 2, 3, and 6, while the factors of 25 are 1, 5, and 25.

The process of writing fractions in lowest terms is called **reducing fractions.** It's useful to do because it's easier to work with and interpret fractions when they're in lowest terms. For example, it's a lot easier to see that $\frac{5}{2}$ is two and a half than it is to see that $\frac{180}{72}$ is two and a half. To reduce a fraction, divide BOTH the numerator and denominator by any common factors.

- $\frac{4}{6} = \frac{4 \div 2}{6 \div 2} = \frac{2}{3}$ The greatest factor common to 4 and 6 is 2.
- $\frac{30}{12} = \frac{30 \div 6}{12 \div 6} = \frac{5}{2}$ The greatest factor common to 30 and 12 is 6. You could also divide both the numerator and denominator by 3, then by 2.

SKILL 2: WRITE FRACTIONS IN DECIMAL FORM

To write a fraction in decimal form, you just need to perform the division, either by hand or with a calculator.

- $\frac{3}{5} = 0.6$

If the decimal doesn't terminate, it's usually appropriate to round. If a certain number of decimal places isn't required, use your judgment based on the context: How accurate would you want your answer to be in a given situation? Note that since you're rounding, the decimal is just an approximation of the exact fractional value.

- $\frac{3}{7} \approx 0.43$

SKILL 3: SOLVE EQUATIONS USING DIVISION

To solve equations using division, use the fact that dividing both sides of an equation by any nonzero number will produce an equivalent equation.

- $7b = 42$ To solve this equation for *b*, divide both sides by 7.

  $b = 6$
- $1.5x = 15$ To solve this equation for *x*, divide both sides by 1.5.

  $x = 10$

PREP SKILLS QUESTIONS

1. Reduce each fraction to lowest terms.

   $\frac{6}{10}$ $\frac{24}{30}$ $\frac{51}{21}$ $\frac{11}{24}$

2. Reduce each fraction to lowest terms.

   $\frac{60}{100}$ $\frac{24}{100}$ $\frac{98}{100}$ $\frac{33}{100}$

3. Write the decimal form of each fraction.

   $\frac{12}{100}$ $\frac{9}{30}$ $\frac{11}{8}$ $\frac{11}{3}$

4. Solve: $12x = 84$

5. Solve: $0.3y = 210$

6. Solve: $0.125x = 47.99$

Lesson 2-1 Giving 110 Percent

LEARNING OBJECTIVES

- ☐ 1. Perform conversions and calculations involving percents.
- ☐ 2. Find percent increase or decrease.
- ☐ 3. Solve problems using percents.
- ☐ 4. Evaluate the legitimacy of claims based on percents.

Photodisc/
Getty Images

I think I'm a kind of a person who works hard at whatever I do, literally from being a waitress to being on television. I always try to give 110 percent to whatever it is I'm doing.
—Teri Hatcher

Coaches are fond of insisting that their players give 110% effort. Players who are good at math will sometimes counter with the fact that it's impossible to give more than you have, which by the way sounds like a good idea if you want to run laps for about a week. Can you really have 110% of something? Since percents play an important role in many, many different areas where studying numbers is useful, everyone who wants to be successful in our society should be armed with a solid understanding of percents: what they mean, how to compute and use them, and how to interpret information that they're providing. In this lesson, we'll review a lot of useful information about percents, information that will come in handy throughout the rest of the course. Pay attention, or you'll be doing pushups until you drop.

0. Write about the most recent time you've come across something involving percents in your daily life.

2-1 Class

Surely you're familiar with the word "percent," but have you ever thought about what it literally means? "Percent" can be translated literally as "per hundred." So 1% = 1/100. For example, the National Coffee Association reported that 78% of Americans who drink coffee do so mostly at home. Of those folks, 61% described the coffee they prefer as "gourmet." This tells us that 78 out of every 100 coffee drinkers consume coffee at home, and that 61 out of 100 described their coffee as gourmet.

Percent Conversions

To work with percents in calculations, we will need to convert them to either decimal or fractional form. But to interpret the answers to calculations, we will want to convert them back into percents.

1. Using the fact that "percent" means "per hundred," write 84% and 9% as fractions with denominator 100.

Here's a simple rule that can be learned from Question 1:

Converting a Percent to a Fraction

To convert a percent to a fraction, drop the percent sign and use the percent number as the numerator of a fraction whose denominator is 100.

If you noticed that the fraction in the first part of your answer to Question 1 could be reduced, good job! When we're working with percents in fraction form, many times it will be easier if the fraction is reduced to lowest terms.

2. Convert each percent to a fraction in lowest terms.

   a. 42%

   b. $37\frac{1}{2}\%$ (Hint: After initially writing as a fraction, multiply by $\frac{2}{2}$ so that the numerator and denominator of your fraction are both whole numbers.)

A lot of times it's especially useful to write a percent in decimal form when doing calculations, so let's see how that works, using what we already know about fractions.

3. We know that we can write 12% as a fraction in the form $\frac{12}{100}$, and 72% in the form $\frac{72}{100}$. Use a calculator to perform the division in each case, writing the original percent in decimal form.

4. Use the result of Question 3 to fill in the blanks in our next rule for converting percents:

Converting a Percent to a Decimal

To convert a percent to a decimal, drop the percent sign and move the decimal point ________ places to the _______.

5. Write each percent in decimal form.

a. 41% **b.** 6.7% **c.** 120%

Sometimes, interpreting the story told by a percentage will require a conversion in the other direction: starting in decimal form, and converting to percent form. Not surprisingly, we accomplish this by doing exactly the opposite of the steps we did for converting from percent to decimal form.

6. The opposite of moving the decimal two places to the left is moving the decimal ________________. The opposite of dropping the percent sign is ________________.

7. Write a procedure for converting a decimal to percent form, then use it to convert 0.34, 0.07, and 1.658 into percent form.

Math Note

Part c of Question 5 shows that our conversion rules work just fine for percentages greater than 100. We'll study the meaning of such a percentage in the Group portion of this lesson.

Did You Get It

Try this problem to see if you understand the concepts we just studied. The answer can be found at the end of the portfolio section.

1. Perform each conversion.
   a. 47.3% to decimal b. 0.24 to percent c. 80% to fraction in lowest terms

In the most common calculations involving percents, the goal is to find a percentage of some quantity. To understand how to do so, consider the following example. You probably know that 50% of 10 is 5. Let's rewrite that statement, then turn it into a calculation:

50% of 10 is 5

$0.5 \times 10 = 5$

When writing a percentage statement in symbols, the word "of" becomes multiplication, and the word "is" becomes an equal sign. Also, we need to change the percent into decimal or fractional form. Let's look at some sample calculations using this idea.

Solved Example 1

Finding a Percentage of a Dinner Bill

I recently went to dinner at Skyline Chili because their chili is really good, and because they were running a fundraiser for the ALS Society. Win-win! On that particular day, 30% of every purchase was being donated to ALS research. Our check for the meal was $25.49. How much of that was donated?

SOLUTION

Here's a GREAT tip when solving problems involving percents: Write the percent statement in the form of a sentence, as described above. In this case, we're asked to find what 30% of $25.49 is, and we can write that in this form:

30% of $25.49 is what amount?

Notice that the calculation on the left side is something we know how to do. We know that 30% can be written in decimal form as 0.30, and that "of" translates into multiplication. So the left side of our statement is

$$0.30 \times \$25.49$$

and using a calculator, we find that the product is $7.65. This is the amount we're looking for, so $7.65 was donated to ALS research.

The great thing about writing percent statements in words and then translating them to symbols is that it can be used to solve several different types of problems. In Solved Example 1, we knew the percentage and the whole amount and were asked for a portion of that whole. But what if we know the percentage and the portion, and need to find the whole? No problem—the same procedure works. It just requires a little algebra. (Very little, as it turns out.)

Solved Example 2

Calculating the Original Price of a Sale Item

A leather store at an outlet mall is offering 70% off a jacket that is sale priced at \$59. What was the original price of the jacket?

SOLUTION

Let's start with a clever observation that can be used to make many problems involving percents easier: If an item is 70% off, then the sale price is 30% of the original price. Did you catch what we did there? We figured out a way to write the problem in the form of a percent statement:

30% of the original price is \$59

Since we don't know the original price, let's just use a letter to represent it for now so that we can write our statement in symbols. I'll say $p =$ the original price, giving me this equation:

$$0.30 \times p = \$59$$

Now I just have to solve that equation to find the original price. Of course, I can do that by dividing both sides by 0.30, which leaves the variable quantity p alone on the left.

$$p = \frac{\$59}{0.30} = \$196.67$$

The original price of the jacket should be \$196.67, so if that's not what the price tag says, someone is being dishonest.

Guess what? The same idea can be used to find a percentage, but it's more straightforward to use division. For example, of the 32 teams in the National Football League, 14 make the playoffs each year. What percentage of teams make the playoffs?

8. Divide the number of teams that make the playoffs by the total number of teams, and write your answer in decimal form.

9. Based on your answer to Question 8, what percentage of teams make the playoffs?

Did You Get It

2. Calculate each amount involving percents.
   a. Find 38% of 400.
   b. \$17.49 is 20% of what amount?
   c. What percentage of 42 is 15?

2-1 Group

If it doesn't make you uncomfortable, exchange the following information with the classmates in your Unit 2 group. This will be your small group for the second unit. It would be a good idea to schedule a time for the group to meet to go over homework, ask/answer questions, or prepare for exams. You can use this table to help schedule a mutually agreeable time.

Name	Phone Number	Email	Available Times

1. Let's look further into the idea of "giving 110% effort." First, explain why a lot of people feel like that saying doesn't really make sense.

2. Convert 110% into decimal form, then multiply the result by 200 to find what 110% of 200 is.

3. Suppose that in a normal half hour of practice, a tennis player expends 280 calories of energy. How many calories would that player need to expend in order to "give 110%" effort?

Did You Get It

3. A real estate developer is looking for investors to help fund a new condo complex. They promise a 145% return on investment in 4 years. If you're interested in investing $30,000 and the developer follows through on that promise, how much money will you get back in 4 years?

Now let's go shopping. If I had a dollar for every time someone has asked me to calculate a discount price for them when a sign advertises a certain percent off, I'd be able to buy us both a very handsome gift indeed. In fact, shopping is one of the most common ways that almost everyone is likely to encounter percents.

Since my wife and I have a Macy's credit card, we get coupons from them in the mail pretty often. On one recent mall outing, we found a red dress with a selling price of $79, and had two coupons to choose from: One offers $15 off any purchase of $50 or more, the other offers 20% off any item. Which was the better choice?

Dave Sobecki

4. Find 20% of $79.

5. Keeping in mind that you just found the amount the price would be discounted, which coupon should Cat have used? Explain.

6. The sales tax in our county is 6.5%. Find the amount of tax on the dress purchase (after applying the coupon that you chose), and the total cost including tax.

Did You Get It

4. A refrigerator has a regular price of $1,149 at Home Depot and $1,219 at Lowe's. Home Depot is offering 15% off, but Lowe's offers $200 off if you put the purchase on store credit. Which is the better option?

7. It's common for stores to advertise a certain percentage off, then say "discount is reflected in the marked price," leaving you to trust that you're getting the percent off that they claimed. Suppose that a refrigerator is listed as 30% off, with a sale price of $749. What would that make the original selling price? (Look back at Solved Example 2 in the Class section if you need help.)

8. A large latte at Shaky's Caffeine Shack usually sells for the low price of $3.50, but today, Shaky has gone goofy and is offering it for just $3.00. Wow! Why not just give them away? The coffee shop across the street is offering 20% off. Which store is offering the better deal?

The percentage you found in Question 8 is called the **percent decrease.** It's the amount that some quantity is decreased as a percentage of the original amount. Similarly, **percent increase** is the amount of increase as a percentage of the original amount. That sounds like a good opportunity for a colored formula box.

Computing Percent Increase and Decrease

When some quantity gets larger, we calculate the percent increase using the formula

$$\text{Percent Increase} = \frac{\text{Amount of increase}}{\text{Original amount}} = \frac{\text{New amount} - \text{Original amount}}{\text{Original amount}}$$

If the quantity gets smaller, we calculate the percent decrease using the formula

$$\text{Percent Decrease} = \frac{\text{Amount of decrease}}{\text{Original amount}} = \frac{\text{New amount} - \text{Original amount}}{\text{Original amount}}$$

In each case, the result is a percentage in decimal form. Percent increase and decrease are also sometimes called "percent change" or "relative change."

9. In January of 2025, the state minimum wage in Alaska was raised from $11.73 to $13.00. In Hawaii it went from $14 to $16. Which state had the higher percent increase?

Did You Get It

5. In an effort to show solidarity with his workers, a CEO offers to take a $300,000 per year pay cut. Last year, his salary was $11 million. Find the percent decrease in pay.

10. On October 2, 2023, shares of stock in Netflix were selling at $380.33, while shares in Dick's Sporting Goods were selling at $106.30 per share. A year later, Netflix stock had gone up to $706.34, and Dick's was up to $205.40. Compare the increase in the two stocks, as well as the percent increase. Which sounds like a better investment to you? Explain.

Piotr Adamowicz/Alamy Stock Photo

We would never, ever accuse any advertisers of being deceptive (wink wink), but let's just say that it's not uncommon for percents to be used incorrectly in marketing materials. Let's look at one such case to close our review of percents.

11. A department store advertised that certain merchandise was reduced 25%. Also, an additional 10% discount card would be given to the first 200 people who entered the store on a specific day. The advertisement then stated that this amounted to a 35% reduction in the price of an item. Explain why the store marketing manager is either a liar or bad at math. Show calculations to back up your explanation.

12. In promotional materials, a company that increased their work force from 1,000 to 1,500 workers claimed that they hired 50% more workers. A rival of theirs bragged about decreasing emissions from 1,500 to 1,000 cubic feet per day, claiming a 33% reduction. Which company is lying? Explain.

2-1 Portfolio

Name ____________________

Check each box when you've completed the task. Remember that your instructor will want you to turn in the portfolio pages you create.

Technology

1. ☐ In this Excel activity, we'll build a spreadsheet that calculates the change and the relative change between two values.

ALEKS Practice

1. ☐ Include any written work from the ALEKS assignment along with any notes or questions about this lesson's content.

Applications

1. ☐ Complete the applications problems.

Reflections

Type a short answer to each question.

1. ☐ What is the literal translation of the word "percent"?
2. ☐ How can an understanding of percents be valuable to you as a shopper?
3. ☐ Do you think you can give 110% effort to doing well in this class? Explain.
4. ☐ Name one thing you learned or discovered in this lesson that you found particularly interesting.
5. ☐ What questions do you have about this lesson?

Looking Ahead

1. ☐ Complete the prep skills for Lesson 2-2.
2. ☐ Read the opening paragraph in Lesson 2-2 carefully and answer Question 0 in preparation for that lesson.

Answers to "Did You Get It?"

1. a. 0.473 **b.** 24% **c.** $\frac{4}{5}$ **2. a.** 152 **b.** \$87.45 **c.** 35.7% **3.** \$43,500
4. \$976.65 at Home Depot vs. \$1,019 at Lowe's **5.** 2.7%

Answers to Prep Skills Questions

1. $\frac{3}{5}, \frac{4}{5}, \frac{17}{7}, \frac{11}{24}$
2. $\frac{3}{5}, \frac{6}{25}, \frac{49}{50}, \frac{33}{100}$
3. $0.12, 0.3, 1.375, 3.\overline{6}$
4. $x = 7$
5. $y = 700$
6. $x = 383.92$

2-1 Applications

Name ______________________________

Many people probably think that those who are good at math always do exact calculations, but we know better: Estimation is actually a really useful tool, especially when it comes to things like percent discount. Suppose that you're shopping on a rack that offers 40% off the lowest ticketed price, and you're interested in a sweater with a ticket price of $47.99.

PhotoLink/Getty Images

1. Given that 40% is being subtracted from the ticket price, explain why that makes the sale price 60% of the ticket price.

2. The ticket price can be rounded easily to $50, so you can estimate the sale price by multiplying 50 by 0.6. The easy way to do this is to multiply 50 by 6, which you can do in your head, then move the decimal one place to the left. Use this idea to estimate the sale price. (For what it's worth, I used this procedure about eight times yesterday while clothes shopping.)

3. Now use the procedures we reviewed in this lesson to find the exact discount price.

4. Would your estimate from Question 2 have been good enough for you to decide whether or not you wanted to buy the item? Discuss.

2-1 Applications

Name ______________________________

5. After a round of layoffs, a company that used to have 1,200 employees now has 900. They then issue a press release saying "In spite of trying economic times, we were able to limit job cuts to just 25% of our job force." A couple of years later, they raise the number of employees back to 1,200. This time, they claim "We were able to increase our job force by more than 33%!" In which press release were they lying? Explain.

6. An employer offers a retirement program in which 9.8% of a worker's salary is deducted from their pay and deposited into an investment fund. The company then adds an additional 12% of the worker's salary to the fund. How much would an employee have deposited into their fund in the first year if their starting salary is $43,000?

7. While shopping at Nordstrom, you pick out two pairs of jeans at $49 each and a pair of shoes on sale for $39. Then you splurge on a $95 sweater. You have two coupons but can only use one. The first offers 40% off any single item. The second offers 15% off your entire purchase. Which is the better choice?

Lesson 2-2 Prep Skills

SKILL 1: COMPUTE A PERCENTAGE

To compute what percentage of a whole a certain amount is, you perform two steps.

Step 1. Divide the portion by the whole. This gives you the percentage in decimal form.

Step 2. Convert that decimal to percent form by moving the decimal two places to the right. This corresponds to multiplying the decimal form by 100.

- 20/50 = 0.4, which is 40%. So 20 is 40% of 50.
- 110/150 ≈ 0.733, which is 73.3%. So 110 is about 73.3% of 150.
- 8/120 ≈ 0.067, which is 6.7%, so 8 is about 6.7% of 120.
- 200/150 ≈ 1.33, which is 133%, so 200 is about 133% of 150.

SKILL 2: COMPUTE A PERCENTAGE OF A GIVEN AMOUNT

To compute a number that represents a certain percentage of a given amount, you perform two steps:

Step 1. Convert the percentage to a decimal by moving the decimal point two places to the left.

Step 2. Multiply the decimal form of the percentage by the given amount.

- What is 20% of 80?
  0.20(80) = 16
  The decimal form of 20% is 0.20. The calculation tells us that 20% of 80 is 16.
- What is 7.5% of 900?
  0.075(900) = 67.5
  The decimal form of 7.5% is 0.075. The calculation tells us that 7.5% of 900 is 67.5.

SKILL 3: FIND AN AMOUNT INCREASED BY A CERTAIN PERCENTAGE

To find an amount increased by a certain percentage, you have two options: You can find the increase and add it to the original amount, or you can multiply the original amount by a growth factor to find the new amount. The growth factor will always be 1 + the percentage in decimal form. The second method takes a bit of thought, but it's quicker and will really come in handy.

- If $300 increases by 4%, what is the new amount?

  Method 1:
  $300 + 0.04($300) — $300 + 4% of $300
  = $300+ $12
  = $312 — If $300 is increased by 4%, the new amount is $312.

  Method 2:
  Growth factor is 1.04 — 104% is 4% more than 100%
  $300(1.04) = $312 — We calculate 104% of $300. If $300 is increased by 4%, the new amount is $312.

PREP SKILLS QUESTIONS

1. What percent of 100 is 40?
2. What percent of 520 is 100?
3. What percent of 12 is 15?
4. What is 30% of 100?
5. What is 85% of 1,200?
6. What is 3.1% of $1,118.13?
7. If an amount of $500 increases by 15%, what is the new amount?
8. If an amount of $12,000 increases by 3.4%, what is the new amount?

Lesson 2-2 Building It Is the Easy Part . . .

LEARNING OBJECTIVES

- ☐ 1. Calculate take-home pay and monthly expenses.
- ☐ 2. Identify necessary expenses and luxuries.
- ☐ 3. Prepare a monthly budget.
- ☐ 4. Prorate long-term expenses to save in advance for them.

Glow Images

A budget tells us what we can't afford, but it doesn't keep us from buying it.
—William Feather

How much money did you spend last month? If you're like most folks, you probably can't answer that question exactly, and if you're like a lot of folks, it would take you a while to even come up with a ballpark guess. By some estimates, American consumers have a combined debt of over $11 trillion. Trillion! That's an almost unimaginable amount of money. Imagine spending a million dollars a day, every day. That would be pretty much impossible, wouldn't it? But if you could, it would still take you over 2,700 YEARS to spend a trillion dollars. This debt problem happened because people spend more money than they make. And one of the reasons that happens is that so few people put forth the effort to track exactly what they make and spend. Budgeting is the process of deciding how much money you can spend on various expenses based on your income: that is, the amount of money you're earning. Learning how to budget your money is a good first step in avoiding massive debt; the hard part is sticking to that budget!

0. Would you be able to make a reasonable estimate of how much total you spend each month? Do you think this would be useful? Discuss.

2-2 Class

The first step in building a reasonable budget is to calculate exactly how much money you have to spend, so we'll begin there.

Solved Example 1

Calculating Annual Pay

Charlie works full-time as an executive assistant, and he has a check for $2,323.13 direct-deposited into his checking account every other Friday. How much money does Charlie make per year?

SOLUTION

This is an interesting question because there are a few different ways you could approach it, depending on how accurate you want your answer to be. We could get a crude estimate by figuring that every other Friday is about twice a month, or 24 times per year. We could then multiply $2,323.13 by 24 to get $55,755.12. But that's too low, because 4 weeks isn't quite a full month (except for February).

It would be more accurate to estimate a year at 52 weeks, which would mean 26 biweekly checks. This time we'd get 26 times $2,323.13, which is $60,401.38—quite a bit more!

The most accurate result would probably come from breaking the paycheck down to a daily amount. He gets $2,323.13 every 2 weeks, or 14 days. So he makes $\frac{\$2{,}323.13}{14} = \$165.94$ per day. Since there are 365 days in a year, we multiply the daily amount by 365 to get $60,568.10.

Most people can tell you their hourly wage or salary, but because of deductions for taxes, insurance, and retirement, a lot of people don't know how much money is actually left for them to spend. This amount is known as **take-home pay,** while the amount earned before deductions is called **gross pay.** Since many common expenses occur on monthly cycles, it's useful to know exactly how much money you bring home per month.

1. Refer to Solved Example 1. What is Charlie's monthly take-home pay? Please describe how you got your answer.

Did You Get It

Try this problem to see if you understand the concepts we just studied. The answer can be found at the end of the portfolio section.

1. Aaliyah works 25 hours a week at a campus bookstore. She gets a weekly paycheck in the amount of $318.70. What is her monthly take-home pay?

By far the biggest difference between money earned and take-home pay for most people is taxes. The percentage of your pay that is withheld for taxes depends on a large variety of factors including income level, where you live, and family status. Nationally, it's hard to pin down an exact average, but standard estimates are that the average person takes home about 70–75% of his or her pay.

Solved Example 2

Calculating Take-Home Pay

If a worker is paid a salary of $47,400 per year and is in a tax bracket that results in 27% deductions, what is their monthly take-home pay?

SOLUTION

Our strategy will be to first find the total take-home for the year, then divide by 12 to get the monthly take-home. There are two ways to attack this. First we'll demonstrate the most straightforward approach, then the quicker and more clever one.

The straightforward approach is to find the amount that gets deducted, which is 27% of the total, then subtract that from the total:

$$27\% \text{ of } \$47{,}400 = 0.27(\$47{,}400) = \$12{,}798$$
$$\$47{,}400 - \$12{,}798 = \$34{,}602$$

Now we divide by 12 to get $\frac{\$34{,}602}{12} = \$2{,}883.50$ for the monthly take-home.

The more clever approach is one that comes in handy a lot in this course. If 27% is taken out of the total pay, then 73% must be left over. (Note that 100% − 27% is 73%.) So the take-home pay is 73% of the total, and we can get that amount in one step. Efficiency is smart.

$$73\% \text{ of } \$47{,}400 = 0.73(\$47{,}400) = \$34{,}602$$

Again, of course, dividing by 12 will give us $2,883.50 per month.

2. The main topic of this lesson is personal budgeting, but organizations—from small charities to the federal government—must also prepare a budget every year. For example, for fiscal year 2023, the president submitted a budget to Congress requesting $6.1 trillion to be spent by the government. Of that amount, $659 billion went toward paying interest on the national debt. What percentage of the budget went to that one expense? (A trillion is a thousand billion, by the way.)

Knowing how much money you bring home is half of budgeting; the other half is tracking your expenses, which is the amount of money you spend. Most people don't realize that expenses that seem minor can really add up.

3. My friend Charles noticed that one of his students came to class every day with two cups of coffee from a well-known coffee chain that isn't exactly famous for their low prices. He asked her if she'd ever thought about how much she spent on that coffee over the course of a year; not surprisingly, she had not. If she paid $3.25 per cup for that coffee every day, how much would she spend on coffee in a year? First, make a guess without doing any calculations. Then calculate the actual amount and describe your reaction.

John Flournoy/McGraw Hill

4. How much would our coffee-drinking friend save over the course of a year if she settled for convenience store coffee at 79 cents per cup?

Did You Get It

2. The average price of a pack of cigarettes in the U.S. in 2024 was about $8.39. How much money would a pack-a-day smoker spend in a year?

We'll call expenses like daily gourmet coffee **luxuries,** which are things you can live without. Everyone wants to be able to spend money on luxuries, but you won't know how much you can comfortably spend until you calculate your **necessary expenses.** These are the basic things you have to pay for: food, shelter, insurance, transportation, clothing, utilities, student loan payments. Some of those expenses are **fixed:** The amount you spend each month is consistent. Housing, insurance, utilities, loans, and transportation usually fall into this category. Other necessary expenses are **variable,** meaning they can change from month to month, like food and clothing.

The key to maintaining a sensible budget is to add up your necessary expenses and subtract them from your income. The result will let you know how much you have to spend on variable expenses, entertainment, and luxuries. It will also let you know if you're able to save any money (which should be one of your goals every month).

5. Tremaine brings home $2,746.79 per month and has fixed expenses shown in the table. How much can he afford to spend on food, clothing, and luxuries?

Expense	Amount
Rent	$675
Car Payment	$299.23
Cell Phone	$79.50
Gas	$110

Expense	Amount
Insurance	$97.50
Utilities	$175
Student Loan	$211.53

6. Tremaine goes out for happy hour drinks and dinner with friends every Friday, and usually spends about $50. On average, what percentage of his budget after fixed expenses is eaten up by these happy hours?

7. In Question 6, you found a percentage. What expenses would you be willing to spend that percentage of your income on? Discuss with your group and make a list.

8. Good news: Tremaine just got a 4% raise. (Looks like he may be spending more than the usual $50 this Friday.) How much more money does that leave him for variable expenses per month?

Did You Get It ?

3. Leslie shares an apartment with two friends, and the rent and utilities are split equally by all three tenants. Find the amount left for variable expenses and luxuries if her monthly income is $1,556.27 and fixed expenses are shown below.

Total rent:	$1,100	Total utilities:	$327
Public transportation:	$88	Cell phone:	$91.50
Insurance:	$48.20	Gym membership:	$38.95

4. Leslie treats her roommates to salads and pizzas from Papa Antonio's every Monday and Thursday, at a cost of $29, including delivery and tip. What percentage of her budget after fixed expenses goes toward these semiweekly gatherings on average?

2-2 Group

The Group portion of this lesson is all about putting together a monthly budget. You can choose one of the three individuals whose financial parameters are provided below.

Choice 1: Ellen

Ellen has a bachelor's degree in sociology and works for her state's department of human services. Her salary is $38,200 per year, and her payroll deductions amount to 26% of her salary. She earns an additional $125 per week tutoring at-risk students at a community center, with 10% of that amount withheld for taxes. Ellen owns a small home and pays $260 per month in property taxes and insurance.

Choice 2: Kai

Kai is a first-year college student taking 12 hours of classes and working 25 hours per week at a coffee shop. He makes $9.75 per hour, and payroll deductions eat up 18% of his pay. Kai lives rent-free at his parents' house, but he pays a share of the utility bill each month. His tuition and books are paid for by a combination of gifts from his parents and student loans. He makes payments on an 8-year-old used car, and he pays for auto insurance as well.

Choice 3: Maribel and Mark

Maribel and Mark are a young married couple with two children. Maribel makes $46,400 per year as an assistant bank manager and has 29% of her pay deducted each month. Mark is a substitute teacher and earns $22,166, with 21% deducted. The family rents a two-bedroom apartment, and they must pay for day care for the kids five days a week. They have two car payments.

1. My group will be making up a budget for _______________.

2. Find the monthly take-home pay for your choice.

3. The most important part of building a budget is probably listing expenses. Think carefully about the person you're budgeting for, and make a list of expenses that your group agrees are reasonable for a person in that financial situation. You might consider doing a little bit of Internet research to come up with typical expenses, or you can ask your instructor or other groups for input. Wouldn't this be a GREAT place to use what you've learned about spreadsheets?

Expense	Amount	Expense	Amount

Total Necessary Expenses: ______________

4. How much money is left in your budget for luxuries and savings?

5. If the total expenses are greater than total income, what are some options they would have? Discuss.

6. If your person's car breaks down, resulting in a $597.43 repair, how does this affect the monthly budget? What do you think some options would be in that case? Discuss.

So far, we've focused on monthly expenses. But some regular expenses occur on an annual or semiannual basis: Car insurance, tuition payments, and property taxes are three examples that come to mind. If you manage your money by spending every penny you have each month, when those occasional large expenses pop up, it can be very hard to find the money to pay for them. The result is usually mounting debt that can take years to overcome.

One way to avoid the negative effects of these long-term expenses is to budget money every month in advance of the due dates, maybe putting it into a separate savings account. Then when the big expense is due, the money is there waiting.

7. Jin pays $337.24 every 6 months for car insurance. He also pays $2,623 for tuition at the beginning of both fall and spring semesters, and he budgets $650 for books and supplies for each semester. If he wants to plan ahead, how much money should he put into savings every month? Explain how you decided on that amount.

Did You Get It ?

5. Dharma and Greg have a combined monthly income of $5,280.94. They want to save ahead for semiannual property taxes ($1,146), semiannual car insurance ($493.21), and annual homeowners insurance ($893.20). What percentage of their income should be saved each month? (Semiannual means twice a year, by the way.)

8. Suppose that instead of starting a savings account with the money he sets aside monthly, Jin decides to just hide that amount of cash each month in his sock drawer. List as many reasons you can think of for why that's a bad decision.

Hopefully, you know that when you save money with a financial institution, they pay you for access to your money. So the amount that you save grows. Good deal. The amount that they pay you is almost always described in terms of percents: more specifically, a percentage of the amount that you've invested. We'll study how this works in depth in the next two lessons.

9. The amount of money that Jin set aside in the very first month will be in his savings account for 5 months, and during that time will have 1.7% of the amount added to it by the bank. How much extra will Jin make on that initial deposit? (Note that this isn't the way a bank would describe savings: We'll learn everything you might want to know about that later.)

10. Before we close the lesson, take a look back at our friend Tremaine from Questions 5–8 in the Class portion. Suppose that disaster strikes: Tremaine's company goes out of business, and his income disappears just like that. Any financial planner worth a nickel advises clients to have a reserve of saved money, often called a **cushion** or **safety net,** in case of such an emergency. Based on his expenses, how much of a cushion would Tremaine need to get through 3 months while searching for a new job?

> **Math Note**
>
> Most financial experts advise having at least 3–6 months worth of expenses in your cushion fund.

Sidelight The State of Retirement Savings in America

I have no doubt that there are some people that would be perfectly happy to work for a living until they die. I don't know any of them, mind you, but they probably exist. Most of us, though, want to retire at some point. There are two things required to retire: you have to live long enough, and you need money. The amount you need depends on way too many factors to give reasonable estimates. One thing I can tell you, though, is that for MANY Americans, the picture is pretty grim.

The following information comes from the 2024 Retirement Confidence Survey.

- 30% of Americans had saved not one single penny for retirement as of 2024.
- Among those that have some retirement savings, 14% have less than $1,000, 21% have less than $10,000, and 32% have less than $25,000.
- Barely half of workers have even TRIED to calculate how much they might need for retirement.
- In spite of these numbers, Americans believe that they need an average of $1.46 million to retire comfortably. The best time to begin saving for retirement is like 20 years ago. The second best time is now!

2-2 Portfolio

Name ______________________________

Check each box when you've completed the task. Remember that your instructor will want you to turn in the portfolio pages you create.

Technology

1. ☐ In this Excel activity, we'll construct a budget.

ALEKS Practice

1. ☐ Include any written work from the ALEKS assignment along with any notes or questions about this lesson's content.

Applications

1. ☐ Complete the applications problems.

Reflections

Type a short answer to each question.

1. ☐ Why is it valuable to put together a monthly budget?
2. ☐ What is the difference between fixed and variable expenses?
3. ☐ How well do you feel you've done in the past at not spending more money than you bring in? Would a budget have helped you in the past?
4. ☐ What changes could you make now that would have a positive impact on your financial situation?
5. ☐ Name one thing you learned or discovered in this lesson that you found particularly interesting.
6. ☐ What questions do you have about this lesson?

Looking Ahead

1. ☐ Complete the prep skills for Lesson 2-3.
2. ☐ Read the opening paragraph in Lesson 2-3 carefully and answer Question 0 in preparation for that lesson.

Answers to "Did You Get It?"

1. I got $1,388.62; I divided by 7 to get daily pay, then multiplied by 30.5 (the average number of days in a month). Your answer might vary somewhat depending on how you calculated it.
2. $3,062.35 **3.** $813.95 **4.** 28.5% **5.** 6.6%

Answers to Prep Skills Questions

1. 40%
2. 19.2%
3. 125%
4. 30
5. 1,020
6. $34.66
7. $575
8. $12,408

2-2 Applications

Name ______________________________

1. If you drink two cans of Diet Coke (90 cents each) every day, how much do you spend on Diet Coke in a year?

2. Which costs more over the course of a year: a medium three-item pizza, which costs $7.25, four times a week, or going out to a fine dining restaurant ($85) twice a month?

3. After graduating with a degree in hospitality management, Edwin is hired to manage the front desk personnel at a resort hotel. His gross monthly pay is $4,142.86, and 29% is deducted for taxes and health insurance. His pay is divided into two checks per month. What is Edwin's monthly take-home pay, and how much is each paycheck worth?

4. Edwin's fixed expenses are shown below. How much money will he have left over every month after paying these fixed expenses?

Expense	Amount
Rent	$845.00
Car Payment	$224.61
Cell Phone	$69.80
Gas	$40.00

Expense	Amount
Insurance	$112.17
Utilities	$93.00
Student Loan	$256.25
Gym	$39.00

2-2 Applications

Name ______________________________

5. What are some of the things that the amount of money you found in Question 4 would likely be used for?

6. Edwin enjoys visiting the casino at the resort where he works. But like most visitors, he tends to walk in with more money than he leaves with. In fact, on average he loses $85 every weekend. What percentage of his budget after expenses is he losing?

7. In many cases, property taxes when you own a home are paid every 6 months, homeowner's insurance is paid once per year, and car insurance is paid every 6 months. One homeowner pays $1,450 in property taxes twice a year, $946 in homeowner's insurance annually, and makes car insurance payments of $294.32 and $335.40 every 6 months. If this homeowner wants to spread these expenses out by putting some money each month into a savings account, how much should they put aside per month?

8. In the last lesson of this unit, we'll study tax brackets. The short version is that if you make more money, you pay a higher percentage of your income in taxes. That's why the percentage of deductions varied for each individual we've studied in this lesson. If you have a job making $42,750, this puts you in the 12% federal tax bracket, and combining state and local taxes, you might end up having 17% deducted. You then get a raise to $48,200, which moves you up a bracket, and now 27% gets deducted. How much of that raise do you actually get to keep?

Lesson 2-3 Prep Skills

SKILL 1: WRITE PERCENTAGES IN DECIMAL FORM

To convert a percent to its decimal form, try remembering that the word "percent" literally means "per hundred."

- $28\% = \frac{28}{100} = 0.28$ — Dividing 28 by 100 has the effect of moving the decimal point two places to the left.
- $3.75\% = \frac{3.75}{100} = 0.0375$ — Dividing 3.75 by 100 has the effect of moving the decimal point two places to the left.

SKILL 2: COMPUTE A PERCENTAGE OF A GIVEN AMOUNT

As we saw in the prep skills for Lesson 2-2, you perform two steps to compute a number that represents a certain percentage of a given amount:

Step 1. Convert the percentage to a decimal by moving the decimal point two places to the left.

Step 2. Multiply the decimal form of the percentage by the given amount.

- What is 40% of 50?
  $0.40(50) = 20$ — The decimal form of 40% is 0.40.
- What is 0.5% of $468?
  $0.005(468) = 2.34$ — The decimal form of 0.5% is 0.005.

SKILL 3: EVALUATE ALGEBRAIC EXPRESSIONS

To evaluate an expression (or a formula), replace all occurrences of the variable or variables with specific numbers. This allows us to find a numeric value.

- Evaluate $2x - 5$ for $x = -3$. — This means to find the value of $2x - 5$ when $x = -3$.
  $2(-3) - 5 = -6 - 5 = -11$ — Replace x with -3 and simplify the result.
- Find P in $P = 2l + 2w$ if $l = 12$ and $w = 5$.
  $P = 2(12) + 2(5)$ — Replace l with 12 and w with 5 and simplify the result.
  $P = 24 + 10$
  $P = 34$

SKILL 4: ENTER FRACTIONAL EXPRESSIONS INTO A CALCULATOR

When using a calculator to enter expressions containing fractions, it's important to remember that a fraction bar indicates grouping symbols around the entire numerator and the entire denominator.

- $\dfrac{67 - 23}{8}$

  $(67 - 23)/8 = 5.5$ — The fraction bar indicates that the entire numerator, (67 – 23), is to be divided by 8.

- $67 - \dfrac{23}{8}$

  $67 - 23/8 = 64.125$ — The fraction bar indicates that 23 is to be divided by 8.

PREP SKILLS QUESTIONS

1. Convert each percent to its decimal equivalent.

   12% 4% 2.99%

2. Find each requested percentage.

   12.3% of 100 3.1% of $1,118.13

3. Find A in $A = PQR$ if $P = 5$, $Q = 120$, and $R = 0.03$.

4. Use a calculator to complete each computation:

   $\dfrac{50 + 40}{120}$ $1{,}000\left(1 + \dfrac{0.06}{5}\right)$

5. Solve the equation: $5 = \dfrac{x}{0.2}$

6. Solve the equation: $4.36 = 58.20(r)(3)$

Lesson 2-3 A Topic of Interest

LEARNING OBJECTIVES

- ☐ 1. Define interest and understand related terminology.
- ☐ 2. Develop simple interest formulas.
- ☐ 3. Use simple interest formulas to analyze financial issues.

Photodisc/Getty Images

An investment in knowledge pays the best interest.
—Benjamin Franklin

The topic of the next two lessons is of interest to anyone who plans to buy a house or a car, has a credit card, invests money, has a savings account—in short, pretty much everyone. This interesting topic is *interest*—a description of how fees are calculated when money is borrowed, and how your money grows when you save. If you prefer not to be separated from your hard-earned money, this is a topic you should be eager to understand well. The key to understanding isn't to just focus on a bunch of formulas: It's about understanding what interest is about and how it's calculated. This will give you the best chance to make sound financial decisions throughout your lifetime. So unless you can grow money (and if you can, want to be best friends?), this should indeed be a topic of interest.

0. How do you think interest will impact your life? How is it impacting your life currently?

2-3 Group

Interest is a fee paid for the use of money. For example, if you borrow money from a bank to buy a car, you not only have to pay back the amount of money that you borrowed, but also an additional amount, called the interest, for the use of the bank's money. On the other hand, if you deposit money in a savings account, the bank will pay you interest for saving money since it will be using your money to provide loans, mortgages, etc. to people who are borrowing money. The stated rate of interest is generally given as a yearly percentage of the amount borrowed or deposited.

There are two kinds of interest. *Simple interest* is paid strictly as a percentage of the original amount of money. *Compound interest* is a percentage of an original amount, as well as a percentage of the new amount including previously calculated interest. We will study simple interest in this lesson, and compound interest in the next.

1. Let's say that you get a $600 tax refund, and you wisely decide to put it into a long-term investment that pays 5% in simple interest each year. How much money would that add to the account each year? (Remember, with simple interest a certain percentage of an original amount is added to the account.)

2. Write out the multiplication you did to get your answer to Question 1. The next goal is to write a general formula for computing the amount of interest earned. Substitute the letter P for the original amount invested ($600) and the letter r for the percentage you multiplied by (written in decimal form, of course).

Did You Get It

Try this problem to see if you understand the concepts we just studied. The answer can be found at the end of the portfolio section.

1. Suppose that you take out a loan to buy new furniture, borrowing $1,700. If the interest charge will be 7.9% of the borrowed amount per year, how much interest will you pay each year?

3. Fill in the following table, which shows the growth of the account over several years, as well as the total amount of interest earned. You'll need a correct answer to Question 1, and remember that this is simple interest, so the interest earned each year is always calculated on the original value of the investment.

Time	Total Interest Earned	Value of Account
Start	–	$600
After 1 year		
After 2 years		
After 3 years		
After 4 years		
After 5 years		
After 10 years		

4. Look carefully at the first and second columns of your table. What is the relationship between the number of years that have passed and the total interest earned? Describe it verbally.

2-3 Class

In the Group portion of this lesson, you actually developed formulas for computing simple interest—whether you realized it or not! Now let's formalize what we discovered.

In order to compute simple interest, we need exactly three pieces of information: the *principal,* the *rate,* and the *time.* The box below describes the terminology we'll use, as well as the letters we'll use to represent the key data.

Interest (I) is a fee charged for the use of money.

Principal (P) is the original amount of money borrowed or placed into a savings account.

Rate (r) is the percent of the principal paid for having money loaned, or earned for investing money. Unless indicated otherwise, rates are given as a percent for 1 year.

Time (t) or term is the length of time that the money is being borrowed or invested. When the rate is given as a percent per year, time has to be written in years.

Future value (A) is the amount of the loan or investment plus the interest paid or earned.

In the work we did in the Group portion, the principal was $P = \$600$; the rate was 5% written as a decimal, which is $r = 0.05$. The amount of time varied when filling out the table of interest and future value.

1. In Question 2 of the Group portion, you should have discovered that the formula for computing the interest in 1 year is $P \times r$. Use this, and your answer to Question 4 of the Group portion, to write a formula for finding the amount of interest I that accumulates in t years:

   $I =$ __________

2. The total value of the account after t years is calculated by adding the total amount of interest to the principal amount. Use the formula you just wrote in Question 1 above to write a formula for the total value in terms of P, r, and t. Remember, we're using the letter A to represent the ending amount and calling it future value.

   $A =$ ______________

3. Use your formula you just wrote in Question 2 to find the future value of your original \$600 investment at 5% interest in 30 years.

Did You Get It

2. Find the future value of a loan if \$3,600 is borrowed for 3 years at a simple interest rate of 8% per year.

4. The right side of your answer to Question 2 should be a sum, meaning that it has two terms. Use some algebra to rewrite the formula so that it has just one term. The result should look like the principal amount (P) times some quantity.

Here's a summary of the formulas we've built to describe simple interest:

Formulas for Computing Simple Interest and Future Value

1. Interest = principal × rate × time:

$$I = Prt$$

2. Future value = principal + interest:

$$A = P + I \quad \text{or} \quad A = P(1 + rt)$$

5. To meet payroll during a down period, United Ceramics Inc. needed to borrow \$2,000 at 4% simple interest for 3 months. Without doing any calculations, make an educated guess as to how much interest you think would be due.

6. Now use formulas to find the interest they paid, and the future value at the end of 3 months. (Hint: How many years does 3 months correspond to?)

Did You Get It

3. Marta needs some quick cash for books at the beginning of spring semester, so she borrows \$600 at 11% simple interest for 2 months. How much interest will she pay, and what is the future value of the loan?

In the case of simple interest, "simple" doesn't mean "not useful." Simple interest is used in many common borrowing situations. In that case, it's typical to pay off the loan in equal monthly installments. That's how loans for things like furniture and electronics usually work. In Solved Example 1, we'll calculate the monthly payment on a simple interest loan.

Solved Example 1

Calculating a Monthly Payment

A biotech start-up borrowed \$68,000 to buy lab equipment at 9% simple interest for 1½ years. If the loan is to be paid off in equal monthly installments over the life of the loan, find the monthly payment.

SOLUTION

The strategy will be to ignore the monthly part first and just calculate the future value of the loan. Once we know that, we can divide by 18 to find the monthly payment. We'll use the future value formula we developed in Question 4 because that allows us to calculate future value all in one step. The formula is $A = P(1 + rt)$, and in this case we have $P = 68{,}000$, $r = 0.09$ (which is the decimal form of the 9% interest rate), and $t = 1.5$, which represents $1\frac{1}{2}$ years.

$$A = 68{,}000(1 + 0.09 \times 1.5) = 77{,}180$$

So the future value of \$77,180 will be equally divided among 18 payments:

$$\frac{77{,}180}{18} \approx 4{,}287.78$$

Rounded to the nearest cent, the monthly payment is \$4,287.78.

Did You Get It

4. The Lookout Restaurant took out a loan for \$5,000. The simple interest rate was 6.5%, and the term of the loan was 3 years. Find the interest, future value, and monthly payment.

Notice that the simple interest formula contains four variable quantities: interest, principal, rate, and time. Because we know how to solve equations from algebra, if we know three of those four quantities for a given account, we can solve to find the one we don't know.

Solved Example 2

Computing the Principal on a Loan

Phillips Health and Beauty Spa is replacing one of its workstations. The interest on a loan secured by the spa was \$93.50. The money was borrowed at 5.5% simple interest for 2 years. What was the amount that they borrowed?

Christy Thompson/ Shutterstock

SOLUTION

When solving any problem like this, it's great practice to first identify the information that was provided. We're told that the interest was \$93.50, so we have $I = 93.50$. The interest rate was 5.5%, giving us $r = 0.055$. The term of the loan was 2 years, so $t = 2$. This gives us a nice start.

Next, we should think about what we're asked to find, which in this case is the amount borrowed, or P in the notation we've been using. Our strategy? Use the simple interest formula, plugging in the stuff we know, and see what we get.

$$I = Prt$$
$$93.50 = P(0.055)(2)$$

This leaves us with an equation that we can solve for P. That's nice, because the principal is exactly what we were asked to find.

$$P = \frac{93.50}{(0.055)(2)} = 850$$

The business borrowed \$850.

2-3 Group (Again)

Solved Example 2 illustrates a really useful framework for solving this type of problem. When you're using an algebraic formula, it's ALWAYS a good idea to clearly identify the information that you're given, as well as what it is you're being asked to find. Then you can plug that information into the known formula and solve the resulting equation.

1. After receiving an inheritance, a newly married couple invested $15,250 for 10 years and received $9,150 in simple interest. What was the rate that the investment paid? Use the framework from Solved Example 2, and show all of your work. Write your answer in the form of a sentence.

2. If that couple had been able to find an investment that paid 9% interest, how much more quickly would they have made that $9,150 in simple interest? Use the framework from Solved Example 2, and show all of your work. Write your answer in the form of a sentence.

Did You Get It ?

5. If you invest $8,000 for 30 months and receive $1,000 in simple interest, what was the rate?
6. A pawn shop offers to finance a guitar costing $750 at 4% simple interest. The total interest charged will be $150. What is the term of the loan and the monthly payment?

If you look at it a certain way, simple interest in a savings account doesn't seem terribly fair. Remember, interest is a fee charged for using someone's money. So when you put money into a bank account, the bank is essentially paying you to borrow your money, because they can use it to increase their business in other areas. So far, so good.

But when you keep that money in the account for a long time, things start to go bad. Let's say that you invest $1,000 into a simple-interest account at 3% interest per year. Then according to the simple interest formula, after 1 year you would have earned $I = (\$1{,}000)(0.03)(1)$, or $30. Don't spend it all in one place! In any case, at the end of that first year, you now have $1,030 in your account, not $1,000. Yet when interest is simple, the amount of interest paid per year remains constant over the life of the investment. Why is this bad for you? Let's see.

3. Forget about the original amount, and just think of the $1,030 in your account at the end of year 1 as a new investment. If you get $30 in interest over the next year, what percentage did you actually get? Show work, please.

4. If you'd actually been paid the full 3% on your $1,030 in the second year, how much more interest would you have earned? Again, show work!

5. Following up on Question 4: How much interest would you earn in the third year if you get 3% on the new amount again?

We'll take up this discussion in the next lesson. For now, refer back to these questions when answering the first question in the Looking Ahead section of the Portfolio.

2-3 Portfolio

Name ______________________________

Check each box when you've completed the task. Remember that your instructor will want you to turn in the portfolio pages you create.

Technology

1. ☐ In this Excel activity, we'll use a spreadsheet to explore the future value of an investment.

ALEKS Practice

1. ☐ Include any written work from the ALEKS assignment along with any notes or questions about this lesson's content.

Applications

1. ☐ Complete the applications problems.

Reflections

Type a short answer to each question.

1. ☐ Why is it reasonable that a bank would pay interest on savings or charge interest on a loan?
2. ☐ What is the key feature of simple interest?
3. ☐ Why is it important to understand the concept of interest?
4. ☐ Name one thing you learned or discovered in this lesson that you found particularly interesting.
5. ☐ What questions do you have about this lesson?

Looking Ahead

1. ☐ Look back at page 98. Write a brief description of why simple interest is somewhat unfair when applied to a savings account.
2. ☐ Complete the prep skills for Lesson 2-4.
3. ☐ Read the opening paragraph in Lesson 2-4 carefully and answer Question 0 in preparation for that lesson.

Answers to "Did You Get It?"

1. \$134.30 **2.** \$4,464 **3.** \$11, \$611 **4.** \$975, \$5,975, \$165.97 **5.** 5%
6. 5 years; \$15

Answers to Prep Skills Questions

1. 0.12, 0.04, 0.0299
2. 12.3, $34.66
3. 18
4. 0.75, 1,012
5. $x = 1$
6. $r = 0.02$

2-3 Applications

Name ______________________________

Some of the most common (and most dangerous) examples of simple interest are short-term payday loans, which some lenders also call "cash advances." (This is not to be confused with a cash advance on a credit card.) Let's say that you're going to get paid in 10 days, and you need some cash for a car repair now. A payday lender might lend you $350 now, and you'll be asked to pay them back when your paycheck comes. Of course, you'll have to pay interest. The median fee charged by these types of lenders is $15 per every $100 borrowed.

1. How much would you have to pay back in 10 days?

2. What is the percent interest you're charged?

Maybe that interest rate doesn't sound crazy high to you. But remember, the length of the loan is only 10 days! Let's see if we can figure out just how much interest they're actually charging.

3. What do you need to multiply 10 by to turn it into 365?

4. If you want to figure out the annual percentage that corresponds to the percentage you found in Question 2 over 10 days, you'd need to multiply by the same number that turns 10 into 365. Do it! What annual percentage is the lender charging?

5. Discuss how good or bad a deal you're getting as the borrower. Should you avoid these types of short-term loans?

2-3 Applications

Name ______________________________

Another example of simple interest is a certificate of deposit, or CD as they're more commonly known. This is an investment where you agree to leave your deposit untouched for a fixed amount of time, and in exchange you are guaranteed a certain percentage interest. Below are some CD rates listed in October of 2024 on Bankrate.com for CDs with a 1-year term. Use the percentage listed in the APY column as the interest rate.

6. For a $2,500 investment, how much more interest would you earn if you chose Limelight Bank over Live Oak Bank?

	APY	Min. deposit	Term
Limelight Bank Limelight CD Member FDIC ★★★★☆ 3.9	4.75%	$1,000	1yr
AMERICA FIRST CREDIT UNION America First Credit Union CD Insured by NCUA ★★★★★ 5.0	4.70%	$500	1yr
LiveOakBank. Live Oak CD Member FDIC ★★★★☆ 4.4	4.60%	$2,500	1yr

7. If your goal was to earn $130 in interest on your $2,500 investment, what interest rate would you need to find?

8. If you got the interest rate that America First Credit Union was offering for a different length of time, how long would you need to invest in order to get that $130 you have your heart set on?

9. As you can see, CDs don't often return a ton of interest. What they DO offer is a guaranteed return. Suppose that you also had the opportunity to invest that $2,500 in a business venture that promised a 12.5% return in 1 year, unless the company failed to turn a profit, in which case you would get nothing at all back. Calculate the amount of interest you'd get, and discuss whether you'd prefer the risky investment over the CD.

Lesson 2-4 Prep Skills

SKILL 1: WRITE PERCENTAGES IN DECIMAL FORM

As we saw in the prep skills for Lesson 2-3, the key to converting a percentage to its decimal form is the fact that the word "percent" literally means "per hundred."

- $21\% = \dfrac{21}{100} = 0.21$ — Dividing 21 by 100 has the effect of moving the decimal point two places to the left.
- $1.25\% = \dfrac{1.25}{100} = 0.0125$ — Dividing 1.25 by 100 has the effect of moving the decimal point two places to the left.

SKILL 2: COMPUTE SIMPLE INTEREST

Remember from Lesson 2-3 that simple interest is found by using the formula $I = Prt$, where I is the interest amount, P is principal, r is the rate, and t is the time in years.

- Investing \$300 at 5.5% simple interest for 4 years will earn $I = (300)(0.055)(4) = \$66$. — Substitute $P = 300$, $r = 0.055$, and $t = 4$ into the formula $I = Prt$ and evaluate.

SKILL 3: INTERPRET EXPONENTS AS REPEATED MULTIPLICATION

An **exponent** is a way to represent repeated multiplication. For example, 4^3 means $4 \times 4 \times 4$.

- $2^5 = 2 \cdot 2 \cdot 2 \cdot 2 \cdot 2 = 32$
- $(-4)^4 = (-4)(-4)(-4)(-4) = 256$ — The parentheses show that the negative is included in the number raised to the exponent.
- $-4^4 = -(4 \cdot 4 \cdot 4 \cdot 4) = -256$ — Only the 4 is affected by the exponent so the result is negative.
- $\left(\dfrac{1}{5}\right)^2 = \dfrac{1}{5} \cdot \dfrac{1}{5} = \dfrac{1}{25}$

SKILL 4: PERFORM CALCULATIONS INVOLVING ORDER OF OPERATIONS

Order of operations is a group of agreed-upon rules that allow us to write calculations involving more than one operation without always having to use parentheses to indicate which operation to do first. The order is as follows, with operations higher on the list being performed before those lower on the list.

(1) Parentheses
(2) Exponents
(3) Multiplication or division
(4) Addition or subtraction

Perform operations in parentheses first. If there are multiple parentheses, work from innermost outward.
For operations that live at the same level, work from left to right.

- $12 + 3 \cdot 4 = 12 + 12 = 24$ — Multiplication before addition
- $4(3)^2 = 4(9) = 36$ — Exponent before multiplication
- $81 \div (12 - 3) = 81 \div 9 = 9$ — Perform the subtraction first because it's in parentheses.

- $52 - 8 \div 4 \cdot 6 = 52 - 2 \cdot 6$
  $= 52 - 12$
  $= 40$

  Division before multiplication because it's to the left

- $(-2(3+1))^2 - 4 = (-2 \cdot 4)^2 - 4$

  Addition in innermost parentheses first, then multiplication inside outer parentheses

  $= (-8)^2 - 4$
  $= 64 - 4$

  Exponent before subtraction

  $= 60$

- $\frac{7-9}{3+5} = \frac{-2}{8} = -\frac{1}{4}$

  In a fraction, the fraction bar acts like grouping symbols: Interpret the fraction as $(7 - 9) \div (3 + 5)$.

SKILL 5: EVALUATE ALGEBRAIC EXPRESSIONS

As we saw in the Prep Skills for Lesson 2-3, you replace all occurrences of the variable or variables with specific numbers to evaluate an expression (or a formula). This allows us to find a numeric value.

- Evaluate $3x^2 - 4$ for $x = 5$.
  $3(5)^2 - 4 = 3(25) - 4$

  Replace x with 5 and simplify the result.

  $= 75 - 4$
  $= 71$

- Find a in $a = br^{n-1}$ if $b = 25$, $r = 2$, and $n = 9$.
  $a = 25(2)^{9-1}$
  $a = 6{,}400$

  Replace b with 25, r with 2, and n with 9 and simplify the result using a calculator.

PREP SKILLS QUESTIONS

1. Write each percent in decimal form.

   47%

   9.4%

2. If you borrow $1,200 for furniture at 11% simple interest for 2 years, how much will you pay in interest?

3. Write the expression below in a shorter form using an exponent:

   $500(1 + 0.03)(1 + 0.03)(1 + 0.03)(1 + 0.03)$

4. Perform each calculation:

   $12[8 + 4(13)]$

   $400\left(1 + \frac{0.05}{12}\right)^{60}$

5. For the expression $a(1 + b)^c$, find the value when $a = 20$, $b = 0.09$, and $c = 6$.

Lesson 2-4 Like a Snowball Rolling Downhill

LEARNING OBJECTIVES

- ☐ 1. Describe how compound interest differs from simple interest.
- ☐ 2. Develop compound interest formulas.
- ☐ 3. Use compound interest formulas to analyze financial issues.

McGraw Hill

My wealth has come from a combination of living in America, some lucky genes, and compound interest.

—Warren Buffett

If you've ever made a snowman, you're familiar with how it starts. You make a little snowball in your hands, then plop it down and start rolling it through the snow. As you do, it picks up more snow on all sides (if you roll it well) and gets bigger. At first, not that much snow is picked up because the original ball is little. But as it gets bigger, it has more surface area to pick up more snow, and it starts to grow very quickly—to the point that before you know it, you need help pushing it to gather more snow. Push that baby down a steep hill, and you could end up with one massive snowball (and maybe getting grounded or sued if something valuable gets in the way). At the very end of the last lesson, we looked into interest being calculated not just on the principal amount, but on interest that had previously been paid as well. As we'll see in this lesson, given enough time this can become the financial equivalent of a snowball rolling downhill, leading to a wonderful place we call "financial independence."

0. Do you think the snowball analogy works for the simple interest we studied in the last lesson? Why or why not?

2-4 Group

We've already studied how an investment can grow when simple interest is calculated; the table below is borrowed from Lesson 2-3. (Borrowed . . . stolen . . . whatever.) In this case, a principal amount of $600 was invested at 5% interest.

Our first goal in this lesson is to recalculate the interest earned and value of the account, this time calculating the next year's interest on the entire value of the account (including previously earned interest), rather than just on the principal amount.

Time	Total Simple Interest Earned	Value of Account
Start	–	$600
After 1 year	$30	$630
After 2 years	$60	$660
After 3 years	$90	$690
After 4 years	$120	$720

1. Compute the amount of interest earned during the second year if that amount is 5% of both the principal amount of \$600 AND the \$30 in interest that was already earned in the first year. Find the new ending value as well. Show work, please.

2. We know from Lesson 2-3 that we can compute the amount of interest earned in 1 year using the formula $I = Pr$. This is using $t = 1$ in the interest formula $I = Prt$. Use this formula with $P = \$630$ to show that you get the same answer you did in Question 1.

3. Use your answer from Question 1 to fill in the row of the table corresponding to "After 2 years." Then repeat your calculations to fill in the remainder of the table, keeping in mind that in each case interest should be calculated on the *new amount at the beginning of the year.*

Time	Total Interest Earned	Value of Account
Start	–	\$600
After 1 year		
After 2 years		
After 3 years		
After 4 years		

If all went well, you should have found that at the end of 4 years, the account would have \$9.30 more when interest is calculated on the new yearly amount, rather than the original principal like we did in the previous lesson. This type of interest is called **compound interest,** and each time the new amount is calculated (in this case, at the end of each year), we say that interest has been **compounded** at that time. Since interest is being compounded once per year in this example, we say that it is interest **compounded annually.**

You're probably not overly impressed by that extra nine bucks, but when we develop some formulas for efficiently computing compound interest, we'll be able to see that something VERY interesting happens in the long run. So let's get to work on developing a formula by looking for a pattern.

Value after 1 year = 600(1 + 0.05) This comes from our formula $A = P(1 + rt)$ from Lesson 2-3.

Value after 2 years = (Value after 1 year)(1 + 0.05) = $\underbrace{[600(1 + 0.05)]}_{\text{Value after 1 year}}$ (1 + 0.05) = $600(1 + 0.05)^2$

4. Now fill in the boxes in the computation to find the value after 3 years:

Value after 3 years = (Value after 2 years)(1 + 0.05) = $\underbrace{\boxed{}}_{\text{Value after 2 years}}$ (1 + 0.05) = $600(1 + 0.05)^{\square}$

Now here comes the key observation of this lesson, so fasten your seat belts. So far, we know the following:

Value after 1 year = 600(1 + 0.05)

Value after 2 years = $600(1 + 0.05)^2$

Value after 3 years = $600(1 + 0.05)^3$

5. Make a conjecture as to the value after 4 years and 5 years. Recall that "conjecture" is a cool math word for "educated guess."

Value after 4 years = ____________

Value after 5 years = ____________

6. Make a conjecture as to the value after t years, where t represents any number of years.

Value after t years = ____________

Congratulations! Your conjecture (you did get it, right?) can be formalized into our first important formula for compound interest.

Formula for Compound Interest Compounded Annually

$$A = P(1 + r)^t$$

where

A is the future value.
P is the principal amount.
r is the interest rate in decimal form.
t is the term of the investment in years.

7. Use the formula for interest compounded annually to find the future value of the original \$600 investment at 5% interest after 10, 20, 30, and 40 years. Then use the simple interest formula from Lesson 2-3 to find the future value if the interest is simple.

Time	Future Value, Compounded Annually	Future Value, Simple Interest
After 10 years		
After 20 years		
After 30 years		
After 40 years		

8. Use the table to draw a graph of the future value for both accounts, then write a verbal description of what you learned from the table and the graph. Make sure you choose an appropriate scale for each axis, and of course label that scale on your graph. Do something to distinguish the two graphs: Use different colors, or make one dashed or thicker. What kind of growth does each account illustrate?

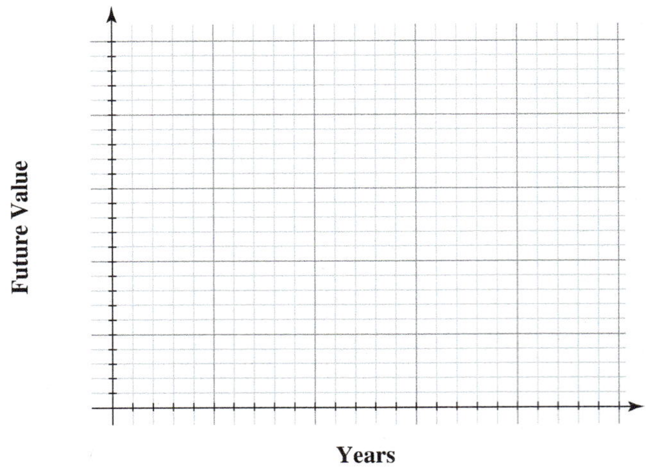

Did You Get It

Try this problem to see if you understand the concepts we just studied. The answer can be found at the end of the portfolio section.

1. For an investment of \$100,000 at 6% interest for 4 years, find (a) the simple interest, and (b) the compound interest if interest is calculated once per year.

2-4 Class

Based on what we've learned so far, if you're the one getting paid interest, compound interest is a MUCH better deal for you than simple interest. But what if you plan to invest your money for less than a year? You'd end up getting a whole lot of nothing, since the interest is compounded for the first time at the end of the first year. This would provide a competitive disadvantage for a banker, as wise investors would prefer more flexibility. At some point, bankers decided that compounding interest more often would make them more attractive to potential customers.

So let's say that interest is being compounded four times per year (or **quarterly**). Going back to our investment of \$600 at 5% interest, the banker would not add 5% to the account four times per year—this would correspond to a far higher rate than the 5% they promised. Instead, they'd divide that rate by 4, and pay 1.25% every 3 months. The first calculation, after 3 months, would look like this:

$$A = 600\left(1 + \frac{0.05}{4}\right)$$

The interest rate t is now divided by four.

1. After 6 months, write a calculation that describes the future value, using the same idea from the Group portion of the lesson. (Hint: How many times has interest been calculated after 6 months?)

2. By the time the first year is up, the interest would have been calculated ________ times, so the total exponent on the parentheses in our future value calculation will be ________, which is the number of times per year that interest has been compounded times the number of years that have passed.

3. If interest is compounded monthly, after 5 years it would be compounded ________ times.

4. If interest is compounded n times per year, after t years it would be compounded ________ times.

5. Now adapt your answer from Question 1 into a formula for calculating the future value of that \$600 investment at 5% interest if interest is compounded n times per year for t years. Your answer to Question 4 will surely come in handy. Remember that the interest rate needs to be divided into n equal portions.

And that, friends, is how you develop a formula for interest compounded a certain number of times per year. Nicely done.

General Formula for Compound Interest

$$A = P\left(1 + \frac{r}{n}\right)^{nt}$$

where

A is the future value.
P is the principal amount.
r is the interest rate in decimal form.
t is the term of the investment in years.
n is the number of times that interest is compounded per year.

Solved Example

Comparing Compounding Periods

Use the compound interest formula to calculate the future value of our \$600 invested at 5% interest compounded quarterly for 10 years, then compare to the future value you calculated in Question 7 of the Group portion using interest compounded annually. Which results in more interest?

Math Note

If you use $n = 1$ in our new compound interest formula, you'll get exactly the formula we already developed for interest compounded annually. So you don't need to think of that as a separate formula. Less stuff to memorize is good.

SOLUTION

As previously mentioned, when using a formula to solve a problem, it's always a good idea to copy down the formula, then identify which variables in the formula we know values for. In this case, our formula is

$$A = P\left(1 + \frac{r}{n}\right)^{nt}$$

We're given a principal amount of $P = \$600$ and an interest rate of $r = 0.05$. Since interest is compounded quarterly, that means that $n = 4$. Finally, the term is 10 years, so $t = 10$. This makes our formula

$$A = 600\left(1 + \frac{0.05}{4}\right)^{4\times10}$$

The best way to perform a complicated calculation like this is all at once, using a graphing calculator or spreadsheet. You can key it in just like it looks, taking care to include parentheses. If you choose to go a different route, order of operations is hugely important. First, do the calculation in parentheses, taking care to do the division before the addition. Then raise the result to the 40th power, and finally multiply by 600. In any case, the future value works out to be \$986.17.

When we calculated the future value for the same principal and interest rate using the formula for interest compounded annually in Group Question 7, we got \$977.34. So compounding four times per year rather than once earned almost \$9 more in interest.

6. Find the future value of the $600 investment at 5% interest using interest **compounded monthly** (12 times per year) and interest **compounded daily** (365 times per year). In each case, find the amount of interest that was earned (not just the future value).

7. Draw a general conclusion about what happens to the amount of interest as the number of times compounding yearly goes up. Then (and this question really requires some thought) see if you can come up with a reason why this makes perfect sense. A true understanding of the difference between simple and compound interest will really help.

Did You Get It

2. Find the interest on an investment of $75,000 that pays 4.5% compounded weekly for 8 years.

Effective Interest Rate

We've now seen that the number of times interest is compounded per year has an effect on how much interest is actually earned. It's not a *huge* effect, mind you, but is at least worth studying. The bottom line is that when interest is compounded more than one time per year, the actual percentage earned is a bit higher than the stated interest rate. That's good if you're the lender, not so good if you're the borrower. In any case, there's a formula that can be used to measure this effect. First, the terminology.

The **effective rate** (also known as the **annual yield**) is the simple interest rate that would yield the same future value over 1 year as the given compound interest rate. The compound interest rate is sometimes known as the **nominal rate.**

Now, the formula.

Formula for Effective Interest Rate

$$E = \left(1 + \frac{r}{n}\right)^n - 1$$

where

E is the effective rate.
n is the number of times that interest is compounded per year.
r is the nominal interest rate.

8. A loan has a nominal interest rate of 4%, and the interest is compounded weekly. Without doing any calculations, would you expect the effective interest rate to be higher or lower than 4%? Explain.

9. Refer to Question 8. Calculate the effective interest rate.

10. Find the future value after 1 year for an investment of $10,000 at 4% compounded weekly.

11. Divide your answer to Question 10 by 10,000. What do you notice?

12. Comparing investments is a very important part of any financial strategy. Which savings account is a better investment: 6.2% compounded daily or 6.25% compounded semiannually? There are different ways you could approach this, so make sure you show all work and explain how you made your decision.

Did You Get It

3. Which is a better investment, 3% compounded monthly or 3.25% compounded semiannually?

Interest Compounded Continuously

The goal of the remainder of this lesson is to see if we can figure out what happens to the compound interest formula when the number of times that interest is compounded per year gets really, really large. Large to the point that in essence, interest is being compounded every instant of every day. Wow.

As you know, the compound interest formula is $A = P\left(1 + \frac{r}{n}\right)^{nt}$, where n is the number of times per year that interest is compounded. In that form, it's REALLY hard to figure out what happens when n gets huge. So we'll need to do some algebra first.

If we do a substitution, using the letter u to take the place of the expression n/r, it turns out that the compound interest formula can be rewritten in the form below. If you're interested in seeing the details, I go through them in a video that you can find either in the video resources for this lesson online, or at this point in the ebook.

$$A = P\left[\left(1 + \frac{1}{u}\right)^{u}\right]^{rt}$$

Remember, we're interested in finding what this formula becomes when n gets bigger and bigger. Sadly, after our little algebra stunt there's no n in the formula anymore. But that's okay: Since $u = \frac{n}{r}$, when n gets bigger and bigger, so does u.

Here's a summary of what we've learned so far: The compound interest formula can be rewritten in the form below, and as we make the number of times that interest is compounded per year get larger and larger, the variable u in that expression gets larger and larger as well. So next we'll study what happens to the expression inside the brackets when u gets larger and larger, and we'll find that we can find a MUCH simpler form for this formula.

$$A = P\left[\left(1 + \frac{1}{u}\right)^{u}\right]^{rt}$$

What happens to this expression when u gets larger and larger?

13. Using a calculator or spreadsheet, fill in the table below. Round to three decimal places. What can you conclude about what happens to the value of $\left(1 + \frac{1}{u}\right)^{u}$ as the number of compounding periods tends to infinity? ("Tends to infinity" is the phrase we use to describe a variable getting larger and larger without bound.)

u	50	100	500	1,000	1,500	2,000	2,500
$\left(1 + \frac{1}{u}\right)^{u}$							

14. The number that you should have found in Question 13 is about 2.718 and is denoted e. (Actually, e is irrational, so its decimal equivalent neither terminates nor repeats: $e \approx 2.718\ldots$) Write the result of the compound interest formula above when the number of times interest is compounded gets larger and larger. In this case, we say that interest is **compounded continuously.** You can think of this as the interest being compounded every instant of every day.

Using Technology: Evaluating Expressions Containing e

To evaluate e^3:

Standard Graphing Calculator: [2nd] [LN] [3] [ENTER]

Standard Scientific Calculator: [3] [2nd] [LN]

Spreadsheet: = EXP(3)

See the Using Tech Videos on Evaluating Expressions Containing e on a Calculator and in Excel for further information.

15. Use the formula you wrote in Question 14 to find the future value after 20 years of a $5,000 account that earns 4% interest compounded continuously, then compare that to the future value of the same account if interest is compounded annually.

16. Finally, let's compare interest compounded continuously to simple interest. Suppose that $50,000 is invested in two accounts: one earns 6% simple interest, the other earns 6% compounded continuously. Fill in the table with the future value of each account after each term, rounded to the nearest dollar. Wouldn't this be a GREAT spot to use a spreadsheet, or at least the table feature on a graphing calculator?

Years	5	10	15	20	25	30
6% simple interest						
6% compounded continuously						

Did You Get It

4. Find the future value of a $40,000 investment with 6.9% interest compounded continuously 25 years after it's started.

2-4 Portfolio

Name ______________________________

Check each box when you've completed the task. Remember that your instructor will want you to turn in the portfolio pages you create.

Technology

1. ☐ In this Excel activity, we'll use a spreadsheet to explore how the number of times interest is compounded affects the future value of an investment.

ALEKS Practice

1. ☐ Include any written work from the ALEKS assignment along with any notes or questions about this lesson's content.

Applications

1. ☐ Complete the applications problems.

Reflections

Type a short answer to each question.

1. ☐ Explain the difference between simple and compound interest.
2. ☐ Why is compound interest a much better deal for you if you're the investor?
3. ☐ What does the term "compounding" refer to?
4. ☐ Name one thing you learned or discovered in this lesson that you found particularly interesting.
5. ☐ What questions do you have about this lesson?

Looking Ahead

1. ☐ Complete the prep skills for Lesson 2-5.
2. ☐ Read the opening paragraph in Lesson 2-5 carefully and answer Question 0 in preparation for that lesson.

Answers to "Did You Get It?"

1. a. \$24,000; **b.** \$26,247.70 **2.** \$32,482.97 **3.** 3.25% semiannually **4.** \$224,500.84

Answers to Prep Skills Questions

1. 0.47, 0.094 **2.** \$264 **3.** $500(1 + 0.03)^4$ **4.** 720, 513.34 **5.** 33.54

2-4 Applications

Name ____________________

Over the last two lessons, we've studied many aspects of interest. The goal of this assignment will be to compare various investments and decide which would be most beneficial. Of course, the amount of money that anyone chooses to invest at any given time is very personal and depends on too many factors to consider in a few lines of text. So we'll pick an amount and go with it, but if you'd prefer to work this assignment with a different amount, check with your instructor to see if that's okay.

Example: An investor has \$12,000 to invest and is willing to keep it invested for up to 5 years. Each of the options in Questions 1–6 is available. Find the future value of each investment.

1. A 1-year certificate of deposit that pays 2.1% simple interest

2. A 5-year certificate of deposit that pays 2.9% interest compounded annually

3. You lend the money to a friend who's starting a business, and they write a contract guaranteeing you a 12% return on your investment, to be paid in full at the end of 4 years, with an option to reinvest if the company is successful.

4. A savings account offers 1.98% interest compounded continuously. You can remove the money any time you like, but plan to keep it in the investment for 5 years.

2-4 Applications

Name ______________________________

5. A savings account that offers 2.18% compounded monthly: you can remove the money any time you like, but plan to keep it in the investment for 5 years. After finding the future value, find the effective interest rate as well.

6. A rival bank to the one in Question 5 offers the same interest rate, but they compound hourly. The plan is also to keep the money invested for 5 years.

7. Review each of the investments that you worked with in Questions 1–6. If you had to choose one of those investments, which would it be? Obviously, you have to discuss your reasoning. Think about the fact that when investing, there are other potential concerns than simply the total amount at the end of the investment.

Lesson 2-5 Prep Skills

SKILL 1: WRITE PERCENTAGES IN DECIMAL FORM

As we have seen in several previous prep skills, the key to converting a percentage to its decimal form is the fact that the word "percent" literally means "per hundred."

- $13\% = \dfrac{13}{100} = 0.13$ — Dividing 13 by 100 has the effect of moving the decimal point two places to the left.
- $4.65\% = \dfrac{4.65}{100} = 0.0465$ — Dividing 4.65 by 100 has the effect of moving the decimal point two places to the left.

SKILL 2: COMPUTE A PERCENTAGE OF A GIVEN AMOUNT

As we saw in the prep skills for Lessons 2-2 and 2-3, you perform two steps to compute a number that represents a certain percentage of a given amount:

Step 1. Convert the percentage to a decimal by moving the decimal point two places to the left.

Step 2. Multiply the decimal form of the percentage by the given amount.

- What is 2.4% of $565?
  $0.024(565) = 13.56$ — The decimal form of 2.4% is 0.024. 2.4% of $565 is $13.56.

SKILL 3: COMPUTE SIMPLE INTEREST

Remember from Lesson 2-3 that the formula for simple interest is $I = Prt$, and the formula for future value is $A = P(1 + rt)$. A represents the future value, P represents the principal, r represents the rate, and t represents the length of time the money is borrowed or invested.

- Find the interest due on a loan of $700 for 4 years at 8% simple interest.
  $I = Prt$
  $I = (\$700)(0.08)(4)$
  $I = \$224$
  — The decimal form of 8% is 0.08. Substitute $700 for P, 0.08 for I, and 4 for t. The total interest on this loan is $224.

- Find the total amount you would have to pay back on a loan of $700 for 4 years at 8% simple interest.
  $A = P(1 + rt)$
  $A = (\$700)(1 + 0.08 \times 4)$
  $A = \$924$
  — The decimal form of 8% is 0.08. Substitute $700 for P, 0.08 for I, and 4 for t. The total amount you would have to pay back on this loan is $924. Notice that this amount is $224 more than the loan value.

PREP SKILLS QUESTIONS

1. Write each percent in decimal form.
   91%
   6.48%

2. What is 1.8% of $11,375?

3. Suppose you borrowed $450 to buy an iPhone. If the interest is 14.7% simple interest for 2 years, how much would you have to pay back?

Lesson 2-5 Buying Stuff Without Money

LEARNING OBJECTIVES

- ☐ 1. Compute payments and charges associated with installment loans.
- ☐ 2. Identify the true cost of a loan by computing APR.
- ☐ 3. Evaluate the costs of buying items on credit.

SW Productions/Brand X Pictures/Getty Images

Before borrowing money from a friend, it's best to decide which you need most.
—Joe Moore

A lot of people celebrate their college graduation by buying themselves a car. Unless you're very wealthy, you won't be plunking down a stack of crisp hundred dollar bills to make that happen—you'll need to get a loan. In this case, you'll be doing what is called **installment buying.** This is when an item is purchased and the buyer pays for it by making periodic partial payments, or installments.

There are natural advantages and disadvantages to installment buying. The most obvious advantage is that it allows you to buy an item that you don't have enough money to pay for, and use it while you're raising that money. The most obvious disadvantage is that you pay interest on the amount borrowed, so you end up paying more for the item—in some cases a lot more. When you buy a car, for example, you know the price of the car, and you're usually told how much the monthly payment will be. If you're buying a good used car for $12,260 with a $3,000 down payment and are told the payment will be $231.50 per month for 48 months, how much will you actually end up paying for that car? Work through this lesson, and soon you will know.

0. Describe at least one experience you've had with borrowing money. Was it a positive or a negative experience?

2-5 Class

Fixed Installment Loans

A **fixed installment loan** is a loan that is repaid in equal payments. Sometimes the buyer will pay part of the cost at the time of purchase. This is known as a **down payment.** Here are some other key terms used to describe installment loans.

The **amount financed** is the amount a borrower will pay interest on.

Amount financed = Price of item − Down payment

The **total installment price** is the total amount of money the buyer will ultimately pay.

Total installment price = Sum of all payments + Down payment

The **finance charge** is the interest charged for borrowing the amount financed.

Finance charge = Total installment price − Price of item

Now let's take a look at the problem posed in this lesson's opening paragraph. The pertinent information is this:

Cost of car: $12,260
Down payment: $3,000
Monthly payment: $231.50
Term of loan: 48 months

> **Math Note**
>
> You're often given quite a bit of information in a problem about finances. Identifying and writing out that information in an organized way, like we did here, is ALWAYS a good idea.

1. What was the amount financed? Refer to the definitions on the previous page if you need to.

2. What's the total amount of all monthly payments?

3. What's the total installment price? (This is the answer to the question "How much money will you actually end up paying for the car?")

4. Find the finance charge, then describe what that amount really represents.

Did You Get It

Try this problem to see if you understand the concepts we just studied. The answer can be found at the end of the portfolio section.

1. If you buy a used car for \$8,200 with a down payment of \$1,000 and 36 monthly payments of \$270, find the amount financed, the total installment price, and the finance charge.

In some cases, it's a good idea to be able to calculate a monthly payment before making a purchase. Not only can this help you to decide whether or not you can afford an item, it gives you a way to make sure you're not getting cheated when you're quoted a monthly payment. So how can you find the monthly payment for a loan? Using a formula, of course. (Although to be fair, there are many loan calculators online.)

Calculating Monthly Payments on a Fixed Installment Loan

A monthly payment M can be calculated using the formula

$$M = \frac{P \cdot \frac{r}{12}}{1 - \left(1 + \frac{r}{12}\right)^{-n}}$$

where

P is the principal value of the loan.
r is the annual percentage rate (APR) in decimal form.
n is the total number of payments.

Memorize this formula, and then eat it. Hold on, we're just kidding—obviously the formula is pretty complicated. You'll need to be especially careful when performing calculations involving this formula. An important note: If you do the calculation in stages, which we don't necessarily recommend, make sure that you don't round too much. That can change the calculation considerably. A reasonable rule of thumb is at least four decimal places.

In order to calculate the monthly payment on an auto loan, we need to know the annual percentage rate, or APR. As we'll see shortly, this isn't exactly the same as the quoted interest rate for a loan. If the APR isn't quoted to you immediately when you start to talk about buying a car, ask for it! Salespeople are fond of steering the conversation toward monthly payments, but what determines how good (or bad) of a deal you're getting is the APR. You should always check local rates in advance of buying a car so you know what rate you should be getting.

5. Have you noticed that car makers never have sales? They're always "events." Weird. Anyhow, for the Honda 2024 President's Day "event," they offered a certified pre-owned CR-V with a special APR of 2.9% for 60 months. If one particular vehicle has a selling price of \$27,295 and a buyer trades in an older car for a down payment of \$4,500, calculate the monthly payment for this loan.

Using Technology: Using the TVM Solver and the PMT Function to Compute Monthly Payments

There are built-in functions in both graphing calculators and spreadsheets to calculate monthly payments, which is pretty nice given how complicated the formula is.

Graphing Calculator

We're going to use a feature of TI-83 and 84 calculators called the TVM solver. The acronym stands for "time value money." First we need to get to the finance menu. This varies by model. If you have a TI-83, you'll want to find FINANCE, which is 2nd x^{-1}. For the TI-83 Plus or TI-84, you just hit the APPS key, then choose 1: Finance.

Now you're on the finance menu.

1. Press 1 to activate the TVM solver. The screen will look like the first screenshot below.
2. Enter the total number of payments for the loan next to N. For monthly payments, this is 12 times the number of years.
3. Enter the interest rate (APR) next to I%. Enter it in percent, not decimal form.
4. Enter the principal value of the loan next to PV. (Note: You'll need to enter it as a negative in order for the monthly payment to end up positive. Think of this as a negative amount of money that you have, since you owe that amount.)
5. Leave 0 next to PMT and FV.
6. Enter the number of payments per year, which for us is 12, next to P/Y. This will automatically enter the same next to C/Y.
7. Move the cursor up to the PMT line, then press ALPHA ENTER to get the SOLVE command. The monthly payment appears next to PMT. The calculations for the loan in Question 5 are shown in the second and third screenshots, before and after the SOLVE command was applied.

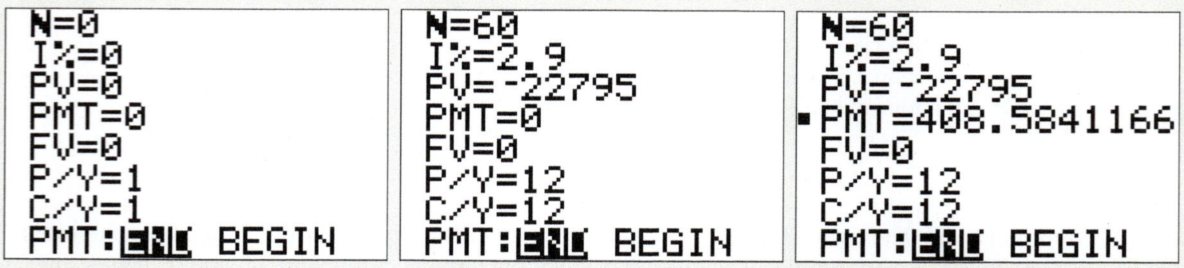

Excel

1. In any blank cell, enter =PMT(
2. As shown in the first screenshot below, Excel prompts you for the information needed. Enter the monthly interest rate (rate) in DECIMAL form. This will be the APR divided by 12.
3. Enter the total number of payments (nper), which is 12 times the number of years.
4. Enter the principal value (pv), then close parentheses. You can ignore [fv] and [type]. As with the calculator, enter the principal value as a negative to get positive payments. The input and the result for the loan in Question 5 are shown in the second and third screenshots. The good news is that the result is the same for calculator and spreadsheet.

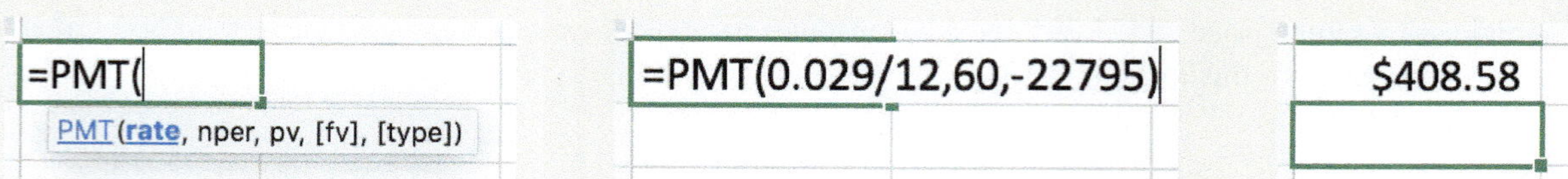

See the Using Tech Videos on Using the TVM Solver and the PMT Function to Compute Monthly Payments for further information.

6. If the loan was based on simple interest, the future value would be $22,795(1 + 0.029 × 5). Calculate this amount, then divide by 60 to find the monthly payment. What do you think accounts for the fact that simple interest makes it HIGHER?

Did You Get It

2. A graphic design pro buys a new iMac for $1,499 with a $200 down payment, and gets manufacturer financing for 5 years at 12% APR. Find: (a) the amount financed, (b) the monthly payment, (c) the total installment price, and (d) the finance charge.

Annual Percentage Rate

Many lenders add upfront fees to a loan and then spread them over the life of the loan. This has the effect of making the actual interest rate that a borrower pays higher than the quoted rate. Because this can get confusing, lenders are required by law to disclose an annual percentage rate, or APR, that reflects the true interest charged. This allows consumers to compare loans with different terms. The mathematical procedures for computing APR are pretty complicated, so we're going to take advantage of technology to compute APR. This is described in the Solved Example.

Solved Example

Computing APR

Elaine is in the market for a good used car, and she finds a Dodge Charger she likes at her friendly local dealer. The dealer is asking $17,499, but Elaine (who's one smart cookie) negotiates the price down to $16,250 with a down payment of $6,000. The dealer claims an interest rate of 7.25% for 48 months and reports the monthly payment as $246.93. Let's help Elaine decide if the dealer is being honest about the interest rate.

SOLUTION

Good news: We can use the TVM solver on a graphing calculator to find APR, so we mostly already know what to do. We can also use Excel, which will require a new function. Let's look at it both ways.

Using the TVM solver, we simply enter the number of payments (48) for N, leave 0 next to I%, enter −10,250 for the principal value (PV), then 246.93 for the monthly payment (PMT), and 12 for the number of payments per year (P/Y). Make sure 0 is entered for future value (FV) since the goal is still to pay off the loan. Then move the cursor to the line we want to calculate

Solved Example *Continued*

(I%) and press **ALPHA** and **ENTER**. The screenshot shows the result: The APR is 7.31%, so the dealer was fudging just a bit when they quoted Elaine an APR of 7.25%.

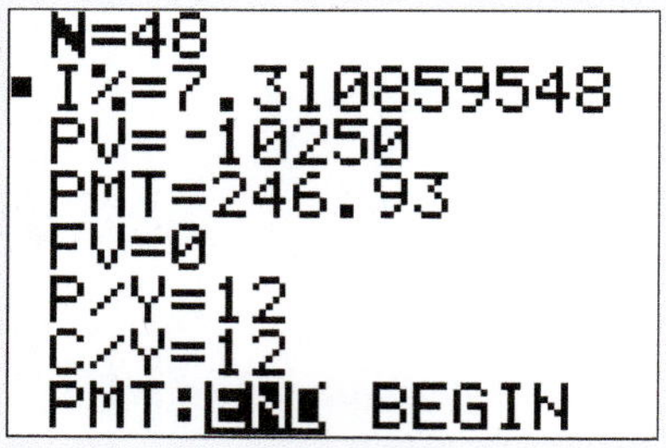

If you're more of a spreadsheet person, the Excel function we need is =RATE. When I enter =RATE(into an empty cell, here's what happens:

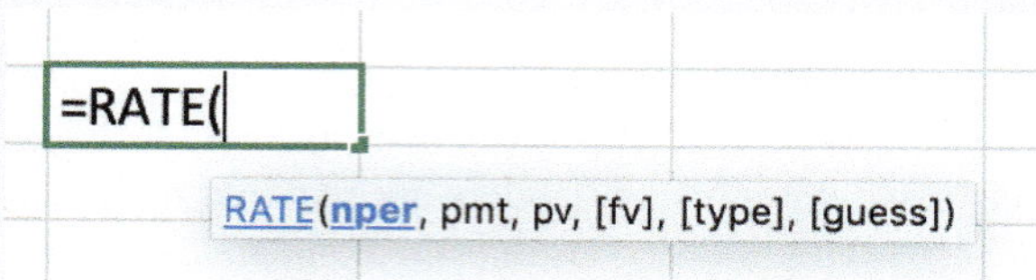

Notice that Excel prompts the inputs it needs. The first input (nper) is the number of payments for the loan. In this case, that's 48. The second input (pmt) is the monthly payment. In this example, the monthly payment is given as \$246.93, so we enter 246.93. Next, pv is the principal value of the loan, which in this case is \$16,250 – \$6,000 = \$10,250. (This is the cost of the car minus the down payment.) We enter this as a negative like we did using the TVM solver. You can ignore the [type] and [guess] inputs. (If you need further guidance on using this function, there's a tech video available in the resources for this lesson.)

The next two graphics are what the cell looks like when we've entered all the required inputs but have not yet hit Enter, and then the result of the calculation.

Wait, what? Is the dealership mistakenly giving us less than 1% interest? If you've ever dealt with a dealership, you know that's unlikely, to say the least. Notice that both our number of payments and our payment are related to months, while APR is an annual percentage. In short, the RATE command in Excel gives us a monthly percentage, not annual. So we just need to multiply by 12: 0.61% × 12 = 7.31%.

Did You Get It

3. Find the APR for the loan we worked with earlier in Questions 1–4.

When you take out a loan for a certain length of time with fixed payments, the usual path is to continue paying for the entire time until you don't owe anymore. But if you have the money to pay off the loan early, you can save some money because you don't end up having to pay the total amount of interest. But how much do you save? Is it worthwhile to pay a big chunk of money to save some on the finance charge? Let's take a look.

The amount of the finance charge that you save when paying off a loan early is called **unearned interest.** There are a couple of methods that are used to calculate unearned interest. We're going to study one called the rule of 78. This is a formula that you can use to calculate it directly.

The Rule of 78

$$u = \frac{fk(k+1)}{n(n+1)}$$

where

u = unearned interest
f = finance charge
k = number of remaining monthly payments
n = original number of payments

As a reminder, here are the specifics of the loan we've been working with:

Cost of car: \$12,260
Down payment: \$3,000
Monthly payment: \$231.50
Term of loan: 48 months

Let's say that we get a nice tax refund and consider paying off the loan a year early to save some interest.

7. Use the rule of 78 to find how much interest you'd save if you pay the loan off with 12 payments remaining. You'll need the finance charge you calculated back in Question 4.

8. In this situation, you would surely want to know how much it would cost to pay off the loan. The first step would be to find the total cost of all the remaining payments after the 36th. So do that.

9. Your answer to Question 8 is what the payoff would be if you weren't getting credit for unearned interest. But you are, so subtracting the unearned interest you found will give you the payoff amount.

10. Discuss the pros and cons of paying off that particular loan early. If you were in that situation, do you think it would have been worthwhile for you to pay it off early?

Did You Get It

4. A $16,500 truck loan is to be paid off in 60 monthly installments of $317.50. The borrower decides to pay off the loan on the 42nd payment. Find the amount of interest saved and the payoff amount.

Sidelight **To Buy or To Lease? That is the Question**

Obvious statement of the day: new cars are expensive. (And used ones aren't particularly cheap, either.) If you don't have a lot of money to make a down payment but you want that new-car smell, leasing may be the way to go. Like an auto loan, with a lease you're making regular equal payments to a financial institution. The difference is simple: after making payments for a set number of months, you don't own the vehicle and you have to give it back. In essence, you're renting the car rather than buying it. So why would you want to do that? There are several reasons.

For one, many leases require little or even no down payment, so you can get a new car without having to save a lot of money in advance. Second, lease payments are usually quite a bit less than those you'd have on a loan (as they should be, since you don't get to keep the car!). Third, most leases last for 2 or 3 years, allowing regular leasers to get a brand new car every few years. And you do have the option of buying the car at a pre-arranged price after the term of the lease expires. The biggest downside to leasing is obvious—you're paying for the car but you don't get to keep it. Also, most leases have mileage limits, and if you go over the number of miles you're allotted, you can end up paying a substantial penalty at the end of the lease. You may pay more for insurance with a leased car as well. In addition, you're committed to making payments for the life of the lease, so if you decide you don't like the car that much or your financial situation changes, you're pretty much out of luck. Depending on what you're looking for and your financial situation, leasing might be the right thing for you. But make sure you understand all of the provisions before signing anything because leases are commitments that are *very* difficult to get out of.

2-5 Group

So far, we've covered examples of closed-ended credit; this is credit with a fixed number of payments and a specific payoff date. We turn our attention now to **open-ended credit,** where there is no fixed number of payments or payoff date. By far the most common example of this is credit cards. Not only are more people using credit cards for convenience, many retail locations don't even take cash anymore, so an understanding of open-ended credit is more important than ever. One of the easiest ways to end up in a really bad place financially is to not understand how open-ended credit works.

The Unpaid Balance Method

One way that banks calculate finance charges on a credit card is the unpaid balance method. In this case, interest is charged only on the balance left over from the previous month.

> **Math Note**
>
> As of the second quarter of 2024, total credit card debt in the U.S. was over 1.1 TRILLION dollars, or a bit less than $9,000 for every household in the country.

Suppose that you had an unpaid balance of $356.75 at the beginning of August and made purchases of $436.50. A payment of $200.00 was made during the month. The interest on the unpaid balance is 1.8% per month. Our goal will be to find the finance charge and the balance on September 1. (The billing cycle for credit cards and loans can begin and end on any day of the month; however, for the examples given in this lesson, we'll assume the cycle ends on the last day of the month.)

1. Find the amount of interest that would be charged on the unpaid balance. Show your work!

2. Use your answer from Question 1 and the information about payments and purchases to find the new balance on the card. Show all work again!

Did You Get It

5. For the month of January, Christina has an unpaid balance of $846.50 from December. Purchases for the month were $532.86 and a payment of $350.00 was made during the month. If the interest on the unpaid balance is 2% per month, find the finance charge and the balance on February 1.

The Average Daily Balance Method

Because a company that issues credit cards doesn't make as much money when users pay off the balance every month, they sometimes charge an annual fee for use of the card. Another way for them to make money from all users is to use the average daily balance method for computing the finance charge. In this method, the balance for each day of the month is used to compute an average daily balance, and interest is computed on that average. So in effect, consumers start paying interest on the day they make a purchase. Ouch!

Calculations using the average daily balance method can be done in a series of steps, as illustrated by the next set of questions. The credit card statement shown here displays more detail on the account we studied in Questions 1 and 2.

August 1	Previous Balance	\$356.75
August 7	Purchases	\$59.95
August 12	Purchases	\$223.75
August 18	Payment	\$200.00
August 24	Purchases	\$152.80

The goal is to find the average daily balance, the finance charge for the month, and the new balance on September 1. The interest rate is 1.8% per month on the average daily balance.

Step 1: Find the balance on each day that a transaction occurred.

3. Write the new balance next to each day listed below in the first blank. Ignore the second blank for now.

August 1 ________ _____

August 7 ________ _____

August 12 ________ _____

August 18 ________ _____

August 24 ________ _____

Step 2: Find the number of days for each balance.

4. How many days was the balance \$356.75? Using the same idea, fill in the second blank next to each balance above with the number of days that the balance was at that amount. (Make sure the total number of days adds up to 31 days, which is the number of days in August.)

Step 3: Multiply each balance by the number of days, and add these products.

5. Copy the first two columns in the table below from your answers on the previous page, then multiply to fill in the last column. Then add your results.

Date	Balance	# of days	Balance × # of days
August 1			
August 7			
August 12			
August 18			
August 24			
		Sum	

Step 4: Divide by the total number of days to get the average daily balance for the month.

6. Divide your sum from Question 5 by the number of days in August. This gives you the average daily balance for the account.

Step 5: Find the finance charge by calculating the amount of interest on the average daily balance.

7. Recall that interest on this card is 1.8% per month. How much interest would be due on the average daily balance?

Step 6: Find the new ending balance by adding the finance charge to the last balance.

8. Find the final balance that would appear on the next statement.

Did You Get It

6. A credit card statement for the month of November showed the following transactions:

November 1	Previous Balance	$937.25
November 4	Purchases	$531.62
November 13	Payment	$400.00
November 20	Purchases	$89.95
November 28	Payment	$100.00

a. Find the average daily balance.
b. Find the finance charge. The interest rate is 1.9% per month on the average daily balance.
c. Find the new balance on December 1.

Sidelight **Money For Nothing?**

Since most credit card companies either use the unpaid balance method, or don't charge interest on new purchases, if you pay off the full balance each month, you never pay any interest. In essence, you're getting to use the bank's money for nothing! But don't shed any tears for the poor banks. They charge the merchant a fee for each transaction. Also, the freedom that comes from buying stuff and walking out the door without handing over any money makes it extremely difficult for people to spend only what they can afford, especially young people using credit cards for the first time. Consider the following statistics:

- The average college senior will graduate with $3,280 in credit card debt in addition to any student loan debt.
- 47% of all credit card users pay interest fees by not paying off the card every month.
- Total credit card debt in the United States in 2024 was over 1.1 *trillion* dollars.
- The average credit card debt per household in 2024 was $8,992. This was up significantly from 2021, when it was $6,270.
- The average household that was carrying a credit card balance in 2024 spent almost $2,000 just in interest on their credit card balance.
- In 2010, over 90% of college students had a credit card, but new regulations on credit card companies targeting students decreased that number by quite a bit. Debit cards are now far more common—debit cards draw from an account, so no money, no purchases. Still, in 2024, 57% of college students had at least one credit card.

If you find yourself spending more than you can pay off, and having to keep a balance on your card, the best approach is probably the one pictured below.

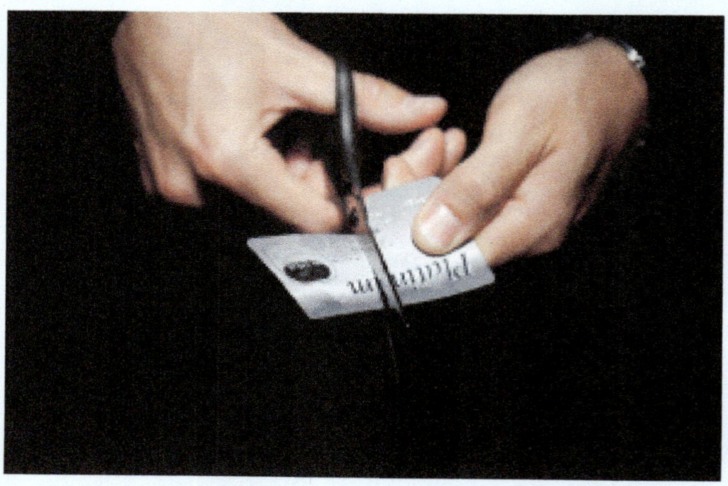

Fuse/Corbis/Getty Images

Each month when you get a credit card statement, there's a minimum payment that must be paid. Before 2003, these minimum payments had shrunk to the point that they didn't even cover the interest for the month, so if a cardholder paid just the minimum payment, the balance would continue to rise. In response to a public outcry, the federal government set new regulations in 2003 requiring credit card companies to set the minimum payment high enough that it covers the interest for the month, as well as part of the principal balance. (The exact minimum payment formula varies some from company to company.)

This is not to say that making minimum payments only is a good idea. If at all possible, you should always pay off the full amount each month. This is the only foolproof way to make sure that you don't get in the habit of spending more money than you can afford. We'll close this lesson by examining the effects of making only minimum payments.

Suppose you have a $2,300 balance on a credit card with an interest rate of 1.72% per month, and the minimum payment for any month is the amount of interest plus 2% of the principal balance, or $20, whichever is higher. If you don't make any more purchases on that card and make the minimum payment for 6 months, how much will you pay down the balance? Let's find out. (By the way, the interest and minimum payment percentages in this problem reflect industry averages in 2024.)

9. This certainly sounds like we'll need to compute some interest. After all, we were told that the interest is 1.72% per month. But sometimes problems contain extraneous information. Read the problem carefully and explain why we don't need to worry about the amount of interest in calculating how much the balance decreases.

10. In fact, what really matters is that the balance is going to decrease by 2% each month. Use this fact to find the new balance after 1 month.

11. Repeat your calculations from Question 10 five more times to find how much the balance would have decreased after 6 months.

12. Just make a guess: How long do you think it would take to pay off the full balance at this rate? Discuss, then ask your instructor to give their thoughts on this question.

Math Note

In the Technology portion of the homework, we'll use Excel to calculate how long it would take to pay off this card if you make minimum payments.

2-5 Portfolio

Name ______________________________

Check each box when you've completed the task. Remember that your instructor will want you to turn in the portfolio pages you create.

Technology

1. ☐ In this Excel activity, we'll build a spreadsheet to analyze how long it would take to pay off a credit card balance when making the minimum payment.

ALEKS Practice

1. ☐ Include any written work from the ALEKS assignment along with any notes or questions about this lesson's content.

Applications

1. ☐ Complete the applications problems.

Reflections

Type a short answer to each question.

1. ☐ Explain the difference between closed-ended and open-ended credit.
2. ☐ What did you learn in this lesson about keeping balances on credit cards?
3. ☐ If making a large purchase, what factors do you need to consider?
4. ☐ Name one thing you learned or discovered in this lesson that you found particularly interesting.
5. ☐ What questions do you have about this lesson?

Looking Ahead

1. ☐ Complete the prep skills for Lesson 2-6.
2. ☐ Read the opening paragraph in Lesson 2-6 carefully and answer Question 0 in preparation for that lesson.

Answers to "Did You Get It?"

1. Amount financed, $7,200; total installment price, $10,720; finance charge, $2,520
2. a. $1,299 **b.** $28.90 **c.** $1,934 **d.** $435 **3.** 9.24%
4. Save $238.28; payoff amount is $5,476.72
5. Finance charge = $16.93; balance on Feb. 1 = $1,046.29 **6. a.** $1,198.69 **b.** $22.78 **c.** $1,081.60

Answers to Prep Skills Questions

1. 0.91, 0.0648
2. $204.75
3. $582.30

2-5 Applications

Name ______________________________

In this assignment, we'll study various options for making a large purchase, including saving money in advance and paying cash, obtaining a fixed installment loan, and using a credit card. The goal is not just to perform calculations, but to think about all aspects and decide which you think is the best choice. We'll use furniture as an example, as it's pretty common to finance furniture purchases when you need to fill a new place. Let's say that the goal is to buy $3,500 worth of furniture.

We'll start with a fixed-installment loan from the furniture store. They're offering 3-year financing with no money down at 11.99% annual interest.

1. Find the monthly payment. You can use the monthly payment formula and calculate by hand if you like, but remember that the TVM solver on a calculator and the PMT function in Excel are good options as well.

2. Find the total installment price and finance charge.

The store also offers 18-month financing with no payments and no interest, but if the balance isn't paid in full at the end of 18 months, 17.99% interest will apply to the entire purchase amount, and will continue on any unpaid balance after 18 months.

3. Let's say you take this option and forget to pay off the final amount before the 18 months is up. And believe me, it's easy to do—they don't exactly go out of their way to remind you of the end date. If you'd paid $100 per month up until that point, how much would you still owe after the interest charge kicks in? Remember, the interest will be calculated on the ENTIRE purchase, not just what's left.

2-5 Applications

Name ______________________________

The next option is to put the entire purchase on a bank-issued Visa card with an annual interest rate of 12.49%. Let's say that you use this particular card strictly for this purchase.

4. In this case, would the amount of interest calculated vary depending on whether the bank used the unpaid balance or the average daily balance method? Explain.

5. This particular card has a minimum monthly payment of 3% of the balance plus any interest accrued during the month. What would your first minimum payment be?

6. In this case, the equation $y = 3{,}500(0.97)^t$ would describe the remaining balance, where t is months after purchase. The graph of this equation is shown below. Use the graph to estimate how long it would take to pay off the purchase making only minimum payments. You can assume that once the balance reaches $100, you'd go ahead and pay off the rest.

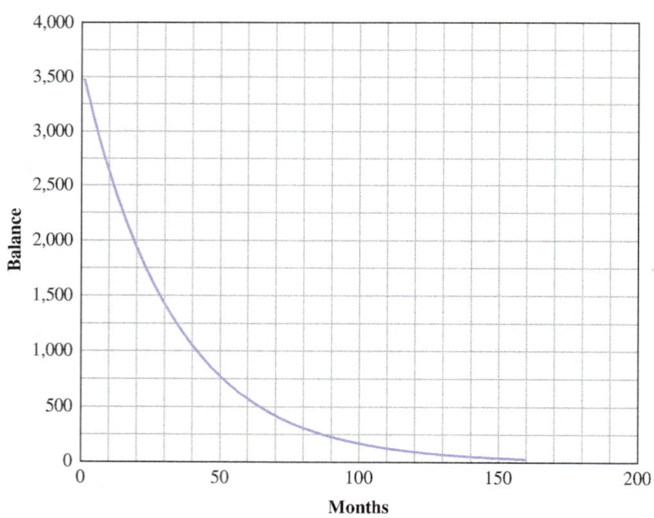

2-5 Applications

Name ______________________________

Suppose that you want to avoid paying interest and decide you'll only buy the furniture when you have the money to pay for it. An **annuity** is basically the opposite of a fixed-installment loan: you deposit a fixed amount each month and receive interest based on the total amount that's been saved. The future value formula is

$$A = \frac{12M\left[\left(1 + \frac{r}{12}\right)^{12t} - 1\right]}{r}$$

where M is the regular monthly payment, r is the annual interest rate in decimal form, and t is the term of the annuity in years.

7. With a monthly payment of $120, what would the future value be if you chose an annuity with a term of 2 years at 4.8% interest?

> **Math Note**
>
> There are great technology options for calculating the future value of an annuity. The TVM solver on a TI–83 or 84 can be used, and there's a future value calculator in Excel as well: the =FV command. A quick Internet search will provide more info, and we've provided a tutorial video in the resources for this lesson.

8. Recalculate the amount if you're willing to raise your monthly payment by $20 per month.

9. Discuss the options you've seen for obtaining the furniture, focusing on which you think sounds like the best choice. You might consider combinations of these choices. Try to consider as many factors as you can.

Lesson 2-6 Prep Skills

SKILL 1: COMPUTE SIMPLE INTEREST

Remember from Lesson 2-3 that the formula for simple interest is $I = Prt$, and the formula for future value is $A = P(1 + rt)$. A represents the future value, P represents the principal, r represents the rate, and t represents the length of time the money is borrowed or invested.

- Find the interest due on a loan of $600 for 8 years at 4.5% simple interest.

  $I = Prt$
  $I = (\$600)(0.045)(8)$
  $I = \$216$

  The decimal form of 4.5% is 0.045. Substitute $600 for *P*, 0.045 for *I*, and 8 for *t*. The total interest on this loan is $216.

- Find the balance in a savings account after 2 years if the account started with $350 and the account pays 2.5% simple interest.

  $A = P(1 + rt)$
  $A = (\$350)(1 + 0.025 \times 2)$
  $A = \$367.50$

  The decimal form of 2.5% is 0.025. Substitute $350 for *P*, 0.025 for *I*, and 2 for *t*. The balance in this savings account after 2 years would be $367.50.

SKILL 2: PERFORM COMPLEX CALCULATIONS

When you perform complex calculations using a calculator, spreadsheet, or other computing device, it is important to pay attention to the intended meaning of the expression. That means looking carefully at exponents and grouping symbols (including fraction bars) that group the entire numerator and denominator separately.

- Use a calculator to evaluate $(1 + 0.035)^{4(8)}$.

  (1 + 0.035)^(4 × 8) ≈ 3.00067

  Note that in the original expression the exponent should be 4 × 8 or 32. Entering the expression as (1 + .035)^4(8) or (1 + .035)^4 × 8 would actually evaluate (1 + .035)^4 and then multiply that result by 8.

- Use a calculator to evaluate $\dfrac{250}{1 + (1 + 0.035)^{4(8)}}$.

  250/(1 + (1 + 0.035)^(4 × 8)) ≈ 62.3954

  Note the need for additional parentheses around the entire denominator.

SKILL 3: EVALUATE ALGEBRAIC EXPRESSIONS

As we saw in the prep skills for Lessons 2-3 and 2-4, you evaluate an expression (or a formula) by replacing all occurrences of the variable or variables with specific numbers to find a numeric value.

- Evaluate $A = P\left(1 + \frac{r}{n}\right)^{nt}$ for $P = 500$,

  $r = 0.0475$, $n = 6$, and $t = 5$.

  $A = 500\left(1 + \frac{0.0475}{6}\right)^{6(5)}$ Replace each variable with the appropriate value and simplify the result.

  $A \approx 633.44$

PREP SKILLS QUESTIONS

1. Compute the simple interest on a \$12,000 loan at 9% interest for 3 years.

2. Compute the future value of a \$3,700 deposit that draws 4.95% simple interest for 18 months.

3. Use a calculator to perform each calculation:

   $8\left(1 - \frac{0.09}{12}\right)^{-24}$

   $\frac{17{,}000}{1 - (1 + 0.03)^{-120}}$

4. Find the value of the formula below if $k = -8.1$, $m = 0.0435$, and $W = 1{,}297.3$. Round your answer to six decimal places.

   $\frac{W\left(\frac{m}{k}\right)}{k - (1 + m)^k}$

Lesson 2-6 Investing in Yourself

LEARNING OBJECTIVES

- ☐ 1. Understand different student loan options.
- ☐ 2. Compute interest and monthly payments on a student loan.
- ☐ 3. Evaluate the effects of capitalizing interest.
- ☐ 4. Analyze various aspects of a mortgage.
- ☐ 5. Compare two mortgages of different lengths.
- ☐ 6. Prepare an amortization schedule.

Digital Vision/SuperStock

If you think education is expensive, try ignorance.
—Derek Bok

Everyone knows that college is expensive. And some folks have opinions on whether or not it's really worth the investment. The numbers, however, say that it absolutely is, since your earning potential with a college degree increases to the point that you'll make back the cost long before retirement. When it comes to credit card debt, the best advice is often "Don't." But that's not a viable option for college costs, since the cost of NOT going to college is far more in the long run than the cost of attending. That makes student loans a necessary evil for most folks. And when you're finished with college and making a living, many of you will make the most expensive and rewarding purchase of your life: a home. If there's one common theme in this unit, it's that understanding the nuances of the financial transactions you're likely to deal with makes you far less likely to end up in a bad situation.

0. Do you think the costs of getting a college degree will be worth it for you personally? Explain.

2-6 Group

Student Loans

There are three basic types of loans that most undergraduates might consider: federal loans made by the government, federal loans made by banks or other lenders but guaranteed by the government, and private loans. Every student's first choice should be the federal loans due to stability: The interest rates are fixed for the life of the loan and are regulated by Congress, so you're unlikely to end up with a nasty surprise somewhere down the line. There are, however, limits on the amount that can be borrowed for one student through federal loan programs, and private loans might be the only alternative for an especially expensive education.

Other than the length of the loan, the big difference between student loans and other installment loans is that the payments are typically deferred until after graduation. This makes perfect sense, as in many cases students aren't working enough to make the money required to keep up with payments. So that's a pretty good deal. But it comes with a catch: In some cases, interest will accumulate on your loan while you're in school even though you're not making any payments. This can increase the principal balance considerably. The loan term can vary depending on the amount borrowed, but the standard term is 10 years, with payments deferred until 6 months after graduation.

Interest on student loans is simple interest: The interest accrues only on the principal balance, not on previously accrued interest. So while the student is still in school, the amount of interest that accrues each month remains constant. The interest is calculated using the simplified daily interest formula, which we can develop using things we know about simple interest.

1. If the principal balance of a loan is $6,500 and the interest rate is 6.8% annually, how much simple interest will accrue in 1 year?

2. Why does it make sense to use 365.25 days in a year for financial calculations?

3. Using Questions 1 and 2, find the daily amount of interest that accrues on that $6,500 loan. Make sure you explain how you got your answer.

4. How much interest accrues in a month? For our purposes, we can use 30.44 for the number of days in a month; this is 365.25 divided by 12.

Math Note

The actual interest accrued in any month will depend on the number of days. We're using 30.44 average days per month to simplify the process just a bit. At worst this will result in a couple dollars of error.

The Simplified Daily Interest Formula

$$\text{Daily interest amount} = \frac{\text{Principal balance} \times \text{interest rate}}{365.25}$$

$$\text{Monthly interest amount} = \text{Daily interest amount} \times 30.44$$

5. For that $6,500 student loan we've been working with: If you start school in August, graduate in May (3 years and 9 months later), and payments begin 6 months after graduation, how many months will pass while interest is accruing?

6. How much total interest accrues before payments begin?

Did You Get It

Try this problem to see if you understand the concepts we just studied. The answer can be found at the end of the portfolio section.

1. Max borrows $11,500 for her last 2 years of college, acquiring a federal student loan at 6.2% interest.
   a. Find the amount of interest that accrues each month.
   b. Find the amount of interest that accrues if she acquires the loan in August, graduates 2 years later, and payments begin 3 months later.

When interest is covered by the government while the borrower is still in school, a student loan is called **subsidized.** If interest is added to the principal while the borrower is in school, the loan is called **unsubsidized.** For the loan in Questions 1–6, we found that $1,878.33 in interest accrued before payments started, so the student would have saved almost $1,900 by acquiring a subsidized loan.

When monthly payments begin on a student loan, they're based on the principal balance at that point in time. These payments can be computed using the following formula, which should look familiar: It's exactly the payment formula we studied in Lesson 2-5. We just thought we'd be nice and recopy it rather than making you page back. You should probably send us a thank-you note, or maybe some homemade cookies. Also, since most student loans have a 10-year term, we took the liberty of adapting the formula to that situation. Again, you should probably send us cookies. Cake would be appreciated, too.

Monthly Payments on a Student Loan

For a 10-year student loan with principal P and interest rate r (written as a decimal), the monthly payment M is given by

$$M = \frac{P \cdot \frac{r}{12}}{1 - \left(1 + \frac{r}{12}\right)^{-120}}$$

Reminder: Perform the calculation in one step if you can. If not, don't round too much. Always use at least four decimal places, and more than that won't kill you. I promise.

7. For the loan in Questions 1–6, what is the principal amount at the time payments begin? (This includes the original amount borrowed, plus the amount of interest that accrued until payments started.)

8. Find the monthly payment on the loan.

> **Math Note**
>
> In the Technology assignment for this lesson, you'll use a spreadsheet to do these calculations. Try it! If you find it helpful, you should use a spreadsheet for the homework as well, provided that you get the okay from your instructor.

We've seen that with a student loan, you have the option of not making any payments until after graduation. When interest on an unsubsidized student loan is not paid during college, we say that the interest is **capitalized.** When interest is capitalized, it adds to the principal balance, which of course results in larger payments. That's what happened in the example that we've been working with so far.

When interest is not capitalized, payments have to be made during college, but only covering the amount of interest. This keeps the principal balance the same until full monthly payments start, which makes those payments smaller.

For the student loan we've been working with, the payment while still in school would simply be the amount of interest that's accruing, which you already calculated. For reference, in the space below write the amount you'd pay in an average 30-day month (see Question 4), as well as the total amount of interest you'd pay before regular monthly payments begin (Question 6).

9. With interest not capitalized, the principal balance of the loan will still be $6,500 when full payments start. Find what those payments will be. How much less is that per month than if interest was capitalized?

In comparing the loans with or without interest capitalized, certainly we'd be interested in finding the difference in the total cost of the loan. Let's check it out.

Solved Example 1

Comparing Loans With and Without Interest Capitalized

We've seen so far that by paying interest on our $6,500 student loan while in school, we'll have a smaller monthly payment when payments begin. Will that result in an overall lower cost over the life of the loan? If so, by how much?

SOLUTION

For the loan with interest capitalized, every dollar you'll pay back comes from the total of the monthly payments. It's a 10-year loan, and we found that the monthly payment will be $96.42. So the total cost of that loan is

$$10 \times 12 \times \$96.42 = \$11{,}570.40$$

Now let's attack the loan with interest not capitalized. In this case, we still have 10 years of monthly payments, at $74.80 per month:

$$10 \times 12 \times \$74.80 = \$8{,}976$$

Wow! That's way cheaper. But wait—we also had to make payments while in college. That amount was $1,878.33. So the total cost of that loan is

$$\$8{,}976 + \$1{,}878.33 = \$10{,}854.33$$

So by making payments while in school, we saved a total of $11,570.40 − $10,854.33 = $716.07.

10. Now, the key question when studying whether capitalizing interest is a good idea: Discuss with your group whether or not you think the amount saved by paying interest while in school would make it worthwhile to scrape together the money to make those interest payments while still a student, then write some of your thoughts.

Did You Get It

2. For the loan in Did You Get It 1:
   a. Find the monthly payment after graduation if interest is not capitalized. (Note that you found how much payments would be while in school in Did You Get It 1.)
   b. Find the full monthly payment after graduation if interest is capitalized.
   c. How much money would be saved by not capitalizing interest?

Home Buying

SW Productions/Brand X Pictures/
Getty Images

For many people, the day they buy their first home is one of the proudest days of their life—and one of the scariest. There is nothing that compares to the feeling of looking at a house and knowing that it's all yours. But the buying process is tremendously intimidating. There are dozens of documents to sign, and the sheer numbers involved are enough to make almost everyone wonder if they're making a colossal mistake.

The most common home loans are paid over a 30-year span. That's a major commitment, and one that nobody should enter into without an understanding of the mathematics that go into the process. In the remainder of this lesson, we will study that math, hopefully helping you to become a well-informed home buyer. Maybe you're thinking, "Dude, I'm a college student. Buying a house is like a million years away." If so, you'll be surprised at how quickly that million years passes. I surely was.

A **mortgage** is a long-term loan where the lender has the right to seize the property purchased if the payments are not made. Homes are the most common items bought using mortgages. The most common mortgage term is 30 years, but they are widely available in terms from 10 to as many as 50 years.

There are several types of mortgages. With a **fixed-rate mortgage,** the rate of interest remains the same for the entire term of the loan. The payments (usually monthly) stay the same. With an **adjustable-rate mortgage,** the rate of interest may fluctuate (i.e., increase and decrease) during the period of the loan. If you opt for an adjustable-rate mortgage, it will start out with a lower interest rate than a fixed-rate loan of the same length.

11. Discuss the pros and cons of fixed and adjustable-rate loans.

12. Discuss the pros and cons of buying a house compared to renting a place to live.

One of the most useful (but frightening) things about home buying is calculating how much money you'll actually pay over the life of a loan. Let's see how that might work for one family.

The Petteys family plans to buy a home for $324,900, and has been offered a 30-year mortgage with a rate of 5.5% if they make a 20% down payment.

13. Calculate the down payment and the amount they'll have to borrow.

14. The monthly payment on the loan will be $1,475.80. How much will the Petteyses actually pay for their $324,900 home? (Don't forget to add in the down payment.)

Math Note

The word "mortgage" comes from a combination of Old French words "mort" (dead) and "gage" (pledge). It is believed the intent was that the debtor pledged the property to secure the loan, and if he or she failed to pay, the property was taken, and was therefore "dead" to the debtor.

Did You Get It

3. The Trissel family agreed on a price of $229,500 for a home. Their company credit union offers a 5.0% 20-year loan with 15% down, resulting in a monthly payment of $1,287.50. Find the down payment amount and the total amount they'll pay for the house.

2-6 Class

When you apply for a home loan, the lender will tell you what your monthly payment will be. And if you're an incredibly trusting sort, that might be good enough for you. For most of us, though, a double-check of the quoted payment would be a big comfort. In addition, knowing how the payment is calculated makes it quite a bit easier to understand the factors that determine how much you'll pay. Better still, it will allow you to compare mortgages with different rates, down payments, and lengths. So let's have a look at computing monthly payments.

1. In the space below, recopy the formula that we used for computing monthly payments on a student loan with a term of 10 years. Then describe the significance of each part of the formula requested below:

   P:

   r:

   12:

2. Where do you think the 120 comes from in the formula? (Hint: Think of 120 as the product of two numbers.)

3. Using your answers to Questions 1 and 2, adapt the student loan monthly payment formula into a general formula for finding the periodic payment on a loan with principal value *P*, annual interest rate *r*, term *t* (in years), and *n* payments per year. *Check with your instructor to make sure that you have the right formula before moving on to the remainder of the lesson!*

> **Math Note**
>
> You can still use the TVM solver to calculate monthly payments. The only difference is that if the number of payments per year is different from 12, you'll enter that number next to P/Y. The PMT function on a spreadsheet works just as it did in Lesson 2-5.

Monthly Payments on a Mortgage

For a mortgage with principal *P*, interest rate *r* (written as a decimal), and *n* payments per year for *t* years, the monthly payment *M* is given by

$$M = \frac{P\left(\frac{r}{n}\right)}{1 - \left(1 + \frac{r}{n}\right)^{-nt}}$$

Now let's use our snazzy new formula to study some variations on the mortgage we've been working with.

Solved Example 2

Comparing Loans With Different Down Payments

Earlier we studied a loan obtained by the Petteys family with these terms: purchase price \$324,900, 20% down payment, 30-year mortgage with a rate of 5.5%. The monthly payment was \$1,475.80. How much higher would their payment be if they'd only made a 10% down payment? (Note that this would raise the interest rate to 5.8%.) How much would this cost them over the life of the loan?

SOLUTION

To calculate a monthly payment we need the principal, the rate, the term, and the number of payments per year. The term is still $t = 30$ years, and the Petteyses will still make $n = 12$ payments per year. The interest rate is now $r = 0.058$. The first calculation we need to do is the new principal. Let's be clever: If they're making a 10% down payment, that means the amount borrowed will be 90% of the purchase price. So we get

$$P = 0.90(\$324{,}900) = \$292{,}410$$

Now we're ready to use the monthly payment formula. Of course, the TVM solver or Excel PMT function that we learned about in Lesson 2-5 are good options as well.

$$M = \frac{292{,}410\left(\frac{0.058}{12}\right)}{1 - \left(1 + \frac{0.058}{12}\right)^{-12\times 30}} = \$1{,}715.72$$

The monthly payment is \$1,715.72. This allows us to find the total cost of this new loan: It's the number of payments (360) times the monthly payment we just found, plus the down payment, which is 10% of the purchase price.

$$360(\$1{,}715.72) + 0.10(\$324{,}900) = \$650{,}149.20$$

In Question 14 of the Group portion, we found that the cost of the loan with 20% down was \$596,268. So with half as much money down, the total cost of the loan goes up by

$$\$650{,}149.20 - \$596{,}268 = \$53{,}881.20$$

4. Now let's examine the term of the loan. If the Petteyses choose a 15-year mortgage with 20% down at 5.5%, what would the monthly payment be?

5. How would their total cost compare to the other two options we studied? Which mortgage do you think is the best choice? Discuss.

Math Note

A Google search for the phrase "mortgage calculator" results in over 60 million hits! There are thousands of pages available that can be used to quickly calculate monthly payments on a mortgage. But doing the calculations by hand will help you to become very familiar with the terms involved in mortgages.

Did You Get It

4. Find the monthly payment for the Trissels (Did You Get It 3) if they choose a 15-year loan instead of 20. How much money would they save overall?

Computing an Amortization Schedule

After securing a mortgage, the lending institution will prepare an **amortization schedule.** This schedule shows what part of the monthly payment is paid on the principal and what part of the monthly payment is paid in interest. Here's the procedure for computing an amortization schedule:

Computing an Amortization Schedule

Step 1: Find the interest for the first month. Use $I = Prt$, where $t = 1/12$. Enter this value in a column labeled Interest.

Step 2: Subtract the interest from the monthly payment to get the amount paid on the principal. Enter this amount in a column labeled Payment on Principal.

Step 3: Subtract the amount of the payment on principal found in step 2 from the principal to get the balance of the loan. Enter this in a column labeled Balance of Loan.

Step 4: Repeat the steps using the amount of the balance found in step 3 for the new principal.

In the next group of questions, we'll use these steps to compute the first 3 months of an amortization table for the Petteyses' loan: principal $259,920, rate 5.5%, monthly payment $1,475.80. A template for the table is on the next page: Fill in the information as you find it.

6. Calculate the interest for the first month on the loan, and put it into the interest column of the table.

7. Subtract the interest from the monthly payment to get the amount paid on the principal, then put that amount in the appropriate column in the table.

Math Note

You'll also be using a spreadsheet to create an amortization table in the Technology assignment. But it will be a lot easier to set up the formulas if you go through the calculations by hand first.

8. Calculate the new balance of the loan by subtracting the payment on principal from the original principal, then put that in the last column.

9. Repeat the previous steps for the next 2 months. Do your calculation in this space, then put the results in the table.

Payment Number	Interest	Payment on Principal	Balance of Loan
1			
2			
3			

10. What does this table say about what the bulk of your payment goes to when you begin making payments on a mortgage? What about near the end of the term of your loan? Explain.

Did You Get It

5. The Trissel loan from Did You Get It 3 has the following parameters: $195,075 principal, 5.0% interest rate, and monthly payment $1,287.50. Complete the first 2 months of an amortization schedule for this loan.

Payment Number	Interest	Payment on Principal	Balance of Loan
1			
2			

After learning about amortization schedules, most people think that mortgages sound like an awful deal: Your early payments go almost completely to interest, so you're paying a LOT of money, and making almost no progress on paying down your loan. But here's the benefit to you: The interest paid on a home loan can be deducted from your income when computing the amount of income tax that you owe. So at least the way mortgages work gives you a bigger break on your income taxes during the early stages of the loan.

Financial advisors have coined two terms for borrowers: "good debt" and "bad debt." Good debt is when you borrow money to buy something that is going to increase in value. Student loans and home loans are pretty much the gold standard in good debt. Houses increase in value, and the amount that you pay for a college education is more than paid back by increased earning potential.

Bad debt, on the other hand, is borrowing to buy something that loses value immediately after purchase. An obvious example is car loans, although for most people this type of bad debt is unavoidable, so we'll coin a new term: "not-so-good debt." Car loans are manageable, but it's important to avoid really long-term loans. Dealers are now offering 60-, 72-, and even 84-month loans. You should think carefully before taking out a 60-month loan, and you should avoid longer loans than that at all costs. If you need to finance a car for that long, you can't afford that car.

The best example of bad debt (or worst example, I guess) is credit card debt. Keeping balances on credit cards is the easiest way to put yourself into a financial hole. Pretty much everything you might buy on credit is instantly worth less than what you paid, and the high interest rates attached to most credit cards make them a financial disaster waiting to happen. The key to managing your financial life is to be an educated borrower and educated consumer. Hopefully these lessons have given you the means to make smart financial decisions when it comes to borrowing.

2-6 Portfolio

Name ______________________________

Check each box when you've completed the task. Remember that your instructor will want you to turn in the portfolio pages you create.

Technology

1. ☐ In this Excel activity, we'll use a spreadsheet to create an amortization table.

ALEKS Practice

1. ☐ Include any written work from the ALEKS assignment along with any notes or questions about this lesson's content.

Applications

1. ☐ Complete the applications problems.

Reflections

Type a short answer to each question.

1. ☐ Describe some of the options you have in student loans.

2. ☐ What are the advantages and disadvantages of a mortgage with a shorter term?

3. ☐ Name one thing you learned or discovered in this lesson that you found particularly interesting.

4. ☐ What questions do you have about this lesson?

Looking Ahead

1. ☐ Complete the prep skills for Lesson 2-7.

2. ☐ Read the opening paragraph in Lesson 2-7 carefully and answer Question 0 in preparation for that lesson.

Answers to "Did You Get It?"

1. a. \$59.42 **b.** \$1,604.34 **2. a.** \$128.83 **b.** \$146.80 **c.** \$552.06
3. Down payment: \$34,425, total amount paid: \$343,425 **4.** \$1,542.64; \$31,324.80
5. Payment 1: Interest = \$812.81, Payment on Principal = \$474.69, Balance = \$194,600.31; Payment 2: Interest = \$810.83, Payment on Principal = \$476.67, Balance = \$194,123.64

Answers to Prep Skills Questions

1. \$3,240
2. \$3,974.73
3. 9.58, 17,504.29
4. 0.7910

2-6 Applications

Name ______________________________

Many mortgage lenders offer a way to lower your interest rate by paying some of the interest up front. This prepaid interest is known in the industry as **points.** Each point corresponds to 1% of the amount borrowed. In short, you're paying a fee to lower your interest rate. The question is whether this is a good deal for you, or for the lender. Let's look into it.

1. A lender offers you a 30-year fixed mortgage of $130,000 at 4% interest with no points. Find the monthly payment. If you're using technology to calculate monthly payments, which is an excellent idea, you should probably indicate that so your instructor knows where your answer came from.

2. You can get the same 30-year mortgage of $130,000 at 3.6% interest by paying 2 points at closing. How much would you have to come up with to cover the 2 points?

3. Find the new monthly payment.

4. How much are you saving each month with the new monthly payment?

5. How long will it take to recover the amount you had to come up with to pay the 2 points? What does this mean in terms of how long you should plan to stay in that home?

2-6 Applications

Name ______________________________

6. If you were to stay in that home long enough to pay off the loan, how much would the mortgage with points save you?

7. Suppose that instead of paying the points, you instead invested that \$2,600 into a long-term fund that paid 2.7% interest compounded monthly. How much would it be worth after 30 years?

8. Discuss which option seems the best to you. Justify your conclusions.

9. If you acquire a \$15,000 student loan at 4.3% interest and have to begin making payments 4 years later, how much would you save in interest if the loan is subsidized?

10. If the loan is not subsidized, the principal of the loan at the time payments start would be the amount borrowed plus the amount of interest you just calculated. Find the monthly payment for both the subsidized and unsubsidized loans. How do they compare?

Lesson 2-7 Prep Skills

SKILL 1: READ INFORMATION FROM A TABLE

Reading information from a table is a valuable skill that, like many things, only improves with practice. Some helpful tips for reading tables include:

- Read the title or the caption to understand what the table is presenting.
- Read the headings for each column and the labels/descriptions for each row to gain more understanding about the data in the table.
- Look for patterns in the data.

SKILL 2: PERFORM COMPUTATIONS INVOLVING PERCENTAGES

In the prep skills for this unit, we have reviewed percentages extensively. Remembering that "percent" literally means "per hundred" is a big key to working with percentages.

- $42\% = \frac{42}{100} = 0.42$

  Converting a percentage to its decimal equivalent requires dividing the percent by 100, which shifts the decimal point two places to the left.

- What percent of 40 is 35?

  $\frac{35}{40} = 0.875 = 87.5\%$

  In decimal form, 35 divided by 40 is 0.875. To convert the decimal to a percentage, multiply the decimal by 100, which shifts the decimal point two places to the right. The result tells us that 35 is 87.5% of 40.

- What is 30% of 60?

  $0.30(60) = 18$

  The decimal form of 30% is 0.30. The calculation tells us that 30% of 60 is 18.

- 24 represents 15% of a number. Find the number.

  $0.15n = 24$

  $n = 160$

  It's helpful to rephrase this question to read "15% of a number is 24." Then set up an equation where n represents the unknown number. Divide both sides by 0.15 to solve the equation. The solution tells us that 24 is 15% of 160.

SKILL 3: SOLVE A FORMULA FOR ONE VARIABLE

Solving a formula for a variable involves the normal steps we use in solving equations to get the desired variable by itself. It really helps to pretend that every letter other than the one you're solving for is a number.

- Solve $y = mx + b$ for x.

  $y = mx + b$ — Subtract b from both sides.

  $y - b = mx$ — Divide both sides by m to isolate x.

  $\frac{y-b}{m} = x$

- Solve $\frac{P_1}{T_1} = \frac{P_2}{T_2}$ for T_2.

  $\frac{P_1}{T_1} = \frac{P_2}{T_2}$ — Multiply both sides by the least common denominator, T_1T_2.

  $P_1T_2 = P_2T_1$ — Divide both sides by P_1 to isolate T_2.

  $T_2 = \frac{P_2T_1}{P_1}$

PREP SKILLS QUESTIONS

1. Below is the nutrition information label for a certain brand of peanut butter.

Serving Size 32g

Amount Per Serving	
Calories 188	Calories from Fat 135
	% Daily Value*
Total Fat 16g	25%
Saturated Fat 3g	17%
Trans Fat 0g	
Cholesterol 0mg	0%
Sodium 147mg	6%
Total Carbohydrate 6g	2%
Dietary Fiber 2g	8%
Sugars 3g	
Protein 8g	
Vitamin A 0% • Vitamin C	0%
Calcium 1% • Iron	3%

a. How many grams of fat are there in one serving?

b. How much sodium is there in three servings?

c. What percent of the daily recommendation of fiber is contained in one serving?

2. Based on the information in the table, what is the recommended daily value for total fat?

3. If you have 20 g of this peanut butter, what percent of a full serving is that?

4. How many grams of protein would you get from that 20 g of peanut butter?

5. Solve the equation $\frac{k}{M} = P$ for M.

Lesson 2-7 A Walk on Wall Street

LEARNING OBJECTIVES

- ☐ 1. Read information from a stock listing.
- ☐ 2. Calculate costs of buying stock, and profit or loss from selling.
- ☐ 3. Study the price-to-earnings ratio, and use it to analyze the value of a stock.
- ☐ 4. Calculate profit from a bond sale.

narvikk/E+/Getty Images

Everyone has the brainpower to make money in stocks. Not everyone has the stomach.

—Peter Lynch

We have seen that the magic of compound interest allows your money to grow considerably over long periods of time. However, interest rates on basic savings accounts are usually quite low, so if you rely on savings alone to build a nest egg, you better hope that you live a very, very long life indeed. Most successful investors grow their money much more quickly using the stock market. In this lesson, we'll learn about the basics of stocks and bonds, and how to get information about the performance of stocks that you might be interested in. When a company files legal papers to become a corporation, it's able to issue **stock.** If you buy shares of stock, you literally become a part owner of the company; for example, if a company issues 1,000 shares of stock and you buy 250 shares, then you own one-quarter of the company. The market can be a very intimidating place, but education holds the key to overcoming that fear and getting a piece of the action.

0. What are your impressions of the stock market at the moment? Be completely honest.

2-7 Class

Basic Definitions

As a shareholder in a company, you own part of that company. When the company makes money, it can choose to distribute part of the profit to its shareholders. This money is called a **dividend.** The stockholder receives a sum of money based on the number of shares of the stock that he or she owns. Sometimes if a company doesn't make a profit or its owners or managers decide to reinvest the money into the company, no dividends are paid. So when choosing whether to buy stock in a certain company, one consideration is whether or not that company typically pays a dividend.

1. If you own stock in a certain company, why are you entitled to a periodic dividend?

2. Describe any advantages or disadvantages of buying stock in a company that pays dividends.

Besides issuing stock, a company can also issue **bonds.** Usually bonds are issued to raise money for the company for start-up costs or special projects. A person who purchases a bond is really lending money to the company. The company, in turn, repays the owner of the bond its original value plus interest. As a general rule, bonds are a safer investment than stocks, but stocks have greater growth potential.

3. Describe the difference between stocks and bonds.

Stocks can be bought and sold on a **stock exchange.** The price of a stock, which varies from day to day (even from minute to minute) depends completely on the amount that investors are willing to pay for it. This can be affected by the profitability of the company, the economy, scandals, even global political concerns. Investors buy or sell stock through a **stockbroker.** Traditionally, this was an individual working for a brokerage firm, but it has become common for people to use online brokers, in which the investor initiates the buying and selling of stocks online. In either case, the brokerage usually charges a fee, called a **commission,** for the service of having their representatives buy or sell the stock at an exchange. Bonds can also be bought and sold like stock.

4. How can an individual buy stock in a company?

5. What are some advantages of investing in the stock market?

Investing in Stocks

When all is said and done, there are two ways to earn money from stocks: buying a stock that pays a regular dividend, and selling shares for more than you paid for them. This means that investors are tremendously interested in the day-to-day price fluctuations of stocks, as well as indicators of how well the company is doing. A **stock table** is a listing of key information; stock tables can be found in newspapers and online financial sites.

The stock table shown below, generated on October 4, 2024, is for a company called the Acme United Corporation, which sounds like the company that makes dynamite and anvils on cartoons, but as near as I can tell is not. Let's see if we can make sense of this mysterious mass of numbers.

Acme United Corporation (ACU) NYSE American

42.12 +0.86 (+2.08%) October 4 4PM EDT

Prev Close:	41.26	Day's Range:	40.88 – 42.21
Open:	41.62	52wk Range:	26.32 – 50.99
Bid:	37.34 x 800	Volume:	68.84
Ask:	44.26 x 3100	Avg Vol (3m):	16,043
1y Target Est:	55.50	Market Cap:	155.632M
Beta:	0.66	P/E (ttm):	8.29
Earnings Date:	Oct 21, 2024	EPS (ttm):	5.08
		Div & Yield:	0.60 (1.42%)

There's a whole lot of information here, and it's going to take some explanation. We suggest that you take notes as you read through the descriptions below, but you might try a quick read-through now, then refer back when you get to Questions 6–11.

At the top, we see the name of the company and the **ticker symbol** (ACU in this case). This is a shorthand symbol assigned by the exchange that represents the stock. The "NYSE American" indicates that this particular stock is traded on the New York Stock Exchange, one of the American stock exchanges. The big bold number at the top is the current price of the stock. The current price isn't like the price of milk at the grocery store: It reflects the most recent price at which the stock has been bought from someone trying to sell. This is followed by how much the price has changed during the current day's trading session.

If you can add and subtract, this makes the **Prev Close** price redundant: Note that this stock has gone up by \$0.86 in this session and is currently worth \$42.12, so of course the closing price on the previous day was \$41.26, which is \$42.12 – \$0.86. (You wouldn't think there'd be many people buying and selling stocks who can't subtract, but I guess anything is possible.) **Open** refers to the price at the time today's session began. This is often different from the previous closing price because many factors can affect what people are willing to pay from day to day. **Bid** and **Ask** are a snapshot of what people have been bidding to buy the stock, and what price current shareholders have been asking for to sell. In this case, \$37.34 × 800 indicates that an offer of \$37.34 for up to 800 shares has been made. \$44.26 × 3100 means that someone has offered up to 3100 shares for sale at \$44.26.

1y Target Est is what analysts are predicting the stock price will be in 1 year. While this is at best an educated guess, it can at least give you some guidance as to whether the stock might be a good buy. **Beta** refers to the **volatility** of the stock, which is a measure of how much the price fluctuates. If the beta is close to 1, then the stock's stability is similar to the rest of the market. Lower numbers indicate that the price is more stable than average. The **Earnings Date** reflects the next time the company is expected to release data on how well the company is doing financially.

The ranges (**Day's Range** and **52wk Range**) indicate the highest and lowest price during the current session, as well as the last full year. The latter is particularly useful, as it gives you a quick snapshot of how the stock has fluctuated, and how its current price compares. **Volume** describes how many shares have been traded in the current session, while **Avg Vol (3m)** provides a 3-month average of shares traded. Lower-volume stocks can be more difficult to buy and sell due to availability.

Market Cap is the total value of the company: It's calculated by multiplying the number of shares that have been issued by the current price, so a little math allows you to calculate how many shares there are. This in turn makes the volume of sales more meaningful. **P/E (ttm)** refers to the P/E (price-to-earnings) ratio. This is a particularly important number, so we'll study that in detail a bit later. **EPS (ttm)** represents earnings per share, or the amount of money the company made for each share of stock that has been issued. Clearly, bigger is better here. **Div & Yield** is the dividend that the company pays for each share, and what percent of the current price that represents. If nothing else, this tells you whether or not the company pays a dividend on shares, but you can also use it for comparison purposes with other stocks.

Now let's put our new knowledge into practice by studying shares of Amazon, also on October 4, 2024.

Amazon.com Inc. (AMZN) NasdaqGS

186.51 +4.55 (+2.50%) October 4 4PM EDT

Prev Close:	181.96	Day's Range:	183.60 – 187.60
Open:	185.71	52wk Range:	118.35 – 201.20
Bid:	186.43 x 200	Volume:	40,862,878
Ask:	186.57 x 100	Avg Vol (3m):	40,261,071
1y Target Est:	219.26	Market Cap:	1.958T
Beta:	1.15	P/E (ttm):	44.41
Earnings Date:	Oct 24, 2024	EPS (ttm):	4.20
		Div & Yield:	N/A (N/A)

6. What is the current price of the stock? What are shareholders trying to sell it currently asking for?

7. What do analysts predict the price will be in 1 year?

8. Is this stock more or less volatile than the market as a whole? How do you know?

9. What's the highest stock price in the last 52 weeks?

10. Does Amazon pay a dividend to shareholders? If so, how much?

11. Use the market cap to find how many shares the company issued.

As previously mentioned, there are two ways to make money from stocks: buy shares of a stock that pays dividends, or buy stock at a low price and sell it at a higher price. But of course, you can't just go buy stock at the corner store—you need to use a brokerage firm, placing an order that is then carried out by representatives at the stock exchange. In exchange for that service, the broker charges a commission, which varies among brokers. Brokers can also make recommendations concerning what stocks to buy and sell, which further justifies their commissions.

The amount that an investor receives from the sale of a stock is called the **proceeds.** The proceeds are equal to the amount of the sale minus the broker's commission. Next, we'll look at some examples of buying and selling stock.

Shares of Apple Computer (AAPL) closed at $12.89 on April 1, 2004. Suppose that an investor bought 600 shares at that price using a broker that charged a 2% commission.

12. Find the amount of commission that was paid.

Bhubeth Bhajanavorakul/Shutterstock

13. Find the total cost of buying the stock.

A **stock split** occurs when a company increases the number of outstanding shares. Most commonly, this is a **2-for-1 split,** in which the number of shares doubles. This can be done for a variety of reasons, but the simplest explanation is that when the price of one share gets particularly high, it tends to scare away smaller investors, and the split effectively halves the price of one share. In any case, the net result is that in an instant, the number of shares owned by shareholders doubles, even though the total value stays the same.

14. Apple did a 2-for-1 split on February 8, 2005, a 7-for-1 split on June 9, 2014, and a 4-for-1 split on August 28, 2020. How many shares would our investor have owned after the third split?

15. Our clever friend the Apple investor could have sold her shares for $223.77 on October 7, 2024. If she had done so, using a brokerage with a 1.5% commission, what would her profit have been? Remember profit is the difference between cost and final proceeds from the sale, which would be selling price minus commission. Needless to say, you should show your work.

16. How much money would our investor have saved by using an online trading site with a flat commission of $19.95 per trade (for both the original purchase and the sale)?

Did You Get It

Try this problem to see if you understand the concepts we just studied. The answer can be found at the end of the portfolio section.

1. a. Amazon stock was available at $1,755.00 on March 13, 2020, and the stock went through a 20-for-1 split on June 6, 2022. If you had bought 10 shares on 3/13/20, using a brokerage with a 1% commission, what would your total cost have been?
   b. If you'd then sold them at the opening price on 10/4/24 (see Questions 6–11), using an online brokerage with a $9.95 flat fee, would you have made or lost money? How much?

2-7 Group

The Price-to-Earnings Ratio For a Stock

The price-to-earnings ratio, commonly known as P/E ratio, is a key indicator of how reasonable the current price of a stock is. It compares the selling price of the stock to the annual amount of money the company has earned per share. If the P/E ratio for a certain stock is 20, that means that the current selling price is 20 times as high as the earnings per share. This tells us that for every $20 you invest in the company, that company will earn $1.

P/E ratio can be a little tricky to interpret. On the one hand, a lower P/E ratio indicates that a stock is priced reasonably, as you need to invest less for the company to make a certain amount of profit. But if investors believe in the growth potential of a company, they may be willing to pay a premium for the stock, which would result in a higher P/E ratio.

To help get some perspective on P/E ratio, it's useful to do some calculations. First, a formal definition:

The P/E Ratio

$$\text{P/E Ratio} = \frac{\text{Current market price}}{\text{Annual earnings per share}}$$

1. For the Amazon stock table from earlier in the lesson, calculate the P/E ratio using the current price and earnings per share. Compare to the P/E ratio listed in the table.

Not every stock table you will encounter has quite the wealth of information that the ones at the beginning of this lesson have. A more abbreviated table might have the P/E ratio but not earnings per share listed, in which case you can rearrange the above formula to calculate earnings per share.

Solved Example 1

Calculating Earnings per Share

On October 7, 2024 stock in Chipotle Mexican Grill, Inc. was $56.47 and the P/E ratio was listed as 55.35. What was Chipotle's earnings per share for the previous year? How does it compare to Amazon's?

SOLUTION

If we know the P/E ratio and the price, we can set up and solve an equation to find the earnings per share.

$$\text{P/E ratio} = \frac{\text{Price}}{\text{Earnings per share}}$$

$$55.35 = \frac{56.47}{\text{Earnings per share}}$$

Solved Example 1 *Continued*

To solve this equation, we can multiply both sides by the earnings per share, then divide both sides by 166.56. This gives us

$$\text{Earnings per share} = \frac{56.47}{55.35} \approx 1.02$$

This shows that Chipotle earned \$1.02 per share, while Amazon earned \$4.20 per share (according to the stock table on page 166).

2. Based on the result of Solved Example 1, which do you think was the better investment at that point in time, Amazon or Chipotle? Describe your reasoning.

Did You Get It

2. a. Use information in the stock table for ACU earlier in the lesson to calculate the P/E ratio by hand, and compare to the P/E ratio in the table. If your result is different, discuss why you think that may be the case.
   b. If a competitor of ACU has a P/E ratio of 21.4 and a market price of \$38.74, calculate the earnings per share.

The table below summarizes some basic rules of thumb when it comes to interpreting the P/E ratio for a stock. Remember that the best approach is to compare a stock's P/E ratio to other companies in the same industry, as current market conditions can have a pretty profound effect on P/E ratio.

P/E Ratio	Common Interpretations
N/A	Company has no earnings; it's also standard for a company with negative earnings (that is, a loss) to have an undefined P/E ratio, even though mathematically a negative P/E ratio does make sense.
0–10	The stock may be undervalued, making it a good buy; alternatively, investors might feel like the earnings are likely to decline. The company may have shown short-term profit from selling assets that will hurt earnings in the long term.
10–17	P/E ratios in this range are historically considered to represent stocks being sold at fair market value.
17–25	The stock may be overvalued, so shop carefully. On the other hand, earnings may have increased since the last official earnings report was released. The company might be up and coming, with earnings expected to increase dramatically at some point.
>25	The stock might be very overpriced at the moment, or the company may be expected to have a big jump in earnings coming up. It could also be the case that the company's most recent earnings are unusually low for that company, with an expectation of recovery.

3. We saw in Solved Example 1 that the P/E ratio for Chipotle was quite high. Come up with some theories as to why that might have been the case.

Investing in Bonds

When an investor buys bonds, the investor is actually loaning money to the company or government entity that issues the bonds. In exchange for lending that money, the investor will receive a fixed return on his or her investment for a given period of time. The ending value of the bond is called the **face value;** this is typically $1,000 but can vary. Bonds also have a **maturity date,** which is the date that the face value is returned to the investor, as well as interest. In some cases, no interest is paid until the maturity date, at which time the full amount of interest is paid. **Coupon bonds,** on the other hand, pay an annual or semi-annual interest payment, known as a **coupon.**

You can think of it this way: If someone offers to sell you a $20 bill for $15, that's a pretty darn good deal. But the catch is that you don't get the $20 until some predetermined time later down the road. That's what a bond is; it's actually pretty simple.

But here's where it can get more interesting. Suppose that you get tired of waiting for the time when you get that $20 bill, and you sell the bond to your neighbor for $18. You've made $3 on your investment and you're done. Your neighbor now waits until the maturity date, cashes in the bond, and makes the other $2 of the original $5 in profit. This is how bonds are bought and sold on the open market. Of course, your neighbor wouldn't pay more than $20 for a bond with a maturity value of $20. But if you were desperate for cash, you might be compelled to unload the bond for less than the $15 you originally paid. You've taken a loss, but at least you got some money out at the time you needed it.

Investors buy bonds for two main reasons. First, they're much safer investments than stocks, even though they have much less opportunity for a big profit. The interest is guaranteed, as long as the company that issued the bond stays in business and has the money to pay its debts. This makes bonds a safe, but not guaranteed, investment. Second, bonds are used to offset losses on stocks when the economy is in a down cycle, because the value of bonds tends to go up when the value of stocks goes down.

Here's a short version of why that makes sense: In our earlier hypothetical example where you're making a $5 profit for waiting to cash in your bond, if the economy is booming in the meantime, prices in general will tend to go up, so the $5 you're getting at some point in the future has less buying power than the $5 you were expecting when you made the investment. This makes the bond in effect worth less. The opposite is also true: If the economy fizzles after you buy that bond, the $5 you eventually get will in effect be worth more than the $5 you thought you were getting, so the bond is worth more. Stocks tend to rise when the economy is doing well and fall when it's not, so the value of bonds usually goes in the opposite direction of the stock market.

Let's look at an example of a bond investment.

Solved Example 2

Investing in Bonds

Suppose that you buy a bond with a face value of $1,000 that was originally issued 18 months ago. The maturity date is 4 years from the time it was issued, and the interest rate is 4% simple interest per year. If you pay $820 for the bond and keep it until the maturity date, what is your profit?

SOLUTION

First we'll need to calculate the amount of interest the bond earns so that we can find your proceeds from the bond. We're told that interest is simple and is 4% per year. So we use our simple interest formula:

$$I = Prt$$

$$I = \$1{,}000 \cdot 0.04 \cdot 4 = \$160 \text{ in interest}$$

With a face value of $1,000 and $160 in interest, the bond will be worth $1,160 at the maturity date, so the profit is $1,160 – $820 = $340.

4. What percentage of your original investment is the profit that we found in Solved Example 2?

Math Note

There are fees and/or commissions for buying bonds as well as stocks. We're omitting fees from this discussion to focus on main issues involved in buying and selling bonds.

5. If you divide your total percent return from Question 4 by the amount of time in years that you hold the bond, you'll get your percent return per year, or **annual** return, which is a key indicator of how well an investment performs. A long-standing rule of thumb on long-term investments in the stock market is that you can expect annual returns in the 6–7% range. How would this bond investment compare?

Did You Get It ?

3. Rashard bought four bonds with face values of $1,000 each, simple interest rate of 5.8% per year, and a maturity date 10 years after they were issued. He paid $4,725 3 years after the bonds were issued. If he keeps the bonds until maturity, find his total profit and his percent return per year.

2-7 Portfolio

Name ______________________________

Check each box when you've completed the task. Remember that your instructor will want you to turn in the portfolio pages you create.

Technology

1. ☐ In this Excel activity, we'll track the values of some stock investments over time.

ALEKS Practice

1. ☐ Include any written work from the ALEKS assignment along with any notes or questions about this lesson's content.

Applications

1. ☐ Complete the applications problems.

Reflections

Type a short answer to each question.

1. ☐ Describe the difference between stocks and bonds.
2. ☐ What are some things you should consider when deciding whether or not to buy stock in a particular company?
3. ☐ What is the advantage of having stock in a company that pays dividends?
4. ☐ Name one thing you learned or discovered in this lesson that you found particularly interesting.
5. ☐ What questions do you have about this lesson?

Looking Ahead

1. ☐ Complete the prep skills for Lesson 2-8.
2. ☐ Read the opening paragraph in Lesson 2-8 carefully and answer Question 0 in preparation for that lesson.

Answers to "Did You Get It?"

1. a. \$17,725.50 **b.** You would have made \$19,406.55.
2. a. 8.29, which matches the table. **b.** \$1.81 **3.** \$1,595; 4.8% per year

Answers to Prep Skills Questions

1. a. 16 **b.** 441 mg **c.** 8%
2. 64 g
3. 62.5%
4. 5 g
5. $M = \frac{k}{P}$

2-7 Applications

Name ________________________________

Use the stock listing for Wal-Mart to answer Questions 1–10.

Wal-Mart Stores Inc. (WMT) - NYSE

60.75 ↓0.08(0.13%) 4:01PM EST

After Hours : **60.75** 0.00 (0.00%) 4:12PM EST

Prev Close:	60.83	Day's Range:	60.50 – 60.97
Open:	60.58	52wk Range:	56.30 – 90.97
Bid:	60.73 x 4300	Volume:	5,296,196
Ask:	60.74 x 2900	Avg Vol (3m):	12,578,300
1y Target Est:	63.63	Market Cap:	194.51B
Beta:	0.444396	P/E (ttm):	13.01
Next Earnings Date:	N/A	EPS (ttm):	4.67
		Div & Yield:	1.96 (3.24%)

1. What is the ticker symbol you'd use to look up stock quotes for Wal-Mart?

2. What was the highest and lowest price that the stock sold for during the last 52 weeks?

3. What was the amount of the dividend per share that the company paid last year?

4. If you own 682 shares, how much in dividends did you make last year?

5. How many shares were traded yesterday?

6. What was the closing price of the stock yesterday?

2-7 Applications

Name ______________________________

7. Find the annual earnings per share.

8. If you purchase 842 shares of Wal-Mart stock at $60.67 per share and the broker's commission is 2%, find the total cost of the purchase.

9. Using information in the table, do a calculation to confirm that the P/E ratio is correct as listed.

10. Based on the information in the table below, do you think Home Depot stock was a better buy than Wal-Mart at that point? Justify your response.

The Home Depot, Inc. (HD) - NYSE

132.68 ↓0.22(0.17%) 4:00PM EST

After Hours : **132.43** ↓0.25 (0.19%) 5: 19PM EST

Prev Close:	132.90	Day's Range:	131.89 – 132.90
Open:	132.25	52wk Range:	92.17 – 135.47
Bid:	132.40 x 300	Volume:	2,607,226
Ask:	132.88 x 100	Avg Vol (3m):	5,027,970
1y Target Est:	141.00	Market Cap:	168.22B
Beta:	0.965301	P/E (ttm):	24.89
Next Earnings Date:	N/A	EPS (ttm):	5.33
		Div & Yield:	2.36 (1.79%)

2-7 Applications

Name ______________________________

11. Compare the two investments below and decide which would have been the better choice, and why.

Investment 1: $10,000 was invested in a 24-month CD that earned 5.1% annual interest compounded monthly.

Investment 2: 1,400 shares of stock in the Lybarger Aviation Company were bought at $7.11 per share using a brokerage with a 0.75% commission rate on both buying and selling stock. Over the 2 years the stock was held, it paid a dividend of $0.48 per share in the first year and $0.36 per share in the second year. The stock was sold through the same brokerage for $7.95 per share.

2-7 Applications

Name ______________________________

12. A bond with face value $10,000, simple interest 6.45%, and term 12 years is originally bought by Jasmine. After 33 months, she sells it to Curtis for $11,400. Curtis then holds onto it for 7 years, eventually selling it to Kim for $14,950. Kim keeps the bond until it matures and cashes it in.
    a. Which investor made the greatest profit?
    b. Which got the greatest percent return on their investment?
    c. Which got the greatest percent return per year?

Lesson 2-8 Prep Skills

SKILL 1: CONVERT DECIMALS TO PERCENT FORM

To convert a decimal to percent form, multiply the decimal by 100. This has the effect of shifting the decimal point two places to the right.

- 0.45 = 0.45(100)% = 45%
- 1.2 = 1.2(100)% = 120%
- 0.033 = 0.033(100)% = 3.3%

Note that each percent form is the decimal form with the decimal point moved two places right.

SKILL 2: COMPUTE A PERCENTAGE OF A GIVEN AMOUNT

As we have seen in previous lessons, you perform two steps to compute a number that represents a certain percentage of a given amount:

Step 1. Convert the percentage to a decimal by moving the decimal point two places to the left.

Step 2. Multiply the decimal form of the percentage by the given amount.

- Calculate 3.2% of $70.
  0.032($70) = $2.24

  The decimal form of 3.2% is 0.032. 3.2% of $70 is $2.24.

- Calculate 24% of the sum of 88 and 43.
  88 + 43 = 131
  0.24(131) = 31.44

  First find the sum of 88 and 43.
  Then find 24% of that sum.

PREP SKILLS QUESTIONS

1. Write each decimal in percent form.

   0.23
   0.794
   0.035

2. What is 10% of $8,603?

3. What is 15% of $23,462.12?

4. Find 20% of the difference between $83,402 and $36,901.

Lesson 2-8 A Taxing Situation

LEARNING OBJECTIVES

- ☐ 1. Understand why we pay taxes.
- ☐ 2. Explain the basic process of paying taxes.
- ☐ 3. Determine the amount of tax due based on taxable income.
- ☐ 4. Complete a 1040 form.

Steven Cole/Photodisc/Getty Images

In this world nothing can be said to be certain, except death and taxes.
—Benjamin Franklin

"I really enjoy paying taxes," said no one. Ever. Of course, if we didn't have to pay taxes, we'd be missing out on some pretty cool stuff, like roads, police, trash pickup, fire fighters, schools, parks . . . Okay, so maybe taxes aren't THAT bad after all. And as our friend Mr. Franklin pointed out, it's not like you get a choice. Taxes are necessary to fund governments at three levels: local, state, and federal. Property taxes, local sales tax, and local income taxes are kept locally to fund services within your community. State income taxes, which are used to fund state governments, vary from state to state depending on revenue a state can generate from other sources, like tourism, oil, or sales tax. Some states have no income tax at all and are funded entirely through other sources. Federal income tax is something that everyone who earns income has to pay (at least in theory. It's not unheard of for wealthy individuals and corporations to beat the system and pay no taxes.). Based on your income and **deductions** (which are characteristics of your financial situation that allow you to subtract from the tax you have to pay), the percentage of your income that you pay can vary considerably. In this lesson we'll study some common aspects of taxation. These are things that every American *should* be aware of, but far too few actually are.

0. How much do you know about federal taxes? Explain.

2-8 Class

How Does the Tax Process Work?

If you earn a paycheck, through a process called **withholding,** taxes are automatically taken from the amount you've earned. The amount of tax that is withheld is based on two main factors: how much you earn, and paperwork you fill out through your employer describing how many other people depend on you financially. The intent is to estimate the amount of tax that you'll owe at the end of the year and pay some of it during each pay period so that you don't end up with a big bill that you can't afford to pay at the end of the year. When it's time to file a tax return, you'll find out if you paid too much, which entitles you to a refund, or too little, which means you need to pay more.

Our tax system is designed so that people who make more money pay a higher percentage of their income in taxes than do those who make less money. The range in which your income falls is known as a **tax bracket,** and the bracket you're in will determine how much of your income you will owe. It's essentially impossible to understand taxes without understanding tax brackets, so that's where we'll begin.

There are four different filing statuses: single, which is self-explanatory; married filing jointly, which is when a married couple files a single return for both; married filing separately, which is when married individuals decide to file separate returns; and head of household, which is for unmarried individuals who financially support others. There are different income levels that qualify you for certain brackets, depending on which filing status applies to you. The following is a table of tax brackets for 2023.

Tax Rate	Single	Married Filing Jointly or Qualifying Surviving Spouse	Married Filing Separately	Head of Household
10%	$0 to $11,000	$0 to $22,000	$0 to $11,000	$0 to $15,700
12%	$11,001 to $44,725	$22,001 to $89,450	$11,001 to $44,725	$15,701 to $59,850
22%	$44,726 to $95,375	$89,451 to $190,750	$44,726 to $95,375	$59,851 to $95,350
24%	$95,376 to $182,100	$190,751 to $364,200	$95,376 to $182,100	$95,351 to $182,100
32%	$182,101 to $231,250	$364,201 to $462,500	$182,101 to $231,250	$182,101 to $231,250
35%	$231,251 to $578,125	$462,501 to $693,750	$231,251 to $346,875	$231,251 to $578,100
37%	$578,126 and up	$693,751 and up	$346,876 and up	$578,101 and up

For each filing status, the range of incomes listed in that column determines which tax bracket an individual falls into. For example, notice that in the second row of the Married Filing Separately column, the range is $11,001 to $44,725, and that row has 12% in the Tax Rate column. This tells us that a married taxpayer filing separately with taxable income between $11,001 and $44,725 is in the 12% tax bracket. (In a while we'll study exactly what that tells us about the amount of tax he or she would pay.)

1. If a person filing as single has taxable income of $58,000, what tax bracket is she in? What about a married couple filing jointly?

Now for the big question: How much would that single person pay in taxes? If you said 22% of $58,000, that's a fine guess. An incorrect one, mind you, but a fine guess nonetheless. So what's the deal? What does the 22% tax bracket mean? The graphic to the right, available on the Internal Revenue Service website (irs.gov) does a nice job of describing the situation.

In short, all taxpayers who are single pay 10% on the first $11,000 in taxable income. They then pay 12% on the next $33,725, which is the amount that falls into the range $11,001 to $44,725. Finally, they pay 22% of the amount left over $44,725. So even though a lot of people think that being in the 22% bracket means they're paying 22% of their income in taxes, that's just not true. They're only paying 22% on the amount over $44,725; the majority of their income is probably being taxed at a much lower rate.

Here's how that works for a **single person with taxable income of $58,000 per year:**

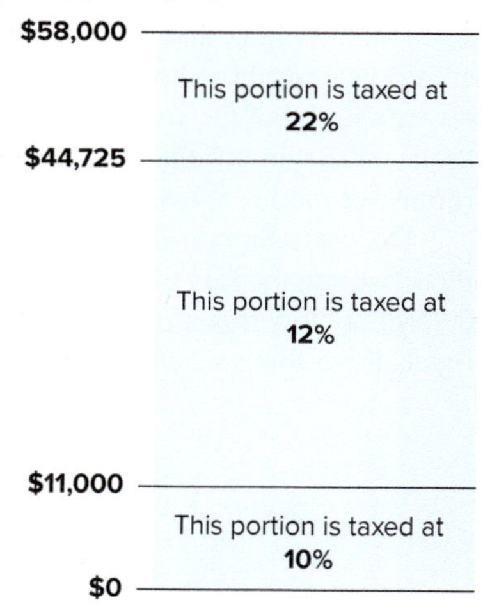

Let's have a look at how we can use the tax bracket to calculate the amount of tax someone would owe.

Solved Example

Computing Tax Owed from a Tax Bracket

Using the information just provided, calculate the amount of tax that our single person would pay on income of $58,000.

SOLUTION

We need to look at the information in the tax bracket line by line, and break our calculation into distinct pieces. First, our taxpayer would pay 10% of the first $11,000 of her income:

$$10\% \text{ of } \$11{,}000 = 0.10 \times \$11{,}000 = \$1{,}100$$

Next, look at the 12% row. Since her income is more than the $44,725 boundary for that bracket, she has to pay 12% on the full amount that falls in that range. We find this by subtracting the upper boundary of that bracket ($44,725) minus the upper boundary of the previous bracket ($11,000):

$$12\% \text{ of } (\$44{,}725 - \$11{,}000) = 0.12 \times \$33{,}725 = \$4{,}047$$

Now we're in the 22% row, so we need to find 22% of the part of her income that's above $44,725. This would be the difference between her $58,000 in income and $44,725. So we get

$$22\% \text{ of } (\$58{,}000 - \$44{,}725) = 0.22 \times \$13{,}275 = \$2{,}920.50$$

Finally, we add up the amounts we've calculated.

$$\$1{,}100 + \$4{,}047 + \$2{,}920.50 = \$8{,}067.50$$

Her total tax owed is $8,067.50.

2. How much would the taxpayer in Solved Example 1 have overpaid if she had misunderstood what the 22% tax bracket means and simply paid 22% of her income in federal tax? With the correct tax calculated in Solved Example 1, what percentage of her income is she paying in taxes?

> **Math Note**
>
> The taxes here are federal only: The amount of tax due at the state and local levels varies widely depending on where you live.

Since many Americans are, shall we say, somewhat math-challenged, the Internal Revenue Service (IRS), the government agency responsible for collecting taxes, publishes tables that let filers look up the amount of tax that they should pay. Their documentation even includes an example with the instructions, in this case for form 1040, which we'll study in a bit.

2023 Tax Table

CAUTION *See the instructions for line 16 to see if you must use the Tax Table below to figure your tax.*

Example. A married couple are filing a joint return. Their taxable income on Form 1040, line 15, is $25,300. First, they find the $25,300-25,350 taxable income line. Next, they find the column for married filing jointly and read down the column. The amount shown where the taxable income line and filing status column meet is $2,599. This is the tax amount they should enter in the entry space on Form 1040, line 16.

Sample Table

At Least	But Less Than	Single	Married filing jointly*	Married filing separately	Head of a household
		Your tax is—			
25,200	**25,250**	2,807	2,587	2,807	2,713
25,250	**25,300**	2,813	2,593	2,813	2,719
25,300	**25,350**	2,819	(2,599)	2,819	2,725
25,350	**25,400**	2,825	2,605	2,825	2,731

3. Use the tax bracket info from earlier in the lesson to calculate the correct tax amount for the married couple described in the example, then compare to the result in the 2023 tax table. If the couple believes the table without doing any calculations, will they pay too much? Too little? By how much?

4. What would happen to you if you fail to pay your taxes? Discuss, and if you're able, maybe do a little bit of Internet research.

Where Do Your Tax Dollars Go?

While paying taxes really hurts, it can help at least some to know exactly what those dollars are accomplishing. There are a handful of websites that illustrate where federal tax dollars go. And for people who stink at percents, which of course does not include you, on some sites you can just enter a taxable income and see how that money gets divided up. But we don't need that . . . we'll just use the percents. The information in the chart is from 2023.

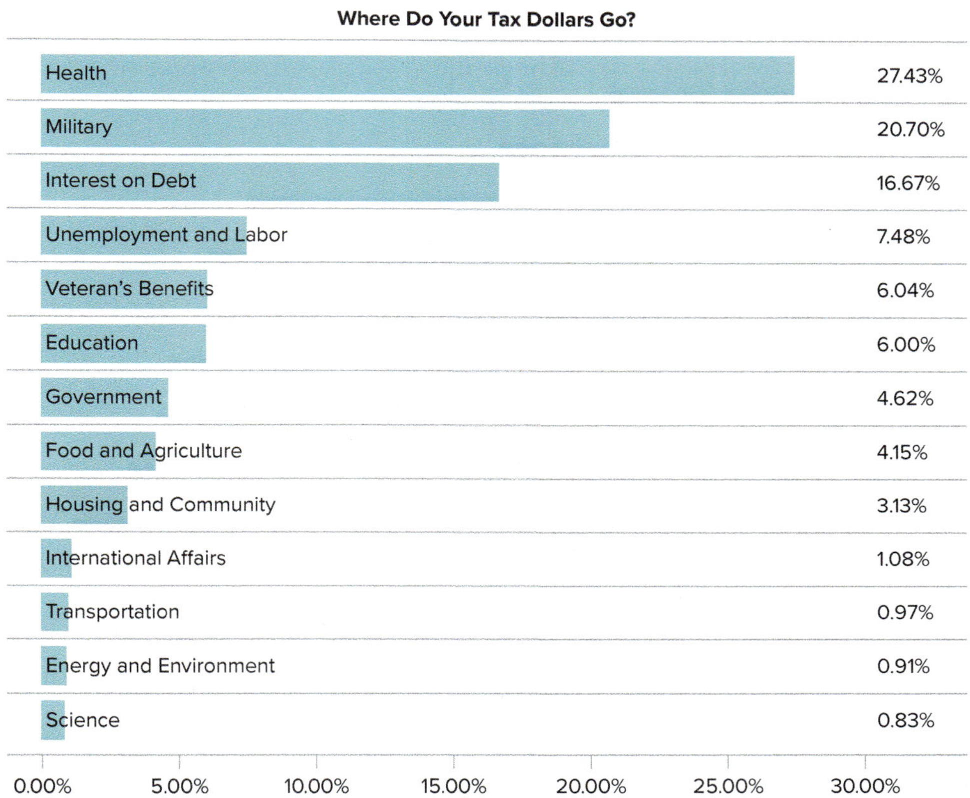

5. If you paid \$6,500 in federal taxes in 2023, how much of your money went to education?

6. For every \$100,000 in income tax revenue collected by the IRS in 2023, how much went just to pay interest on the national debt?

Did You Get It

Try these problems to see if you understand the concepts we just studied. The answer can be found at the end of the portfolio section.

1. Find the tax bracket for a person with status head of household if she has a taxable income of \$97,800.
2. Calculate the amount of federal tax that person would owe. What percentage of her taxable income would that be?

2-8 Group

Tax Forms: A Lesson in Following Directions

Many people are intimidated into submission by filling out tax forms. To be fair, you don't want to be making mistakes on forms this important, but with a bit of education and a solid ability to follow written instructions, completing basic tax forms is very doable. In this case, we're going to study federal form 1040, which is the standard tax form that most taxpayers will have to submit by April 15 of each year.

On pages 189 and 190, you'll find a copy of the 1040 form for 2023, and we'll practice filling out the form. But first, we should discuss how you get information about your income. For many folks, this will come from two sources: a W-2 form, which reports your wages, and a 1099-INT, which reports interest income. First, we'll look at some documents for a guy named Darin. Darin is 29 years old and single. He has no dependents, and nobody can claim him as a dependent on their taxes.

a Employee's social security number XXX-XX-XXXX	OMB No. 1545-0008	Safe, accurate, FAST! Use IRS e-file	Visit the IRS website at www.irs.gov/efile
b Employer identification number (EIN)		1 Wages, tips, other compensation 58,584.78	2 Federal income tax withheld 8,941.46
c Employer's name, address, and ZIP code Kramerica Industries 129 West 81st Street, Apartment 5B New York, NY 10029		3 Social security wages 66,482.21	4 Social security tax withheld 2,132.11
		5 Medicare wages and tips 66,482.21	6 Medicare tax withheld 963.99
		7 Social security tips	8 Allocated tips
d Control number		9	10 Dependent care benefits
e Employee's first name and initial Last name Suff. Darin Roberts 18 Newman Ave, Apt. 4B New York, NY 10012		11 Nonqualified plans	12a See instructions for box 12
		13 Statutory employee [X] Retirement plan [ ] Third-party sick pay [ ]	12b
		14 Other	12c
f Employee's address and ZIP code			12d

15 State	Employer's state ID number	16 State wages, tips, etc.	17 State income tax	18 Local wages, tips, etc.	19 Local income tax	20 Locality name
NY	47-325941	58,584.78	1,762.12	66,482.21	1,329.66	NYC

Form **W-2** **Wage and Tax Statement**

Department of the Treasury—Internal Revenue Service

Copy B—To Be Filed With Employee's FEDERAL Tax Return.
This information is being furnished to the Internal Revenue Service.

☐ VOID ☐ CORRECTED

PAYER'S name, street address, city or town, state or province, country, ZIP or foreign postal code, and telephone no. Great Eastern Savings and Loan 400 West 79th Street New York, NY 10029		Payer's RTN (optional) 1 Interest income $ 749.36	OMB No. 1545-0112 Form **1099-INT**	**Interest Income**
		2 Early withdrawal penalty $		**Copy 1**
PAYER'S federal identification number	RECIPIENT'S identification number XXX-XX-XXXX	3 Interest on U.S. Savings Bonds and Treas. obligations $		**For State Tax Department**
RECIPIENT'S name Darin Roberts		4 Federal income tax withheld $	5 Investment expenses $	
		6 Foreign tax paid $	7 Foreign country or U.S. possession	
Street address (including apt. no.) 18 Newman Ave, Apt. 4B		8 Tax-exempt interest $	9 Specified private activity bond interest $	
City or town, state or province, country, and ZIP or foreign postal code New York, NY 10012		10 Market discount $	11 Bond premium $	
	FATCA filing requirement ☐	12	13 Bond premium on tax-exempt bond $	
Account number (see instructions)		14 Tax-exempt and tax credit bond CUSIP no.	15 State 16 State identification no.	17 State tax withheld $ $

Form **1099-INT** www.irs.gov/form1099int Department of the Treasury - Internal Revenue Service

Next we see a W-2 form for Archibald, a college student who works part time while taking classes. Archibald is also single and has no dependents, and nobody can claim him as a dependent, either. He received $3,000 in scholarship money in 2018. Because he's not currently pursuing a degree, that money is considered taxable income.

a Employee's social security number: XXX-XX-XXXX	OMB No. 1545-0008	Safe, accurate, FAST! Use IRS e-file. Visit the IRS website at www.irs.gov/efile
b Employer identification number (EIN)	**1** Wages, tips, other compensation: 23,667	**2** Federal income tax withheld: 3,518
c Employer's name, address, and ZIP code Heisman Marketing 1975 Lane Ave. Columbus, OH 43018	**3** Social security wages: 23,667	**4** Social security tax withheld: 943.12
	5 Medicare wages and tips: 23,667	**6** Medicare tax withheld: 247.01
	7 Social security tips	**8** Allocated tips
d Control number	**9**	**10** Dependent care benefits
e Employee's first name and initial / Last name / Suff. Archibald Griffen 45 Buckeye Way Columbus, OH 43085	**11** Nonqualified plans	**12a** See instructions for box 12
	13 Statutory employee [X] / Retirement plan [ ] / Third-party sick pay [ ]	**12b**
	14 Other	**12c**
f Employee's address and ZIP code		**12d**

15 State	Employer's state ID number	16 State wages, tips, etc.	17 State income tax	18 Local wages, tips, etc.	19 Local income tax	20 Locality name
OH	11-004141	23,667	852.19	23,667	356.97	Colum

Form **W-2 Wage and Tax Statement** — Department of the Treasury—Internal Revenue Service

Copy B—To Be Filed With Employee's FEDERAL Tax Return.
This information is being furnished to the Internal Revenue Service.

The goal of the Group portion of this lesson is to fill out the 1040 form for either Darin or Archibald. You can choose which, and when filling out the form you'll indicate which you chose by filling in the name field at the top of the form. All the information you need is either on the W-2 and 1099 forms provided, or was in the description of your chosen client. You will, however, need to use the tax bracket information from earlier in the lesson to find the amount of tax that your client owes.

Form 1040 has many instructions directly on the form, so working together you should be able to navigate the form successfully. We've added some additional instructions in purple right on the form. Since both clients have no other sources of taxable income and will be taking standard exemptions, any mention of "Schedules" on the form do not apply. After you've filled out your group's 1040 form, switch with another group so that they can perform an audit and see how you did.

Form **1040** Department of the Treasury—Internal Revenue Service
U.S. Individual Income Tax Return **2023** OMB No. 1545-0074 IRS Use Only—Do not write or staple in this space.

For the year Jan. 1–Dec. 31, 2023, or other tax year beginning ________, 2023, ending ________, 20 ____ See separate instructions.

Your first name and middle initial	Last name		**Your social security number**
If joint return, spouse's first name and middle initial	Last name		**Spouse's social security number**
Home address (number and street). If you have a P.O. box, see instructions.		Apt. no.	**Presidential Election Campaign** Check here if you, or your spouse if filing jointly, want $3 to go to this fund. Checking a box below will not change your tax or refund. ☐ You ☐ Spouse
City, town, or post office. If you have a foreign address, also complete spaces below.	State	ZIP code	
Foreign country name	Foreign province/state/county	Foreign postal code	

Filing Status Check only one box.
☐ Single
☐ Married filing jointly (even if only one had income)
☐ Married filing separately (MFS)
☐ Head of household (HOH)
☐ Qualifying surviving spouse (QSS)
If you checked the MFS box, enter the name of your spouse. If you checked the HOH or QSS box, enter the child's name if the qualifying person is a child but not your dependent: ________

You can ignore these two boxes.

Digital Assets At any time during 2023, did you: (a) receive (as a reward, award, or payment for property or services); or (b) sell, exchange, or otherwise dispose of a digital asset (or a financial interest in a digital asset)? (See instructions.) ☐ **Yes** ☐ **No**

Standard Deduction **Someone can claim:** ☐ You as a dependent ☐ Your spouse as a dependent
☐ Spouse itemizes on a separate return or you were a dual-status alien

Age/Blindness **You:** ☐ Were born before January 2, 1959 ☐ Are blind **Spouse:** ☐ Was born before January 2, 1959 ☐ Is blind

Dependents (see instructions):
If more than four dependents, see instructions and check here . . ☐

(1) First name	Last name	(2) Social security number	(3) Relationship to you	(4) Check the box if qualifies for (see instructions): Child tax credit	Credit for other dependents
				☐	☐
				☐	☐
				☐	☐
				☐	☐

Income

Attach Form(s) W-2 here. Also attach Forms W-2G and 1099-R if tax was withheld.
If you did not get a Form W-2, see instructions.

1a	Total amount from Form(s) W-2, box 1 (see instructions)	1a
b	Household employee wages not reported on Form(s) W-2	1b
c	Tip income not reported on line 1a (see instructions)	1c
d	Medicaid waiver payments not reported on Form(s) W-2 (see instructions)	1d
e	Taxable dependent care benefits from Form 2441, line 26	1e
f	Employer-provided adoption benefits from Form 8839, line 29	1f
g	Wages from Form 8919, line 6	1g
h	Other earned income (see instructions)	1h
i	Nontaxable combat pay election (see instructions) 1i	
z	Add lines 1a through 1h	1z

Lines 1b through i, 3-8, 10, and 13 don't apply to our taxpayers, except for 1g which applies only to Archibald.

Attach Sch. B if required.

2a	Tax-exempt interest	2a	b	Taxable interest	2b
3a	Qualified dividends	3a	b	Ordinary dividends	3b
4a	IRA distributions	4a	b	Taxable amount	4b
5a	Pensions and annuities	5a	b	Taxable amount	5b
6a	Social security benefits	6a	b	Taxable amount	6b

Standard Deduction for—
- Single or Married filing separately, $13,850
- Married filing jointly or Qualifying surviving spouse, $27,700
- Head of household, $20,800
- If you checked any box under *Standard Deduction*, see instructions.

c	If you elect to use the lump-sum election method, check here (see instructions) ☐	
7	Capital gain or (loss). Attach Schedule D if required. If not required, check here ☐	7
8	Additional income from Schedule 1, line 10	8
9	Add lines 1z, 2b, 3b, 4b, 5b, 6b, 7, and 8. This is your **total income**	9
10	Adjustments to income from Schedule 1, line 26	10
11	Subtract line 10 from line 9. This is your **adjusted gross income**	11
12	**Standard deduction or itemized deductions** (from Schedule A)	12
13	Qualified business income deduction from Form 8995 or Form 8995-A	13
14	Add lines 12 and 13	14
15	Subtract line 14 from line 11. If zero or less, enter -0-. This is your **taxable income**	15

For Disclosure, Privacy Act, and Paperwork Reduction Act Notice, see separate instructions. Cat. No. 11320B Form **1040** (2023)

Form 1040 (2023) — Page **2**

Tax and Credits

Lines 17, 19-21, 23, and 26-32 do not apply here.

To find the tax owed on line 16, you'll need to calculate the tax on the amount on line 15 using what we learned earlier in the lesson.

16	**Tax** (see instructions). Check if any from Form(s): 1 ☐ 8814 2 ☐ 4972 3 ☐ ________		16
17	Amount from Schedule 2, line 3		17
18	Add lines 16 and 17		18
19	Child tax credit or credit for other dependents from Schedule 8812		19
20	Amount from Schedule 3, line 8		20
21	Add lines 19 and 20		21
22	Subtract line 21 from line 18. If zero or less, enter -0-		22
23	Other taxes, including self-employment tax, from Schedule 2, line 21		23
24	Add lines 22 and 23. This is your **total tax**		24

Payments

If you have a qualifying child, attach Sch. EIC.

25	Federal income tax withheld from:		
a	Form(s) W-2	25a	
b	Form(s) 1099	25b	
c	Other forms (see instructions)	25c	
d	Add lines 25a through 25c		25d
26	2023 estimated tax payments and amount applied from 2022 return		26
27	Earned income credit (EIC)	27	
28	Additional child tax credit from Schedule 8812	28	
29	American opportunity credit from Form 8863, line 8	29	
30	Reserved for future use	30	
31	Amount from Schedule 3, line 15	31	
32	Add lines 27, 28, 29, and 31. These are your **total other payments and refundable credits**		32
33	Add lines 25d, 26, and 32. These are your **total payments**		33

Refund

Direct deposit? See instructions.

34	If line 33 is more than line 24, subtract line 24 from line 33. This is the amount you **overpaid**		34
35a	Amount of line 34 you want **refunded to you**. If Form 8888 is attached, check here ☐		35a
b	Routing number ________ **c** Type: ☐ Checking ☐ Savings		
d	Account number ________		
36	Amount of line 34 you want **applied to your 2024 estimated tax**	36	

Amount You Owe

37	Subtract line 33 from line 24. This is the **amount you owe**. For details on how to pay, go to *www.irs.gov/Payments* or see instructions		37
38	Estimated tax penalty (see instructions)	38	

Third Party Designee

Do you want to allow another person to discuss this return with the IRS? See instructions ☐ **Yes.** Complete below. ☐ **No**

Designee's name | Phone no. | Personal identification number (PIN)

Sign Here

Under penalties of perjury, I declare that I have examined this return and accompanying schedules and statements, and to the best of my knowledge and belief, they are true, correct, and complete. Declaration of preparer (other than taxpayer) is based on all information of which preparer has any knowledge.

Joint return? See instructions. Keep a copy for your records.

Your signature	Date	Your occupation	If the IRS sent you an Identity Protection PIN, enter it here (see inst.)
Spouse's signature. If a joint return, **both** must sign.	Date	Spouse's occupation	If the IRS sent your spouse an Identity Protection PIN, enter it here (see inst.)
Phone no.	Email address		

Paid Preparer Use Only

Preparer's name	Preparer's signature	Date	PTIN	Check if: ☐ Self-employed
Firm's name			Phone no.	
Firm's address			Firm's EIN	

Go to *www.irs.gov/Form1040* for instructions and the latest information. Form **1040** (2023)

Did You Get It

3. A taxpayer with status married filing jointly has no dependents other than her spouse. Her federal wages are \$81,112, and she also had \$1,462.12 in interest income. What would you write in the following lines on her 1040 form?

   a. 9 b. 12 c. 15 d. 16

2-8 Portfolio

Name ______________________________

Check each box when you've completed the task. Remember that your instructor will want you to turn in the portfolio pages you create.

ALEKS Practice

1. ☐ Include any written work from the ALEKS assignment along with any notes or questions about this lesson's content.

Applications

1. ☐ Complete the applications problems.

Reflections

Type a short answer to each question.

1. ☐ Why do we have to pay taxes?
2. ☐ What is a tax bracket? What is the point of tax brackets?
3. ☐ What is a 1040 form used for?
4. ☐ Name one thing you learned or discovered in this lesson that you found particularly interesting.
5. ☐ What questions do you have about this lesson?

Looking Ahead

1. ☐ Complete the prep skills for Lesson 3-1.
2. ☐ Read the opening paragraph in Lesson 3-1 carefully and answer Question 0 in preparation for that lesson.

Answers to "Did You Get It?"

1. 24% **2.** $15,266; 15.6%
3. Line 9: $82,574.12; Line 12: $27,700; Line 15: $54,874.12; Line 16: $6,144.89

Answers to Prep Skills Questions

1. 23%, 79.4%, 3.5%
2. $860.30
3. $3,519.32
4. $9,300.20

2-8 Applications

Name ____

Everyone fears the tax auditor—that's a government employee who checks over tax returns for accuracy. This is your chance to play auditor. Check the 1040 filed by the deputy sheriff whose W-2 and 1099 forms are below. The completed 1040 is on page 195. Rick is filing jointly for himself and his wife.

a Employee's social security number	XXX-XX-XXXX
OMB No. 1545-0008	
Safe, accurate, FAST! Use IRS e-file	Visit the IRS website at www.irs.gov/efile
b Employer identification number (EIN)	
c Employer's name, address, and ZIP code	King County Sherriff 411 W. Main Ave Cynthiana, KY 41031
d Control number	
e Employee's first name and initial / Last name / Suff.	Richard Grimes 11 Apocalypse Circle Cynthiana, KY 41031
f Employee's address and ZIP code	
1 Wages, tips, other compensation	92,513
2 Federal income tax withheld	14,319
3 Social security wages	96,303
4 Social security tax withheld	3,601.11
5 Medicare wages and tips	96,303
6 Medicare tax withheld	952.00
7 Social security tips	
8 Allocated tips	
9	
10 Dependent care benefits	
11 Nonqualified plans	
12a See instructions for box 12	
12b	
12c	
12d	
13 Statutory employee / Retirement plan / Third-party sick pay	Statutory employee: X
14 Other	

15 State	Employer's state ID number	16 State wages, tips, etc.	17 State income tax	18 Local wages, tips, etc.	19 Local income tax	20 Locality name
KY	21-006390	96,303	3,774	96,303	917	Cynthi

Form **W-2** **Wage and Tax Statement**

Department of the Treasury—Internal Revenue Service

Copy B—To Be Filed With Employee's FEDERAL Tax Return.
This information is being furnished to the Internal Revenue Service.

☐ VOID ☐ CORRECTED

PAYER'S name, street address, city or town, state or province, country, ZIP or foreign postal code, and telephone no.	Kentucky State Employee Credit Union 142 Poplar St Cynthiana, KY 41031
Payer's RTN (optional)	
OMB No. 1545-0112	Form 1099-INT
1 Interest income	$ 1,104.61
2 Early withdrawal penalty	$
PAYER'S federal identification number	
RECIPIENT'S identification number	XXX-XX-XXXX
3 Interest on U.S. Savings Bonds and Treas. obligations	$
RECIPIENT'S name	Richard Grimes
Street address (including apt. no.)	11 Apocalypse Circle
City or town, state or province, country, and ZIP or foreign postal code	Cynthiana, KY 41031
4 Federal income tax withheld	$
5 Investment expenses	$
6 Foreign tax paid	$
7 Foreign country or U.S. possession	
8 Tax-exempt interest	$
9 Specified private activity bond interest	$
10 Market discount	$
11 Bond premium	$
FATCA filing requirement	☐
12	
13 Bond premium on tax-exempt bond	$
Account number (see instructions)	
14 Tax-exempt and tax credit bond CUSIP no.	
15 State	
16 State identification no.	
17 State tax withheld	$ $

Interest Income

Copy 1

For State Tax Department

Form **1099-INT** www.irs.gov/form1099int Department of the Treasury - Internal Revenue Service

2-8 Applications

Name ______________________________

1. Using the tax bracket table on page 182, calculate the tax due on the taxable income shown on Rick's 1040. Then compare to the amount shown in the IRS tax table to the right. Is Rick better off using his amount, or the IRS amount?

If line 15 (taxable income) is—		And you are—			
At least	But less than	Single	Married filing jointly *	Married filing sepa-rately	Head of a house-hold
		Your tax is—			
68,500	**68,550**	10,383	7,783	10,383	8,777
68,550	**68,600**	10,394	7,789	10,394	8,788
68,600	**68,650**	10,405	7,795	10,405	8,799
68,650	**68,700**	10,416	7,801	10,416	8,810
68,700	**68,750**	10,427	7,807	10,427	8,821
68,750	**68,800**	10,438	7,813	10,438	8,832
68,800	**68,850**	10,449	7,819	10,449	8,843
68,850	**68,900**	10,460	7,825	10,460	8,854
68,900	**68,950**	10,471	7,831	10,471	8,865
68,950	**69,000**	10,482	7,837	10,482	8,876

2. List any mistakes you found on the 1040, and write how they should be corrected. If you need extra space for calculations, use the blank space on the next page.

3. Would our friend Rick have paid too much or too little in taxes without your corrections? By how much?

2-8 Applications

Name ______________________________

4. Using the corrected tax that Rick owed after you finished your audit, how much of Rick's taxes will go to paying interest on the national debt?

Form **1040** Department of the Treasury—Internal Revenue Service
U.S. Individual Income Tax Return **2023** OMB No. 1545-0074 IRS Use Only—Do not write or staple in this space.

For the year Jan. 1–Dec. 31, 2023, or other tax year beginning ________, 2023, ending ________, 20____ See separate instructions.

Your first name and middle initial	Last name	Your social security number
Richard	Grimes	XXX XX XXXX
If joint return, spouse's first name and middle initial	Last name	Spouse's social security number
Lori	Grimes	XXX XX XXXX

Home address (number and street). If you have a P.O. box, see instructions.		Apt. no.
11 Winding Way		
City, town, or post office. If you have a foreign address, also complete spaces below.	State	ZIP code
Cynthiana	KY	41031
Foreign country name	Foreign province/state/county	Foreign postal code

Presidential Election Campaign Check here if you, or your spouse if filing jointly, want $3 to go to this fund. Checking a box below will not change your tax or refund. ☐ You ☐ Spouse

Filing Status
Check only one box.

- ☐ Single
- ☑ Married filing jointly (even if only one had income)
- ☐ Married filing separately (MFS)
- ☐ Head of household (HOH)
- ☐ Qualifying surviving spouse (QSS)

If you checked the MFS box, enter the name of your spouse. If you checked the HOH or QSS box, enter the child's name if the qualifying person is a child but not your dependent: ____________

Income

Attach Form(s) W-2 here. Also attach Forms W-2G and 1099-R if tax was withheld.

If you did not get a Form W-2, see instructions.

1a	Total amount from Form(s) W-2, box 1 (see instructions)	1a	
b	Household employee wages not reported on Form(s) W-2	1b	
c	Tip income not reported on line 1a (see instructions)	1c	
d	Medicaid waiver payments not reported on Form(s) W-2 (see instructions)	1d	
e	Taxable dependent care benefits from Form 2441, line 26	1e	
f	Employer-provided adoption benefits from Form 8839, line 29	1f	
g	Wages from Form 8919, line 6	1g	
h	Other earned income (see instructions)	1h	
i	Nontaxable combat pay election (see instructions) 1i		
z	Add lines 1a through 1h	1z	

Attach Sch. B if required.

2a	Tax-exempt interest	2a		b Taxable interest	2b	
3a	Qualified dividends	3a		b Ordinary dividends	3b	
4a	IRA distributions	4a		b Taxable amount	4b	
5a	Pensions and annuities	5a		b Taxable amount	5b	
6a	Social security benefits	6a		b Taxable amount	6b	

Standard Deduction for—
- Single or Married filing separately, $13,850
- Married filing jointly or Qualifying surviving spouse, $27,700
- Head of household, $20,800
- If you checked any box under *Standard Deduction,* see instructions.

c	If you elect to use the lump-sum election method, check here (see instructions) ☐		
7	Capital gain or (loss). Attach Schedule D if required. If not required, check here ☐	7	
8	Additional income from Schedule 1, line 10	8	
9	Add lines 1z, 2b, 3b, 4b, 5b, 6b, 7, and 8. This is your **total income**	9	
10	Adjustments to income from Schedule 1, line 26	10	
11	Subtract line 10 from line 9. This is your **adjusted gross income**	11	
12	**Standard deduction or itemized deductions** (from Schedule A)	12	
13	Qualified business income deduction from Form 8995 or Form 8995-A	13	
14	Add lines 12 and 13	14	
15	Subtract line 14 from line 11. If zero or less, enter -0-. This is your **taxable income**	15	

For Disclosure, Privacy Act, and Paperwork Reduction Act Notice, see separate instructions. Cat. No. 11320B Form **1040** (2023)

2-8 Applications

Name ______________________________

Form 1040 (2023) Page **2**

Section	Line	Description	Box	Amount	Box	Amount
Tax and Credits	16	**Tax** (see instructions). Check if any from Form(s): **1** ☐ 8814 **2** ☐ 4972 **3** ☐ ______			16	
	17	Amount from Schedule 2, line 3			17	
	18	Add lines 16 and 17			18	
	19	Child tax credit or credit for other dependents from Schedule 8812			19	
	20	Amount from Schedule 3, line 8			20	
	21	Add lines 19 and 20			21	
	22	Subtract line 21 from line 18. If zero or less, enter -0-			22	
	23	Other taxes, including self-employment tax, from Schedule 2, line 21			23	
	24	Add lines 22 and 23. This is your **total tax**			24	
Payments	25	Federal income tax withheld from:				
	a	Form(s) W-2	25a			
	b	Form(s) 1099	25b			
	c	Other forms (see instructions)	25c			
	d	Add lines 25a through 25c			25d	
If you have a qualifying child, attach Sch. EIC.	26	2023 estimated tax payments and amount applied from 2022 return			26	
	27	Earned income credit (EIC)	27			
	28	Additional child tax credit from Schedule 8812	28			
	29	American opportunity credit from Form 8863, line 8	29			
	30	Reserved for future use	30			
	31	Amount from Schedule 3, line 15	31			
	32	Add lines 27, 28, 29, and 31. These are your **total other payments and refundable credits**			32	
	33	Add lines 25d, 26, and 32. These are your **total payments**			33	
Refund	34	If line 33 is more than line 24, subtract line 24 from line 33. This is the amount you **overpaid**			34	
	35a	Amount of line 34 you want **refunded to you.** If Form 8888 is attached, check here ☐			35a	
Direct deposit? See instructions.	b	Routing number ______ **c** Type: ☐ Checking ☐ Savings				
	d	Account number ______				
	36	Amount of line 34 you want **applied to your 2024 estimated tax**	36			
Amount You Owe	37	Subtract line 33 from line 24. This is the **amount you owe.** For details on how to pay, go to *www.irs.gov/Payments* or see instructions			37	
	38	Estimated tax penalty (see instructions)	38			

Third Party Designee Do you want to allow another person to discuss this return with the IRS? See instructions ☐ **Yes.** Complete below. ☐ **No**

Designee's name | Phone no. | Personal identification number (PIN)

Sign Here

Under penalties of perjury, I declare that I have examined this return and accompanying schedules and statements, and to the best of my knowledge and belief, they are true, correct, and complete. Declaration of preparer (other than taxpayer) is based on all information of which preparer has any knowledge.

Joint return? See instructions. Keep a copy for your records.

Your signature	Date	Your occupation	If the IRS sent you an Identity Protection PIN, enter it here (see inst.)
Spouse's signature. If a joint return, **both** must sign.	Date	Spouse's occupation	If the IRS sent your spouse an Identity Protection PIN, enter it here (see inst.)
Phone no.	Email address		

Paid Preparer Use Only

Preparer's name	Preparer's signature	Date	PTIN	Check if: ☐ Self-employed
Firm's name			Phone no.	
Firm's address			Firm's EIN	

Go to *www.irs.gov/Form1040* for instructions and the latest information. Form **1040** (2023)

Math in Paying For an Education REVISITED

Jasper White/Image Source

1. Multiplying the yearly amount by 4, we get \$108,600 needed for 4 years.
2. With 25% paid for, that leaves 75% we're responsible for, minus \$5,000 a year we're contributing up front: 0.75 (\$108,600) – 4(\$5,000) = \$61,450 we need to borrow.
3. If we pay interest while in school, the principal at the time payments start will remain \$61,450. Use the student loan payment formula with $P = 61{,}450$, and $r = 0.068$: the results give us monthly payments of \$707.17. The total of all payments is 120 × \$707.17 = \$84,860.40. Now calculate the simple interest paid while in school: $P = \$61{,}450$, $r = 0.068$, $t = 4$, so $I = \$61{,}450 \times 0.068 \times 4 = \$16{,}714.40$. (Note: This assumes that you borrowed the entire amount up front, which is not the smartest idea in the world, but simplifies the calculation. More likely, you'd borrow one year at a time which would lower the total interest.) Adding the two amounts, we get a total of \$84,860.40 + \$16,714.40 = \$101,574.80 in total payments. Subtracting the amount borrowed, we find that total interest is \$101,574.80 – \$61,450 = \$40,124.80.
4. If we capitalize the interest, upon graduation our principal will be \$61,450 + \$16,714.40 = \$78,164.40. This time the monthly payment formula yields \$899.52. Multiply this by 120 payments to get total payments of \$107,942.40. Subtract the principal of \$61,450 to get total interest of \$46,492.40.
5. According to the table, the median salary with a bachelor's degree is \$30,108 higher than with a high school diploma. In Question 4, we found that the total amount paid on loans is \$107,942.40; add the \$20,000 contributed while in school, and the total cost of the degree is \$127,942.40. Finally, divide this amount spent by the extra \$30,108 we'll make each year to find that it will take 4.25 years to make back the amount spent. Here's some great news: in the last edition of this book, using numbers from 2018, the time to catch up was 5.1 years, so the numbers are moving further in favor of education.

Unit 3
Place Your Bets

View Stock/age fotostock

Outline

Math in Gambling

The fact that you're reading this sentence means that you're probably taking a math class right now. But maybe not . . . you could be an instructor evaluating the book, or maybe an editor looking for mistakes (unsuccessfully, no doubt). Still, I would be willing to bet that you're taking a math class. The word "probably" indicates a certain likelihood of something happening, and that basic idea is the topic of this unit. We call the study of the likelihood of events occurring *probability*.

Probability is one of the most useful concepts in math because being able to anticipate the likelihood of events can be useful in so many different areas. Games of chance, business and investing, sports, and weather forecasting are just a few samples from an essentially limitless list of applications. What are the chances of your team winning the championship? Should you take an umbrella to the golf course today? Will stock in a company you're keeping an eye on go up or down? Is that new job offer a good opportunity, or a disaster waiting to happen? Every day you make decisions regarding possible events that are governed at least in part by chance. The more you know about the likelihood of events, the more informed your decisions are likely to be.

We titled this chapter Place Your Bets not because everything we deal with will involve traditional gambling games, but to encourage you to think about the fact that almost everything we do is a gamble to some extent. And in that regard, almost everything we do relates to probability in some way. Have you ever gone outside during a thunderstorm? Not worn a seatbelt? Texted while driving? Smoked? Eaten a fatty diet? Ridden a motorcycle? Flown on a plane? Sped up to beat a red light?

In each of those instances, you were gambling, and not with something silly like casino chips. You were literally gambling with your life as the stakes. To be fair, the odds were DRAMATICALLY in your favor, and the fact that you're not currently deceased proves that you're on a heck of a winning streak. But the more you learn about probability, the more likely you'll be to give some deeper thought to the consequences of actions. And that's what we in academia call "getting an education."

So let's talk about gambling, both traditional and otherwise. You can think about these questions now, make some educated guesses, and maybe discuss them with your group. Eventually, you can come back and answer all of them, and see how accurate your guesses were.

1. Just last week the Powerball lottery reached a jackpot of $1.5 billion. Billion. With a "b." If you bought ten tickets, were you more likely to win the jackpot or get struck by lightning? Make a guess, then look up the odds of winning the jackpot, and your probability of getting struck by lightning, on the Internet. Which is more likely?
2. What does texting while driving do to the probability that you'll be in a fatal accident? Again, make an educated guess, then do some research to see how you did.
3. A handful of gambling scenarios is provided below. In each case, find the expected value (that is, the average amount a person would win or lose) if placing the bet 100 times. Then rank the scenarios from best to worst in terms of your likelihood of winning or losing money. It's a GREAT idea to rank them before you do the calculations. You may be surprised at the results.

   a. At a church fair, you bet $1 and roll two dice. If the sum is 2, 3, 11, or 12, you get back your dollar plus four more. On any other roll, you lose.

   b. In a casino, you bet $1 on 33 at a roulette table. There are 38 possible numbers that can come up. If you win, you get your dollar back, plus 35 more.

   c. You buy a $1 ticket to a multistate lottery. If you match all six numbers, including the Mega Ball, you win the $20 million jackpot. If not, you lose. There are 175,711,536 possible combinations, and only one of them will be a winner.

   d. You bet $1 on flipping a coin with your roommate. Heads, you win, tails, your roommate wins.

For answers, see Math in Gambling Revisited on page 346.

Lesson 3-1 Prep Skills

SKILL 1: REDUCE FRACTIONS

The process of writing fractions in lowest terms is called reducing fractions. It's useful to do because it's easier to work with and interpret fractions when they're in lowest terms. For example, it's a lot easier to see that $\frac{5}{2}$ is two and a half than it is to see that $\frac{180}{72}$ is two and a half. To reduce a fraction, divide BOTH the numerator and the denominator by any common factors.

- $\frac{9}{12} = \frac{9 \div 3}{12 \div 3} = \frac{3}{4}$ The greatest factor common to 9 and 12 is 3.
- $\frac{60}{48} = \frac{60 \div 12}{48 \div 12} = \frac{5}{4}$ The greatest factor common to 60 and 48 is 12.

SKILL 2: CONVERT BETWEEN FRACTIONS, DECIMALS, AND PERCENTS

To write a fraction in decimal form, you just need to perform the division, either by hand or with a calculator.

- $\frac{3}{5} = 0.6$

If the decimal doesn't terminate, it's usually appropriate to round. If a certain number of decimal places isn't required, use your judgment based on the context: How accurate would you want your answer to be in a given situation? Note that since you're rounding, the decimal is just an approximation of the exact fractional value.

- $\frac{3}{7} \approx 0.43$

To write a decimal in percent form, move the decimal two places to the right.

- $0.435 = 43.5\%$

To write a percent in decimal form, move the decimal two places to the left.

- $8.2\% = 0.082$

To write a percent in fraction form, write the fraction as $\frac{\text{Percent}}{100}$ and reduce. This makes sense because "percent" literally means "per hundred."

- $70\% = \frac{70}{100} = \frac{7}{10}$

SKILL 3: INTERPRET "AT MOST" AND "AT LEAST"

Interpreting words that indicate inequalities can be confusing. Reading them carefully and thinking about the meaning is the best way to interpret these phrases accurately. ***Less than*** (<) and ***greater than*** (>) are straightforward because they mean exactly what they say. ***At most*** is another way of expressing ***less than or equal to*** (≤). For example, if you say you can fit at most 5 people in your car, you are saying you can fit 5 or fewer (less than or equal to 5) people in your car. ***At least*** is another way of expressing ***greater than or equal to*** (≥). For example, if you say you need at least $20 to take a date to the movie theater, you are saying you need $20 or more (greater than or equal to $20).

PREP SKILLS QUESTIONS

1. Write each fraction in lowest terms.

   $\frac{12}{16}$

   $\frac{65}{80}$

2. Write each percent in decimal form.

   60%

   12.5%

3. Write each fraction in percent form.

   $\frac{3}{5}$

   $\frac{12}{16}$

4. Write each decimal in percent form.

   0.13

   0.058

5. Write the whole numbers between 1 and 20 that are at least 15, and those that are at most 8.

Lesson 3-1 So You're Saying There's a Chance . . .

Ingram Publishing/age fotostock

LEARNING OBJECTIVES

- ☐ 1. Understand key terminology in the study of probability.
- ☐ 2. Compute and interpret theoretical and empirical probabilities.
- ☐ 3. Compare theoretical and empirical probability.

A scientist worthy of a lab coat should be able to make original discoveries while wearing a clown suit, or give a lecture in a high squeaky voice from inhaling helium. It is written nowhere in the math of probability theory that one may have no fun.

—Eliezer Yudkowsky

Walking into a casino without knowing anything about probability is kind of like going to a stick fight without a stick—you're likely to take a beating. Casinos aren't in the business of losing money, and the games are designed so that most people lose more than they win. But an understanding of what is likely to happen in a given situation can give you an advantage over other players, giving you a better chance of walking out the door with some cash in your pockets. (Although if you learn enough about probability, you might decide that staying away from casinos is your best approach.)

The study of probability originated in an effort to understand games of chance, like those that use coins, dice, and playing cards. Generally speaking, probability is simply a number that describes how likely an event is to occur. We will use games of chance to illustrate the ideas, but will eventually see that probability has many applications beyond simple games.

0. How likely do you think it is that you'll get a C or better in this course? How would you describe that likelihood using a number?

3-1 Class

In this lesson, we'll examine the basic concepts involved in studying **probability.** In short, the probability of something occurring is a number that represents how likely it is to occur. You can think of it as the percent chance of something happening. The major difference between percent chance and probability is that we write probability as a number between zero and one. So in that regard, it's like percent chance with the percentage written in decimal or fractional form.

1. A forecaster says that there's a 70% chance of rain today. What is the probability of rain?

Sample Spaces

Processes such as flipping coins, rolling dice, or drawing cards from a deck are called probability experiments.

A **probability experiment** is a process that leads to well-defined results called **outcomes.** An outcome is the result of a single trial of a probability experiment.

Let's illustrate the terminology using a simple example.

Flipping a coin: A probability experiment
One flip of a coin: A single trial of that probability experiment
Coin landing heads up: An outcome of the experiment

In a probability experiment, we can predict what outcomes are possible, but we can't make any definite statements about which one will occur for any given trial. We say that the outcomes occur at **random.** In any probability experiment, the set of all possible outcomes is called the **sample space.**

Solved Example 1

Finding the Sample Space for a Probability Experiment

List out the sample space for each probability experiment.

a. Flipping a coin

b. Rolling a standard die

c. Answering a true or false question

d. Flipping two coins

> **Math Note**
>
> Sample spaces are defined to be sets, and math folks usually indicate that something is a set by enclosing it in curly brackets: { }. So that's what we did. But many instructors will probably be okay if you just list all the outcomes.

SOLUTION

In each case, our job is to make a list of all the things that can possibly occur for the given experiment.

a. When a coin is flipped, it either lands heads up or tails up. So the sample space is {heads, tails}.

b. A standard die has six sides numbered from 1 to 6, and each number can be rolled. The sample space is {1, 2, 3, 4, 5, 6}.

c. This could be interpreted in a couple of different ways. There are two possible answers, so the sample space could be {true, false}. You could also consider that every answer is either right or wrong and write the sample space as {right, wrong}.

d. The key here is to think of the experiment as a sequence of two flips. You could get either heads or tails on the first. Then, once you know what came up first, you could also get heads or tails on the second. So the sample space is {heads then heads, heads then tails, tails then heads, tails then tails}.

2. Write the sample space for the probability experiment of identifying a student's grade in a chemistry course.

When studying probability, it's pretty common to think about several outcomes of an experiment that are related in some way. For example, when rolling a die, we might be interested in whether the result is odd.

3. Write all outcomes of rolling one die that are odd.

You just wrote an example of an *event.*

An **event** is a collection of one or more outcomes in a sample space. Some events, like rolling a 3 with one die, have only one outcome. Others, like rolling an odd number, have more than one outcome.

There's a subtle distinction between an outcome and an event: An outcome is a single occurrence, while an event can contain a number of outcomes. So rolling 3 with a single die is an outcome, but it can also be considered an event. Rolling an odd number, on the other hand, is an event made up of three outcomes but is not itself an outcome.

Let's look at that distinction more closely. Look back at the sample space for flipping two coins in Solved Example 1.

Photodisc/Getty Images

Experiment: Draw a card. Sample space: 52 cards. Outcome: Draw the ace of spades. Event: Draw an ace.

4. How many results match getting heads on the first flip and tails on the second?

That's an outcome.

5. How many results match getting one head and one tail?

That's an event.

Identify each scenario in Questions 6–8 as an event, an outcome, or both.

6. When rolling two dice, getting 3 on the first die and 5 on the second.

7. When rolling two dice, getting a total of 10.

8. When flipping a coin 10 times consecutively, getting tails 8 times.

Did You Get It

Try this problem to see if you understand the concepts we just studied. The answer can be found at the end of the portfolio section.

1. A baseball team has 12 pitchers on the roster, and the manager, who evidently is intent on losing his job, decides to pick the starting pitcher for tonight's game randomly. What is the sample space? Write an example of an outcome and an example of an event.

Theoretical Probability

Now we're ready to specifically define what is meant by probability. The first type we'll study is called **theoretical probability.** The goal is to identify all of the possible outcomes in a sample space and calculate the probability, or likelihood, of an event occurring without actually performing experiments. There is one key assumption we make in theoretical probability: that every outcome in a sample space is equally likely. For example, when a single die is rolled, we assume that each number is equally likely to come up. When a card is chosen from a deck of 52 cards, we assume that each card has the same probability of being drawn.

9. You'd have to be awfully bored to flip a coin 100 times, but if you did, about how many times do you think it would land tails side up?

10. What percentage of the time would you expect the coin to land tails up?

11. Write a fraction with the number of flips in the denominator and the number of times you're expecting it to land tails up in the numerator. What does this have to do with your answer to Question 10?

Congratulations: You've just computed your first theoretical probability. You didn't actually flip a coin 100 times: You just used reasoning ability to decide what the likelihood of getting tails is. That's what makes it a theoretical probability. You developed a theory by thinking, as opposed to gathering data experimentally. (We'll get to the experimental part in a bit.)

Calculating a Theoretical Probability

The probability of an event E in a sample space S occurring is given by

$$P(E) = \frac{\text{The number of outcomes in E}}{\text{The number of outcomes in S}} = \frac{n(E)}{n(S)}$$

For Questions 12–14, a single six-sided die is rolled. Find the probability of each event. If you need to, look back at the sample space from Solved Example 1, part b for reference.

12. Rolling a 2.

Jules Frazier/Photodisc/Getty Images

13. Rolling a number less than 5.

14. Rolling an odd number.

15. If THREE coins are flipped, write the sample space. Remember that Tails, Heads, Heads is a different outcome than Heads, Heads, Tails even though both result in one tail and two heads (which, by the way, sounds like a bunny rabbit from a horror story).

Brandon Laufenberg/E+/Getty Images

For Questions 16–18, three coins are flipped. Find the probability of each event.

16. Two tails

17. At least one tail

18. At most one tail

Did You Get It

2. There are 75 balls in a bingo hopper. The ones numbered 1–15 are labeled with B, the ones from 16–30 with I, from 31–45 N, from 46–60 G, and from 61–75 O. Find the probability of these events:
   a. Drawing N42
   b. Drawing N11
   c. Drawing a number less than 50
   d. Drawing an odd number
3. In the game of rock, paper, scissors, two people play, and it's equally likely that you will win, lose, or tie. Find the probability of
   a. Losing twice in a row.
   b. Winning at least once in two tries.
   c. Having the same outcome twice in a row.

Solved Example 2

Describing Properties of Probability

For each observation, explain why it makes sense based on our understanding of probability.

a. Probability can never be greater than one or less than zero.

b. Probability is one when an event is certain to occur and zero when an event can't occur.

c. If you add the probabilities for all outcomes in any sample space, the result has to be one.

SOLUTION

a. Remember that in essence, probability is a fraction: the number of outcomes in a given event divided by the number of outcomes in the sample space. A fraction can only be more than one if the numerator is bigger than the denominator, but that can't happen here—the denominator is the TOTAL number of outcomes, so no event can have more outcomes than the entire sample space. Conclusion: The fraction can't be more than one. Next, a fraction can only be less than zero if either the numerator or denominator (but not both) is. But they're both numbers of outcomes in this case, and neither can be negative.

b. The only way an event can be certain to occur is if it contains every possible outcome. In that case, the event is the whole sample space, and the number of outcomes in the event and the sample space are equal. So dividing them to get the probability will always result in one. The only way an event cannot possibly occur is if it doesn't have any outcomes in it, in which case the numerator of the probability is zero.

c. If you add the fractions for all possible outcomes, they'll all have the same denominator (the total number of outcomes). That gives us a bunch of fractions with a common denominator. The numerators, on the other hand, will consist of the number of all possible outcomes, which gets us back to a fraction with the same numerator and denominator. That makes the sum of those probabilities one.

3-1 Group

If it doesn't make you uncomfortable, exchange the following information with the classmates in your Unit 3 group. This will be your small group for the third unit. It would be a good idea to schedule a time for the group to meet to go over homework, ask/answer questions, or prepare for exams. You can use this table to help schedule a mutually agreeable time.

Name	Phone Number	Email	Available Times

Probabilities are usually expressed as fractions or decimals between (and including) zero and one. But occasionally we'll express probabilities as percents. For example, when the probability of an event is 1/2, there's a 50% chance that it'll occur. If an event has probability close to zero, it's very unlikely to occur, and if the probability is close to one, it's very likely to occur.

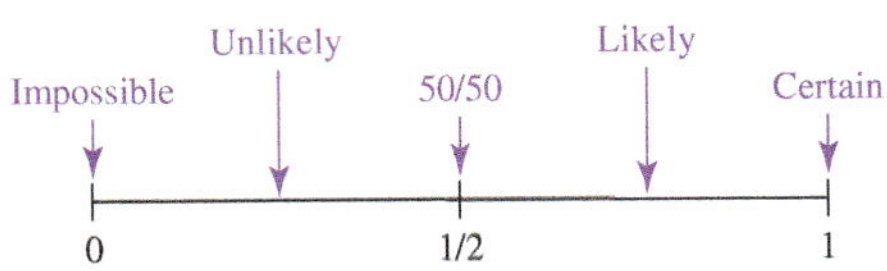

pikepicture/Shutterstock

1. Describe exactly what you think it means when a weather forecaster says that there's a 10% chance of rain today.

A lot of people think that a 10% chance of rain means that it will rain for 10% of the day, but that's not correct. The forecaster is expressing a probability—how likely he or she thinks it is that there will be SOME measurable precipitation in that area today. So in essence, it means that if you keep track of 10 days with a 10% chance of rain, you would expect rain on one of those 10 days.

In Questions 2–5, estimate the probability of each event occurring in your life. You don't have to do any calculations; the idea is just to think about the size of a probability compared to how likely something is to occur.

2. Passing this course

3. Living past the age of 100

4. Winning a million-dollar lottery

5. Doing tonight's homework

Sometimes it's useful to find the probability that a given event will NOT occur. Here's an example.

6. What's the probability of rolling a 2 with one die?

7. How many outcomes are in the event of NOT rolling a 2?

8. Use your answer to Question 7 to find the probability of NOT rolling a 2.

When you roll one die, you will either get 2, or you will not. If we call the event of rolling a 2 E, then we call the event of not rolling a 2 E', and we read it as "E complement" or "the complement of E."

For an event E in a sample space, the set of all outcomes that are not in E is called the **complement** of E.

9. Describe the complement of each event in detail. Don't just put "not" in front. Think hard! The full answer may not be the obvious one.

a. Passing this course

b. Winning $1,000,000 when you buy a lotto ticket

Given the fact that all probabilities for an experiment have to add up to 1, it makes perfect sense that the probability of rolling 2 plus the probability of not rolling 2 is 1. This line of reasoning gives us a general rule:

The Probability of a Complement

For any event E, if E' is the event "E does not occur," then

$$P(E') = 1 - P(E)$$

10. Of the next 32 trials on the docket in a county court, 5 are homicides, 12 are drug offenses, 6 are assaults, and 9 are property crimes. If jurors are assigned to trials randomly, what's the probability that a given juror won't get a homicide case? Describe two different ways you could calculate this probability.

Did You Get It ?

4. For the docket in Question 10, what's the probability that a juror gets assigned to a case that isn't a drug offense?

Empirical Probability

The second approach to probability we will study is computed using experimental data, rather than counting equally likely outcomes. For example, suppose that out of the last 100 days, you've gone to the gym 60 times. It would be reasonable to guess that the probability of you going to the gym on any given day is about 60/100, or 0.6.

This type of probability is called **empirical probability,** and is based on *observed frequencies*—that is, the number of times a particular event has occurred out of a certain number of trials. In this case, the observed frequency of days going to the gym is 60, the observed frequency of not going is 40, and the total number of trials is $60 + 40 = 100$.

11. On any given day, there are exactly two possibilities: You can go to the gym, or not go to the gym. Why would this fact and theoretical probability not be useful in deciding the probability that you'll go to the gym?

When calculating theoretical probabilities, we assumed that all outcomes were equally likely. But there are plenty of situations where that assumption would be faulty. In that case, we often perform an experiment to estimate probabilities.

12. If you flip a paper or plastic cup, it can land on its top, its bottom, or its side. Are these outcomes equally likely? Explain.

13. Estimate the probability of each outcome.

The way that we estimated the probability that you go to the gym on any given day is a good model for how to compute empirical probabilities in general:

Calculating an Empirical Probability

The empirical probability of an event E is given by

$$P(E) = \frac{\text{Observed frequency of the event}}{\text{The number of trials}} = \frac{f}{n}$$

14. Describe how you could design an experiment to develop empirical probabilities for flipping a cup.

15. Now let's perform that experiment. Your professor will supply your group with a cup. Flip it from about waist level and record how it lands in the table below. Complete as many trials as you feel like you need in order to get accurate empirical probabilities, and then compute those probabilities.

Position	Put a Mark for Each Occurrence	Add the Marks
Land on top		
Land on bottom		
Land on side		

The table you used to organize the results of your experiment is an example of a **frequency distribution.** This is a table that lists categories that we call **classes** (in this case the outcomes of our experiment) and frequencies for each class.

Earlier, we talked about the number of times you might go to the gym in 100 days.

16. Complete a frequency distribution for going to the gym based on the data provided earlier.

Result (Class)	Observed Frequency
Gym	
No gym	
Total	

17. In a random sample of 500 people, 210 had type O blood, 223 had type A, 51 had type B, and 16 had type AB. Set up a frequency distribution for the data.

Result (Class)	Observed Frequency

When a blood transfusion is needed in an emergency medical situation, you might think that you'd have to receive the same blood type as your own, but that's actually not the case. This chart describes which types can donate, and receive, various blood types.

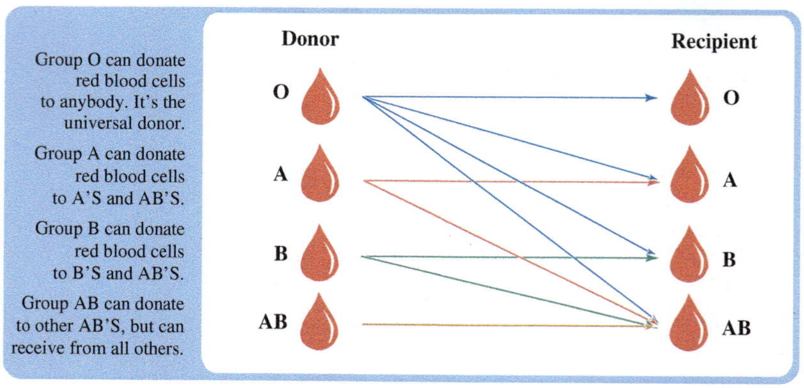

18. What's the probability that a randomly selected patient can receive only one type of blood? Use the chart and the frequency distribution from Question 17.

19. If an accident victim has type B blood, and an EMT on site is the only person willing to donate blood, what's the probability that it will be compatible with the victim?

20. If a person with blood type A gets a transfusion, what's the probability that the donor was not type A?

21. Find the probability that a randomly selected donor's blood can be given to patients of more than just one blood type.

Did You Get It ?

5. A 2-lb bag of Hershey's miniatures contains 33 milk chocolate bars, 27 Krackel bars, 26 Mr. Goodbars, and 19 Special Dark bars. Set up a frequency distribution and find the probability that a bar chosen from a randomly selected bag is
   a. A Mr. Goodbar.
   b. A Krackel or Special Dark.
   c. Not a milk chocolate bar.

Sidelight Probability and Your Fears

All of us at one time or another have thought about dying. Some people have fears of dying in a plane crash or dying from a shark attack. Plane crashes, in particular, are feared far more than reality should dictate–statistically, it's 125 times safer to fly across the United States than to drive. Statisticians who work for insurance companies (called actuaries) also calculate probabilities for dying from other causes. For example, based on deaths in the United States, here are the probabilities of dying from a variety of frightening causes:

Motor vehicle accident	1 in 7,000
Shot by a gun	1 in 10,000
Accident while walking across the street	1 in 60,000
Lightning strike	1 in 3 million
Shark attack	1 in 100 million

The death risk for various diseases is much higher:

Heart attack	1 in 400
Cancer	1 in 600
Stroke	1 in 2,000

Clearly, the chances of dying from diseases are much higher than the chances of dying from accidents. (For the record, this is intended to ease some of your fears, not encourage you to act recklessly. Your chances of dying in a motor vehicle accident go up by an awful lot if you're sending texts and speeding while you drive.)

We'll close the lesson with a comparison between the two types of probability we've studied, theoretical and empirical. The key thing to remember is that in theoretical probability, we use a sample space with a definite number of outcomes and assume that all of them are equally likely. This makes theoretical probability a bad option for something like the probability of a team winning a given game, because in most cases it would be silly to assume that who wins a game is random.

Empirical probability, on the other hand, uses observed frequencies and a total number of trials to compute a probability. A perfectly reasonable question is, "How many trials are needed to get an accurate probability?" The answer, of course, is seven. Just kidding. There IS no answer to that question. All we can say is that the more trials you do in an experiment, the more likely it is that the probabilities you calculate are true representatives of the situation.

The distinction between theoretical and empirical probability can sometimes be pretty subtle, and it may depend on the exact probability you're interested in. For example, suppose that you buy a bag of M&M's, which is always a good idea, and find that of the 50 candies in the bag, 10 are brown. You could conclude that the probability of drawing a brown M&M from *that particular bag* is 10/50, or 0.2, and that would be a theoretical probability. "But wait," you say, "isn't it based on observed frequencies?" Well, yes, but since you know for a fact that there are 50 total outcomes and 10 of them satisfy the event "picking a brown candy," then you're using our theoretical probability formula.

But if, under the same circumstances, you conclude that there's a 20% chance that the first M&M you pick out of the *next bag you buy* will be brown, now it's an empirical probability, because you're using the observed frequencies from that one bag to predict probabilities for a general situation. So you might find it helpful to think "Are we using the number of all potential outcomes?" If we are, you're probably looking at a theoretical probability.

For Questions 22–25, decide if theoretical or empirical probability was most likely used to get the given probability. Justify your answer, of course.

22. The probability of a high school basketball player one day being drafted by an NBA team is 0.000408.

23. The probability of being dealt one pair in five-card poker is 0.422569.

Math Note

Each of these probabilities was found by doing a Google search for the string "The probability of."

24. The probability of being in a fatal accident if you make two car trips a day for one year is $\frac{1}{13{,}750}$.

25. The probability of winning some prize when buying one New York State Lottery scratch-off ticket is 0.216.

3-1 Portfolio

Name ______________________________

Check each box when you've completed the task. Remember that your instructor will want you to turn in the portfolio pages you create.

Technology

1. ☐ In this Excel activity, we'll calculate empirical probabilities from a simulated data set representing numerous rolls of a 15 sided die.

ALEKS Practice

1. ☐ Include any written work from the ALEKS assignment along with any notes or questions about this lesson's content.

Applications

1. ☐ Complete the applications problems.

Reflections

Type a short answer to each question.

1. ☐ Describe the connection between the probability of an event occurring and percent chance that it will occur.
2. ☐ Explain how theoretical probability differs from empirical probability.
3. ☐ When would you be likely to use empirical methods to calculate a probability rather than theoretical methods?
4. ☐ Name one thing you learned or discovered in this lesson that you found particularly interesting.
5. ☐ What questions do you have about this lesson?

Looking Ahead

1. ☐ Complete the prep skills for Lesson 3-2.
2. ☐ Read the opening paragraph in Lesson 3-2 carefully and answer Question 0 in preparation for that lesson.

Answers to "Did You Get It?"

1. Sample space: all 12 pitchers. An outcome would be choosing one specific pitcher. An event would be something like choosing a left-handed pitcher.

2. a. $\frac{1}{75}$ **b.** 0 **c.** $\frac{49}{75}$ **d.** $\frac{38}{75}$ **3. a.** $\frac{1}{9}$ **b.** $\frac{5}{9}$ **c.** $\frac{1}{3}$

4. $\frac{5}{8}$ **5. a.** 0.248 **b.** 0.438 **c.** 0.686

Answers to Prep Skills Questions

1. $\frac{3}{4}, \frac{13}{16}$

2. 0.6, 0.125

3. 60%, 75%

4. 13%, 5.8%

5. 15, 16, 17, 18, 19; 2, 3, 4, 5, 6, 7, 8

3-1 Applications

Name ______________________________

1. On *The Price Is Right* game show, to get into the showcase at the end, contestants spin a wheel with 20 equally sized spaces, numbered 5 cents, 10 cents, 15 cents, and so on, up to $1.00. If the contestant spins once, find the probability of each event below.

Valerie Macon/Getty Images

a. $1.00

b. An amount less than 50 cents

c. An amount greater than 75 cents

d. An amount that is divisible by five

e. Which type of probability (empirical or theoretical) are you using to answer these questions? Explain.

2. If there are only 50 lottery tickets for the Big Game, 1 of which is a winning ticket, and you buy 7 of those tickets at random, what is the probability that you'll win the super jackpot? Which type of probability (empirical or theoretical) did you use to answer these questions? Explain.

3-1 Applications

Name ______________________________

3. In April of 2024, the Pew Research Center reported that 70% of Americans say that access to in vitro fertilization (IVF) is a good thing. If an American is selected at random, find the probability that he or she will feel that access to IVF is a bad thing, or have no opinion on the issue. Which type of probability (empirical or theoretical) did you use to answer this question? Explain.

4. There were 12,355 single-bias hate crime offenses reported in the United States in 2023. These crimes were divided into categories, with the following breakdown: 6,307 were based on racial or ethnic bias, 2,882 on religious bias, 2,296 on sexual orientation bias, 202 on bias against those with disabilities, 126 on gender bias, and 542 on gender identity bias.

   a. Make a frequency distribution for these data.

Class	Frequency

Now use your frequency distribution to find the probability of each event occurring for a year after 2023 in parts b–d.

   b. Crime was motivated by sexual orientation bias.

   c. Crime was motivated by bias against race/ethnicity or religion.

   d. Crime was not motivated by bias against religion or those with disabilities.

   e. Which type of probability (empirical or theoretical) are you using to answer these questions? Explain.

3-1 Applications

Name ______________________________

Each year, the Yale Program on Climate Change Communication releases results of an extensive survey on attitudes toward climate change. The table below summarizes results of one sample group responses to the question "Do you think that global warming is happening?" in 2010 and in 2023.

Response	2010	2023
Yes	571	821
No	200	182
Don't know	230	137
Total	1,001	1,140

5. If a respondent to either survey is chosen randomly, what's the probability that he or she felt that global warming was not happening at the time?

6. What's the probability that a respondent in 2023 didn't answer yes?

7. How much did the probability of a respondent saying that global warming was happening increase from 2010 to 2023?

8. Which type of probability (empirical or theoretical) are you using to answer these questions? Explain.

Lesson 3-2 Prep Skills

SKILL 1: MULTIPLY STRINGS OF INTEGERS

When multiplying strings of integers without a calculator, use the fact that multiplication is commutative and associative to multiply pairs of factors that produce convenient results where possible.

- $4 \cdot 7 \cdot 25 = 7 \cdot 100$
  $= 700$

  Multiply 4 and 25 first because $7 \cdot 100$ is easier to calculate than $28 \cdot 25$.

- $7 \cdot 6 \cdot 5 \cdot 4 = 42 \cdot 20$
  $= 840$

  If you just started multiplying left to right, you would quickly get to a point where it was quite challenging to do without a calculator; $42 \cdot 20$ is more manageable.

SKILL 2: COMPUTE THEORETICAL PROBABILITIES

As we saw in Lesson 3-1, the probability of an event E in a sample space S occurring is given by

$$P(E) = \frac{\text{The number of outcomes in } E}{\text{The number of outcomes in } S} = \frac{n(E)}{n(S)}$$

For the following examples, remember that the sample space for flipping three coins is {TTT, TTH, THH, THT, HTT, HTH, HHT, HHH}. The number of outcomes in this sample space is 8.

- Find the probability of getting one tail.

$$P(\text{one tail}) = \frac{\text{The number of outcomes resulting in one tail}}{8} = \frac{3}{8}$$

The three outcomes resulting in one tail are THH, HTH, and HHT.

- Find the probability of getting at least two tails.

The phrase "at least two tails" means getting two or more tails from the three flips (2 or 3 tails).

$$P(\text{at least two tails}) = \frac{\text{The number of outcomes resulting in at least two tails}}{8} = \frac{4}{8} = \frac{1}{2}$$

The four outcomes resulting in 2 or 3 tails are TTT, TTH, THT, and HTT.

PREP SKILLS QUESTIONS

1. Perform each multiplication without a calculator.

   $3 \cdot 2 \cdot 5$

   $4 \cdot 3 \cdot 5 \cdot 2$

2. The math department at my school offers a total of 27 different classes, 3 of which are calculus classes. If someone is goofy enough to pick a math class randomly, what's the probability that they'll end up in a calculus class?

3. If you flip a coin twice, what is the probability of it landing heads up both times?

4. What's the probability of rolling a 5 with one die?

5. What's the probability of rolling a number less than 5 with one die?

Lesson 3-2 Make It Count

LEARNING OBJECTIVES

- ☐ 1. Describe how counting techniques are useful in probability theory.
- ☐ 2. Use tree diagrams and tables to determine sample spaces and compute probabilities.
- ☐ 3. Develop and use the fundamental counting principle.

Photodisc/MedioImages/Getty Images

Why does my brain insist on counting the steps every time I walk up a flight of stairs? I just can't help myself. There's something about my mind that always wants to keep counting.

—Rachel Nichols

For centuries, people have tried a wide variety of techniques, some of them pretty bizarre, to try and influence the sex of their children. (Look it up sometime—it's fascinating.) The truth is, without the aid of cutting-edge science, you don't get to choose. But that doesn't stop many young couples from planning the type of family they hope to have. Suppose that one couple would like to have three children, but they definitely want to have at least one boy and one girl. What is the probability that they'll get their wish without having to go beyond three kids?

Believe it or not, the subject of this lesson is counting. Seriously. When working with theoretical probabilities, we know that we need to decide on the sample space for an event, and then find how many individual outcomes are in that event. When situations start to get complicated, it might not always be apparent how to do so. That's where developing counting techniques will be a big help.

0. How did counting play a role in our study of probability in Lesson 3-1?

3-2 Class

Alright, let's start counting. One, two, three, four … just kidding. That's not the kind of counting we're talking about. The first technique we will study involves diagramming all of the possible outcomes for probability experiments, like the one referenced in the opening paragraph above. The specific type of diagram we'll study is called a *tree diagram*.

A **tree diagram** is a diagram where we use branches, similar to a flow chart, to list the outcomes of two or more probability experiments that are done in sequence.

While it's not the most sensitive thing we can do, we can regard the birth of each child for the couple above as a probability experiment: Each child is either a boy or a girl. When you start to look at three such experiments in succession, it's helpful to diagram the possible outcomes.

To draw a tree diagram, we start with branches emanating, or starting from, a single point to show the possible outcomes for the first experiment. Next, we show the outcomes for the second experiment using branches emanating from the end of each branch that was used for the first experiment, and so on.

Solved Example

Drawing and Using a Tree Diagram

Draw a tree diagram for the possible sexes of three children in a family, and use it to write out the sample space. Use G to indicate the outcome of having a girl and B to indicate the outcome of having a boy.

SOLUTION

To draw our diagram, we'll have to think in stages: Each child is a stage in the process. The first child will be a girl or a boy, so the tree diagram begins with two branches:

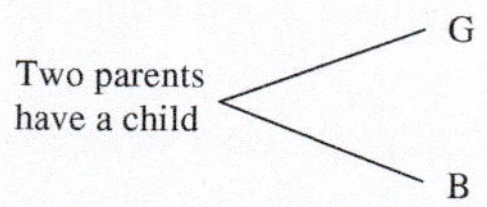

Once the first child's sex is known, at the next stage (meaning the birth of the next child), there are again two possibilities starting with each of the possibilities for the first child.

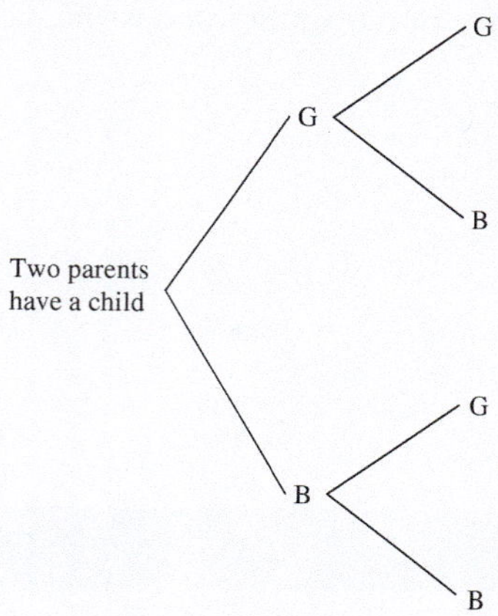

Now we repeat the process for the four possibilities at stage two, and we have our tree diagram.

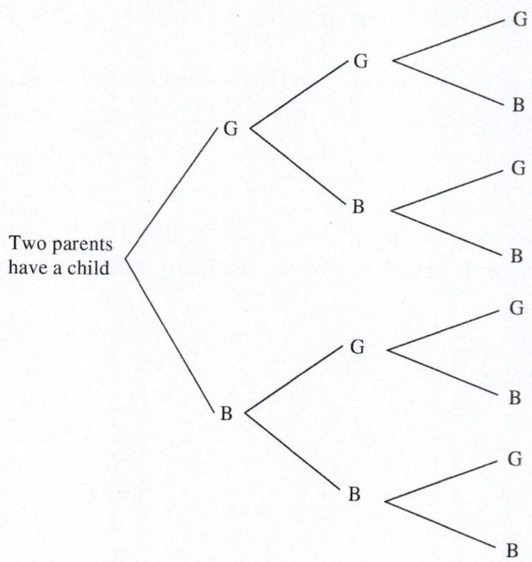

Solved Example *Continued*

To find the sample space, we'll trace along each branch of the tree diagram from left to right. If we start at the far left and trace along the top branch at each stage, we get GGG, which tells us that one outcome is having three girls. If we follow the top branch at the first and second stages, then choose the lower branch at the last stage, we get the outcome GGB. Following along all branches in this way, we get the sample space:

{GGG, GGB, GBG, GBB, BGG, BGB, BBG, BBB}

Notice that the number of outcomes, eight, is exactly the number of choices at the last stage of the process: This is a nice way to check and make sure you have the right number of outcomes in your sample space when using a tree diagram.

1. Use the tree diagram from Solved Example 1 to find the probability that a family with three children will have exactly two boys.

> **Math Note**
>
> Computing probability here requires assuming that there's a 50–50 chance of getting a boy every time. That's not exactly true, but birth statistics show that it's pretty close.

2. Find the probability that a family with three children will have at least two girls.

3. Look back at the opening paragraph. What is the percent chance that the couple will get their wish?

4. Describe how being able to count different possibilities helps us in computing probability.

Did You Get It

Try these problems to see if you understand the concepts we just studied. The answers can be found at the end of the portfolio section.

1. A soda machine dispenses both Coke and Pepsi products, in both 12-ounce cans and 20-ounce bottles. For each brand, it has a regular cola, diet cola, and lemon-lime drink. Use a tree diagram to find the sample space for the experiment of choosing one drink at random from this machine.
2. Suppose the soda machine goes berserk and starts dispensing drinks randomly. If you want a diet cola, what is the probability that you'll get one?

In order to collect information for a student survey, a researcher classifies students according to class rank (freshman, sophomore), scholarship status (full, partial, none) and where the student lives when not on campus (in state, out of state). A folder for each classification is then made up (e.g., freshman/partial scholarship, in state).

5. How many different outcomes do you think make up the sample space here? Discuss and see if you can come up with an answer without using a tree diagram.

6. Find the sample space for the folders using a tree diagram.

Write sample space here

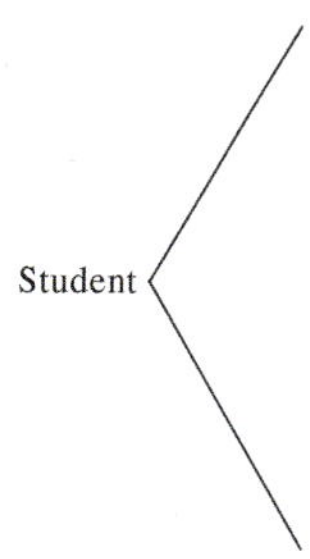

7. If a folder is selected at random, find the probability that it includes students with a full scholarship.

8. If a folder is selected at random, find the probability that it includes students who live out of state and do not have a partial scholarship.

9. If a folder is selected at random, find the probability that it includes students who are in state freshmen.

Did You Get It

3. A coin is flipped, and then a die is rolled. Use a tree diagram to find the probability of getting heads on the coin and an even number on the die.

Tables

Another way of listing out a sample space is by making a table. Consider the sample space of selecting a card from a standard deck of 52 cards. (The cards are assumed to be shuffled to make sure that the selection occurs at random.) There are four suits—hearts, diamonds, spades, and clubs, and 13 cards of each suit consisting of the denominations ace (A), 2, 3, 4, 5, 6, 7, 8, 9, 10, and 3 picture or face cards—jack (J), queen (Q), and king (K). The sample space is shown in the table below.

	A	2	3	4	5	6	7	8	9	10	J	Q	K
♥	A♥	2♥	3♥	4♥	5♥	6♥	7♥	8♥	9♥	10♥	J♥	Q♥	K♥
♦	A♦	2♦	3♦	4♦	5♦	6♦	7♦	8♦	9♦	10♦	J♦	Q♦	K♦
♠	A♠	2♠	3♠	4♠	5♠	6♠	7♠	8♠	9♠	10♠	J♠	Q♠	K♠
♣	A♣	2♣	3♣	4♣	5♣	6♣	7♣	8♣	9♣	10♣	J♣	Q♣	K♣

A card is drawn from an ordinary deck. In Questions 10–12, use the sample space shown above to find the probabilities of getting

10. A jack.

11. A black card with a number between 2 and 7.

12. A 3 or a diamond.

13. Describe why having the sample space listed out in table form is so useful for probability questions like this one.

3-2 Group

Let's continue our study of using tables to help us with probability by looking at an example where we need to construct our own table. When two dice are rolled, how many outcomes are in the sample space?

To get a handle on this question, let's pretend that one die is red and the other is green, and let's say that we roll the red die first, then the green one. If we get 1 on the red die, there are six possible overall outcomes: 1 on red and 1 on green, 1 on red and 2 on green, 1 on red and 3 on green, and so on. If we represent each possible outcome with an ordered pair, like (1, 3), then we can systematically list the possibilities in table form.

1. Complete the sample space table below for rolling a red die followed by a green one. The first few outcomes have been done for you because that's the way we roll.

Green Die

Red Die	1	2	3	4	5	6
1	(1, 1)	(1, 2)	(1, 3)			
2						
3						
4						
5						
6						

Now that we have our sample space, we can calculate tons of probabilities. Notice, by the way, that the sum of the two dice gives us the total for the roll, so the outcome (1, 3) would be called "rolling a 4." Find each probability for a roll of two dice:

2. Rolling an 8.

3. Rolling doubles (the same number on each die).

4. Rolling less than 5.

Did You Get It

4. Two dice are rolled. Use your sample space from Question 1 to find the probability of
   a. Rolling a 9.
   b. Rolling an even number.
   c. Rolling a number greater than 6.

After getting a cool new job, you naturally want a new apartment, and furniture to go along with it. The hip furniture boutique around the corner has the couch you want in either leather or micro suede, and each comes in your choice of four colors: tan, gray, red, and black.

atproduction/123RF

5. Create a tree diagram for this situation to determine how many different couch options you have to choose from.

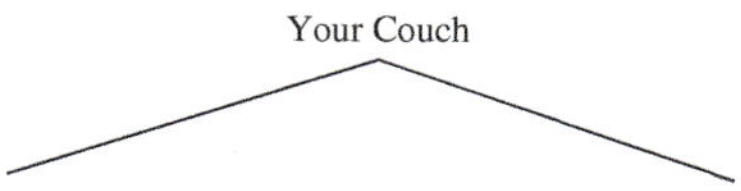

6. How many choices of couch are there? USING YOUR TREE DIAGRAM, explain why it makes perfect sense that the number of total choices is the product of the number of fabric choices and the number of color choices.

7. Here's a way to connect the first two topics of this lesson. The table below can be used to list all of the couch choices. Explain why this also justifies that the total number of choices is the product of the number of fabric choices and the number of color choices.

Fabric \ Color	Tan (T)	Gray (G)	Red (R)	Black (B)
Leather (L)	LT	LG	LR	LB
Micro Suede (M)	MT	MG	MR	MB

8. More choices is better, right? Suppose the furniture salesperson calls you and points out that your couch also comes in a version with an end recliner, and one without. Based on what we've learned so far, make a conjecture as to how this will change the number of options. Explain.

Math Note

Are more choices *really* better? There's some interesting research that says maybe not. Do a Google search for "choice paralysis" if you're interested.

9. Add this extra layer of choices to your tree diagram on the previous page. Does this confirm your answer to Question 8? Explain.

We've now discovered an idea so important in counting that it's literally called "The Fundamental Counting Principle."

The Fundamental Counting Principle

In a sequence of n procedures, if the first can occur in k_1 ways, the second in k_2 ways, the third in k_3 ways, and so on, the total number of ways the sequence of procedures can occur is

$$k_1 \cdot k_2 \cdot k_3 \cdot \cdots \cdot k_n$$

IMPORTANT NOTE: The fundamental counting principle can only be used when the occurrence of the first event in no way affects the occurrence of the second event, which in turn does not affect the occurrence of the third event, and so on.

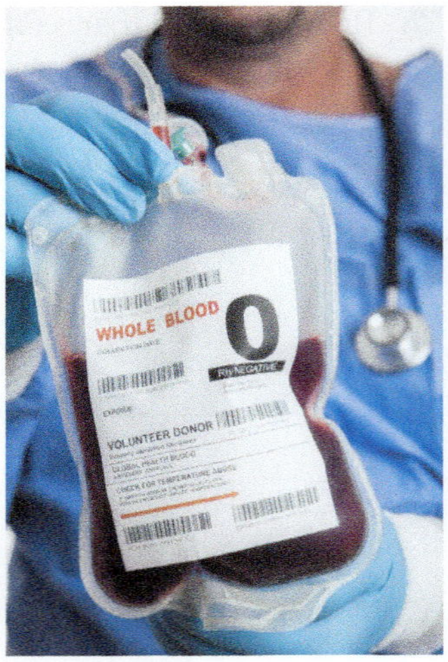

DieterMeyrl/iStock/Getty Images

The basic blood types are represented by letters: A, B, AB, and O. There is also an Rh factor represented by either positive or negative. If a local blood bank labels donations according to type, Rh factor, and biological sex of the donor, it would be useful to look at how many different ways a blood sample can be labeled, as this would have a significant impact on preparing for the labeling.

10. How many different blood types are listed?

11. How many different Rh factors are possible?

12. How many different biological sexes are likely to be listed?

13. Based on these answers and the fundamental counting principle, how many different ways can the blood sample be labeled?

14. If you choose one of the labels randomly, what is the probability that it corresponds to male and AB-negative?

15. In reality, the probability of a randomly selected person in the U.S. population being a male with AB-negative blood is about 0.0035. What does that fact and your answer to Question 14 tell you about blood type?

16. A company with 412 employees is designing new ID badges for its employees. Each badge will be coded with a four-character sequence, with each character chosen from the letters A, B, C, D, and E. Will there be enough badges so that everyone gets a different code? Use the fundamental counting principle to decide. Note that letters can be used more than once.

17. Will there be enough badges to go around if no letter can appear more than once in the code? (Hint: Think about the number of choices at each stage.)

18. You could absolutely use a tree diagram to list out the possible codes. Explain why that's a tremendously bad idea.

Did You Get It ?

5. A discount paint manufacturer plans to make several different paints. The categories include

   Color: Red, blue, white, black, green, brown, yellow
   Type: Latex, oil
   Texture: Flat, semigloss, high-gloss
   Use: Outdoor, indoor
   How many different kinds of paint can be made?

6. The lock on a storage facility is controlled by a keypad containing digits 1 through 5.

   How many three-digit codes are possible if digits can be repeated?
   How many three-digit codes are possible if digits cannot be repeated?

3-2 Portfolio

Name ______________________________

Check each box when you've completed the task. Remember that your instructor will want you to turn in the portfolio pages you create.

ALEKS Practice

1. ☐ Include any written work from the ALEKS assignment along with any notes or questions about this lesson's content.

Applications

1. ☐ Complete the applications problems.

Reflections

Type a short answer to each question.

1. ☐ What is meant by the term "counting" in this lesson?
2. ☐ Explain how counting techniques are important to the study of probability.
3. ☐ Why is the fundamental counting principle more widely applicable than tree diagrams?
4. ☐ Is the fundamental counting principle useful for empirical or theoretical probabilities? Why?
5. ☐ Name one thing you learned or discovered in this lesson that you found particularly interesting.
6. ☐ What questions do you have about this lesson?

Looking Ahead

1. ☐ Complete the prep skills for Lesson 3-3.
2. ☐ Read the opening paragraph in Lesson 3-3 carefully and answer Question 0 in preparation for that lesson.

Answers to "Did You Get It?"

1. Sample space: {Coke, can, cola; Coke, can, diet cola; Coke, can, lemon-lime; Coke, bottle, cola; Coke, bottle, diet cola; Coke, bottle, lemon-lime; Pepsi, can, cola; Pepsi, can, diet cola; Pepsi, can, lemon-lime; Pepsi, bottle, cola; Pepsi, bottle, diet cola; Pepsi, bottle, lemon-lime}

2. $\frac{1}{3}$ **3.** $\frac{1}{4}$ **4. a.** $\frac{1}{9}$ **b.** $\frac{1}{2}$ **c.** $\frac{7}{12}$ **5.** 84

6. 125 repeated, 60 not repeated

Answers to Prep Skills Questions

1. 30, 120

2. $\frac{1}{9}$

3. $\frac{1}{4}$

4. $\frac{1}{6}$

5. $\frac{2}{3}$

3-2 Applications

Name ______________________________

1. Use the sample space for a standard deck of cards (refer back to page 229 if necessary) to answer each question.

   a. We know that there are 52 cards in a standard deck. Explain how you can use the fundamental counting principle to arrive at this same number.

   b. Before the cards are dealt for a game of poker, what's the probability that the 31st card in the deck is a spade?

   c. After 5 cards have been dealt to each of six players, with eight spades among those dealt, what's the probability that the 31st card in the deck is a spade?

2. Akira wants to buy his first new car and can select one option from each category:

Model	Engine Type	Color
Ford Edge	Hybrid	Burnt Copper
Honda Civic	Gasoline	Cobalt blue
Toyota Corolla		Metallic green

   a. Using the fundamental counting principle, how many outcomes should there be in the sample space?

3-2 Applications

Name ____________________

b. Draw a tree diagram and write out the sample space for all possible choices.

In parts c–g, find the probability that the car, if chosen at random,

c. Has a gasoline engine.

d. Is a burnt copper Ford Edge.

e. Is a metallic green hybrid Honda Civic.

f. Is cobalt blue.

g. Is either a metallic green or cobalt blue Toyota Corolla.

3-2 Applications

Name ______________________________

3. In between classes, Jade plays a game of online Monopoly on her laptop. Using the sample space for rolling two dice that you created in the Group portion of this lesson, find the probability that when Jade rolls the two dice, she gets a

a. Sum of 5.

b. Sum of 7 or 11.

c. Sum greater than 9.

d. Sum less than or equal to 5.

e. Three on one die or on both dice.

f. Sum that is odd.

g. Prime number on one or both dice.

h. Sum greater than 1.

3-2 Applications

Name ______________________________

While making up his schedule for spring semester, Tom complains that he doesn't have very many choices of schedule because of the general education requirements he has to meet. His advisor tells Tom that he has to take one course from each of English (three choices), history (five choices), math/stats (five choices), computer science (four choices), and general science (six choices). Does Tom have a legitimate gripe? Let's examine this in Questions 4–7.

4. If every possible course is available at the time he's registering, how many possible schedules can he choose from (disregarding when the classes meet)?

5. If Tom hasn't met the prerequisites for two of the math/stat courses and three of the general science courses, by how much does this reduce his number of possible schedules?

6. Good news! The prerequisite issue was a mistake on his transcript, so all courses are back in play. But when trying to schedule, Tom finds that all but one of the English courses is closed, as are two history courses and one general science course. How many schedules does he have to choose from now?

7. In an unprecedented effort to make the general education requirements more accessible, the dean of Tom's college decides to double the number of acceptable courses in each of those five areas. What effect does this have on the number of possible schedules?

Lesson 3-3 Prep Skills

SKILL 1: USE THE FUNDAMENTAL COUNTING PRINCIPLE

As we saw in Lesson 3-2, in a sequence of n procedures, if the first can occur in k_1 ways, the second in k_2 ways, the third in k_3 ways, and so on, the total number of ways the sequence of procedures can occur is $k_1 \cdot k_2 \cdot k_3 \cdot \cdots \cdot k_n$.

- How many different outfits can you put together if you have 3 pairs of pants, 5 shirts, and 2 pairs of shoes? (Assume a pair of pants is only one item and that you wear shoes as a matching pair.)
  $3 \cdot 5 \cdot 2 = 30$

  Multiply the number of ways you can pick pants, the number of ways you can pick shirts, and the number of ways you can pick shoes. You can put together 30 different outfits from these clothes.

SKILL 2: COMPUTE THEORETICAL PROBABILITIES

As we saw in Lesson 3-1, the probability of an event E in a sample space S occurring is given by

$$P(E) = \frac{\text{The number of outcomes in } E}{\text{The number of outcomes in } S} = \frac{n(E)}{n(S)}$$

- Find the probability that a randomly generated three-digit PIN consisting only of numbers will be 000.

$$P(\text{Code is } 000) = \frac{1}{10 \cdot 10 \cdot 10}$$
$$= \frac{1}{1{,}000}$$

There is only one way to have a PIN of 000. Using the fundamental counting principle, there are $10 \cdot 10 \cdot 10 = 1{,}000$ different three-digit PIN codes.

SKILL 3: DIVIDE OUT COMMON FACTORS

Dividing out common factors involves the same skill as reducing fractions. When you reduce $\frac{10}{15}$ to $\frac{2}{3}$, you divide the numerator and the denominator by 5. It can be helpful to think of $\frac{10}{15}$ as $\frac{2 \cdot 5}{3 \cdot 5}$, and then divide both the numerator and denominator by 5. Sometimes this is shown by crossing out the common factors $\frac{2 \cdot \cancel{5}}{3 \cdot \cancel{5}} = \frac{2}{3}$.

But DON'T THINK OF THIS AS JUST CROSSING OUT STUFF THAT LOOKS ALIKE. Think of it as what it is: division.

Write each fraction in reduced form.

- $$\frac{5 \cdot 3 \cdot 7}{5 \cdot 7} = \frac{\cancel{5} \cdot 3 \cdot \cancel{7}}{\cancel{5} \cdot \cancel{7}}$$
  $$= \frac{3}{1} = 3$$

  Divide numerator and denominator by the common factors of 5 and 7.

- $$\frac{4 \cdot 3 \cdot 2 \cdot 1}{5 \cdot 4 \cdot 3 \cdot 2 \cdot 1} = \frac{\cancel{4} \cdot \cancel{3} \cdot \cancel{2} \cdot \cancel{1}}{5 \cdot \cancel{4} \cdot \cancel{3} \cdot \cancel{2} \cdot \cancel{1}}$$
  $$= \frac{1}{5}$$

  Divide out the common factors of 4, 3, 2, and 1.

PREP SKILLS QUESTIONS

1. How many possible outcomes are there if you flip a coin then roll one six-sided die?
2. When eating at his campus cafeteria, Deshaun can choose from six sandwiches, four sides, and six drinks. If he wants a different combination every day, how many days can he eat without repeating a meal?
3. There are 38 spaces on a roulette wheel, and 18 of them are colored red. What's the probability of getting red on one spin?
4. There are also 18 black spaces on a roulette wheel. If you repeatedly bet on a space that is neither red nor black, what percent of the time will you be likely to win?
5. Write each fraction in lowest terms:

$$\frac{7 \cdot 6 \cdot 5 \cdot 4 \cdot 3 \cdot 2 \cdot 1}{3 \cdot 2 \cdot 1}$$

$$\frac{10 \cdot 9 \cdot 8 \cdot 7 \cdot 6 \cdot 5 \cdot 4 \cdot 3 \cdot 2 \cdot 1}{8 \cdot 7 \cdot 6 \cdot 5 \cdot 4 \cdot 3 \cdot 2 \cdot 1}$$

Lesson 3-3 Combining Forces

LEARNING OBJECTIVES

- ☐ 1. Understand how combinatorics are useful in probability theory.
- ☐ 2. Distinguish between permutations and combinations.
- ☐ 3. Find the number of permutations and combinations of *n* objects.
- ☐ 4. Find the number of permutations and combinations of *n* objects chosen *r* at a time.

Darren Greenwood/Design Pics

Most discoveries even today are a combination of serendipity and of searching.
—Siddhartha Mukherjee

Ah, the high school locker: providing students the thrill of fumbling with a combination lock for over 100 years. The combination to my locker was 3-24-32. If I had absentmindedly tried 24-3-32—a reasonably common occurrence given that school started at 7:30 and I am NOT a morning person—would that have opened the locker? Of course not. The point of a combination lock is that you need an exact string of numbers in the exact correct order. Anything other than that, and you're spinning that wheel at least twice to the right and trying again. In the last lesson, we learned that being able to count up possibilities is a really important part of computing probabilities. Now it's time to expand our counting repertoire. The key idea of this lesson is deciding whether or not order matters when counting the number of outcomes in an event or a sample space. And we'll eventually find out that combination locks are completely misnamed.

0. Write an example of something where order is crucial, and an example where order doesn't matter.

3-3 Group

Two couples go out to dinner, and as you might expect they are given a table for four.

1. How many different people could choose to sit in the first seat?

2. Once that person is seated, how many different people could sit in the second seat?

3. Then how many in the third seat?

4. And the fourth?

5. Now use the fundamental counting principle to calculate the number of ways that Cat, Dave, Nikki, and Brian can sit at that table.

6. Using Question 5 as a model, how many different ways could a group of 18 friends sit at a table with 18 chairs? Write out the calculation, but you don't need to perform it.

Dave Sobecki

Kind of annoying to write, wasn't it? (You can be honest.) Fortunately, there's a notation that will allow us to write calculations of this sort concisely: It's called *factorial notation,* and it plays a really important role in the next few counting methods we'll study.

Factorial Notation

The symbol we'll use to represent the computation you did in Question 5 is $4!$, which we read "Four factorial." To find $n!$, you simply multiply all of the whole numbers from n down to 1.

7. Fill in the blanks, using the first two lines for reference if you need to.

$1! = 1 = 1$

$2! = 2 \cdot 1 = 2$

$3! = \_\_\_ \cdot \_\_\_ \cdot \_\_\_ = \_\_\_$

$4! = \_\_\_ \cdot \_\_\_ \cdot \_\_\_ \cdot \_\_\_ = \_\_\_$

$5! = \_\_\_ \cdot \_\_\_ \cdot \_\_\_ \cdot \_\_\_ \cdot \_\_\_ = \_\_\_$

Math Note

The reason we advised against performing the calculation in Question 6? Eighteen factorial is 6,402,373,705,728,000.

Using Technology: Computing Factorials

The factorial command is on the MATH-PRB menu on a TI-84. To compute 5! easily:

5 MATH; then use arrow keys to select PRB and press 4 ENTER

In Excel, enter =FACT(5).

See the Using Tech video on Computing Factorials for further information.

8. Look back at your answers to Questions 5 and 6. Using factorial notation, fill in the blanks:

The number of ways to arrange 4 people around a table is _______.

The number of ways to arrange 18 people around a table is _______.

An arrangement of n distinct objects in a specific order is called a **permutation** of the objects.

9. In the case of the two couples meeting for dinner, each different way they could be seated is called a permutation, and we would say that the number of permutations of four objects (people) in this case is _______. (See Question 8.)

10. Now you should be able to fill in the blank in the colored box below, which provides our first counting formula in this lesson.

Permutations of n Objects

The number of permutations of n distinct objects using all of the objects is _______.

11. There are seven horses in a race and a large jackpot for anyone who can pick the finishing order of all seven exactly right. Find the number of possible orders in which the horses can finish.

12. What is the probability of picking the exact correct finishing order?

Did You Get It

Try this problem to see if you understand the concepts we just studied. The answer can be found at the end of the portfolio section.

1. A hospital floor nursing supervisor needs to assign five nurses to five different rotations. How many different ways can they be assigned?

So far, in calculating permutations, we've used all of the available objects. But what if only some of them are selected?

13. Most horse tracks offer a trifecta bet, in which the goal is to predict the exact order of the first three finishers. For the seven-horse race in Questions 11 and 12:

a. How many different horses could finish first?

b. Once you know which horse finished first, how many could finish second?

c. So we know who finished first and second. How many horses could finish third?

d. Use the fundamental counting principle to find the number of possible ways to pick the trifecta in a seven-horse race.

Now let's go back to our original two couples enjoying a night out. Suppose that the only available table when they arrive has six chairs. Here's how we calculate the number of possible seating arrangements:

6	·	5	·	4	·	3
Chairs the 1st person can choose		Chairs the 2nd person can choose		Chairs the 3rd person can choose		Chairs the 4th person can choose

This is not a factorial because we didn't continue down to one, but it kind of looks like one. If we're clever, we can figure out how to write this calculation in terms of factorials. First, we're going to multiply by one, because of course we can without changing the value.

$$6 \cdot 5 \cdot 4 \cdot 3 \cdot \frac{2 \cdot 1}{2 \cdot 1}$$

Okay, so we multiplied by one in a strange way, but that fraction is still a perfectly good way to say one. Now we'll do some very basic rewriting:

$$\frac{6 \cdot 5 \cdot 4 \cdot 3}{1} \cdot \frac{2 \cdot 1}{2 \cdot 1} = \frac{6 \cdot 5 \cdot 4 \cdot 3 \cdot 2 \cdot 1}{2 \cdot 1}$$

14. Presto! The numerator and denominator are both factorials now. Write this expression as a single fraction with a factorial in the numerator and another factorial in the denominator.

Now the key observation: The point of that calculation was to find the number of ways we can arrange 4 objects chosen from a set of 6. Notice that in the numerator you have 6!, and in the denominator, you have (6 – 4)!. You've just discovered our next important counting formula. Nice! Is it me, or are you getting smarter?

Permutations of *n* Objects Taken *r* at a Time

The arrangement of *n* objects in a specific order using *r* of those objects without replacement is called a **permutation** of *n* objects taken *r* at a time. It is written as $_nP_r$ and is calculated using the formula

$$_nP_r = \frac{n!}{(n-r)!}$$

15. An iPhone allows users to choose either a four- or six-digit lock code. How many four-digit codes are possible if you decide you don't want any number repeated?

16. How many more choices do you have if you DO allow numbers to be repeated? (Hint: Think fundamental counting principle.)

What if we're interested in a permutation of *n* objects taken *n* at a time, like the original question about four diners picking four chairs? Ideally, the above formula should still work, but we already know that for a permutation of *n* objects, the number of possibilities is *n*!. Let's see what happens if we try our new formula:

$$_nP_n = \frac{n!}{(n-n)!} = \frac{n!}{0!}$$

This presents us with a problem: What exactly does 0! even mean? In terms of how we defined factorials, it doesn't make sense. So . . .

17. In order to make the above formula provide the answer that we already know is correct (*n*!), what should we define 0! to be?

In case you didn't catch that, we've decided that 0! is defined to be 1. This means we don't have to know two different permutation formulas. Instead, we can always use the formula for finding a permutation of n objects chosen r at a time. If all of the objects are chosen, you'll get $n!$ / 0! and if you know that 0! = 1, you get our original permutation formula ($n!$).

Using Technology: Computing Permutations

The permutation command is on the MATH-PRB menu on a TI-84. To compute ${}_{10}P_4$:

10 MATH; then use arrow keys to select PRB and press 2 4 ENTER

In Excel, enter =PERMUT(10,4).

See the Using Tech video on Computing Permutations for further information.

Did You Get It

2. In a lottery game, 30 numbered ping-pong balls are put in a bin, and 5 are randomly chosen. To win the jackpot, you need to match all 5 numbers in the order in which they were drawn. How many different winning orders are there?

Summary of Permutations

These are the key elements that distinguish when we need to use a permutation formula:

- Finding the number of ways to arrange a certain number of objects.
- Order matters.
- Once an object is used, it can't be used again.

3-3 Class

Suppose that after waiting in line overnight, you manage to snag the last three tickets to a big concert. Sweet! The bad news is that you can only take two of your four housemates. How many different ways can you choose the two friends that get to go?

This sounds a little bit like a permutation problem, but there's a key difference: The order in which you choose two friends doesn't make any difference. They either get to go or they don't. So choosing Ruth and Ama is exactly the same as choosing Ama and Ruth.

When order matters in a selection, we call it a permutation, but when order is not important, we call the selection a *combination*.

People images/Getty Images

A selection of n distinct objects without regard to order is called a **combination** of the objects.

To clarify, let's study the difference between permutations and combinations.

1. Given four housemates, Ruth, Elaine, Ama, and Jasmine, list all the ways you can choose two of them to attend the concert with you. Is this a permutation or a combination? (Use R, E, A, and J to represent housemates.)

2. What if seating is assigned, and you're not only choosing who gets to attend, but also which of two seats each will sit in? List all the ways you can choose now. Is this a permutation or a combination?

3. There are _______ permutations, but only ______ combinations.

If all went well, the number of combinations you found was half the number of permutations. Here's why this makes perfect sense: After you picked your two guests, there were exactly two ways to arrange them. So by ignoring order, you essentially cut the number of possibilities in half. This provides an important fact:

> *The number of combinations of a certain number of objects is the number of permutations divided by the number of orders you can arrange those objects in.*

4. How many different ways can you order r objects?

Now we have a formula for finding a number of combinations.

Combinations of *n* Objects Taken *r* at a Time

The arrangement of n objects in any order using r of those objects without replacement is called a **combination** of n objects taken r at a time. It is written as ${}_nC_r$, and it is calculated using the formula

$$ {}_nC_r = \frac{{}_nP_r}{r!} = \frac{n!}{(n-r)!r!} $$

Notice that there are two ways to think of this formula: as the number of permutations divided by the number of orders, or "from scratch," as n factorial divided by the product of $(n - r)$ factorial and r factorial.

5. Use the combination formula to find the number of ways you can choose two of four housemates to go to the concert with you. Does this agree with the number of combinations you wrote out in Question 1?

Using Technology: Computing Combinations

The combination command is on the MATH-PRB menu on a TI-84. To compute ${}_4C_2$:

4 MATH then use arrow keys to select PRB and press 3 2 ENTER

In Excel, enter =COMBIN(4,2).

See the Using Tech video on Computing Combinations for further information.

Did You Get It

3. How many combinations of eight objects are there taken five at a time?

6. While studying abroad one semester, Tran is required to visit 10 different cities. He plans to visit 3 of the 10 over a long weekend. How many different ways can he choose the 3 to visit? Assume that distance isn't a factor.

Math Note

Some people use the terminology "10 choose 3" to describe the combination ${}_{10}C_3$, and it's also represented by $\binom{10}{3}$.

Did You Get It

4. An instructor posts a list of 12 group projects to her website. Every group is required to do four projects at some point during the semester. How many different ways can a group choose the four projects they want to do?

In some cases, it's useful to use the combination rule in conjunction with the fundamental counting principle. Let's look at some problems that illustrate specific situations.

Solved Example

Finding a Number of Possible Committees

At one school, the student government consists of seven women and five men. How many different committees can be chosen that consist of three women and two men?

SOLUTION

This problem pretty much has to have something to do with combinations since we're choosing a certain number of objects from a larger group, and the order doesn't matter. The thing that makes it a bit more complicated is that we're making TWO selections: three women and two men.

Let's focus on the ladies first. We need to choose three women from seven, and order doesn't matter, so this is ${}_7C_3$.

$${}_7C_3 = \frac{7!}{(7-3)!3!} = 35$$

There are 35 ways to choose the three women on the committee. Next, we choose two men from five, which is ${}_5C_2$.

$${}_5C_2 = \frac{5!}{(5-2)!2!} = 10$$

There are 10 ways to choose the two men on the committee. Now we think: Once the women have been chosen, does this have ANY effect on which men are chosen? Nope. This is exactly the situation that we developed the fundamental counting principle for: a sequence of two procedures that don't affect each other. So we can use the fundamental counting principle and multiply the 35 ways to choose the women and the 10 ways to choose men to get 350 total committees that can be chosen.

Here's the next problem to solve:

To raise money for a charity event, a campus honor society plans to sell a calendar featuring tasteful photos of some of the more popular professors on campus. They will need to choose six models from a pool of finalists that includes nine women and six men. How many possible choices are there if they want to feature at least four women?

7. How many different choices could they make if they decide on exactly four women? Note that this question is very similar to the Solved Example.

8. So four women and two men is one possible composition of the calendar. There are two others: Find the number of different choices for each of those. Then decide what to do with your results to solve the calendar problem, and solve it.

Did You Get It

5. A four-person crew for the international space station is to be chosen from a candidate pool of 10 Americans and 12 Russians. How many different crews are possible if there must be at least 2 Russians?

This is a good time to talk about the significance of the words *and* and *or* when applying counting techniques. In the Solved Example, we needed a committee with three women AND two men; this led us to conclude that we could think of it as a two-stage process and multiply using the fundamental counting principle. In the calendar problem, we end up choosing between three possible options: four women OR five women OR six women. In this case, we found the number of possibilities for each and added them.

In using counting to continue our study of probability, of course it's important to know the formulas. But the formulas do no good unless you know *which* formula to use in a given situation. In the Applications, you'll be asked to identify whether a given problem matches the fundamental counting principle, the permutation formula, or the combination formula. In preparation for that, you should end this lesson by reviewing the work we've done. Make sure you're clear on exactly what each formula is used for.

9. Oh, one final question: Why did we say earlier that combination locks are misnamed?

3-3 Portfolio

Name ______________________________

Check each box when you've completed the task. Remember that your instructor will want you to turn in the portfolio pages you create.

Technology

1. ☐ In this Excel activity, we'll calculate the probabilities of winning a few different games of chance.

ALEKS Practice

1. ☐ Include any written work from the ALEKS assignment along with any notes or questions about this lesson's content.

Applications

1. ☐ Complete the applications problems.

Reflections

Type a short answer to each question.

1. ☐ What is the difference between permutations and combinations?
2. ☐ Why are we studying these counting techniques in this unit?
3. ☐ Describe how to compute a factorial.
4. ☐ Name one thing you learned or discovered in this lesson that you found particularly interesting.
5. ☐ What questions do you have about this lesson?

Looking Ahead

1. ☐ Complete the prep skills for Lesson 3-4.
2. ☐ Read the opening paragraph in Lesson 3-4 carefully and answer Question 0 in preparation for that lesson.

Answers to "Did You Get It?"

1. 120 **2.** 17,100,720 **3.** 56 **4.** 495 **5.** 5,665

Answers to Prep Skills Questions

1. 12 **2.** 144 **3.** $\frac{9}{19}$ or about 0.47

4. About 5.3% **5.** 840, 90

3-3 Applications

Name ____

In preparation for this assignment, it would be a good idea for you to review the three major counting rules that we've studied.

Rule	Description	Formula
Fundamental counting principle	The number of ways a sequence of *n* events can occur. Each event is unaffected by the others. The first event can occur in k_1 ways, the second in k_2 ways, and so on.	$k_1 \cdot k_2 \cdot k_3 \cdot \cdots \cdot k_n$
Permutation rule	The number of ways we can choose *r* objects from a set of *n*, with order important.	${}_nP_r = \frac{n!}{(n-r)!}$
Combination rule	The number of ways we can choose *r* objects from a set of *n*, without regard to order.	${}_nC_r = \frac{n!}{(n-r)!r!}$

For Questions 1–4, decide if each selection is a permutation or a combination, and briefly explain your choice.

1. From a class of 25 students, a group of 5 is chosen to give a presentation.

2. A starting pitcher and catcher are picked from a 12-person intramural softball team.

3. A 5-digit passcode is chosen from the numbers 0 through 9.

4. A gardener picks 4 vegetable plants for a garden from 10 choices.

3-3 Applications

Name ______________________________

For Questions 5–7, write the appropriate counting method to use, then answer the question.

5. A campus pizzeria offers regular crust, thin crust, or pan pizzas. You can get either white or red sauce. The owner is kind of eccentric, and only sells pizzas with one topping, chosen from pepperoni, sausage, ham, onions, and ground beef. Determine the number of possible choices of pizza.

6. Suppose that a chef is choosing from 20 toppings to make a gourmet pizza, and he plans to choose 6 of them. How many different ways can he do so if you're keeping track of the order in which he adds them?

7. Let's be real, though . . . when you're putting toppings on a pizza, it doesn't really matter what order you choose. So how many different pizzas can the chef make?

For the remaining questions, find the number requested and show your work.

8. A major league baseball team has 25 players on the active roster. How many choices does a manager have for batting order, listing the nine starters from 1 through 9?

9. In 5-card poker, each player is dealt 5 cards (go figure) from a standard deck of 52 cards. How many different hands can be dealt?

10. At the movies, Shana wants to get snacks for her friends. How many ways can she choose three types of candy and two types of soda from the eight types of candy and five types of soda available?

3-3 Applications

Name ______________________________

11. Steve wants to download new music into his phone from iTunes. How many ways can Steve pick 2 rock songs, 3 alternative songs, and 3 hip-hop songs from a list of 8 rock songs, 6 alternative songs, and 10 hip-hop songs?

12. A college basketball team has 14 women on the roster. In how many ways can the coach choose a lineup featuring five different positions?

13. An inspector with the Nuclear Regulatory Commission is tasked to visit five nuclear plants this month, randomly chosen from three in Ohio, four in New York, and five in Pennsylvania. How many different ways can he choose the plants to visit if at least three will be in Pennsylvania?

14. For the new fall season, a network president has 11 shows in development, and six openings in the prime-time schedule. In how many ways can she arrange new shows to fit into the schedule?

15. Think about the definitions of what ${}_nP_r$ and ${}_nC_r$ actually mean (NOT the formulas for computing them). Explain why both have to be zero when r is greater than n.

16. Now look at the formulas for ${}_nP_r$ and ${}_nC_r$. Can you use either of them if r is bigger than n? Why or why not?

Lesson 3-4 Prep Skills

SKILL 1: COMPUTE COMBINATIONS

As we saw in Lesson 3-3, the arrangement of *n* objects in any order using *r* of those objects without replacement is called a combination of *n* objects taken *r* at a time. It is written as ${}_nC_r$ and is calculated using the formula

$$ {}_nC_r = \frac{{}_nP_r}{r!} = \frac{n!}{(n-r)!r!} $$

- Find the number of combinations of 9 objects taken 2 at a time.

$$ {}_9C_2 = \frac{9!}{(9-2)!2!} $$

Substitute 9 and 2 into the formula for computing combinations and simplify the result.

$$ = \frac{9 \cdot 8 \cdot 7!}{7! \cdot 2!} $$

9! can be written as $9 \cdot 8 \cdot 7!$, allowing us to divide out 7!.

$$ = \frac{9 \cdot 8}{2 \cdot 1} $$

$$ = 36 $$

There are 36 combinations of 9 objects taken 2 at a time.

SKILL 2: COMPUTE PERMUTATIONS

As we saw in Lesson 3-3, the arrangement of *n* objects in a specific order using *r* of those objects without replacement is called a permutation of *n* objects taken *r* at a time. It is written as ${}_nP_r$ and is calculated using the formula

$$ {}_nP_r = \frac{n!}{n-r!} $$

- Find the number of permutations of 11 objects taken 3 at a time.

$$ {}_{11}P_3 = \frac{11!}{(11-3)!} $$

Substitute 11 and 3 into the formula for computing permutations and simplify the result.

$$ = \frac{11!}{8!} $$

$$ = \frac{11 \cdot 10 \cdot 9 \cdot 8!}{8!} $$

11! can be written as $11 \cdot 10 \cdot 9 \cdot 8!$, allowing us to divide out 8!.

$$ = 11 \cdot 10 \cdot 9 $$

$$ = 990 $$

There are 990 permutations of 11 objects taken 3 at a time.

SKILL 3: USE THE FUNDAMENTAL COUNTING PRINCIPLE

As we have seen in previous lessons, in a sequence of n procedures, if the first can occur in k_1 ways, the second in k_2 ways, the third in k_3 ways, and so on, the total number of ways the sequence of procedures can occur is $k_1 \cdot k_2 \cdot k_3 \cdot \cdots \cdot k_n$.

- A box lunch is put together by selecting a sandwich, a side, a cookie, and a drink. If there are 5 options for sandwiches, 4 types of sides, 2 types of cookies, and 3 drinks, how many different box lunches can be created?
  $5 \cdot 4 \cdot 2 \cdot 3 = 120$

  There are 120 different box lunches that can be created from these available options.

SKILL 4: COMPUTE THEORETICAL PROBABILITY

As we have seen in previous lessons, the probability of an event E in a sample space S occurring is given by $P(E) = \dfrac{\text{The number of outcomes in } E}{\text{The number of outcomes in } S} = \dfrac{n(E)}{n(S)}$

- There are 20 marbles in a bag and 7 of them are red. Find the probability of drawing a red marble from this bag.
  $P(\text{Red}) = \dfrac{7}{20} = 0.35$

  There are 7 red marbles in the bag out of 20 total marbles. The probability of drawing a red marble is 7/20.

PREP SKILLS QUESTIONS

1. Find each combination.

   The number of combinations of 11 objects taken 4 at a time.

   ${}_8C_3$

2. In how many different orders can 5 distinct objects be arranged?

3. Find the number of permutations of 12 objects taken 9 at a time.

4. Compute ${}_7P_3$.

5. If you have 7 choices of entree, 5 choices of side, and 3 choices of dessert, how many different meals could you have?

6. A tech company has 144 employees, 90 of whom have a bachelor's degree. There are 32 with master's degrees and 4 PhD's. What's the probability that a randomly selected employee has a bachelor's?

7. What's the probability that a randomly selected employee has an associate's or no college degree?

Lesson 3-4 Too Good to Be True?

LEARNING OBJECTIVES

- ☐ 1. Recognize probability problems where permutations are useful and where combinations are useful.
- ☐ 2. Use permutations to calculate probabilities.
- ☐ 3. Use combinations to calculate probabilities.

George Doyle/Stockbyte/Getty Images

If you're playing a poker game and you look around the table and can't tell who the sucker is, it's you.

—Paul Newman

Sometimes a friendly game of poker can become less friendly when someone seems just a bit too lucky. Suppose one player in such a game gets dealt all four aces in one hand. Would you suspect that something fishy was going on? What is the probability of that happening?

Initially, we used tree diagrams and tables to find sample spaces and the number of outcomes in certain events. This was a pretty good strategy, but when the number of possibilities gets larger, diagrams can get out of hand. Fortunately, we know about counting techniques that are tailor-made for answering questions about probability. If our job is to find how many ways something can happen, the combination and permutation rules from Lesson 3-3 will be our best friends. Our general game plan will be to use these rules to find the number of outcomes that satisfy a certain event, as well as the total number of outcomes in the sample space. Then we can divide the first number by the second to obtain the probability of the event occurring.

0. Make a guess as to the probability of drawing four aces in a five-card hand without doing any calculations. We'll find out later how well you did.

3-4 Class

So now we're pretty good at counting up outcomes, and we have a good understanding of how to compute probabilities. That means we're ready to combine those skills to answer more complicated questions involving probability. Just to make sure you're ready to go, let's do a quick review.

1. Describe how to find the probability of an event based on number of outcomes.

2. When will we need to use permutation rules? What about combination rules?

Marc Romanelli/Blend Images LLC

Solved Example 1

Using Counting to Compute a Probability

Like most math classes, you probably got one choice of book in this course. Sorry. But in other areas, there's sometimes a book list, and students have to choose which ones they want to read. Let's say that a student named Kalani is taking a humanities class with a reading list that contains 10 biographies and five books on current events. She's required to read three books of her choice at some point during the semester. Being a fan of probability, Kalani decides to roll the dice and pick her three books randomly. What she doesn't realize is that her professor won't look very kindly on a student who totally avoids biographies. So how likely is it that Kalani will inadvertently hurt her grade? Find the probability that she gets no biographies.

SOLUTION

We know that we'll need to count outcomes, so our first task is to decide if we're dealing with permutations or combinations. The key question, of course, is "Does order matter?" The statement of the problem simply refers to the total number of books Kalani is choosing—it says nothing about the order in which she reads them. So we're looking at combinations, not permutations.

In order to get no biographies, Kalani would have to end up with three books on current events, and there are five to choose from. The number of ways to choose three books from five with order unimportant is ${}_5C_3$, which is 10. You can get this using the nCr command on a calculator or by using the formula 5!/(3! 2!).

Now for the entire sample space. Ultimately, we're choosing 3 books from 15 total, again with no interest in the order. This can be done in ${}_{15}C_3$ ways, which works out to be 455 ways. Now it's a simple probability calculation: There are 455 outcomes in the sample space, and just 10 of them result in three books on current events, so the probability of getting no biographies is

$$P(\text{No biographies}) = \frac{10}{455} \approx 0.022$$

3. Discuss how likely you think it is that Kalani won't have to worry about reading all current events books and getting downgraded by her instructor.

Did You Get It ?

Try this problem to see if you understand the concepts we just studied. The answer can be found at the end of the portfolio section.

1. There are 12 women and 8 men in a seminar course. If the professor chooses five-person groups at random, what is the probability that the first group chosen will consist of all women?

Now back to that poker game . . . In Questions 4–9, our goal is to find the probability of getting four aces when drawing 5 cards from a standard deck of 52 cards.

Pixtal/SuperStock

4. First, we need to know how many five-card hands have four aces in them. The key observation is that there are only four aces in the deck, so we need to get them all. How many ways can this happen?

5. After we have all four aces, we still need one more card, right? How many ways can we choose a fifth card once we have all four aces?

6. How many five-card hands have all four aces?

7. Will we be using permutations or combinations to find the total number of five-card hands? Why?

8. The total number of five-card hands comes from choosing __________ cards from __________ possibilities.

9. Find the number of five-card hands, and the probability of getting four aces when drawing five cards from a standard deck. Write your answer as a fraction.

Chances are you would prefer a decimal answer to Question 9 in order to get an idea of how small the probability of drawing 4 aces is. Computing that decimal answer will likely raise a key point about rounding: You can lose a lot of important information if you restrict yourself to rounding to a certain number of decimal places. To deal with this issue, sciencey-type folks developed the idea of *significant digits*. In short, strings of zeroes that act as placeholders are not considered to be significant because they don't provide specific information. For example, a typical grain of sand weighs about 0.004401 grams. If we round that to two decimal places, we get 0.00, and we conclude that sand is weightless. Of course, that's silly. The zeroes at the beginning of 0.004401 are not significant—they're just needed as placeholders because the number is so small.

In science classes, you'd probably get pretty in-depth on the concept of significant digits, but for our purposes we'll just call any digits (zero or non-zero) that come after an initial string of zeroes **significant digits.** So the weight of a grain of sand is 0.0044 rounded to two significant digits and 0.00440 rounded to three significant digits.

Solved Example 2

Rounding to a Given Number of Significant Digits

Round each number to the given number of significant digits.

a. 0.000916; two significant digits

b. 0.008603; three significant digits

c. 2.513; two significant digits

SOLUTION

a. The three zeroes after the decimal point are placeholders and are not significant. So the first significant digit is the 9, and the second is 1. The next digit after 1 is more than 4, so we round up and get 0.00092.

b. The first three significant digits are 860. The next digit is less than 5, so we round down and get 0.00860. Remember, zeroes CAN BE significant digits if they come after nonzero digits.

c. Since there are no zeroes to act as placeholders, all displayed digits are significant, and we need two digits. The number rounds to 2.5.

10. Find a decimal equivalent for the probability you found in Question 9 rounded to two significant digits.

11. Now the key question: If your friend comes up with four aces, is she cheating? Discuss, using the probability you found as evidence.

Did You Get It ?

2. Suppose the deck of cards in Questions 4–9 has all 32 cards with numbers less than 10 removed, so that only 10s, jacks, queens, kings, and aces remain. Now what is the probability of getting 4 aces when drawing 5 cards?

Sidelight **Unlikely Events Are Very Likely to Occur**

When the probability of some event is low, like say 0.05, it's common and reasonable to guess that event most likely will not occur. But if you look at it a certain way, events that are incredibly unlikely happen all the time. Let's use poker as an example. When you're dealt five cards, there's just a 1 in 13 chance that the first card you are dealt will be a 5, because there are 13 denominations. That's a probability of about 0.077. Now think about this: regardless of what denomination your first card ends up being, there was the same probability—just 0.077—of getting it. But of course the first card has to be *something,* so every time that first card is dealt, something with probability 0.077 is *guaranteed to happen.*

Now let's extend that out to all five cards. As we saw in Question 9, there are 2,598,960 possible hands you can be dealt. That means the probability of getting any particular hand is 0.00000038. Yet every single time that you are dealt five cards, you have to get one possible hand, so something with that spectacularly low probability will occur with probability 1. Wild!

Finally, let's talk about the most unlikely lottery win ever, which happened to you even if you don't realize it. That's the genetic lottery, which resulted in you being born as you. Genetics tells us that the number of possible DNA combinations, which determine all of your characteristics and make you who you are, is on the order of 3 followed by 614 zeros. This is far larger than the total number of stars in the universe. Of all those possible combinations, only one of them could have resulted in you, an event so unlikely that the probability is for all intents and purposes zero. And yet somehow that happened. And continues to happen thousands of times across the world every single day. Next time you're feeling a little down, think about how incredibly lucky you were to have ever been born, and maybe you'll feel just a little special.

3-4 Group

For the next few probability questions, we're going to provide a little less guidance than we did on the last two. Fortunately, you have group mates to work with, so we're sure you'll be just fine.

F Schussler/PhotoLink/
Photodisc/Getty Images

This lock has 40 numbers on it, from 0 to 39. In Questions 1–4, our goal is to find the probability that if the combination to unlock it consists of 3 numbers, it will contain the numbers 1, 2, and 3 in some order. Assume that numbers can't be repeated in the combination.

1. Does the order matter? Do we want to use permutations or combinations?

2. How many different combinations for the lock contain 1, 2, and 3?

3. How many different combinations are there total?

4. Write the probability that the combination contains 1, 2, and 3 in some order as a fraction, and as a decimal rounded to three significant digits.

Did You Get It

3. A different "permutation" lock has letters from A through L on it, and the combination consists of four letters with no repeats. What is the probability that the combination is I, J, K, and L in some order?

Ah, politics. It used to be that the easiest way to get in a fight with someone was to make fun of their mom. Now, all you have to do is disagree with someone on a political issue or candidate. We've become a frighteningly polarized society in that regard, and the makeup of committees among political bodies has become a very common issue.

5. In one town, the city council consists of seven Republicans and three Democrats. A committee of three is being selected by random draw to revise the town's election laws. Based on the overall composition of the council, what would be a reasonable makeup for the committee?

6. Are we dealing with permutations or combinations here? Why?

7. How many different ways can the Republican(s) on the committee be chosen, based on the composition you chose?

8. How many different ways can the Democrat(s) be chosen?

9. Look back at your answer to Question 5. Does it contain the word "and" or "or"? Does that word indicate that you should multiply or add to find the number of committees that match the composition you specified?

10. Find the number of ways a committee can be chosen that matches your composition.

11. Find the probability that the committee will work out the way you think it should.

Did You Get It

4. A store has six different fitness magazines and three different news magazines. Find the probability that a customer picking three magazines randomly gets at least two fitness magazines.

12. You're in a room with 29 other people and someone says "I'll bet you fifty bucks that two people in this room have the same birthday." Would you take the bet? Why or why not?

3-4 Class (Again)

The goal of the remainder of this lesson, and of the tech assignment, is to study this question to see if you made a wise decision in either taking, or turning down, the bet. It's not a particularly simple question, so we'll start out with a case that's simpler and see if we can generalize.

1. It turns out it's easier to first find the probability that everyone in the room has a different birthday, so we'll begin there. If there are only three people in the room, how hard can it be? In that case, there are 365 choices for the first person's birthday (assuming it's not a leap year). If the second person has a different birthday, there are 364 choices, then 363 for the third. So the number of ways that all three can have different birthdays is

_______________. (Write as a product, and don't multiply it out.)

2. The total number of possible birthdays for the three people, on the other hand, is

_______________, or ______.

3. Use your answers from the last two questions to find the probability that all three people have different birthdays. Find a decimal equivalent rounded to three significant digits.

4. What is the probability that at least two of the people have the SAME birthday? (Hint: This event is the complement of everyone having a different birthday.)

5. Look carefully at the calculation of the number of ways that all three can have different birthdays (Question 1). If it doesn't look familiar, you probably need to review permutations. That was a hint. Write that calculation as a permutation in the form ${}_nP_r$.

6. Summary so far: With three people in the room, the probability that at least two have the same birthday can be written as: (Fill in the blanks.)

$$1 - \frac{{}_{365}P_{\_}}{365^{\_}}$$

7. Now here's the key to the entire problem: We can repeat all of the steps we've done for any other number of people in the room, and the only thing it changes is where the number of people appeared in your answer to Question 6. If there are 30 people in the room, you should now be able to find the probability that at least 2 have the same birthday. That's the question we started this activity with. Find the probability now.

8. Now would you take the bet? Why or why not?

9. In the technology portion of the homework, you'll use Excel to study the probability that at least two people in a room have the same birthday for many different room populations. In order to do so, you'll want to have a general formula for that probability when there are k people in the room. Write that here, and save it for reference when you start the tech assignment.

3-4 Portfolio

Name ______________________________

Check each box when you've completed the task. Remember that your instructor will want you to turn in the portfolio pages you create.

Technology

1. ☐ In this Excel activity, we'll use the formula we developed at the end of this lesson to examine the probability that two people in a room have the same birthday based on a varying number of people in the room.

ALEKS Practice

1. ☐ Include any written work from the ALEKS assignment along with any notes or questions about this lesson's content.

Applications

1. ☐ Complete the applications problems.

Reflections

Type a short answer to each question.

1. ☐ Now that we've completed this lesson, reflect back on why we spent so much time in the early parts of this unit learning to count. Write some of your thoughts.
2. ☐ What was the point of defining significant digits in this lesson? How did understanding the concept help us to study probability?
3. ☐ Discuss the birthday problem we closed the lesson with. What do you find surprising about it? Why?
4. ☐ Name one thing you learned or discovered in this lesson that you found particularly interesting.
5. ☐ What questions do you have about this lesson?

Looking Ahead

1. ☐ Complete the prep skills for Lesson 3-5.
2. ☐ Read the opening paragraph in Lesson 3-5 carefully and answer Question 0 in preparation for that lesson.

Answers to "Did You Get It?"

1. $\frac{33}{646} \approx 0.051$ **2.** $\frac{1}{969} \approx 0.00103$ **3.** $\frac{1}{495} \approx 0.002$ **4.** $\frac{65}{84} \approx 0.77$

Answers to Prep Skills Questions

1. 330, 56
2. 120
3. 79,833,600
4. 210
5. 105
6. 0.625
7. 0.125

3-4 Applications

Name ______________________________

1. In a math class with 18 students, there are 11 majors and 7 non-majors. Four students are randomly picked to present a demonstration on the use of graphing calculators. Find the probability that the group consists of

a. All non-majors.

b. Two majors and two non-majors.

2. A chef is choosing from 12 different entrees for an important banquet, 3 of which contain spinach. The guests will have a choice of 4 entrees. If the chef chooses those 4 at random, find the probability that

a. At least one has spinach. (Hint: Complements will help.)

Purestock/SuperStock

b. Three have spinach.

c. All four have spinach.

3-4 Applications

Name ______________________________

3. An engineering company has four openings, and the applicant pool consists of six database administrators and eight network engineers. If the hiring is done without regard for the specific qualifications of the applicants, find the probability that the four hired will be

   a. All network engineers.

   b. Three database administrators and one network engineer.

In one lottery game, contestants pick five numbers from 1 through 40 and have to match all five for the big prize (in any order). Questions 4–6 refer to this game. Round your answers to two significant digits.

4. What's the probability you'll win if you buy one ticket? (Hint: Your chances are NOT good.)

5. You'll get twice your money back if you match three of the five numbers. If you buy two tickets, what's the probability of matching three out of five numbers?

3-4 Applications

Name ______________________________

6. To celebrate Fourth of July week, the lottery commission is offering a $1 million bonus if a winner of the big prize matches the numbers in the order in which they were drawn. What's the probability of this happening if you buy one ticket?

Think you can guess your way into someone's garage by guessing the garage door code? Well let's just see about that.

David Sobecki

7. What is the probability that you can guess my four-digit garage code in 10 random guesses if repeat digits are not allowed? Give your answer as a fraction.

8. What is the probability that you can guess my four-digit garage code in 10 random guesses if repeat digits are allowed? Write as a fraction.

9. If you don't have any repeated numbers in your four-digit code, and someone suspects that, how much easier are you making it for them to guess the code?

10. Let's go back to no repeats. What's the probability of guessing a six-digit code in 10 tries? What does that suggest about security?

Lesson 3-5 Prep Skills

SKILL 1: DIVIDE FRACTIONS AND REDUCE THE RESULT

To divide two fractions, multiply the first fraction by the reciprocal of the second. $\frac{a}{b} \div \frac{c}{d} = \frac{a}{b} \cdot \frac{d}{c} = \frac{ad}{bc}$

- $\frac{2}{3} \div \frac{1}{2} = \frac{2}{3} \cdot \frac{2}{1} = \frac{4}{3}$

  Rewrite the division problem as a multiplication problem. Then multiply. This result is already simplified.

- $\frac{4}{15} \div \frac{2}{5} = \frac{4}{15} \cdot \frac{5}{2}$

  $= \frac{\cancel{4}^{2}}{\cancel{15}_{3}} \cdot \frac{\cancel{5}^{1}}{\cancel{2}_{1}} = \frac{2}{3}$

  Rewrite the division problem as a multiplication problem. Then multiply. When multiplying fractions, it is often helpful to reduce by dividing away common factors before actually multiplying.

SKILL 2: COMPUTE THEORETICAL PROBABILITIES

As we have seen in previous lessons, the probability of an event E in a sample space S occurring is given by

$$P(E) = \frac{\text{The number of outcomes in } E}{\text{The number of outcomes in } S} = \frac{n(E)}{n(S)}$$

In a large bowl in the break room, there are 12 Snickers, 5 Milky Ways, 11 Kit Kats, and 8 Twix candy bars.

- If a candy bar is randomly chosen from the bowl in that break room, find the probability that it is a Snickers.

  $P(\text{Snickers}) = \frac{12}{36} \approx 0.33$

  Out of the 36 candy bars in this bowl, there are 12 Snickers. The probability of drawing a Snickers is 12/36.

- If a candy bar is randomly chosen from the bowl in that break room, find the probability that it is a either a Kit Kat or a Twix.

  $P(\text{Kit Kat or Twix}) = \frac{11+8}{36} = \frac{19}{36} \approx 0.53$

  Out of the 36 candy bars in this bowl, there are 11 Kit Kats and 8 Twix. The probability of drawing either a Kit Kat or a Twix is 19/36.

SKILL 3: CALCULATE A SUM OF PRODUCTS

When calculating expected value, a topic you'll cover in this lesson, you'll need to simplify expressions containing a sum of products like $6 \cdot \frac{1}{5} + 5 \cdot \frac{3}{5} + 2 \cdot \frac{1}{5}$. Simplifying an expression like this one amounts to following order of operations. First multiply, then add fractions with common denominators.

- $6 \cdot \frac{1}{5} + 5 \cdot \frac{3}{5} + 2 \cdot \frac{1}{5}$

  Multiply.

  $= \frac{6}{5} + \frac{15}{5} + \frac{2}{5}$

  Note that we don't want to reduce $\frac{15}{5}$ because we want to keep the common denominator.

  $= \frac{23}{5} = 4.6$

  Add fractions with common denominators. Now that we have an exact fraction, we can divide to get a decimal equivalent or approximation.

PREP SKILLS QUESTIONS

1. Perform the division. Make sure your answer is in lowest terms.

$$\frac{\frac{1}{5}}{\frac{3}{5}}$$

$$\frac{\frac{8}{13}}{\frac{10}{13}}$$

2. What is the probability of rolling a total of 6 with two dice? What about a total greater than 6?

3. There are 2 seniors, 1 junior, 7 sophomores, and 11 freshmen in a math course. If one student is randomly picked for extra credit, what's the probability that it's either a freshman or a senior?

4. Find the value of the following expression in both fractional and decimal forms:

$$-10 \cdot \frac{4}{9} + 2 \cdot \frac{2}{9} + 4 \cdot \frac{2}{9} + 15 \cdot \frac{1}{9}$$

Lesson 3-5 Odds and Ends

LEARNING OBJECTIVES

- ☐ 1. Distinguish between odds and probability.
- ☐ 2. Compute and interpret the odds in favor of and odds against an event.
- ☐ 3. Compute odds from probability and vice versa.
- ☐ 4. Develop a procedure for finding expected value.
- ☐ 5. Compute and interpret expected values.

Tammy Ljungblad/Kansas City Star/Tribune News Service via Getty Images

Think of success as a game of chance in which you have control over the odds. As you begin to master concepts in personal achievement, you are increasing your odds of achieving success.

—Bo Bennett

The Kansas City Chiefs won the Super Bowl on February 11, 2024, and by the time the ink dried on newspapers reporting the victory, oddsmakers in Las Vegas had listed the odds against the Chiefs winning it again in 2025 as 15 to 2. But what exactly does that mean? The term "odds" is used all the time in describing the likelihood of something happening, but a lot of people don't understand exactly what a given set of odds means. Odds are used by casinos, racetracks, and other gambling establishments to determine the payoffs when bets are made or lottery tickets are purchased. They're also used by insurance companies in determining the amount to charge for premiums. In a very real sense, odds are a way of stating what you expect will happen. So in this lesson, we'll study odds and expectations.

0. What exactly do you think it means to say that the odds against the Chiefs winning the Super Bowl are 15 to 2?

3-5 Class

The formulas for computing odds are similar to the formula we've been using for theoretical probability, and shortly we will see a strong connection between the two concepts.

If an event E has a favorable outcomes and b unfavorable outcomes, then

- The **odds in favor** of event E occurring are $\frac{a}{b}$ (also written as $a:b$).
- The **odds against** event E occurring $\frac{b}{a}$ (also written as $b:a$).

Don't think of favorable as "good" and unfavorable as "bad." We're describing the odds of some event occurring, so we call any outcome that satisfies that event favorable and any outcome that doesn't satisfy that event unfavorable. If we were studying the odds of getting a sharp stick in the eye while walking in the woods, we would call getting a sharp stick in the eye favorable even though clearly it would not be a good thing.

Odds can be expressed as a fraction or a ratio.

1. Write the odds against Kansas City repeating as Super Bowl champion as a fraction, and as a ratio.

2. Write the odds in favor of Kansas City repeating as Super Bowl champion as a fraction, and as a ratio.

> **Math Note**
>
> In common usage, the phrase "the odds of" means the same thing as "the odds against." Odds are also commonly written with a hyphen rather than a colon: 8–1, for example.

3. By setting Kansas City's odds at 15 to 2, the oddsmakers were predicting that if the season were played 17 times, Kansas City would win the Super Bowl ______ times, and not win it _______ times.

Did You Get It

Try this problem to see if you understand the concepts we just studied. The answer can be found at the end of the portfolio section.

1. What are the odds in favor of rolling a prime number sum with a roll of two dice? What are the odds against? The table showing possible rolls of two die is reproduced here for reference.

	1	2	3	4	5	6
1	(1, 1)	(1, 2)	(1, 3)	(1, 4)	(1, 5)	(1, 6)
2	(2, 1)	(2, 2)	(2, 3)	(2, 4)	(2, 5)	(2, 6)
3	(3, 1)	(3, 2)	(3, 3)	(3, 4)	(3, 5)	(3, 6)
4	(4, 1)	(4, 2)	(4, 3)	(4, 4)	(4, 5)	(4, 6)
5	(5, 1)	(5, 2)	(5, 3)	(5, 4)	(5, 5)	(5, 6)
6	(6, 1)	(6, 2)	(6, 3)	(6, 4)	(6, 5)	(6, 6)

4. Based on the answer to Question 3, write the probability of Kansas City winning the Super Bowl and the probability of Kansas City not winning the Super Bowl in fraction form. Then compare your answers with the formula below. Careful! Since this is a question about probability, think about the total number of outcomes.

P(winning) =

P(not winning) =

Formula for Probability in Terms of Odds

If the odds in favor of an event E are $a{:}b$, then the probability that the event will occur is

$$P(E) = \frac{a}{a+b}$$

This formula makes perfect sense based on our probability formulas. Remember that a and b represent the number of favorable and unfavorable outcomes, so $a + b$ is the TOTAL number of outcomes, and we're back to number of outcomes that satisfy an event divided by the total number of outcomes.

5. Fill in the blanks to make this a true statement: P(winning) = ______ – P(____________).

6. Explain why your answer to Question 5 makes perfect sense based on a fact about probability that we studied earlier in this unit.

7. For the probabilities in Question 4, compute and simplify $\frac{P(\text{winning})}{P(\text{not winning})}$. What does your answer represent in terms of odds?

8. Now compute $\frac{P(\text{not winning})}{P(\text{winning})}$. What does your answer represent in terms of odds?

The results of Questions 7 and 8 can be summarized in these formulas:

Formulas for Odds in Terms of Probability

If the probability that an event E will occur is $P(E)$, then

$$\text{Odds in favor of } E = \frac{P(E)}{1 - P(E)} \text{ and Odds against } E = \frac{1 - P(E)}{P(E)}$$

Now that we've developed formulas relating odds and probability, it's worth thinking about whether the formulas are actually the best approach. Let's look at a couple of examples.

Solved Example

Relating Odds and Probability

a. According to the National Safety Council, the odds in favor of dying due to injury at some point in your life are about 10:237. Find the probability of dying from injury.

b. The probability of getting exactly one pair in a five-card poker hand is 0.423. Find the odds in favor of getting exactly one pair, and the odds against.

SOLUTION

a. First let's use the formula on page 280. With odds of 10:237, we should use the formula with $a = 10$ and $b = 237$:

$$P(\text{Dying}) = \frac{10}{10 + 237} = \frac{10}{247}$$

But what if you forgot that formula? Or never learned it in the first place? If you know what odds really mean, you should be able to figure out the probability anyhow. Since the odds are 10:237, we know that means there are 10 favorable outcomes and 237 unfavorable. Common sense then dictates that there are 247 total outcomes. Ten of them represent dying, so we can just use our formula for theoretical probability to get that the probability of dying is 10/247. Personally, I think that's a lot easier than memorizing a formula, but do what you have to.

By the way, this problem is a great reminder of the fact that in the study of odds, favorable means "satisfies a given outcome," not "good." You'd be hard-pressed to find many people who think that dying from an injury at some point is a good thing.

b. Again, we'll start with the formula approach, then do the cooler thing, which is figure it out.

$$\text{Odds in favor} = \frac{P(\text{Getting one pair})}{1 - P(\text{Getting one pair})} = \frac{0.423}{1 - 0.423} = \frac{0.423}{0.577}$$

This is kind of an awkward answer: a fraction with two decimals in it. We can make it look a whole lot better by multiplying the numerator and denominator both by 1,000, giving us

$$\text{Odds in favor} = \frac{423}{577}$$

We wanted to keep our answer in fraction form because that's the way that odds are written. Also, this makes it nice and easy to find the odds against getting one pair: It's just the reciprocal of the odds in favor, so

$$\text{Odds against} = \frac{577}{423}$$

Now the "figure it out" approach. If the probability of getting one pair is 0.423, we can write that in fraction form as 423/1,000. This tells us that if there are 1,000 outcomes, 423 represent getting one pair, while 577 represent not getting one pair (because $1{,}000 - 423 = 577$). That makes the odds in favor 423/577 and the odds against 577/423. Formulas are great for a lot of things, but understanding and figuring stuff out is ALWAYS better if you can do it.

Did You Get It

2. According to the American Cancer Society, the probability of an American female developing some type of cancer at some point in her life is about 1/3. Find the odds in favor of and against an American woman developing cancer.

3. When two dice are rolled, the odds in favor of getting a sum of 9 are 1:8. Find the probability of not getting a sum of 9 when two dice are rolled.

9. On the morning of the 2024 Kentucky Derby, Fierceness was listed as the favorite, with odds of 5:2. Write a sentence or two describing what this means.

10. If someone had bet $10 on Fierceness to win, about how much money do you think they would have gotten back if he had in fact won the race? (For what it's worth, he did not. Sigh.) Justify your answer.

Sidelight **Odds and Betting**

In almost any gambling enterprise, from Vegas to a church raffle, odds and expected value (which we'll learn about next) are used to determine what the payouts will be. Let's use the game of roulette as an example. There are 38 partitions on a roulette wheel, with a small ball that is equally likely to land in any of them. Players make various bets on where the ball will land. If you simply bet on a particular number, the odds against you are 37:1. If you win, the casino will pay 35 times what you bet. So if you bet a dollar on each spin, on average, in 38 spins, you would lose 37 times and win once, which would put you three dollars in the hole. That doesn't sound like the recipe for the casino making a lot of money, but when you consider the large number of people in a typical casino, and the fact that they're placing multiple bets at frequent intervals, it starts to make more sense.

This is where it becomes clear that state lotteries are an awful bet. In the Ohio lottery's "Classic Lotto" game, for example, the odds in favor of matching 5 of 6 numbers are 1:54,021. So to make the game completely fair, the payout should be in the neighborhood of $54,000 on a $1 ticket. The actual payout? Just $1,500! Any gambler or statistician will tell you that lotteries offer the worst odds of just about any game of chance. (In fact, you're at least 50 times more likely to be struck by lightning than win a multi-million dollar lottery.) So why do so many people play lotteries? For one, a complete lack of understanding of probability. But the biggest reason is probably that the massive payouts grab people's attention and make it seem like it's worth the risk of losing money. In almost every case though, no matter how huge the payout is in a lottery, it's still many, many times smaller than the odds of winning would dictate. In fact, the irony of this is that if a privately owned casino tried to run a game with payouts anywhere near that low compared to the odds of winning, the states would shut them down in about 5 minutes. Yes, the same states operate lotteries with those dreadful payouts. Go figure.

3-5 Group

Expected Value

If you play lottery games regularly (something we completely advise against, by the way), how much money do you expect to win? Obviously, this is not the same question as "How much money do you hope to win?" The concept of **expectation,** or **expected value,** is used to determine the result that would be expected in some sort of gamble. It's used not only for games of chance, but in areas like insurance, management, engineering, and others. In short, the expected value of some procedure or experiment is the average result you'd expect to get in the long term, if you kept repeating that procedure over and over.

Here's the key thing to remember as we study expected value: *It only makes sense for events that have numerical outcomes.* For example, rolling a die has a numerical outcome (1 through 6), and expected value can be used to determine what the average long-run result is likely to be. But it doesn't make sense to ask what the long-term average of flipping a coin is. (There's nothing between heads or tails, while an average roll of something like 2.7 at least has some meaning.)

Now let's illustrate a procedure for computing the expected value of a game, and then we'll summarize that procedure for future reference.

Suppose you pay a dollar to roll two dice. If you roll 5 or 6, you get your dollar back plus two more just like it. If not, you get nothing and like it. The goal will be to find the amount of money you can expect to win or lose if you play this game 100 times. First, we'll find the probability of winning and losing. We know that there are 36 possible rolls of two dice. (Refer back to the table at the end of Did You Get It 1 if you need to.)

1. How many outcomes result in 5? List them.

2. How many outcomes result in 6? List them.

3. What's the probability of winning the game?

4. What's the probability of losing the game?

5. Based on Questions 3 and 4, on average, you can expect to win _______% of the time.

6. So if you play the game 100 times, on average you'd expect to win _______ times and lose _______.

7. With each win, you make _______, and with each loss, you lose _______.

8. Multiply the number of times you expect to win by the amount you'd win each time.

9. Multiply the number of times you expect to lose by the amount you'd lose each time.

The next three questions are the key points in this part of the lesson.

10. After playing the game 100 times, find how much you would expect to win or lose total. Show how you got your answer.

11. Divide your answer to Question 10 by 100, and use your result to answer the next question.

12. You would expect to ________ (win/lose) on average ___________ every time you play this game. This amount is the **expected value** of the game.

Did You Get It

4. On a roulette wheel, there are 38 slots, 18 of which are colored red. If you bet $5 on red and win, you get $10 back. If red doesn't come up, you lose your $5. Find the expected value of playing the game 100 times. If you need help, follow Questions 3–12, which built the procedure for finding expected value.

Your answer to Question 12 might seem really puzzling if you think about it. How can you lose 25 cents on one trial of the game? The issue is that you can't: You either win $2 or lose $1. And that illustrates a hugely important point in understanding expected value: *The expected value never tells you what will happen on any given outcome; it tells you what the average result will be in the long run.* That's a key difference.

The key part of the procedure that we built for computing expected value was finding the probability of each outcome, then multiplying by the outcome itself to find out how much we'd win or lose on that particular outcome. We then need to add the results to get the expectation for all outcomes.

If we apply that same procedure and extend to cases with more than two possible outcomes, we get a general formula for the expected value of a probability experiment.

Formula for Expected Value

The expected value for the outcomes of a probability experiment is

$$E = X_1 \cdot P(X_1) + X_2 \cdot P(X_2) + \cdots + X_n \cdot P(X_n)$$

where the X's correspond to the numerical outcomes and the $P(X)$'s are the corresponding probabilities of the outcomes.

Verbally, the expected value formula tells us to multiply each outcome by its probability, then add the results.

13. When a single die is rolled, make a guess as to what the expected value of the roll will be.

To describe all of the outcomes in a probability experiment, we can build a **probability distribution.** A probability distribution consists of a list of all outcomes and the corresponding probabilities for a probability experiment. A quantity that can vary randomly from one individual trial to another, like the number on a die, is called a **random variable.** Why? Because it's random, and it can vary. See, math DOES make sense!

14. To help us calculate the expected value, complete the probability distribution below, listing the possible outcomes and their probabilities:

Number on Die (X)	1					
$P(X)$	$\frac{1}{6}$					

15. Using information from your table and the formula in the box above, calculate the expected value of one roll.

Did You Get It

5. If seven cards are numbered with integers from −2 to 4, then placed into a box and picked out at random, find the expected value.

While gambling games are the typical go-to examples for computing expected value, they're not by any means the only application. Let's look at the world of automobile insurance. If you think of it, buying insurance is kind of like gambling: You're betting that something bad will happen to your car. If nothing happens, you've paid the insurance company for nothing. But if something does happen, when they pay for repairs, you're basically "winning" the bet.

Suppose that a driver named Karl pays $1,100 per year on insurance for his SUV. According to insurance industry stats I found online, the average motorist has a 3.9% chance of getting in an accident in any given year, and the average collision claim paid by insurance companies is $5,313. Our goal will be to find the expected value if Karl buys 1 year's worth of car insurance.

16. If Karl does get in an accident and his insurance company pays the average claim, what is Karl's net gain? Don't forget the amount he paid for insurance.

17. If Karl remains accident-free for the year, what is his net loss?

18. Fill in the probability distribution, and use it to find Karl's expected value.

Result	Outcome (Gain or Loss)	Probability
Accident		
No accident		

19. Based on the expected value you calculated, is Karl paying too much for car insurance? Explain.

A favorite game for small children at fairs and carnivals is a game some folks call Pick-a-Duck. It's a simple game, which makes it ideal for little ones. A small pool is full of rubber ducks. Children get to pick one of the ducks and their prize is determined by the color of the dot on the bottom of the duck. Of course, the probability of winning a certain prize is based on how many ducks are in the pool, and how many have each color.

Brian Mercer

In one such game, you pay $1 to select one of 50 ducks from the pool and prizes are determined by the following list:

Number of Ducks	Outcome
20 ducks with red dots	Lose the money you paid to play*
10 ducks with orange dots	Win a sticker valued at $0.05
10 ducks with green dots	Win a small toy valued at $0.10
6 ducks with blue dots	Get to play again
4 ducks with purple dots	Win a stuffed animal valued at $4.00

*Then threaten to cry unless Mom lets you play again.

20. Fill in the probability distribution.

Color of Dot	Net Gain X	Probability $P(X)$
Red		
Orange		
Green		
Blue		
Purple		

21. Find the expected value of playing the game once.

22. Is the owner of the game making money? Explain.

Did You Get It

6. The profit made by a small ski resort, not surprisingly, depends largely on the seasonal weather. In a season with more than 75 inches of snow, it makes an average of $250,000. If snowfall is between 40 and 75 inches, the average profit is $160,000, and if snowfall is less than 40 inches, it loses $70,000. The resort gets over 75 inches of snow 40% of years, between 40 and 75 inches 45% of years, and less than 40 inches 15% of years. Find the resort's expected yearly profit.

In gambling games, if the expected value of the gain is 0, the game is said to be a **fair game**. If the expected value of the gain of a game is positive, then the game is in favor of the player. That is, the player has a better-than-even chance of winning. If the expected value of the gain is negative, then the game is said to be in favor of the house. That is, in the long run, the players will lose money.

23. The expected value of different casino games varies, but the expected values of every game in every casino have one thing in common. What is it?

Math Note

In American roulette, there are 22 different types of bets you can place, but it's interesting to note that all but one of them have the exact same expected value: −$0.053 on a $1 bet. (Betting on 0, 00, 1, 2, and 3, called a five-number bet, is worse, at −$0.079.)

A final note: The computation for expected value can be programmed into a spreadsheet. In fact, that's what you'll be doing in the technology assignment. There's also a command for computing expected value on a TI-84 graphing calculator. There's a document in the online resources for this lesson that walks you through that process.

3-5 Portfolio

Name ____________________

Check each box when you've completed the task. Remember that your instructor will want you to turn in the portfolio pages you create.

Technology

1. ☐ In this Excel activity, we'll compute the expected value for rolling a single 6-sided die. Then we'll compute the expected value for rolling two 6-sided dice.

ALEKS Practice

1. ☐ Include any written work from the ALEKS assignment along with any notes or questions about this lesson's content.

Applications

1. ☐ Complete the applications problems.

Reflections

Type a short answer to each question.

1. ☐ I just looked up a sports betting website online, and they have the Washington Commanders listed as a 21/10 favorite to make the football playoffs this year. Explain what that means, including relating it to probability.
2. ☐ How is it possible for the expected value of a game to be losing 65 cents if the only outcomes are win \$2 or lose \$1?
3. ☐ Name one thing you learned or discovered in this lesson that you found particularly interesting.
4. ☐ What questions do you have about this lesson?

Looking Ahead

1. ☐ Complete the prep skills for Lesson 3-6.
2. ☐ Read the opening paragraph in Lesson 3-6 carefully and answer Question 0 in preparation for that lesson.

Answers to "Did You Get It?"

1. In favor 5:7; against 7:5 **2.** In favor 1:2, against 2:1 **3.** $\frac{8}{9}$ **4.** –$26.32
5. 1 **6.** $161,500

Answers to Prep Skills Questions

1. $\frac{1}{3}, \frac{4}{5}$

2. $\frac{5}{36}, \frac{7}{12}$

3. $\frac{13}{21}$

4. $-\frac{13}{9}$, or $-1.\overline{4}$

3-5 Applications

Name ______________________________

1. A charity is running an auction. One thousand tickets are sold at $2 each for four prizes of $100, four of $50, four of $25, and four of $10. What are the odds in favor of winning $100?

2. What are the odds against winning any prize at all?

3. What is the expected value if you buy one ticket? Fill in the probability distribution to help organize the needed information.

Outcome	Gain	Probability
Lose		
Win $10		
Win $25		
Win $50		
Win $100		

4. What does your answer to Question 3 mean?

3-5 Applications

Name ____

5. A stock you bought 2 years ago with high hopes is now selling for less than you paid, and things look grim for the company. Do you sell, or hold on and hope it will come back to the original price before you sell? A model economists use for such situations is a game no one wants to play. Suppose you have a choice: Lose $100, or take a 50–50 chance between losing nothing and losing $300. Which do you choose? Find the expected value for each strategy.

There's an old joke among math and stats professors that lotteries are a tax on people who are bad at math. Let's take a look at one lottery game and use what we've learned about odds and expected value to study how good an idea it would be to play the game. The image below is the back of a $5 scratch-off ticket from the Illinois Lottery.

- SEND WINNING TICKETS TO ILLINOIS LOTTERY, P.O. BOX 19080, SPRINGFIELD, IL, 62794-9080.
- Tickets, transactions and winners are subject to all Lottery procedures, rules, directives and state law. Official game rules are on file at the Illinois Lottery.
- Approximate overall odds of winning (including break-even prizes) are 1 in 4.05.
  Approximate Prize odds are: $5 - 1 in 12.00; $10 - 1 in 8.57; $15 - 1 in 120.00; $20 - 1 in 48.00; $25 - 1 in 150.00; $30 - 1 in 300.00; $50 - 1 in 300.00; $100 - 1 in 250.00; $200 - 1 in 3,000.00; $1,000 - 1 in 6,667; $2,000 - 1 in 60,000; $5,000 - 1 in 160,800; $10,000 - 1 in 268,000; $200,00 - 1 in 1,340,000.

Courtesy of Brian Mercer

6. Some of the data above are described as "odds." Are they really odds? Describe exactly what these numbers represent, using terminology from this lesson.

3-5 Applications

Name ______________________________

7. Fill in the probability distribution below by using the "odds" on the back of the ticket to calculate probabilities for each numerical outcome (from the perspective of a player, not the lottery commission). When looking at the outcomes, remember that you have to pay $5 for the ticket. Write probabilities in fractional form. The first probability has been filled in for you. We strongly suggest using a spreadsheet to do the calculations for this question and the next one.

Result	Outcome	Probability
Lose	−$5	$1 - \frac{1}{4.05}$
Win $5	$0	
Win $10		
Win $15		
Win $20		
Win $25		
Win $30		
Win $50		

Result	Outcome	Probability
Win $100		
Win $200		
Win $1,000		
Win $2,000		
Win $5,000		
Win $10,000		
Win $200,000		

8. Use your table to find the expected value of buying one ticket. Again, a spreadsheet would be a great idea.

Result	Outcome	Probability	X·P(X)

Result	Outcome	Probability	X·P(X)

3-5 Applications

Name ____________________

9. Carefully explain exactly what your answer to Question 8 represents. If you don't have an answer for Question 8, make an educated guess as to what you think the expected value would be and explain what it would mean.

10. How much profit would the state expect to make if they sold 5,000,000 of these tickets?

11. Some people think that lotteries are a bad idea because they encourage gambling among people who can least afford to lose money. Others feel like the revenue that states make from lotteries is important to education and other social services. What are your thoughts?

Lesson 3-6 Prep Skills

SKILL 1: COMPUTE THEORETICAL PROBABILITIES

As we have seen in previous lessons, the probability of an event E in a sample space S occurring is given by $P(E) = \frac{\text{The number of outcomes in } E}{\text{The number of outcomes in } S} = \frac{n(E)}{n(S)}$.

For the following examples, remember that the sample space for rolling one die is {1, 2, 3, 4, 5, 6}. The number of outcomes in this sample space is 6.

- Find the probability of rolling an odd number.

$$P(\text{odd}) = \frac{3}{6} = \frac{1}{2}$$

The three outcomes resulting in an odd number being rolled are 1, 3, and 5.

- Find the probability of rolling a number greater than 2.

$$P(\text{greater than 2}) = \frac{4}{6} = \frac{2}{3}$$

There are four possible outcomes where a number greater than 2 is rolled: 3, 4, 5, and 6.

SKILL 2: ADD FRACTIONS WITH THE SAME DENOMINATOR

When adding or subtracting fractions, first check whether the fractions have a common denominator. If all fractions have the same denominator, then we add or subtract the numerators and keep the common denominator.

$\frac{a}{c} + \frac{b}{c} = \frac{a+b}{c}$ and $\frac{a}{c} - \frac{b}{c} = \frac{a-b}{c}$

- $\frac{3}{7} + \frac{2}{7} = \frac{3+2}{7} = \frac{5}{7}$

The fractions have a common denominator of 7. Perform the addition by adding the numerators and keeping the common denominator.

- $\frac{5}{10} - \frac{3}{10} = \frac{5-3}{10} = \frac{2}{10} = \frac{1}{5}$

The fractions have a common denominator of 10. Perform the subtraction by subtracting the numerators and keeping the common denominator.

The result can be reduced.

SKILL 3: SOLVE BASIC EQUATIONS

Solving equations involves performing strategic simplification and algebraic steps that produce equivalent equations to isolate the variable. The simplest of those steps involves adding quantities to and subtracting quantities from both sides of the equation.

- Solve $x - 3 = 2$.
  $x - 3 + 3 = 2 + 3$
  $x = 5$

  Isolate x by adding 3 to both sides of the equation. $x - 3 = 2$ and $x = 5$ are considered equivalent equations. The solution is $x = 5$. This means both sides of the original equation will have the same value when $x = 5$, making the equation a true statement.

- Solve $x + 1 = 8$.
  $x + 1 - 1 = 8 - 1$
  $x = 7$

  Isolate x by subtracting 1 from both sides of the equation. $x + 1 = 8$ and $x = 7$ are considered equivalent equations. The solution is $x = 7$. This means both sides of the original equation will have the same value when $x = 7$, making the equation a true statement.

PREP SKILLS QUESTIONS

1. After winning a drawing at a social club, one of the members gets to pick an envelope out of a bin. There are 40 envelopes in the bin: One has $1,000 in it, 2 have $50,10 have $20, and the rest contain blank scraps of paper. What is the probability that she'll pick an envelope with some cash in it?

2. What's the probability that she'll win more than $20?

3. Find the sum: $\frac{2}{11} + \frac{7}{11}$

4. Find the sum: $\frac{3}{16} + \frac{9}{16}$

5. Compute: $\frac{27}{80} + \frac{13}{80} - \frac{9}{80}$

6. Solve: $0.36 = 0.28 + 0.39 - P$

Lesson 3-6 An Exclusive Club

LEARNING OBJECTIVES

- ☐ 1. Distinguish between events that are and are not mutually exclusive.
- ☐ 2. Develop addition rules for finding probabilities of "or" events that are and are not mutually exclusive.
- ☐ 3. Use the addition rules to calculate probabilities.

Tupungato/iStockphoto/Getty Images

When politicians start talking about large groups of their fellow Americans as "enemies," it's time for a quiet stir of alertness. Polarizing people is a good way to win an election, and also a good way to wreck a country.
—Molly Ivins

As you know if you pay any attention at all to the news, ours is a country that is becoming increasingly divided by political affiliation. As the animosity between the two major parties grows, our lawmakers become more predictable. On most issues, you can have a pretty good idea of which side a politician will take just by seeing if they have a D or an R after their title. Of course, other characteristics come into play as well, and some political reporters make a living trying to forecast how our elected officials are going to react to the key issues of the day.

As you may know, every member of Congress declares a single party affiliation, either Democrat, Republican, or independent. Suppose that one commentator feels like a particular issue before Congress is most likely to be supported by women and Democrats. It would be useful to find the probability that any given representative is either female or a Democrat. At first thought, it seems reasonable to add up the number of women and the number of Democrats in Congress, then divide by the total number of representatives. But if you think about it, this won't work—those who are both female and Democrats will get counted twice. Developing rules for more complicated probability situations, like this one, will allow us to analyze more problems in a complex world.

0. What is it about the probability question described in the opening paragraphs that make it more complicated than the questions we've dealt with so far?

3-6 Group

The situation described in the opening paragraphs would be a lot simpler if the commentator were interested in the probability of a representative being either a Republican or an independent. The key difference is that everyone has to be either one or the other. (Of course, that's not the case when the two categories are women and Democrats.) This distinction leads to our first important new term:

> Two events are called **mutually exclusive** if they can't both occur at the same time. Another way of saying this is that the two events have no outcomes in common.

1. The example of finding the probability of a congressperson being a woman or a Democrat illustrates events that are not mutually exclusive. Explain why. Extra props if your answer identifies specific outcomes.

For Questions 2–4, assume that a random member of Congress is being chosen, and decide whether or not the given events are mutually exclusive. Please justify your answer.

2. The representative is not a Democrat, the representative is an independent.

3. The representative is from a state west of the Mississippi River, the representative is a man.

4. The representative has been in Congress for more than 10 years, the representative is serving their first 2-year term in Congress.

Did You Get It

Try this problem to see if you understand the concepts we just studied. The answer can be found at the end of the portfolio section.

1. If student government picks students at random to win free books for a semester, determine whether or not the two events are mutually exclusive.
   a. The winner is a sophomore or a business major.
   b. The winner is a junior or a senior.

A type of diagram borrowed from set theory is often used to illustrate sample spaces in probability. **Venn diagrams** use circles to portray relationships between sets of objects, in this case people. Each circle represents a set, with the objects in the set, or the number of objects in the set, written inside. If two circles overlap, the region that's common to both circles represents objects that are in both sets. We'll study Venn diagrams in much greater depth in Unit 6.

In this lesson, we'll first use a Venn diagram to develop a rule for working with mutually exclusive events in probability. In 2024, there were 49 Republican senators, 47 Democratic, and 4 independents. The Venn diagram below illustrates this composition, with the orange circle representing Republicans and the blue circle representing independents. The 47 written outside of the circles represents the number of senators who are neither Republican nor independent (that is, the Democrats).

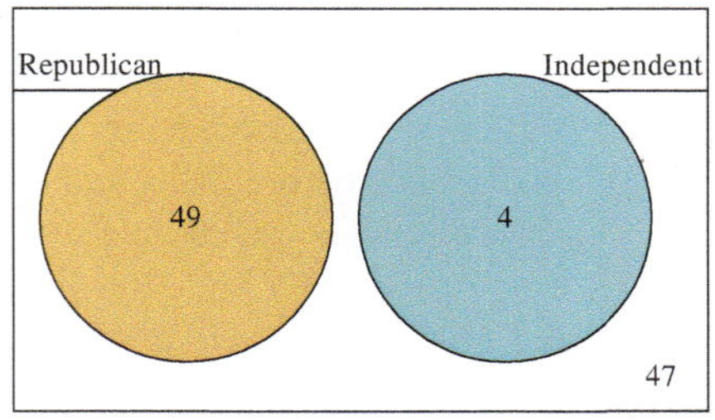

Since a senator can't be both Republican AND independent, notice that the two circles don't overlap at all. This indicates that the conditions of being a Republican senator and being an independent senator are mutually exclusive.

5. Find the probability of a senator being Republican.

6. Find the probability of a senator being independent.

7. How many senators are either Republican or independent? Use this to find the probability that a senator is either Republican or independent.

8. Describe the relationship between the probabilities found in Questions 5, 6, and 7, then use the Venn diagram to explain why that makes perfect sense when events are mutually exclusive.

Addition Rule for Probability (Mutually Exclusive Events)

When two events A and B are mutually exclusive, the probability that A or B will occur is

$$P(A \text{ or } B) = P(A) + P(B)$$

9. In the House of Representatives, there are 435 total representatives. Eight are from Maryland, 14 are from North Carolina, 12 are from New Jersey, and 11 are from Virginia. Find the probability that a randomly chosen representative is from North Carolina or Virginia.

10. In 2024, if you picked a random senator, you would have had a 0.09 probability of getting one who had been in the Senate since before 2000, and a 0.67 probability of getting one who took office sometime after 2010. What is the probability that you chose a senator who had been in office either since before 2000 or took office after 2010?

Did You Get It

2. When the ace of spades is removed from a standard deck of cards, if one card is randomly drawn, the probability of getting a spade is $\frac{4}{17}$ and the probability of an ace is $\frac{1}{17}$. What is the probability of drawing an ace or a spade?

11. There's a key difference between Questions 9 and 10. You can answer Question 9 two different ways, while Question 10 requires our first addition rule. Describe the two ways you can find the probability in Question 9.

12. Can our addition rule be extended to more than two events? Suppose that you draw one card from a standard deck. Find three probabilities: the probability that the card is a club, the probability that it's a heart, and the probability that it's a diamond.

13. Recall from Lesson 3-1 that for any event E, if E' is the event "E does not occur," then $P(E') = 1 - P(E)$. Use this to find the probability that the card is not a spade.

14. How does your answer to Question 13 relate to the three probabilities in Question 12? Explain why this makes sense and what it indicates about the addition rule for more than two events. This should be a reasonably in-depth answer.

3-6 Class

You can probably guess what's next: What about events that are not mutually exclusive? We can turn to our friend the Venn diagram again to develop a rule for finding probabilities involving "or" when events can have outcomes in common.

Before we get started, it's useful to talk about the word "or" itself. That simple two-letter word can actually have two different interpretations. A lot of restaurant menus say something like "All entrees come with soup or a salad." This pretty much always means that you can have one or the other, but not both. In math we tend to favor a more permissive approach—at our place, you'd be welcome to have soup and salad if you like. So when we use the word "or," we mean "one or the other, or both."

This diagram illustrates the number of women and Democrats in the U.S. Senate in 2024.

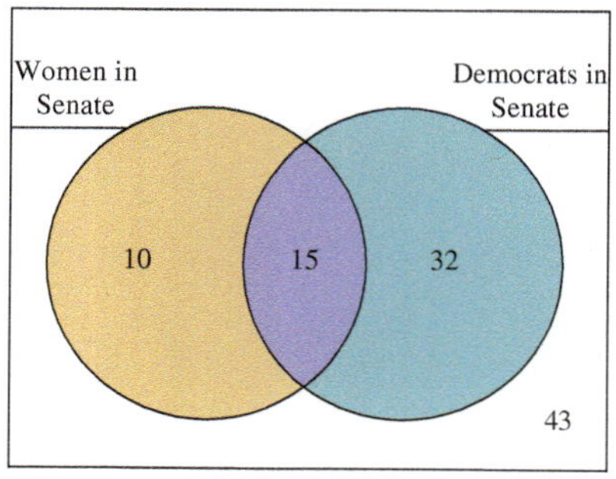

1. How many senators are women? (Pay attention to the entire left circle!)

2. How many are Democrats?

3. Write the probability of a senator being a woman, and the probability of a senator being a Democrat.

4. If we try to use our first addition rule, we'd conclude that the probability of a senator being a woman or a Democrat is $0.25 + 0.47 = 0.72$. Based just on the numbers in the Venn diagram, how can we be sure that this is incorrect?

5. Explain why using our addition rule doesn't work here. Make sure your answer refers to the Venn diagram! I didn't draw that baby for nothing.

6. Fill in the blanks:

The number of senators who are either female or Democrats = the number who are ______________

PLUS the number who are ________________ MINUS the number who are ________________.

7. Fill in numbers in this calculation, using your answer to Question 6 and the Venn diagram.

$$P(\text{Woman or Democrat}) = \frac{\text{Number who are women or Democrat}}{100} = \frac{\_\_ + \_\_ - \_\_}{100}$$

8. Now we're going to use some algebra: split the fraction in the calculation above into three separate fractions with denominator 100, then describe why you're allowed to do that. It might help to look at it backward, starting with three separate fractions and going back to one fraction.

$$P(\text{Woman or Democrat}) = \frac{\_\_ + \_\_ - \_\_}{100} = \frac{\_\_}{100} + \frac{\_\_}{100} - \frac{\_\_}{100}$$

9. Describe what the three fractions at the end of this calculation represent in terms of probability.

If all went well (and if it didn't you should ask for help!), you've just discovered our second addition rule for probability. You're pretty smart.

Addition Rule for Probability (General)

When two events *A* and *B* are not mutually exclusive, the probability that *A* or *B* will occur is

$$P(A \text{ or } B) = P(A) + P(B) - P(A \text{ and } B)$$

Solved Example

Using the General Addition Rule

Architect of the Capitol

At the time I'm writing this, in the House of Representatives, there are 102 women, 13 of whom are Republican. There are 435 total representatives, and 199 of those are Republican. If one representative is randomly chosen for a TV interview, what's the probability that the representative is either a woman or a Republican?

SOLUTION

On any "or" probability question, it's a good idea to start out by deciding if the events are mutually exclusive. In this case, they aren't because the statement flat-out says that there are some female Republicans. So we definitely need the general addition rule, which applies to events that are not mutually exclusive. We'll need three probabilities to use the rule: the probability that a randomly selected representative is a woman, the probability that he or she is a Republican, and the probability that both of those things happen.

$$P(\text{Woman}) = \frac{102}{435} \quad P(\text{Republican}) = \frac{199}{435} \quad P(\text{both}) = \frac{13}{435}$$

Now we can directly apply the addition rule.

$$P(\text{Woman or Republican}) = P(\text{Woman}) + P(\text{Republican}) - P(\text{both})$$

$$= \frac{102}{435} + \frac{199}{435} - \frac{13}{435} = \frac{288}{435} \approx 0.66$$

10. At Big Tony's Pizzas and Loans, 45% of customers order pizzas with either sausage or pepperoni. If 31% get pizzas with pepperoni and 26% with sausage, find the probability that the next person to call will order both sausage and pepperoni. (Hint: This will require the general addition rule and some ingenuity.)

Did You Get It

3. When drawing a single card from a standard deck of cards, what's the probability that the card you draw is either a face card or a heart?

11. We now have two addition rules. Or do we? Look carefully at the general addition rule on page 304. Explain why it will also work when two events are mutually exclusive, and in that regard is the only addition rule you need to know.

We've now studied two distinctly different situations involving two events: The key difference between them is the word that connects the descriptions of the two events: "and" or "or."

12. If we're trying to find the probability of one event AND another, what operation do we use on the individual probabilities?

13. If we're trying to find the probability of one event OR another, what operation do we use on the individual probabilities?

14. When a lot of information is provided verbally, it's often helpful to organize it when problem solving. In a hospital there are 86 nurses and 41 physicians. Seventy nurses and 14 physicians are females. If a staff person is selected, find the probability that the subject is a nurse or a male. Organizing the information in the table below should probably help.

Staff	Male	Female	Total
Doctor			
Nurse			
Total			

Did You Get It ?

4. In one class, there are 15 freshmen and 10 sophomores. Six of the freshmen are education majors and 4 of the sophomores are education majors. If a student is selected at random, find the probability that the student is a sophomore or an education major.

15. Earlier, we thought about an addition rule for three mutually exclusive events. What about three events that are NOT mutually exclusive? That turns out to be quite a bit more complicated. It will require the three-circle Venn diagram below. For three events, *A*, *B*, and *C*, see if you can make a conjecture on a formula for finding the probability of *A* or *B* or *C* occurring.

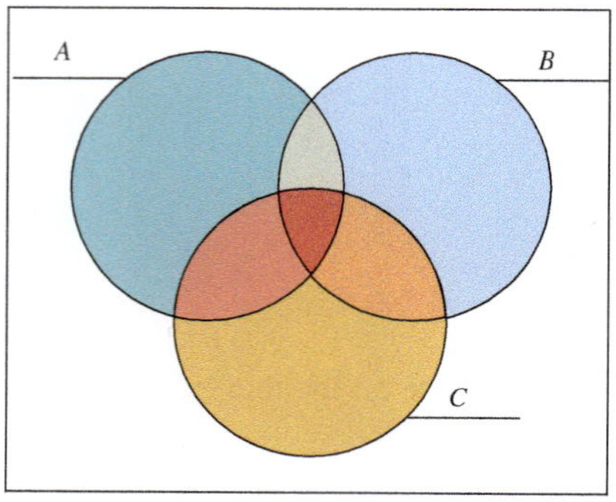

3-6 Portfolio

Name ______________________________

Check each box when you've completed the task. Remember that your instructor will want you to turn in the portfolio pages you create.

ALEKS Practice

1. ☐ Include any written work from the ALEKS assignment along with any notes or questions about this lesson's content.

Applications

1. ☐ Complete the applications problems.

Reflections

Type a short answer to each question.

1. ☐ How do you decide if two events are mutually exclusive?
2. ☐ Explain why deciding if events are mutually exclusive is important in probability. Make sure you describe the types of probability problems where this comes up.
3. ☐ Describe the two addition rules in your own words.
4. ☐ Name one thing you learned or discovered in this lesson that you found particularly interesting.
5. ☐ What questions do you have about this lesson?

Looking Ahead

1. ☐ Complete the prep skills for Lesson 3-7.
2. ☐ Read the opening paragraph in Lesson 3-7 carefully and answer Question 0 in preparation for that lesson.

Answers to "Did You Get It?"

1. a. Not mutually exclusive **b.** Mutually exclusive **2.** $\frac{5}{17}$ **3.** $\frac{11}{26}$ **4.** $\frac{16}{25}$

Answers to Prep Skills Questions

1. $\frac{13}{40}$ or 0.325
2. $\frac{3}{40}$, or 0.075
3. $\frac{9}{11}$
4. $\frac{3}{4}$
5. $\frac{31}{80}$
6. $P = 0.31$

3-6 Applications

Name ______________________________

Rank	Program	Network	Avg. Viewers (Thousands)
1	NFL Sunday Night Football	NBC	19,838
2	NFL Monday Night Football	ABC	11,986
3	Tracker	CBS	10,799
4	NCIS	CBS	9,660
5	Young Sheldon	CBS	9,101
6	FBI	CBS	8,960
7	Chicago Fire	NBC	8,794
8	Blue Bloods	CBS	8,449
9	60 Minutes	CBS	8,266
10	Ghosts	CBS	8,206
11	Chicago Med	NBC	8,203
12	Chicago PD	NBC	7,959
13	The Equalizer	CBS	7,900
14	NCIS: Hawai'i	CBS	7,788
15	Elsbeth	CBS	7,454
16	FBI: Most Wanted	CBS	7,389
17	Fire Country	CBS	7,332
18	FBI: International	CBS	7,208
19	The Voice (Monday)	NBC	7,143
20	Law & Order: SVU	NBC	6,905

This table shows the top 20 highest-rated American network TV shows for the 2023–2024 season. The last column is the average number of viewers in thousands. You'll be using this table throughout the Applications, so memorize it and then eat it. Or I guess you could just refer back to it. That probably makes more sense.

1. If one of the top 20 shows is chosen at random, what is the probability that it was on NBC or CBS?

2. If you choose one top 20 show randomly, what is the probability that it was on CBS or had more than 9 million viewers on average?

3. When choosing a random top 20 show, what's more likely: that it was on CBS, or that it was on ABC or NBC? Justify by calculating probabilities, please.

3-6 Applications

Name ______________________________

4. In the rankings of all shows for the season, the probability of a show being on each network was: 0.222 for ABC, 0.243 for CBS, 0.139 for the CW, 0.208 for Fox, and 0.1875 for NBC. Find the probability that one randomly selected network show was not on NBC or CBS.

5. Back to the information from the table. Write an example of an event involving "or" based on the table where two or more events ARE mutually exclusive. Write a sentence or two to justify your choice.

6. Write an example of an event involving "or" based on the table where two or more events ARE NOT mutually exclusive. Write a sentence or two to justify your choice.

7. There were 144 total shows listed, and 35 of them were on CBS. Find the probability that a show selected at random from the entire rankings list was on CBS or was ranked in the top 20.

8. Find the probability that a show selected at random from the entire rankings list was on CBS or had more than 8 million viewers.

Lesson 3-7 Prep Skills

SKILL 1: MULTIPLY FRACTIONS

When multiplying fractions, multiply the numerators and multiply the denominators separately, then reduce the result if possible: $\frac{a}{b} \cdot \frac{c}{d} = \frac{ac}{bd}$.

Often, dividing away common factors before actually multiplying can simplify the process.

- $\frac{3}{10} \cdot \frac{4}{5} = \frac{12}{50}$
  $= \frac{6}{25}$

  Multiplying the numerators results in $3 \cdot 4 = 12$. Multiplying the denominators results in $10 \cdot 5 = 50$. Then reduce the result $\frac{12}{50}$ by dividing away the common factor of 2 from the 12 and the 50.

- $\frac{3}{10} \cdot \frac{4}{5} = \frac{3}{\cancel{10}_{5}} \cdot \frac{\cancel{4}^{2}}{5}$
  $= \frac{6}{25}$

  Note that 4 from the numerator of the second fraction shares a common factor of 2 with the 10 from the denominator of the first fraction. Dividing away that common factor first makes the multiplication slightly easier and eliminates the need for reducing the fraction at the end. Either process will produce the correct answer, but this second method is usually quicker and easier.

SKILL 2: COMPUTE THEORETICAL PROBABILITIES

As we have seen in previous lessons, the probability of an event E in a sample space S occurring is given by $P(E) = \frac{\text{The number of outcomes in } E}{\text{The number of outcomes in } S} = \frac{n(E)}{n(S)}$.

- Calculate the probability of winning a raffle if you buy 3 of the 400 tickets sold for a local school fundraiser.

  $P(\text{Win}) = \frac{3}{400} = 0.0075$

  You have 3 chances out of 400 to win.

SKILL 3: COMPUTE PROBABILITY FROM A DECK OF CARDS

A standard deck contains 52 cards. Consider the sample space of selecting a card from a standard deck of 52 cards. (The cards are assumed to be shuffled to make sure that the selection occurs at random.) There are four suits—hearts, diamonds, spades, and clubs, and 13 cards of each suit consisting of the denominations ace (A), 2, 3, 4, 5, 6, 7, 8, 9, 10, and 3 picture or face cards—jack (J), queen (Q), and king (K). The sample space is shown in the following table.

	A	2	3	4	5	6	7	8	9	10	J	Q	K
♥	A♥	2♥	3♥	4♥	5♥	6♥	7♥	8♥	9♥	10♥	J♥	Q♥	K♥
♦	A♦	2♦	3♦	4♦	5♦	6♦	7♦	8♦	9♦	10♦	J♦	Q♦	K♦
♠	A♠	2♠	3♠	4♠	5♠	6♠	7♠	8♠	9♠	10♠	J♠	Q♠	K♠
♣	A♣	2♣	3♣	4♣	5♣	6♣	7♣	8♣	9♣	10♣	J♣	Q♣	K♣

A card is drawn from an ordinary deck. Use the sample space shown above to find the probabilities of getting

- An 8.

$P(8) = \frac{4}{52} = \frac{1}{13}$ There are four 8s in a standard deck of cards.

- A face card.

$P(\text{Face}) = \frac{12}{52} = \frac{3}{13}$ There are 12 face cards in a standard deck of cards.

SKILL 4: FIND PROBABILITY FROM DATA IN TABLE FORM

When information is displayed in a table, carefully read the row and column labels to identify what is being displayed. Note whether the totals are given for each row and for each column. If they aren't, there's a pretty good chance that you'll need to compute those totals yourself. The table at right displays the sex and size of dogs in a local shelter.

	Large Dog (L)	Small Dog (S)	Total
Female (F)	4	3	7
Male (M)	7	2	9
Total	11	5	16

- Find the probability of randomly selecting a small female dog from this shelter.

$P(SF) = \frac{3}{16}$ There are 3 small female dogs out of the 16 total dogs in this shelter.

- Find the probability of randomly selecting a male dog from this shelter.

$P(M) = \frac{9}{16}$ There are 9 male dogs out of the 16 total dogs in this shelter.

PREP SKILLS QUESTIONS

1. Perform each multiplication. Write your answer in lowest terms.

   $\frac{1}{3} \cdot \frac{6}{11}$ $\frac{9}{10} \cdot \frac{3}{10}$ $\frac{1}{5} \cdot \frac{2}{3} \cdot \frac{3}{5}$

2. A fast-food restaurant has 11 meal choices on its value menu. If a friend runs to pick up some food and chooses your meal randomly, what's the probability that you'll get your favorite?
3. When drawing a single card randomly from a deck, find the probability that the card is a 10.
4. Find the probability that the card is a heart.
5. Find the probability that the card is a red face card.
6. The table shows light conditions for November 4, 2024, in the Cincinnati area, as well as a month later. Find the probability that a randomly selected hour on one of those days had sunny conditions.

Date	Hours Sunny	Hours Cloudy	Hours Dark
Nov. 4	6.2	4.05	13.75
Dec. 4	1.7	7.8	14.5

Lesson 3-7 Independence Day

Chip Simons/Creatas Images/
Getty Images

LEARNING OBJECTIVES

- ☐ 1. Distinguish between events that are and are not independent.
- ☐ 2. Develop multiplication rules for finding probabilities of "and" events that are and are not independent.
- ☐ 3. Use the multiplication rules to calculate probabilities.
- ☐ 4. Define, compute, and interpret conditional probabilities.

The five separate fingers are five independent units. Close them and the fist multiplies strength.

—James Cash Penney

Even after learning about probability, a lot of people are tempted to play lottery games, looking for that one big score that will put them on easy street. Suppose that you play a daily game in which you try to match a randomly selected four-digit number. On Monday, the number is 2471. On Tuesday, the number is again 2471. Would you play 2471 on Wednesday? People are usually split on this question: some say "There's no way it's going to come up three days in a row!" while others say "That number's hot! Go with it!"

The logic in both of those arguments is faulty, however, assuming that the game isn't rigged and the numbers really are chosen at random. When Wednesday's numbers are being drawn, they have no idea what may have been drawn on Monday and Tuesday. Those two results can't jump into the future and affect what happens on Wednesday's drawing. So it's neither a good thing nor a bad thing to play 2471 on Wednesday. It's just a thing.

Let's consider a related question: What's the probability that 2471 will come up three times in a row? In this lesson, we'll study rules for finding the probability of consecutive events occurring. In every case, we'll have to decide if any event affects the outcome of the ones that follow.

0. Would you play 2471 on the third day? Why or why not?

3-7 Group

The key idea in the opening paragraphs is that when finding probabilities of consecutive events, it's important to decide whether any of the later events are in some way affected by the ones that already occurred. One of the terms we use to describe this condition is *independence*.

Two events *A* and *B* are **independent** if the fact that *A* occurs has no effect on the probability of *B* occurring.

1. Based on our description above, are the lottery numbers that are drawn on Tuesday and Wednesday independent events or not? Why?

On the other hand, when the occurrence of the first event changes the probability of the occurrence of the second event, the two events are said to be *dependent.* For example, suppose a card is drawn from a deck and not replaced, and then a second card is drawn. Probabilities for the second card are all different now; the first card can't be picked, so the remaining sample space has only 51 outcomes.

Two events A and B are **dependent** if the outcome of A has some effect on the probability of B occurring.

Suppose each day a professor picks a student at random to work out a homework problem on the board.

2. If the student is selected regardless of who already went, the events

   are ______________.

3. If the student is selected from the pool of students who havent gone yet, the events

   are ______________.

moodboard/Cultura/Getty Images

In Questions 4–9, the goal is to decide if the two events described are independent or dependent.

4. Rolling a die and getting a 6, and then rolling a second die and getting a 3

5. Drawing a card from a deck, not replacing it, and then drawing a second card

6. Parking in a no-parking zone and getting a parking ticket

7. Drawing a card from a deck and getting a queen, replacing it, and drawing a second card and getting a queen

8. Selecting a lottery ball from a tumbler, not replacing it, and then drawing a second ball

9. Flipping a coin and then flipping it again

10. Speaking of flipping a coin twice consecutively: What's the probability of getting tails on the first flip? What's the probability of getting tails on the second flip?

11. Complete the sample space for two flips of a coin: {HH, ____, ____, ____}.

12. What is the probability of getting tails twice? This could be written as P(Tails AND Tails).

13. How do your answers to Question 10 relate to your answer to Question 12? Make a conjecture on a rule for finding the probability of two events occurring consecutively when those events are independent.

If you did well, your conjecture from Question 13 is our first multiplication rule for probability.

The Multiplication Rule for Independent Events

When two events A and B are independent, the probability of both occurring is

$$P(A \text{ and } B) = P(A) \cdot P(B)$$

14. A coin is flipped and a die is rolled. Find the probability of getting heads on the coin and a 4 on the die. (Are you sure that the multiplication rule applies?)

Math Note

You could find the probability in Question 14 using standard sample space methods, but as outcomes get more complicated, it can be extremely difficult to list sample spaces. That's why multiplication rules are so helpful.

Did You Get It

Try this problem to see if you understand the concepts we just studied. The answer can be found at the end of the portfolio section.

1. Talk about a bad day: If the probability of your alarm not going off is 1/20, and the probability of getting a ticket on your way to work is 1/200, find the probability that both will happen in the same morning.

As part of a psychology experiment on perception and memory, colored marbles are picked from a small vase. The vase contains three red marbles, two green marbles, and five white marbles. A marble is picked and its color is noted, then it's replaced. A second marble is picked and its color is noted. Find the probability of each event in Questions 15–18.

15. Picking two green marbles. This could be written as *P*(green and green).

16. Picking a green marble and then a white marble

17. Picking a red marble and then a green marble

18. Picking a white marble, then a red marble, and then a green marble

19. Did you make up a new rule in Question 18? Explain.

Did You Get It

2. As part of a card trick, a card is drawn from a deck and replaced; then a second card is drawn. Find the probability of getting a queen and then an ace.
3. Given that the probability of rain on any given day in March in Daytona Beach is 1/5, find the probability that
   a. It rains three straight days in March.
   b. It rains on March 10 and 12, but not March 11.

   Assume that weather on any day is independent of the others.

3-7 Class

We know when a card is picked from a deck and not replaced, this changes the probabilities for what can happen when a second card is drawn. Let's explore exactly how the probabilities are affected.

Solved Example

Studying Probabilities for Dependent Events

Cards are being drawn one at a time from a standard deck.

a. Find the probability that the first card is a queen.

b. If the first card is a queen, find the probability that the second card is also a queen.

c. Find the relative change between the probability of drawing the first queen and drawing the second queen. What does this mean?

SOLUTION

a. This is just a quick review of theoretical probability. There are 52 cards in the deck and 4 of them are queens, so

$$P(\text{Drawing a queen}) = \frac{4}{52}, \text{ or about } 0.077$$

b. Now we have to think a bit. One of the queens is out of the deck, so there are only 3 left to pick from. And there are now only 51 cards total. So this time we get

$$P(\text{Drawing a second queen}) = \frac{3}{51}, \text{ or about } 0.059$$

c. Recall that to find the relative difference between two numbers, you divide the difference between them by the original value.

$$\text{Relative difference} = \frac{0.059 - 0.077}{0.077} \approx -0.23$$

This means that you're 23% less likely to draw a queen on the second draw if a queen was already drawn on the first.

Part c of the Solved Example illustrates probably the key point in understanding why independence of events matters: The occurrence of the first event had a really significant effect on the probability of the second event occurring.

1. Now let's change the situation. Suppose instead of 1 deck, there were 1,000 standard decks, all mixed together. How many queens are in this gigantic deck? How many are left after drawing one?

Math Note

I just did a quick calculation and found that a stack of 1,000 decks of standard playing cards would be about 52 feet tall. Why would I do this? Because I can, and I wondered.

2. Do you think that not replacing the queen in this case will significantly affect the probability on the second draw? Explain.

3. Well, let's just see about that. Find the probability of drawing a first queen from 1,000 decks, and round to three significant digits.

4. Now find the probability of drawing a second queen, assuming you don't replace the first one. What do you notice?

This illustrates a useful idea in computing probabilities in our world: If the sample space is really large, choosing a handful of objects without replacement technically makes the choices dependent, but the effect on probability is so small that we can pretend the choices are independent, making the calculations simpler. We'll use this idea in the next group of questions.

According to the National Center for Education Statistics, 85% of students entering college receive some form of financial aid–loans, grants, or scholarships. Our goal for Questions 5–8 is to find the probability that four randomly chosen students entering college all receive some kind of aid.

5. For any individual student, what is the probability that he or she receives financial aid? Find the probability, then write your answer in the form of a formal probability statement in the form $P(\text{event}) = \text{probability}$.

6. Explain why the four choices are technically not independent events.

7. Explain why the four choices can be treated as independent events.

8. Treating the four choices as independent and using the multiplication rule for independent events, find the probability that all four will be receiving financial aid. Again, write your answer in the form of a formal probability statement, this time using the word AND.

Did You Get It

4. According to the *Princeton Review,* just 29% of college-bound students reported that their stress level wasn't either high or very high. If three college-bound students are picked at random, find the probability that all three rate their stress as high or very high.

When we are interested in finding the probability of consecutive events that are dependent, we can still use the multiplication rule, but with a minor modification. For example, let's go back to drawing two cards at random from a standard deck. We know that the probability of drawing a queen on the first draw is 4/52 and on the second draw is 3/51.

9. What does multiplication provide as the probability of getting a queen on both the first and second draws?

The procedure we developed in Question 9 is a multiplication rule that can be used for any two events, whether they're independent or not.

The Multiplication Rule for Any Two Events

For two events A and B, the probability of both occurring is

$$P(A \text{ and } B) = P(A) \cdot P(B \text{ given that } A \text{ has already occurred})$$

10. An appliance store gets a shipment of 25 plasma TVs, and 3 of them are defective. If two of the TVs are chosen at random, find the probability that both are defective. (The first TV is not replaced after it's tested because that would be silly.)

Math Note

Notice that if the two events are independent, then the probability of B given that A has already occurred is just the regular probability of B occurring, and we get our first multiplication rule back. Good deal.

Did You Get It

5. The 2024 NCAA men's basketball tournament field had (among 68 teams) 8 teams from the Big 12 Conference and the SEC, 6 each from the Big Ten and Mountain West, and five from the ACC. If you were randomly assigned 2 teams in a dorm pool, find the probability that
   a. Both were from the Big 12.
   b. The first was from the ACC and the second was from the Big Ten.

Conditional Probability

We know that to find the probability of two dependent events occurring, it's important to find the probability of the second event occurring given that the first has already occurred. We call this the **conditional probability** of event B occurring given that event A has occurred, and denote it $P(B|A)$.

Now that we have a symbol to represent the probability that an event B has occurred given that an event A already occurred, it makes sense to write a formula for calculating it.

11. Solve the general multiplication rule in the box above to find $P(B|A)$.

Computing Conditional Probability

The probability that a second event B occurs given that a first event A has occurred can be found by dividing the probability that both events occurred by the probability that the first event has occurred. The formula is

$$P(B|A) = \frac{P(A \text{ and } B)}{P(A)}$$

Hate crimes are defined to be crimes in which the victim is targeted because of one or more personal characteristics, such as race/ethnicity, religion, or sexual orientation. These types of crimes have been historically underreported, and there were no specific legal protections for at-risk groups until the late 1960s. The 2009 Hate Crimes Prevention Act was a positive step, but there has been an alarming increase in these crimes in the last several years. The table below lists the motivation for certain hate crimes as reported by the FBI for 2023. If you've been the victim of a hate crime, you're not alone. Don't be afraid to speak up and seek support. You can find a number of helpful resources at https://victimconnect.org/learn/types-of-crime/hate-crimes/.

Motivation	Crimes against Persons	Crimes against Property	Crimes against Society	Total
Race/Ethnicity	5,137	1,787	125	**7,049**
Religion	1,316	1,585	66	**2,967**
Sexual Orientation	1,574	812	16	**2,402**
Total	**8,027**	**4,184**	**207**	**12,418**

In Questions 12–14, write your answer as both an unreduced fraction and a decimal rounded to three significant digits.

12. Find the probability that a hate crime was a crime against persons. We'll call this $P(A)$.

13. Find the probability that a hate crime was motivated by race or ethnicity. We'll call this $P(B)$.

14. Find the probability that a hate crime was against persons and was motivated by race or ethnicity. Use an appropriate notation for this event in your answer.

15. Use your unreduced fractional answers and the conditional probability formula to find the probability that a hate crime was motivated by race or ethnicity given that it was a crime against persons. Write your answer as a decimal rounded to three significant figures.

16. Use your decimal approximations and the conditional probability formula to find the probability that a hate crime was motivated by race or ethnicity given that it was a crime against persons. How does your answer compare to Question 15?

17. Find the probability that a hate crime was against property given that it was motivated by the victim's sexual orientation. Don't use the conditional probability formula: Just use information in the table.

Did You Get It

6. Based on the data in the table on page 321, find the probability that
   a. A crime was motivated by either race/ethnicity or religion given that it was a crime against society.
   b. A crime was against persons given that it was motivated by religion or sexual orientation.

3-7 Portfolio

Name ______________________________

Check each box when you've completed the task. Remember that your instructor will want you to turn in the portfolio pages you create.

Technology

1. ☐ In this Excel activity, we'll compute the probability of drawing two consecutive queens from a varying number of decks of cards.

ALEKS Practice

1. ☐ Include any written work from the ALEKS assignment along with any notes or questions about this lesson's content.

Applications

1. ☐ Complete the applications problems.

Reflections

Type a short answer to each question.

1. ☐ Describe the difference between dependent and independent events, and give an example of each.
2. ☐ What is meant by the term conditional probability?
3. ☐ When can you treat dependent events as if they were independent? Why is it reasonable to do so?
4. ☐ Name one thing you learned or discovered in this lesson that you found particularly interesting.
5. ☐ What questions do you have about this lesson?

Looking Ahead

1. ☐ Complete the prep skills for Lesson 3-8.
2. ☐ Read the opening paragraph in Lesson 3-8 carefully and answer Question 0 in preparation for that lesson.

Answers to "Did You Get It?"

1. 1/4,000 **2.** 1/169 **3. a.** 1/125 **b.** 4/125 **4.** 0.358 **5. a.** 0.0123 **b.** 0.0066 **6. a.** 0.923 **b.** 0.538

Answers to Prep Skills Questions

1. $\frac{2}{11}, \frac{27}{100}, \frac{2}{25}$

2. $\frac{1}{11}$

3. $\frac{1}{13}$

4. $\frac{1}{4}$

5. $\frac{3}{26}$

6. About 0.165

3-7 Applications

Name ____

In Questions 1–4, decide whether the events are independent or dependent, and explain your answer.

1. Flipping a coin and drawing a card from a deck

2. Getting a raise in salary and buying a new car

3. Driving on ice and having an accident

4. Having a large shoe size and having a high IQ

Use this information for Questions 5 and 6: A telecommunications company has six satellites, two of which are sending a weak signal. Assume two are picked at random without replacement.

5. Find the probability that both are sending a weak signal.

6. Find the probability that one satellite sends a strong signal and the other sends a weak signal.

3-7 Applications

Name ______________________________

In Questions 7–9, find each probability assuming three cards are drawn from an ordinary deck and not replaced.

7. Getting three jacks

8. Getting an ace, a king, and a queen in order

9. Getting a club, a spade, and a heart in order

A survey of 200 college students shows the average number of text messages sent per month. Find each probability in Questions 10–13.

	Less than 1,000	1,000–1,999	2,000–2,999	3,000 or more
Men	56	18	10	16
Women	61	18	13	8

10. The student sent less than 1,000 texts given that it was a woman.

3-7 Applications

Name ______________________________

11. The student sent more than 3,000 texts given that it was a man.

12. The student was a woman given that they sent between 1,000 and 1,999 texts.

13. The student was a man given that they sent between 1,000 and 2,999 texts.

Use this information for Questions 14–16. On one large campus, 18% of students surveyed said that they spend less than an hour a night studying, 72% of students surveyed said they were in a math class, and 13% of students surveyed said that they spend less than an hour a night studying and that they're in a math class.

14. If three students are picked at random, what's the probability that all three spend less than an hour a night studying?

15. What's the probability that only the first two of the three students picked spend less than an hour a night studying?

16. Find the probability that a randomly selected student spends less than an hour a night studying given that the student is in a math class.

3-7 Applications

Name ______________________________

17. Classify each statement as sensible or silly, and briefly explain your answer.

a. I flipped a coin five times in a row and got heads, so I'm willing to bet $100 that it will be tails on the next flip.

b. There's a 30% chance of rain tomorrow and there's a 50–50 chance of my only class getting canceled, so there's a 35% chance I'll be able to go golfing without missing class. (By the way, I'm not going to go golfing if it rains.)

c. According to duilawblog.com, the probability of being convicted when charged with DUI in California is 0.794. If 50% of those convicted get at least 48 hours in jail, the probability of spending at least 48 hours in jail if charged with DUI in California is 1.294.

Lesson 3-8 Prep Skills

SKILL 1: COMPUTE COMBINATIONS

As we saw in Lesson 3-3, the arrangement of n objects in any order using r of those objects without replacement is called a combination of n objects taken r at a time. It is written as ${}_nC_r$ and is calculated using the formula

$${}_nC_r = \frac{{}_nP_r}{r!} = \frac{n!}{(n-r)!r!}$$

- Find the number of combinations of six objects taken two at a time.

$${}_6C_2 = \frac{6!}{(6-2)!2!}$$

Substitute 6 and 2 into the formula for computing combinations and simplify the result.

$$= \frac{6 \cdot 5 \cdot 4!}{4! \cdot 2!}$$

6! can be written as $6 \cdot 5 \cdot 4!$, allowing us to divide out 4!.

$$= \frac{6 \cdot 5}{2 \cdot 1}$$

$$= 15$$

There are 15 combinations of 6 objects taken 2 at a time.

SKILL 2: COMPUTE POWERS OF A FRACTION

When raising a fraction to a power, it's useful to remember the exponent rule for fractions, $\left(\frac{a}{b}\right)^n = \frac{a^n}{b^n}$.

In words, this says that to raise a fraction to a power, apply that power to each of the numerator and denominator separately.

- $\left(\frac{2}{3}\right)^4 = \frac{2^4}{3^4}$

$$= \frac{16}{81} \approx 0.20$$

Raise the numerator and the denominator to the 4th power and simplify.

SKILL 3: INTERPRET A TREE DIAGRAM

As we saw in Lesson 3-2, a tree diagram is a diagram where we use branches, similar to a flow chart, to list the outcomes of two or more probability experiments that are done in sequence.

The tree diagram below shows the outcomes for selecting a colored ball from a bag containing a red, a green, and a purple ball and then flipping a coin.

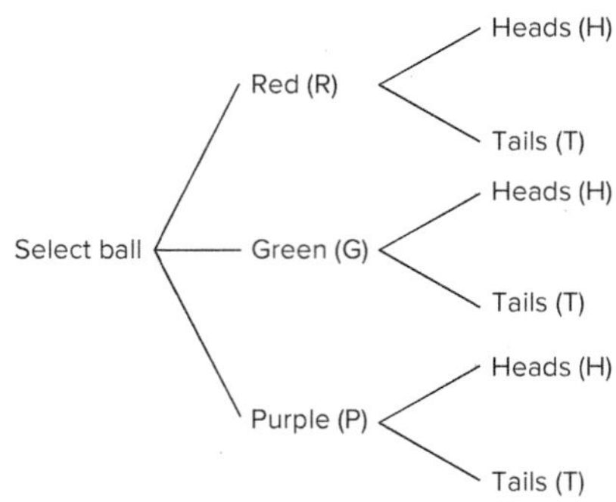

- List the sample space for this tree diagram. {RH, RT, GH, GT, PH, PT}

  There are six possible outcomes in this sample space, which we find by tracing along each branch of the tree diagram.

- Find the probability of drawing a green ball and having the coin land on tails.

  $P(\text{Green, Tails}) = \frac{1}{6}$

  There is only one outcome with a green ball and the coin landing on tails.

PREP SKILLS QUESTIONS

1. Compute each combination.

   ${}_5C_5 \qquad {}_{12}C_8$

2. Compute each power. Write your answer in fraction form, then in decimal form rounded to three significant digits.

   $\left(\frac{1}{2}\right)^3 \qquad \left(\frac{4}{5}\right)^9$

3. The tree diagram illustrates the outcomes for the experiment of flipping two coins then playing one game of rock, paper, scissors (where each option of win, lose, or tie is equally likely). What's the probability of getting one head and one tail on the coins and not losing rock, paper, scissors?

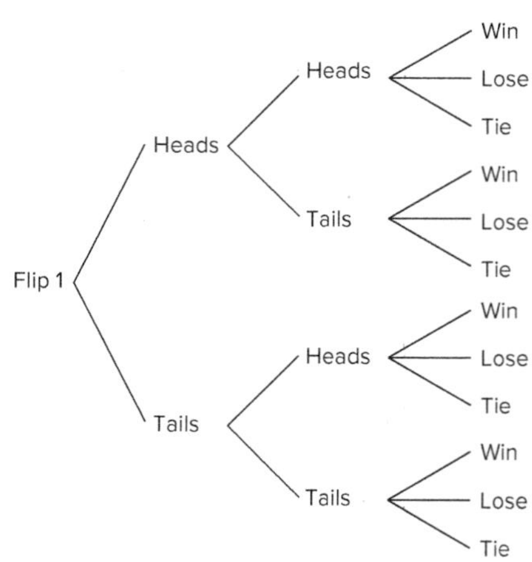

Lesson 3-8 Either/Or

LEARNING OBJECTIVES

- ☐ 1. Identify binomial experiments.
- ☐ 2. Compute and interpret probabilities of outcomes in a binomial experiment.
- ☐ 3. Compute cumulative binomial probabilities.

HBSS/Fancy Photography/Getty Images

When you wake up every day, you have two choices. You can either be positive or negative; an optimist or a pessimist. I choose to be an optimist. It's all a matter of perspective.

—Harvey Mackay

Many years ago, I might have read the quote above and said "Bah! That's a bunch of psychobabble." Then one day I started to get tired of being crabby every morning, and wondered maybe if I *pretended* to be in a good mood, I might actually *feel* that way. And I'll be darned if it didn't work. It didn't change my life overnight, but I've become a believer in the notion that on most days, we have the power to choose whether to be in a good mood or a sour one. It's an either/or proposition, so why not choose the more pleasant outcome?

What does this have to do with a unit on probability? Many probability problems involve situations that have just two outcomes. When a baby is born, it will be either male or female. When a coin is flipped, it will land either heads or tails. When the New York Yankees play, they either win or they lose. The classmate you think is cute will either go out with you or will not. Other situations can be reduced to two outcomes. For example, medical procedures can be classified as either successful or unsuccessful. An answer to a multiple choice exam question can be classified as right or wrong even though there may be four answer choices. Since either/or situations like this are common, it's worth a bit of our time and effort to study probability in this setting.

0. On any given day, you can be in a good mood or a bad mood. Does that mean the probability of being in a good mood is 1/2? Explain.

3-8 Class

Situations like the ones mentioned above are called *binomial experiments.* This is the only type of probability question we'll study in this lesson, so we should begin by clearly defining exactly what types of questions qualify.

A **binomial experiment** is a probability experiment that satisfies the following requirements:

1. Each trial has only two outcomes, or outcomes that can be reduced to two outcomes. These outcomes can be considered either a success or a failure.
2. The outcomes must be independent of each other.
3. There must be a fixed number of independent trials.
4. The probability of a success must remain the same for all trials of the experiment.

In Questions 1–4, decide whether or not each is a binomial experiment, and briefly explain your choice.

1. Drawing a card from a deck and seeing what suit it is

2. Answering a question on a true-false test

3. Asking 100 people whether or not they smoke

4. Drawing cards at random from a deck without replacement and deciding if they are red or black cards

Did You Get It

Try this problem to see if you understand the concepts we just studied. The answer can be found at the end of the portfolio section.

1. Decide whether or not each experiment is a binomial experiment.
   a. Picking a colored ball with replacement from an urn containing three balls of different colors, and seeing if the chosen ball is orange
   b. Picking a number from a bingo machine
   c. Drawing a card at random from a deck with replacement and noting its color, red or black
   d. Rolling a die and getting a 3

Here's a more specific example of a binomial probability experiment. You're one of four finalists in a raffle, and each finalist is represented by a single colored ball thrown into a box. The colors are red, black, white, and green (you). A ball is picked from the box and its color is recorded, then it's mixed back into the box and a second ball is picked. If your green ball is picked twice, you win a 2-year lease on a Fiat convertible. As each ball is chosen, you don't care what color it is if it isn't green. All you're interested in is whether the chosen ball is green or not. The tree diagram for two trials of this experiment is shown below.

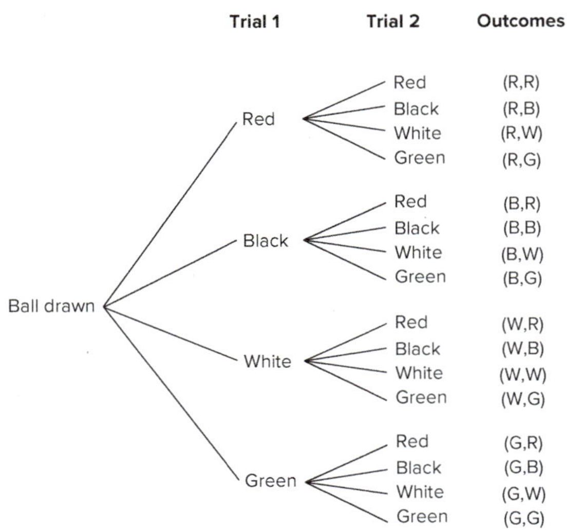

5. How many outcomes are in the sample space?

6. How many outcomes result in no green balls being picked? List them.

7. How many outcomes result in one green ball being picked? List them.

8. How many outcomes result in you driving away in a brand-new Fiat? List them.

9. Use your answers from Questions 5–8 to fill in the probability distribution.

Number of Green Balls Picked	Probability
0	
1	
2	
Sum of probabilities	

Let's understand how the experiment can be considered a binomial experiment. We'll make sure it meets the four conditions of a binomial experiment.

10. Even though four colors can be chosen in each trial, why are there only two outcomes that we're interested in? Explain.

11. Are the outcomes independent of each other? Why?

12. Is there a fixed number of trials? What is that fixed number?

13. Explain how the probability of success remains the same for each trial. What is the probability of success and the probability of failure on each trial?

Next, we're going to make some observations that will help you to develop a formula for the probability of events in binomial experiments. We're going to use the following letters:

n is the total number of trials in a binomial experiment.
x is the number of successes.
p is the probability of success on any given trial.
q is the probability of failure on any given trial.

Now our key observations:

- If there are x successes in n trials, there must be $n - x$ failures. And if the probability of success is p, the probability of failure is $1 - p$.
- Since the trials are independent, if the probability of success on one trial is p, the probability of three successes in a row is $p \cdot p \cdot p$, or p^3; the probability of five successes in a row is $p \cdot p \cdot p \cdot p \cdot p = p^5$; and the probability of x successes is p^x.
- Using the same reasoning, the probability of $n - x$ failures is q^{n-x}.

Now let's build that formula. The goal is to find the probability of getting x successes in n trials of a binomial probability experiment.

14. First, note that we're interested in the probability of getting x successes AND $n - x$ failures. And we know the probability of x successes and $n - x$ failures (from our key observations). Use what we learned about probabilities for "and" events to write the probability of getting x successes and $n - x$ failures in n trials.

15. The issue with that formula is that it provides the probability of one particular combination of failures and successes. But we don't know (or care, really) WHICH trials result in successes. We're only interested in the total number of successes. So you'll need to multiply your answer from Question 14 by the number of ways to get x successes in n trials. Do that and you've developed the key formula for this lesson.

Needless to say, after reading the information in the colored box you should make sure that you wrote the right formula in Question 15. Then you should make sure that you understand the formula, and what exactly it calculates.

The Binomial Probability Formula

When conducting a binomial probability experiment, the probability of getting x successes in n trials is

$$P(x) = {}_nC_x \cdot p^x \cdot q^{n-x}$$

where p is the probability of a success on one trial, and q is the probability of failure on one trial. Note that q is $1 - p$.

Now let's get back to trying to win you that convertible. Remember, we were drawing a ball from a box with four balls, colored red, green, black, and white. The ball was then replaced and another was drawn.

Solved Example

Computing a Binomial Probability

You're one of four finalists in a raffle, and each finalist is represented by a single colored ball thrown into a box. The colors are red, black, white, and green (you). A ball is picked from the box and its color is recorded, then it's mixed back into the box and a second ball is picked. If your green ball is picked twice, you win a 2-year lease on a Fiat convertible. Use the binomial probability formula to find the percent chance that you'll win.

SOLUTION

As we've already seen, this qualifies as a binomial probability experiment. For each of two trials, "success" means the green ball gets drawn. What we want is the probability of two successes in two trials. Here's the information we know:

$n = 2$
$x = 2$
$p = 1/4$
$q = 3/4$

Now we just need to plug this info into our binomial probability formula:

$$P(\text{Two successes}) = {}_2C_2 \cdot \left(\frac{1}{4}\right)^2 \left(\frac{3}{4}\right)^{2-2} = 1 \cdot \frac{1}{16} \cdot 1 = \frac{1}{16}$$

This is the probability, but we were asked for the percent chance. So we convert 1/16 to decimal form, which is 0.0625. You have a 6.25% chance of winning the car.

16. If your color is picked once, you at least get 100 bucks and a bag of muffins. What's the probability of that happening?

17. How do the answers to the Solved Example and Question 16 compare to the results you got from the tree diagram? Why is it worthwhile to compare them?

18. Suppose that the morning after your birthday, you remember that you have a 20-question true/false quiz in your early class. Uh oh! Completely unprepared and a little woozy, you decide to guess on every question. Use the binomial probability formula to find the probability that you'll get 12 out of 20 right, and just barely pass. Begin by identifying values for all variables in the formula.

> **Math Note**
>
> There are 20 trials in this experiment, so to say the very least it would be cumbersome to draw a tree diagram. To say the very most, it would be completely ridiculous. The last level of the tree would have over a million branches.

Did You Get It

2. If you take a 10-question multiple choice quiz, with four choices for each question, and completely guess on every one, what's the probability of getting exactly 6 questions right? (Only one of the choices is the right answer.)

Not surprisingly, both graphing calculators and spreadsheets have commands for computing binomial probabilities.

Using Technology: Computing Binomial Probabilities

TI-84 Plus

You could certainly calculate directly using the combination command we already know multiplied by p^x and q^{n-x}. But graphing calculators have a built-in command for computing binomial probabilities directly. The **binompdf(*n*,*p*,*x*)** command can be found by pressing **2nd VARS,** which is the **DISTR** menu. Then scroll down to binompdf. (Don't confuse this with binomcdf, which we'll use later.) The screen shot shows how this works for finding the probability of guessing 12 out of 20 true/false questions correctly.

```
binompdf(20,1/2,
12)
        .1201343536
```

Notice that the inputs in parentheses correspond to the number of trials (n), then the probability of success (p), and finally the number of successes (x).

See the Using Tech video on Computing Binomial Probabilities with a Graphing Calculator for further information.

Using Technology: Computing Binomial Probabilities

Excel

The Excel command is =BINOM.DIST. When you start typing =BINOM.DIST in Excel here's what appears, as a guide to help you:

=BINOM.DIST(number_s, trials, probability_s, cumulative), where

- number_s = number of successes (This is our *x*.)
- trials = number of trials (*n*)
- probability_s = the probability of success (*p*)
- cumulative = false; this tells Excel that you want the probability of EXACTLY *x* successes.

The sample below also shows the calculation for guessing 12 of 20 true/false questions correctly.

A1 *fx* =BINOM.DIST(12,20,1/2,FALSE)

	A	B	C	D
1	0.1201344			

See the Using Tech video on Computing Binomial Probabilities in Excel for further information.

3-8 Group

We were able to compute the probability of getting exactly 12 questions right when guessing on a 20-question quiz. But wouldn't it be more reasonable to ask a question like "What's the probability of getting AT LEAST 12 questions right?" You'd be a whole lot more likely to wonder how likely you were to pass, rather than to get exactly 12 right. You'll answer that question in the homework. For now, we'll look at a similar, but simpler, question.

Of five physical therapists who work at a rehab center, three have master's degrees and two have doctorates. Each therapist is equally likely to be assigned to a patient on any given visit. Suppose Tom has five sessions scheduled in the next 2 weeks.

1. Find the probability that he gets a therapist with a doctorate twice.

In Questions 2–5, we'll work on finding the probability that he gets a therapist with a doctorate less than two times.

2. List the possible number of times he could get a therapist with a doctorate if it is less than two times.

3. Are these outcomes mutually exclusive? Explain.

4. In that case, what should we do with the probabilities?

5. Find the probability using the binomial probability formula and the operation you specified in Question 4.

Did You Get It

3. If a different patient at the rehab center in Questions 1–5 has seven appointments scheduled, find the probability that she gets a therapist with a master's degree
   a. Every time.
   b. At least five times.

Questions 1–5 show that we can use the binomial probability formula and the addition rule for independent events to answer questions like "What's the probability of getting at least 12 questions right when guessing on a 20-question true/false quiz?" But doing so requires calculating several individual probabilities: the probability of getting exactly 12 right, then 13, 14, 15, 16, 17, 18, 19, and 20. Who has the time (or the patience) for that? I know that I don't, and I'm guessing that you probably do not either.

Fortunately, technology can save the day. Graphing calculators and spreadsheets have functions that are specifically designed to find the probability of x successes OR LESS in a binomial experiment. This is known as a **cumulative probability.**

Using Technology: Computing Cumulative Binomial Probabilities

TI-84 Plus

The **binomcdf(*n*,*p*,*x*)** command can be found by pressing **2nd VARS,** which is the **DISTR** menu. Then scroll down to binomcdf. This calculates the probability of getting *x* successes or less in *n* trials. The screen shot shows how this works for finding the probability of zero or one therapists with a doctorate (Questions 1–5).

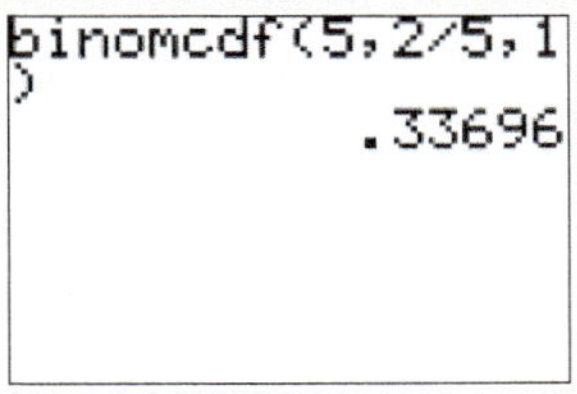

Notice that the inputs in parentheses correspond to the number of trials (5), then the probability of success (2/5), and finally the number of successes or less that we're interested in (1).

Excel

To find a cumulative binomial probability, use the same command we already learned: =BINOM.DIST, followed by the number of successes, the number of trials, and the probability of success on each trial in parentheses. The only difference is that you finish the string in parentheses with true rather than false. This tells the formula to find the probability of *x* successes OR LESS.

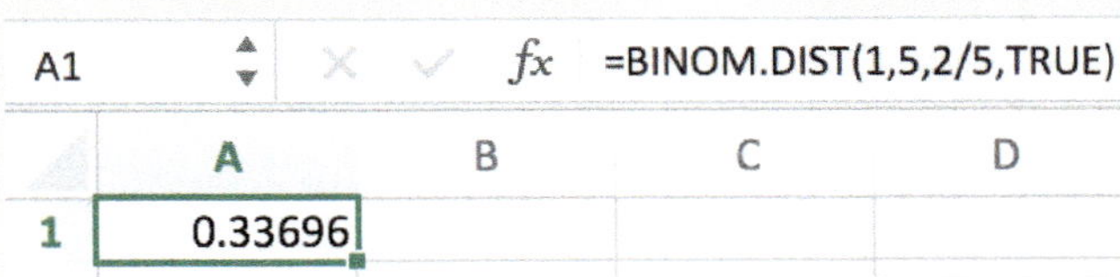

See the Using Tech video on Computing Cumulative Binomial Probabilities for further information.

We'll close the lesson with some practice on using technology to find binomial probabilities.

We can only use the binomial probability formula for a series of trials that are independent. As we saw in Lesson 3-7, when subjects are chosen from a sample and not replaced, the choices aren't independent, but if the sample is really large, the lack of replacement has a negligible effect on probability. So we can treat the choices as independent even though they technically aren't, which comes in handy when using the binomial probability formula.

According to the Edumed.org, 14% of college undergrads don't have health insurance. Suppose that 10 undergrads are surveyed at random.

6. Find the probability that half of them will be uninsured. Why can we use our binomial probability formula?

7. Find the probability that less than seven will be uninsured.

8. Find the probability that three or less will be uninsured.

9. What's the probability that four or more will be uninsured?

Question 9 brings up a really useful point. The cumulative binomial commands are designed to find the probability of a certain number of successes or LESS. If you're asked to find the probability that a certain number of successes or MORE occurs, you can use a cumulative binomial command, then find a complement.

Did You Get It

4. In a large community, it was determined that 44% of the residents use the public library at least once a year. If 10 people are picked randomly, find the probability that
   a. Exactly 2 of them have used the library during the last year.
   b. At least 5 of them have used the library during the last year.

3-8 Portfolio

Name ____

Check each box when you've completed the task. Remember that your instructor will want you to turn in the portfolio pages you create.

Technology

1. ☐ In this Excel activity, we'll calculate a binomial probability distribution and a cumulative binomial probability distribution.

ALEKS Practice

1. ☐ Include any written work from the ALEKS assignment along with any notes or questions about this lesson's content.

Applications

1. ☐ Complete the applications problems.

Reflections

Type a short answer to each question.

1. ☐ How can we recognize when a probability experiment qualifies as a binomial experiment?
2. ☐ Give a brief description of how an experiment with several outcomes can be reduced to one with two outcomes.
3. ☐ In a binomial experiment, given the probability of any trial being a success, explain how to find the probability of a failure.
4. ☐ Name one thing you learned or discovered in this lesson that you found particularly interesting.
5. ☐ What questions do you have about this lesson?

Looking Ahead

1. ☐ Complete the prep skills for Lesson 4-1.
2. ☐ Read the opening paragraph in Lesson 4-1 carefully and answer Question 0 in preparation for that lesson.

Answers to "Did You Get It?"

1. a. Yes **b.** No **c.** Yes **d.** Yes **2.** About 0.016
3. a. About 0.028 **b.** About 0.420 **4. a.** About 0.084 **b.** About 0.470

Answers to Prep Skills Questions

1. 1,495
2. $\frac{1}{8}$ or 0.125; $\frac{262{,}144}{1{,}953{,}125}$ or 0.134
3. $\frac{1}{3}$

3-8 Applications

Name ______________________________

In Questions 1–6, decide if the experiment or question described is or is not a binomial experiment, then explain your reasoning. You don't need to find probabilities.

1. Keeping track of the number of customers in a week that get pepperoni on their pizza

2. Recording the numerical results of rolling two dice 20 different times

3. Playing 10 different people in tennis and recording whether or not you win

4. Drawing 12 different cards from a standard deck and recording whether or not each is a spade

5. Finding the probability that three out of the next five men you run into are 6′3″ or taller, given that 3.9% of men in America are taller than 6′2″.

6. Playing 25 rounds of rock, paper, scissors against the same person and recording the result (If you're not familiar with the game, you can find the rules online very easily.)

3-8 Applications

Name ______________________________

7. Of all identity thefts, 27% are solved in 1 day or less. If 10 cases are studied, find the probability that exactly 5 are solved in 1 day or less.

8. Of all people who do banking, 16% prefer to use ATMs. Of 20 people who are banking customers, find the probability that exactly 4 prefer to use the ATM.

9. The research group Statcounter reported that in 2024, 58.8% of smartphones in operation in the U.S. were iPhones. If this is still accurate, what's the probability that less than 7 of the next 10 smartphone users you see have iPhones?

10. A Telenav survey found that one in five people surveyed would rather go shoeless than cell phone-less for a week. Find the probability that if 25 people are surveyed, between 20% and 40% of them would be more willing to give up their shoes than their phones (include 20% and 40% of respondents).

3-8 Applications

Name ______________________________

11. Various studies have indicated that about 40% of Americans have at least mild aviophobia, which is the fear of flying. About 5% have a severe enough case that they're incapable of getting on a plane. If 50 Americans are randomly selected, what is the probability that less than 20 will have at least some fear of flying?

12. In a recent survey, 2% of the people surveyed said that they would keep their current job if they won a multi-million-dollar lottery. If 20 people are chosen randomly, find the probability that 3, 4, or 5 of them would keep their job.

13. If you take a true/false quiz with 20 questions and totally guess on every one, find the probability that you pass (get at least 60% right).

14. After taking a 10-question multiple choice quiz where there are four choices on each question (and only one is correct), a classmate who scored 9 out of 10 right says they didn't study, didn't even read the questions, and randomly guessed on all 10. Would you believe him? Use what we've learned in this lesson to back up your answer.

Math in Gambling REVISITED

View Stock/age fotostock

1. The odds against winning the Powerball jackpot are 292 million to 1. The probability of getting struck by lightning is 1/600,000. You are WAY WAY WAY more likely to get struck by lightning. By the way, think about 292 million to 1 for a second. Your chance of winning Powerball with one ticket is only slightly better than your chance of having your name chosen out of a hat (a really big hat) containing the names of *every human being in the United States.*
2. The probability of getting killed in an accident if you text and drive varies by year, but estimates are that you're AT LEAST four times as likely to get killed on the road if you text. Not very smart, is it?
3. **a.** Using the table for rolling two dice on page 231, the probability of rolling 2, 3, 11, or 12 is $\frac{1}{6}$, meaning the probability of losing is $\frac{5}{6}$. So the two outcomes are +\$4 with probability $\frac{1}{6}$, and −\$1 with probability $\frac{5}{6}$. The expected value of each trial is then

$$+4 \cdot \frac{1}{6} + (-1) \cdot \frac{5}{6} = -\frac{1}{6}$$

In 100 trials, you would expect to lose $\$100\left(\frac{1}{6}\right)$, or \$16.67.

b. The probability of winning \$35 is $\frac{1}{38}$, and the probability of losing \$1 is $\frac{37}{38}$. The expected value of each spin is

$$+35 \cdot \frac{1}{38} + (-1) \cdot \frac{37}{38} = -\frac{2}{38} = -\frac{1}{19}$$

In 100 trials, you would expect to lose $100\left(\frac{1}{19}\right)$ or \$5.26.

c. Using the same idea as scenarios a and b, the expected value of each ticket is

$$+20{,}000{,}000 \cdot \frac{1}{175{,}711{,}536} + (-1) \cdot \frac{175{,}711{,}535}{175{,}711{,}536}$$

$$\approx -\$0.8862$$

Buying 100 tickets, you would expect to lose \$100(−0.8862), or \$88.62.

d. You would win \$1 with probability $\frac{1}{2}$, and lose \$1 with probability $\frac{1}{2}$, so the expected value of each flip is

$$1 \cdot \frac{1}{2} + (-1) \cdot \frac{1}{2} = 0$$

In 100 flips you would expect to break even.

The best by far is flipping a coin, followed by roulette, the church fair dice game, and, bringing up the rear by quite a bit, the multistate lottery. Notice that under the best of circumstances, you are likely to break even!

Unit 4
Statistically Speaking

James W. Hall

Outline

Math in Sociology

Broadly defined, sociology is the study of human behavior within society. One important branch is criminology. This is not about investigating crimes, but rather studying patterns of criminal behavior and their effect on society. One of the main tools used by sociologists is statistics. This important area of math allows researchers to study patterns of behavior objectively, by analyzing information gathered from a mathematical perspective, not a subjective one.

It all begins with the gathering of data. Data are measurements or observations that are gathered for a study of some sort. Let's look at an example. There's an old adage in police work that violent crime increases with the temperature. A statement like "My cousin Jed is a cop, and he told me that there are more violent crimes during the summer" is not evidence, nor is it data. It's the subjective opinion of one guy. In order to study whether this phenomenon is legitimate, we would need to gather information about the number of violent crimes at different times during the year, as well as temperature information, and study those numbers objectively to see if there appears to be a connection. This is one of the most important topics in this chapter.

Data on its own isn't good for very much unless we develop methods for organizing, studying, and displaying that data. These are the methods that make up the bulk of this chapter. After discussing methods of gathering data, we'll learn effective methods for organizing and displaying data so that it can be presented in a meaningful, understandable way. We will then turn our attention to techniques of analyzing data that will help us to unlock the secrets of what sets of data may be trying to tell us.

As for math in sociology, the table shows the number of homicides committed in each month of 2023 in Chicago, along with the average high temperature for that month in degrees Fahrenheit. After finishing the chapter, you'll be able to analyze the data and answer the following questions. For now, discuss the table with your group and see if you can come to a consensus as to whether the claim about violent crime being connected to temperature seems to be legitimate.

Month	Avg. High Temp.	Homicides
1 (Jan.)	37.0	37
2	42.0	49
3	45.4	49
4	61.7	44
5	73.1	57
6	80.6	55
7	83.8	62
8	82.4	58
9	76.9	49
10	64.1	36
11	50.4	42
12	43.9	42

Source: Chicago Sun-Times

1. Draw a bar graph for these data, with months on the horizontal axis and number of homicides on the vertical. Do you notice any trends?
2. There are four different numbers that might be considered an "average" for the number of homicides: mean, median, mode, and midrange. Find each for the data. Which do you think is most revealing? Explain.
3. Based on a graph of the data called a scatter plot, does there appear to be a relationship between temperature and homicide rate?
4. Can you conclude, using a measure called the correlation coefficient, with 95% certainty that there is in fact a relationship between the data sets? What about with 99% certainty?
5. Do the results indicate that warmer temperatures cause people to behave in a more violent manner? Discuss. A one-sentence answer here won't cut it: This is a really important question.

For answers, see Math in Sociology Revisited on page 533.

Lesson 4-1 Prep Skills

SKILL 1: READ INFORMATION FROM A TABLE

Reading information from a table is a valuable skill that, like many things, only improves with practice. Some helpful tips for reading tables include:

- Some tables include a title and/or a caption. Read the title or the caption to understand what the table is presenting.
- Some tables have headings for each column and labels for each row. Read the headings for each column and the labels/descriptions for each row to gain more understanding about the data in the table.
- As you examine the data in a table, look for patterns in the data.
- Some tables will not have headings at all and may provide a description of the data before or after the data. Read the description carefully to pick out units and other key information.

PREP SKILLS QUESTIONS

The cost in thousands of dollars of all houses sold in one town for the last week are below.

146	197	112	124	133	219	245	150	165	179
119	133	202	107	198	348	167	209	124	123

1. How many houses sold for less than $125,000?
2. What was the cost of the most expensive house that sold?
3. What was the cost of the least expensive house that sold?
4. How many houses sold last week?
5. How many houses had a sale price between $150,000 and $175,000?

Lesson 4-1 Crunching the Numbers

LEARNING OBJECTIVES

- ☐ 1. Explain the difference between a population and a sample.
- ☐ 2. Compare and contrast different sampling methods.
- ☐ 3. Organize data with frequency distributions.
- ☐ 4. Analyze data with stem and leaf plots.

Chad Baker/Photodisc/Getty Images

Errors using inadequate data are much less than those using no data at all.
—Charles Babbage

One of the big issues being discussed over the last 20 years or so in higher education is grade inflation—the perception that college students in general are getting much higher grades than they used to. But perception and reality are often two different things, so how could we decide if this is actually taking place? And if so, what are some possible reasons? Are students just getting smarter? Are professors lowering their standards?

In order to examine questions like this, it would be very valuable to gather some **data,** which are measurements or observations that are gathered for an event under study. The term "data age" is coming into vogue to describe this era because of the vast amount of data that is available to anyone with Internet access and a web browser. But the expense of having access to so much data is that knowing how to gather data appropriately, and how to organize them into a useful form, is more important than it's ever been. So that's what we'll study in this lesson.

0. Do you think having access to so much data is a good thing or a bad thing? Explain.

4-1 Group

If it doesn't make you uncomfortable, exchange the following information with the classmates in your Unit 4 group. This will be your small group for the fourth unit. It would be a good idea to schedule a time for the group to meet to go over homework, ask/answer questions, or prepare for exams. You can use this table to help schedule a mutually agreeable time.

Name	Phone Number	Email	Available Times

As we begin our study of statistics, it would seem reasonable to have a really good idea of what that actually means. **Statistics** is a branch of math that involves collecting, organizing, summarizing, and presenting data and drawing general conclusions from those data. If we're talking about grade inflation, then relevant data might be average grade point averages for students in several different years. The table below shows some GPA data from the National Center for Education Statistics.

Year	Average Undergraduate GPA
1990	2.81
1993	2.85
1996	2.84
2000	2.89
2004	2.96
2008	2.94
2012	2.94
2016	2.93
2020	3.15

In statistics, we *start* with a question we'd like to answer, like "Is grade inflation a real phenomenon?" *Then* we gather data and try to interpret them in order to answer that question.

1. Write a verbal description of the information provided in the table. Do you think it supports the idea of grade inflation?

2. How do you think these numbers were calculated? What students' GPAs were considered? Discuss.

ALEX S/Shutterstock

The ideal way to gather GPA data would be to add the GPAs of every college student in America during each year, then divide by the number of students. Needless to say, that's not particularly realistic. Aside from legal and privacy ramifications, in any given year there are more than 20 million students enrolled in American colleges, and by the time all of that GPA data were gathered and recorded, many of them would be contemplating retirement. Instead, GPA data are obtained from a *sample* of college students.

A **population** consists of all subjects under study. A **sample** is a representative subgroup or subset of a population.

3. In the case of the GPA data we've been studying, the population is ____________________.

4. The sample is ____.

In this case, GPA data from a sample of 170 different colleges were used to compute the average. The thing that makes the data compelling is that the sample is representative of the population as a whole. The schools chosen vary in size, cost, and geographic location. Some are public, some are private.

5. Discuss what exactly is meant by a representative sample, and why it's so important in answering questions using statistics.

Sampling Methods

We will study four basic sampling methods that can be used to obtain a representative sample: random, systematic, stratified, and cluster.

In order to obtain a **random sample,** every subject of the population has to have an equal chance of being selected. The best way to obtain a random sample is to use a list of random numbers. These can be generated from a variety of Internet sources, or from a computer or calculator. Subjects in the population are numbered, and then the ones to be studied are picked by using the corresponding random numbers.

> **Math Note**
>
> The word "data" is plural, so we say "data are," not "data is." The singular version is "datum."

6. If you were going to do a survey with every student at your campus as the population, describe how you might obtain a random sample to survey. More specific is better.

A **systematic sample** is taken by numbering each member of the population and then selecting every kth member, where k is some natural number. For example, researchers might have numbered all of the colleges willing to release grade information and chosen every tenth one. In this case, $k = 10$. When using systematic sampling, it's important that the starting number is chosen randomly.

7. Let's get back to the campus survey mentioned in Question 6. Describe how you might develop a systematic sample.

When a population is divided into groups where the members of each group have similar characteristics (like large public schools, large private schools, small public schools, and small private schools) and a certain number of members from each group are chosen at random, the result is called a **stratified sample.** For example, the grade inflation researcher might have decided to choose five schools from each of those groups. Medical researchers often stratify samples by dividing subjects into sex groups, age groups, racial groups, or many other categories. They would then randomly choose a similar number of subjects from each group to compose the stratified sample.

8. Describe how you'd put together a stratified sample for that campus survey we've been talking about.

When an existing group of subjects that represent the population is used for a sample, it is called a **cluster sample.** For example, suppose we were interested in studying hunger in Ohio. There are 88 counties in the state, so we could randomly pick a dozen of those counties. We could then choose a random sample of cities in each selected county to gather data on the number of food insecure families in that city. In the case of the grade inflation study, researchers might look for existing groups that colleges fall into. There are several different organizations that provide accreditation to colleges in the U.S. If a couple of those organizations were chosen randomly, then schools from each of those groups could be surveyed for GPA data. The accreditation organizations would be the clusters.

It can be a little tricky to distinguish between stratified and cluster samples, so you might find it helpful to think of it this way: If the entire population is divided into distinct categories and some subjects are chosen from each category, that's a stratified sample. If the subjects are chosen from just a small handful of existing groups, that's a cluster sample.

9. Describe how you might develop a cluster sample for our campus survey.

Math Note

In the last lesson of this unit, we'll study some ways that samples can be poorly chosen, either intentionally or not, which can invalidate the conclusions of a study.

Think about four different sampling methods the folks researching grade inflation could have used. For Questions 10–13, write which of the sampling methods we've studied was used for each situation. Explain your choice.

10. The researcher randomly selected three states and then asked all colleges willing to release grade information from those states.

11. The researcher numbered all of the colleges willing to release grade information and chose every 10th one.

12. The researcher numbered all of the colleges willing to release grade information and chose several of them using a random number generator.

13. The researcher grouped the schools by 2-year state schools, 4-year state schools, 2-year private schools, and 4-year private schools. The researcher then decided to choose five schools from each of those groups.

14. Rank the four sampling methods in Questions 10–13 from best to worst in terms of how representative you think the sample would be. Explain your rankings.

Did You Get It

Try this problem to see if you understand the concepts we just studied. The answer can be found at the end of the portfolio section.

1. To study the number of credit hours taken by a typical student, Shawna asks the registrar to provide email addresses for 10 freshmen, 10 sophomores, 10 juniors, and 10 seniors. From each group, she asks for one whose student ID ends in 0, one whose ID ends in 1, and so forth.
   a. What method of sampling did she use?
   b. Do you think the sample will be representative?

4-1 Class

Frequency Distributions

The data collected for a statistical study are called **raw data.** Raw data can be really difficult to interpret; in order to describe situations and draw conclusions, we need to organize the data in a meaningful way. Two methods that we will use are *frequency distributions* and *stem and leaf plots.* First, we'll study **categorical frequency distributions,** which are a way to organize data that are divided into distinct categories, like sex, your class standing in school, or conferences for college football teams.

A categorical frequency distribution is a table in which you list different categories, then count up how many individuals in a sample or population fall into each category and list those frequencies. In Questions 1 and 2, you'll create a categorical frequency distribution.

1. Twenty-five volunteers for a medical research study were given a blood test to obtain their blood types. The data are shown below. What would the categories for this data be? Why?

A	B	B	AB	O
O	O	B	AB	B
B	B	O	A	O
AB	A	O	B	A
A	O	O	O	AB

2. Build a categorical frequency distribution by writing in your categories, then putting tally marks in the second column as you count the number of individuals in each category.

Categories	Tally Marks	Frequency

3. Use your distribution to write some conclusions on blood types in the population. Do you think that there's enough data for your conclusions to be reliable?

Did You Get It ?

2. A health-food store recorded the types of vitamin pills 35 customers purchased during a 1-day sale. Construct a categorical frequency distribution for the data.

C C C A D E C E E A B D C E C E C C

C D A B B C C A A E E E E A B C B

Another type of frequency distribution that can be constructed uses numerical data and is called a **grouped frequency distribution.** In a grouped frequency distribution, the numerical data are divided into **classes.** For example, if you gathered data on the current grade percentage for people in your class, there's a chance that no two people have exactly the same percentage. In that case, everyone would be in a separate category, and each category would have one score in it. This, of course, would make the frequency distribution no more useful than the original list of raw data!

So it would be reasonable to group people into grade ranges, like 50–59 percent, 60–69 percent, 70–79 percent, and so forth. In the 50–59 percent class, we call 50 the **lower limit** and 59 the **upper limit.**

4. With these ranges, where would we put someone with a current grade of 69.8%?

As Question 4 illustrates, the way we choose classes is dependent on how the data values are rounded. The percentage classes we chose earlier would be fine if all the grades were rounded to the nearest percent. But if grades were rounded to the tenth of a percent we'd want to make the classes 50–59.9, 60–69.9, and so on.

Constructing a Grouped Frequency Distribution

Step 1: Decide how many classes you want to divide the data into. Between 5 and 15 is acceptable, depending on the number of data values in the set.

Step 2: Determine the width of each class:

$$\text{Class width} = \frac{\text{largest data value} - \text{smallest data value}}{\text{number of classes}}$$

Then round this value up to the next biggest whole number.

Step 3: Use the lowest number in the data set or a "nice" number lower than that number for the lower limit of the first class.

Step 4: Add the class width to the first lower class limit to get the second lower class limit. Do this for all classes.

Step 5: Find the upper class limit for each class by subtracting one from each lower class limit. (If the data is not rounded to the nearest whole number, you may need to adjust the upper class limits as described in the paragraph above the colored box.)

Step 6: Tally up the number of data values in each class, then put the frequencies next to each class in a table.

We'll illustrate by organizing the following data, which are the record high temperatures for each of the 50 states in degrees Fahrenheit.

State	Temp.	State	Temp.	State	Temp.	State	Temp
Alabama	112	Indiana	116	Nebraska	118	South Carolina	113
Alaska	100	Iowa	118	Nevada	125	South Dakota	120
Arizona	128	Kansas	121	New Hampshire	106	Tennessee	113
Arkansas	120	Kentucky	114	New Jersey	110	Texas	120
California	134	Louisiana	114	New Mexico	122	Utah	117
Colorado	115	Maine	105	New York	108	Vermont	107
Connecticut	106	Maryland	109	North Carolina	110	Virginia	110
Delaware	110	Massachusetts	107	North Dakota	121	Washington	120
Florida	109	Michigan	112	Ohio	113	West Virginia	112
Georgia	112	Minnesota	115	Oklahoma	120	Wisconsin	114
Hawaii	100	Mississippi	115	Oregon	119	Wyoming	115
Idaho	118	Missouri	118	Pennsylvania	111		
Illinois	117	Montana	117	Rhode Island	104		

Source: National Climatic Data Center

Using Technology: Using the Sort Tool in Excel and a Calculator

You may realize that it is a lot easier to build a grouped frequency distribution if the data is sorted.

1. Enter the data.
2. Select the cells containing the data.
3. Click on the **Data** tab.
4. Click on **Sort.**
5. Select options and click OK.

To sort data on a TI-84 calculator:

1. Enter the data in list L1. (Access L1 by first pressing STAT and ENTER.)
2. Press 2nd, STAT,), ENTER to access the **SortA(** feature.
3. Press 2nd, STAT, ENTER to input L1 after **SortA(.**
4. Press the) key and then press ENTER.
5. Go back to L1 and the data you entered should be sorted.

See the Using Tech Videos on Using the Sort Feature in Excel and a Calculator for further information.

In Solved Example 1, we'll ask you to build a grouped frequency distribution. To make that process much easier, we've used Excel to sort the data from the table by temperature.

	A	B
1	**State**	**Temp.**
2	Alaska	100
3	Hawaii	100
4	Rhode Island	104
5	Maine	105
6	Connecticut	106
7	New Hampshire	106
8	Massachusetts	107
9	Vermont	107
10	New York	108
11	Florida	109
12	Maryland	109
13	Delaware	110
14	New Jersey	110
15	North Carolina	110
16	Virginia	110
17	Pennsylvania	111
18	Alabama	112
19	Georgia	112
20	Michigan	112
21	West Virginia	112
22	Ohio	113
23	South Carolina	113
24	Tennessee	113
25	Kentucky	114
26	Louisiana	114
27	Wisconsin	114
28	Colorado	115
29	Minnesota	115
30	Mississippi	115
31	Wyoming	115
32	Indiana	116
33	Illinois	117
34	Montana	117
35	Utah	117
36	Idaho	118
37	Iowa	118
38	Missouri	118
39	Nebraska	118
40	Oregon	119
41	Arkansas	120
42	Oklahoma	120
43	South Dakota	120
44	Texas	120
45	Washington	120
46	Kansas	121
47	North Dakota	121
48	New Mexico	122
49	Nevada	125
50	Arizona	128
51	California	134

5. Why would a categorical frequency distribution be a bad idea for these data? (Hint: What would the categories be?)

Solved Example 1

Building a Grouped Frequency Distribution

Organize the record high-temperature data from the table into a grouped frequency distribution.

SOLUTION

Let's follow along the steps listed in the colored box:

Step 1: Decide how many classes we want. The lowest temperature in the table is 100 and the highest is 134, for a span of 34 degrees. With 50 data values, somewhere in the neighborhood of 6 to 8 classes sounds reasonable to me. Any more than that and we'd probably have many classes with just a couple values in them. It occurred to me that the span (34) is close to 35, and 35 divides evenly by 7, so I choose 7 classes.

Step 2: Find the class width. Here's the calculation:

$$\text{Class width} = \frac{\text{largest data value} - \text{smallest data value}}{\text{number of classes}} = \frac{134 - 100}{7} \approx 4.9$$

Rounding up we find that each class will cover a range of 5 degrees.

Solved Example 1 *Continued*

Step 3: Decide on a lower limit for the first class. The lowest data value is 100, and if we start there, our classes will have nice easy boundaries, so let's do it.

Step 4: Find the lower limit for all the other classes. Starting at a lower limit of 100 and adding the class width (5) repeatedly, we get 100, 105, 110, 115, 120, 125, 130.

Step 5: Find upper limits for each class. All of the data values are whole numbers, so we won't need a decimal part on our class limits. In this case, that makes the upper limits start at 104 and increase by the class width: 104, 109, 114, 119, 124, 129, 134.

Step 6: Set up a table and build the grouped frequency distribution. We set up the table with the class limits that we chose, then go through the data values one by one, putting a tally mark in the appropriate class for each value. Finally we add the tally marks to complete the distribution.

Class	Tally Marks	Frequency
100–104	III	3
105–109	~~IIII~~ III	8
110–114	~~IIII~~ ~~IIII~~ ~~IIII~~	15
115–119	~~IIII~~ ~~IIII~~ III	13
120–124	~~IIII~~ III	8
125–129	II	2
130–134	I	1

6. Write down at least three observations you can make from analyzing your frequency distribution.

7. The data below represent the team record for wins in a single season by all current Major League Baseball teams. Organize the data into a grouped frequency distribution. You may find it easier to sort the data first.

Team	Wins	Team	Wins	Team	Wins
Arizona	100	Houston	107	Philadelphia	102
Atlanta	106	Kansas City	102	Pittsburgh	110
Baltimore	109	Los Angeles Angels	100	San Diego	98
Boston	108	Los Angeles Dodgers	111	San Francisco	107
Chicago Cubs	116	Miami	92	Seattle	116
Chicago White Sox	100	Milwaukee	96	St. Louis	106
Cincinnati	108	Minnesota	102	Tampa Bay	100
Cleveland	111	New York Mets	108	Texas	96
Colorado	92	New York Yankees	114	Toronto	99
Detroit	104	Oakland/Las Vegas	107	Washington	98

Did You Get It

3. The online homework system I use in my calculus classes allows me to track how much time students spend working on homework. In my two Calc 2 sections last semester, here are the average number of hours spent per week for each student. Construct a grouped frequency distribution for the data.

1	2	6	7	12	13	2	6	9	5
18	7	3	15	15	4	17	1	14	5
4	16	4	5	8	6	5	18	5	2
9	11	12	1	9	2	10	11	4	10
9	18	8	8	4	14	7	3	2	6

Stem and Leaf Plots

Another way to organize data is to use a stem and leaf plot (sometimes called a stem plot). Each data value or number is separated into two parts. For a two-digit number such as 53, the tens digit, 5, is called the **stem,** and the ones digit, 3, is called its **leaf.** For the number 72, the stem is 7, and the leaf is 2. For a three-digit number, say 138, the first two digits, 13, are used as the stem, and the third digit, 8, is used as the leaf. For values rounded to the tenths place, like 8.4, you can use the value to the left of the decimal place as stem and the tenths place as leaf, as in our next example. In any case, the very last digit is used as the leaf, and what comes before is the stem. Note that we'll include a key with our plot that clarifies what the stems and leaves represent.

The following data are the percentage of days with at least 80% cloud cover at some point during daylight hours in 50 American cities:

City	Percent days with clouds	City	Percent days with clouds	City	Percent days with clouds	City	Percent days with clouds
Atlanta, Georgia	41%	Fresno, California	27%	Milwaukee, Wisconsin	48%	Richmond, Virginia	44%
Austin, Texas	37%	Grand Rapids, Michigan	56%	Minneapolis, Minnesota	46%	Rochester, New York	55%
Baltimore, Maryland	42%	Hartford, Connecticut	48%	Nashville, Tennessee	43%	Sacramento, California	27%
Birmingham, Alabama	42%	Honolulu, Hawai	26%	New Orleans, Louisiana	40%	Salt Lake City, Utah	38%
Boston, Massachusetts	45%	Houston, Texas	44%	New York, New York	36%	San Antonio, Texas	39%
Buffalo, New York	57%	Indianapolis, Indiana	49%	Oklahoma City, Oklahoma	36%	San Francisco, California	29%
Charlotte, North Carolina	42%	Jacksonville, Florida	39%	Orlando, Florida	36%	Seattle, Washington	62%
Chicago, Illinois	48%	Kansas City, Missouri	41%	Philadelphia, Pennsylvania	44%	St Louis, Missouri	45%
Cleveland, Ohio	55%	Las Vegas, Nevada	20%	Phoenix, Arizona	19%	Tampa, Florida	33%
Columbus, Ohio	52%	Los Angeles, California	28%	Pittsburgh, Pennsylvania	56%	Tulsa, Oklahoma	37%
Dallas, Texas	36%	Louisville, Kentucky	47%	Portland, Oregon	61%	Washington, DC	45%
Denver, Colorado	33%	Memphis, Tennessee	41%	Providence, Rhode Island	45%		
Detroit, Michigan	51%	Miami, Florida	32%	Raleigh, North Carolina	41%		

Source: National Conference of State Legislators

Solved Example 2

Building a Stem and Leaf Plot

Draw a stem and leaf plot for the cloudy day data provided.

SOLUTION

The logical choice of stems here is the first digit in each percentage, which means the second digit will be the leaves. Notice that the digits 1, 2, 3, 4, 5, and 6 appear in the first digits, so those will be our stems. We then go through the data values one by one and write each second digit under leaves next to the appropriate stem.

Stems	Leaves
1	9
2	7 6 0 8 7 9
3	7 6 3 9 2 6 6 6 8 9 3 7
4	1 2 2 5 2 8 8 4 9 1 7 1 8 6 3 0 4 5 1 4 5 5
5	7 5 2 1 6 6 5
6	1 2

Key: 1 | 9 means 19%

Notice the key at the bottom: this is really important for letting the reader know exactly what the stems and leaves represent.

Solved Example 2 *Continued*

There's nothing wrong with our plot at the moment, but to make it a bit more organized, it's customary to rearrange the leaves in increasing order, so our final plot looks like this:

Stems	Leaves
1	9
2	0 6 7 7 8 9
3	2 3 3 6 6 6 6 7 7 8 9 9
4	0 1 1 1 1 2 2 2 3 4 4 4 5 5 5 5 6 7 8 8 8 9
5	1 2 5 5 6 6 7
6	1 2

Key: 1 | 9 means 19%

Math Note

The online homework for this lesson requires that the leaves are ordered from smallest to biggest, so you should get in the habit of doing that whenever you draw a stem and leaf plot.

Pro tip: If you have the data in Excel, putting it in order from lowest to highest will make building the stem and leaf plot much quicker.

8. Write at least three observations about the cloudy day data from looking at the stem and leaf plot in Solved Example 2.

Rodney Todt/Alamy Stock Photo

Deciding on stems and leaves for the cloudy day data was fairly simple because there was a natural choice of stems (the first digit of each percentage) that gave us a reasonable number of them (6). That's not always the case: for the high-temperature data earlier in this lesson, the natural choice of stem would seem to be the first two digits. But here's what the stem and leaf plot would look like in that case.

Stems	Leaves
10	0 0 4 5 6 6 7 7 8 9 9
11	0 0 0 0 1 2 2 2 2 3 3 3 4 4 4 5 5 5 5 6 7 7 7 8 8 8 8 9
12	0 0 0 0 0 1 1 2 5 8
13	4

Key: 13 | 4 means 134

9. Explain why this stem and leaf plot isn't really very helpful in organizing the data.

In general, it's a good idea to choose stems so that there are between 5 and 15 categories. In the case of the high-temperature data, we'd get a better organization if we divided the categories in half, so that the first category covers 100 to 104, the second 105 to 109, and so on. To signify this, we can use the symbols $10^{(0)}$ to represent a stem of 10 with leaves that are digits from zero to four, and $10^{(5)}$ to represent a stem of 10 with leaves that are digits from five to nine.

10. Complete the stem and leaf plot for the high-temperature data on page 359. We've helped you start by including the first four stems. Don't forget to include a key.

Stems	Leaves
$10^{(0)}$	
$10^{(5)}$	
$11^{(0)}$	
$11^{(5)}$	
$12^{(0)}$	
$12^{(5)}$	
$13^{(0)}$	

11. Describe why this stem and leaf plot makes it easier to analyze the organized data than the preceding one.

Did You Get It ?

4. It's no secret that gas prices have fluctuated wildly in the last 20 years, but from 1980–1999, they were surprisingly stable. According to the Energy Information Administration, the data below represent the average price (in cents) per gallon of regular unleaded gas for those years. Draw a stem and leaf plot of the data.

119	131	122	113	112	86	90	90	100	115
114	113	111	111	115	123	123	123	106	117

12. Compare grouped frequency distributions and stem and leaf plots. How are they similar? Which do you think makes it easier to analyze a set of data?

4-1 Portfolio

Name ____

Check each box when you've completed the task. Remember that your instructor will want you to turn in the portfolio pages you create.

Technology

1. ☐ In this Excel activity, we'll build a frequency distribution based on blood type data.
2. ☐ In this Excel activity, we'll build a frequency distribution based on record high temperatures.

ALEKS Practice

1. ☐ Include any written work from the ALEKS assignment along with any notes or questions about this lesson's content.

Applications

1. ☐ Complete the applications problems.

Reflections

Type a short answer to each question.

1. ☐ What is the point of sampling in statistics? What's the difference between a sample and a population?
2. ☐ What makes cluster sampling different from stratified?
3. ☐ Why is it bad to have too few or too many categories in a grouped frequency distribution?
4. ☐ Name one thing you learned or discovered in this lesson that you found particularly interesting.
5. ☐ What questions do you have about this lesson?

Looking Ahead

1. ☐ Complete the prep skills for Lesson 4-2.
2. ☐ Read the opening paragraph in Lesson 4-2 carefully and answer Question 0 in preparation for that lesson.

Answers to "Did You Get It?"

1. a. Stratified sample **b.** The last digit of a student ID number is likely to be totally random, so this should lead to a representative sample.

2.

Type	Tally Marks	Frequency
A	卌 I	6
B	卌	5
C	卌卌 II	12
D	III	3
E	卌 IIII	9

3.

Class	Tally Marks	Frequency
1–3	卌卌	10
4–6	卌卌 IIII	14
7–9	卌卌	10
10–12	卌 I	6
13–15	卌	5
16–18	卌	5

4.

Stems	Leaves
8	6
9	0 0
10	0 6
11	1 1 2 3 3 4 5 5 7 9
12	2 3 3 3
13	1

Key: 13 | 1 means 131

Answers to Prep Skills Questions

1. 6
2. \$348,000
3. \$107,000
4. 20
5. It depends on if you do or do not include \$150,000: either 2 or 3.

4-1 Applications

Name ______________________________

1. Twenty-five fans of reality TV were asked to rate four shows, and the data below reflect which one each rated highest. (S = Survivor, D = Dancing with the Stars, B = Big Brother, A = The Amazing Race)

S	S	B	A	D	B	S	D	A	B	A	A	S
B	A	D	B	A	S	A	D	B	B	S	A	

a. Construct a frequency distribution for the data.

b. Which show did those 25 fans rate the highest?

2. The data shown are the cigarette taxes per pack in cents imposed by each state as of July 22, 2024.

68	435	320	129	351	170	200	203	153	60
200	210	57	110	200	64	535	333	62	303
200	450	298	108	304	180	45	260	141	120
115	134	100	200	68	178	44	425	170	252
194	37	136	375	17	270	160	57	308	60

Source: American Lung Association

a. Construct a grouped frequency distribution for this data using seven classes with the first one being 0–69.

b. Which class contains the highest number of values? Write a sentence explaining what this means using units on any numbers used.

4-1 Applications

Name ______________________________

3. As an experiment in a botany class, plants are placed in a greenhouse, and their growth in centimeters after 25 days is recorded, with the results shown below in a stem and leaf plot.

Stems	Leaves
1	2
2	0 3
3	2 5 8 8 9
4	1 3 3
5	0 1 2 3 3 5 8 9 9

Key: 1|2 means 12

a. How many plants grew 203 cm?

b. How many plants grew 38 cm?

c. How many plants had their growth measured and recorded?

d. What was the largest growth recorded for any plant?

4. The National Insurance Crime Bureau reported that these data represent the number of registered vehicles per car stolen for 35 selected cities in the United States. For example, in Miami, 1 automobile is stolen for every 38 registered vehicles in the city. (The data have been rounded to the nearest whole number.)

38	53	53	56	69	89	94
41	58	68	66	69	89	52
50	70	83	81	80	90	74
50	70	83	59	75	78	73
92	84	87	84	85	84	89

a. Build a stem and leaf plot for the data.

Stems	**Leaves**

4-1 Applications

Name ____________________

b. Build a grouped frequency distribution for the data, where the classes correspond to the stems that you chose.

c. A **histogram** is very much like a bar graph that represents the information in a frequency distribution. There are some key differences between a typical bar graph and a histogram. First, the width of the bars has meaning and is representative of the class width. Second, the bars are drawn right next to each other with no space in between. Create a histogram from the grouped frequency distribution above.

Vehicles Registered per Stolen Vehicle

Class

> **Math Note**
>
> Histograms are really important in the next lesson. We'll learn that the way in which the horizontal axis is labeled is also usually different for histograms.

d. Which class contains the most values? Explain what this tells us.

e. Which class would you want to have your city listed in? Why?

4-1 Applications

Name ____________________

f. Of the three methods you used to represent the data, which was the easiest to interpret in your opinion? Explain.

Lesson 4-2 Prep Skills

SKILL 1: PERFORM COMPUTATIONS INVOLVING PERCENTS

Working with percentages has come up often in Prep Skills. Remembering that "percent" literally means "per hundred" is a big key to working with percentages.

- $3.2\% = \frac{3.2}{100} = 0.032$

  Converting a percentage to its decimal equivalent requires dividing the percent by 100, which shifts the decimal point two places to the left.

- What percent of 20 is 12?
  $\frac{12}{20} = 0.6 = 60\%$

  In decimal form, 12 divided by 20 is 0.6. To convert the decimal to a percentage, multiply the decimal by 100, which shifts the decimal point two places to the right. The result tells us that 12 is 60% of 20.

- What is 40% of 60?
  $0.40(60) = 24$

  The decimal form of 40% is 0.40. The calculation tells us that 40% of 60 is 24. The word "of" represents multiplication, so we get 0.4 times 60.

- 24 represents 15% of a number. Find the number.
  $0.15n = 24$
  $n = 160$

  It's helpful to rephrase this question to read "15% of a number is 24." Then set up an equation where n represents the unknown number. Divide both sides by 0.15 to solve the equation. The solution tells us that 15% of 160 is 24.

SKILL 2: FIND THE MIDPOINT OF TWO NUMBERS

Finding the midpoint of two numbers amounts to finding the arithmetic mean (average) of those two numbers.

- Find the midpoint of 12 and 28.

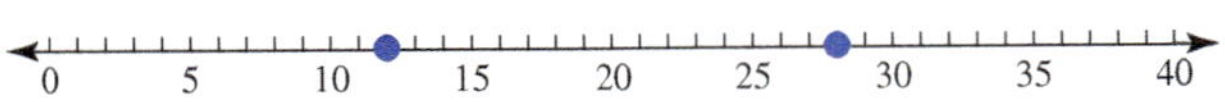

  The midpoint of 12 and 28 is found by calculating the mean of these two numbers.

  $\frac{12 + 28}{2} = \frac{40}{2} = 20$

  Notice that 20 is midway between 12 and 28 on the number line.

SKILL 3: ADD AND SUBTRACT A FIXED NUMBER

Adding and subtracting a fixed number is a really useful skill when setting up certain graphs in this unit. The key is to focus and be consistent—and also trust yourself to do the calculations by hand! Sure, you can use a calculator or spreadsheet if you have to. But it will be a big time saver if you get used to doing this on your own so that you can set up the scale for a graph and move on to focusing on the information that the graph is illustrating. Try Prep Skill Questions 4 and 5, and if you struggle, make up your own similar examples and practice until you feel comfortable.

SKILL 4: PLOT POINTS

When plotting points, we use a system of graphing called a rectangular coordinate system. The horizontal axis is usually called the x axis and the vertical axis is usually called the y axis. The point where the two axes meet is called the origin. A single point in the rectangular coordinate system is expressed by using two numbers separated by a comma and enclosed in parentheses. The two numbers we use to locate a point are called coordinates. For the point (2, −3), the first number, 2, is called the *x* coordinate, and the second number, −3, is called the *y* coordinate. To plot the point (2, −3), start at the origin and count 2 units to the right and 3 units down, paying careful attention to the scale used.

PREP SKILLS QUESTIONS

1. What is 27% of 360?

2. There were 93 people in my high school graduating class. Sixty-two went on to college. What percent went on to college?

3. What number is halfway between 8.5 and 16.5 on a number line?

4. Starting with 0.75, add 0.5 repeatedly until doing so one more time would result in a number greater than 8. List all of your results.

5. Expand your list of numbers from Question 4 by also subtracting 0.5 from 0.75 one time.

6. Plot each point on the axes. Make sure you indicate an appropriate scale on each axis!

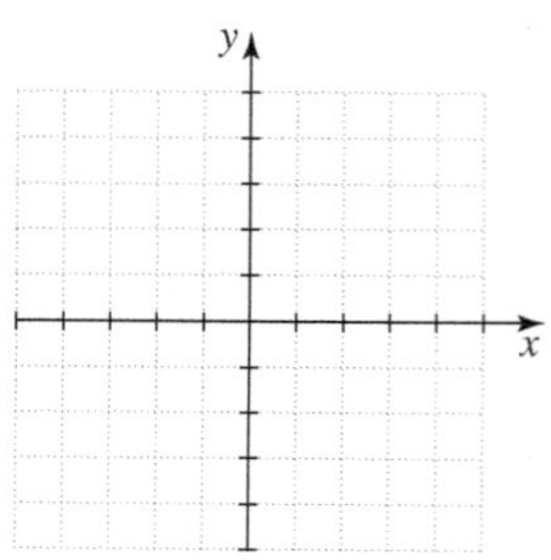

A. (−12, 100)
B. (18, 0)
C. (−6, −60)
D. (4, −10)

Lesson 4-2 Picture This

LEARNING OBJECTIVES

- ☐ 1. Draw and interpret bar graphs from frequency distributions.
- ☐ 2. Draw and interpret pie charts from frequency distributions.
- ☐ 3. Draw and interpret histograms and frequency polygons from frequency distributions.
- ☐ 4. Draw and interpret time series graphs.

Siegrist Photography/Shutterstock

The greatest moments are those when you see the result pop up in a graph or in your statistics analysis—that moment you realize you know something no one else does and you get the pleasure of thinking about how to tell them.

—Emily Oster

Now we've gathered some data, and we think maybe some of it has an interesting story to tell. How can we most effectively present that data? In Lesson 4-1, we displayed data in table form, using frequency distributions, and using stem and leaf plots. All are perfectly valid, but . . . they don't exactly pop. The data in the table below, supplied by the FBI, shows the number of hate crimes reported in the U.S. for the years 2008–2023.

Year	2008	2009	2010	2011	2012	2013	2014	2015
Hate crimes	7,783	6,604	6,628	6,222	5,796	5,928	5,479	5,850

Year	2016	2017	2018	2019	2020	2021	2022	2023
Hate crimes	6,121	7,175	7,120	7,314	7,759	10,840	11,634	11,862

Now let's look at the same information presented in graphic form:

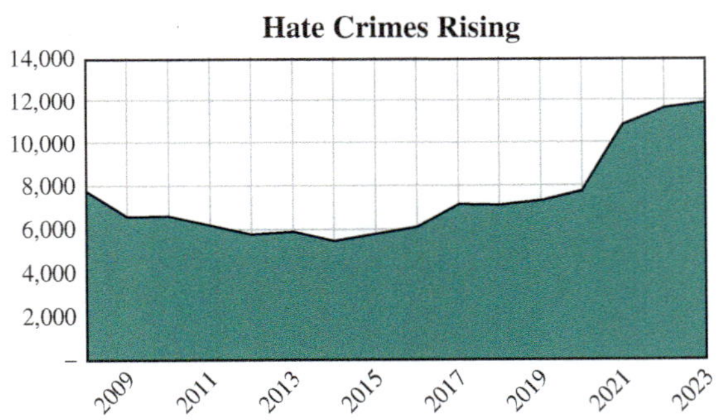

The information is the same, but with the graph, just a quick glance is enough to see that hate crimes decreased significantly from 2008 to 2014, started to increase in 2015, and then spiked in 2020, reaching a new high in 2023. Simply put, if you want your data to really catch someone's eye, you can't do much better than a nice graphical representation of the data. In Lesson 1-2, we studied interpreting information from graphs that had been supplied for us. But those graphs don't just appear out of thin air. In this lesson, we'll study several methods for turning data into graphical form.

0. Name some places where you've seen data represented in graphical form.

4-2 Group

The first type of frequency distribution we studied in Lesson 4-1 was the categorical frequency distribution. When data are representative of certain categories, rather than numerical, we often use pie charts or bar graphs to illustrate the data. We'll start with a pie chart.

Pie charts are most appropriate for data where every individual datum fits into just one category. Consider two different surveys about candidates for the presidency: If the question is "Which candidate is your favorite?" everyone will give exactly one response. So if we calculate the percentage of respondents that chose each candidate, those percentages will add to 100%, and a pie chart would be a good choice for illustrating the survey results. But if the question is "What candidates do you find acceptable?" many respondents will list more than one candidate, and the percentages would add to more than 100%. That type of data is more appropriately illustrated with a bar graph.

1. When measured in degrees, a full circle measures 360°. If 50% of the respondents to our candidate survey chose one particular candidate, how many degrees would make up his or her slice of a pie chart? What if the candidate got 25% of the responses?

> **Math Note**
>
> We studied reading the types of graphs in this lesson back in Unit 1. Now we'll focus more on creating them to represent data that have been gathered.

2. Describe how you got the degree measures in Question 1.

Drawing a Pie Chart

Step 1: Find the percentage of the circle covered and number of degrees corresponding to each slice. If f is the frequency for each class and n is the sum of all frequencies, then $\frac{f}{n}$ is the percentage in decimal form, and

$$\text{Degrees} = \text{Percentage in decimal form} \times 360°$$

Step 2: Using a protractor or special graph paper, graph each section on the circle using the angles you calculated.

Step 3: Label the chart so that it's clear what each slice represents.

3. The marketing firm Deloitte Retail conducted a survey of 80 grocery shoppers. The frequency distribution below represents the responses to the survey question "Which grocery store is your favorite?" Complete the table below and use that information to create a pie chart to represent these data.

Store	Frequency	Percent in Decimal Form	Degrees
Publix	13		
Trader Joe's	29		
Fareway	23		
Aldi	10		
Other	5		

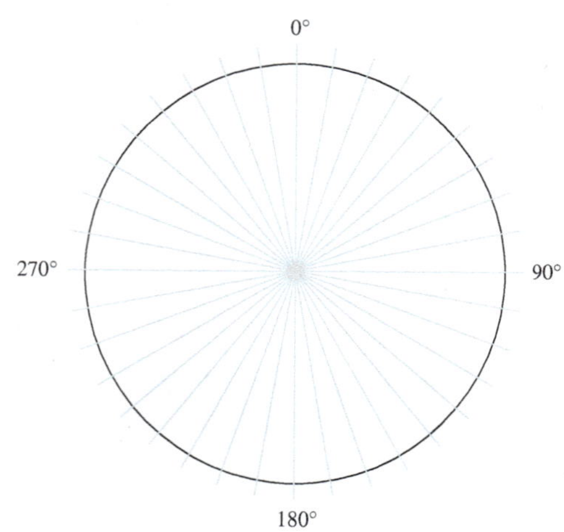

Using Technology: Creating a Pie Chart

To create a pie chart in Excel:

1. Type the category names in one column (or row).
2. Type the category values in the next column (or row).
3. Use the mouse to drag and select all the data in those two columns (or rows).
4. With the appropriate cells selected, click the **Insert** tab, then choose **Chart** and Pie chart. There are a few different styles you can experiment with, but starting with the simplest is a good idea.

You can add titles, change colors, and adjust other formatting by right-clicking on certain elements or by using the options on the **Charts** menu. Try some options and see what you can learn!

See the Using Tech Video on Creating a Pie Chart in Excel for further information.

The purpose of a pie chart is to show the relationship of the parts to the whole by visually comparing the sizes of the sections.

4. What can you observe by looking at the pie chart for the grocery store survey?

5. Let's revisit the marketing firm's survey of 80 grocery shoppers. Suppose the frequency distribution below represents the responses to the survey question "Which grocery stores have you shopped at in the last month?" Would a pie chart make sense for these data? Why or why not?

Store	Frequency
Publix	50
Trader Joe's	40
Fareway	28
Aldi	33
Other	65

While a pie chart is used to compare parts to a whole, a bar graph is used for comparing parts to other parts. That makes a bar graph an outstanding choice for the second grocery store survey—the intent is to compare the number of people who shop at each store to each other, rather than to the total frequency, which isn't terribly significant in this setting.

Drawing a Bar Graph

Step 1: Draw and label the axes. If you're drawing a vertical bar graph, the responses go on the horizontal axis and the frequencies on the vertical. For a horizontal bar graph, reverse that.

Step 2: Draw vertical bars with heights that correspond to the frequencies, or horizontal bars with lengths that correspond to the frequencies.

6. Draw a vertical bar graph illustrating the data from the survey about which grocery stores shoppers had visited in the last month.

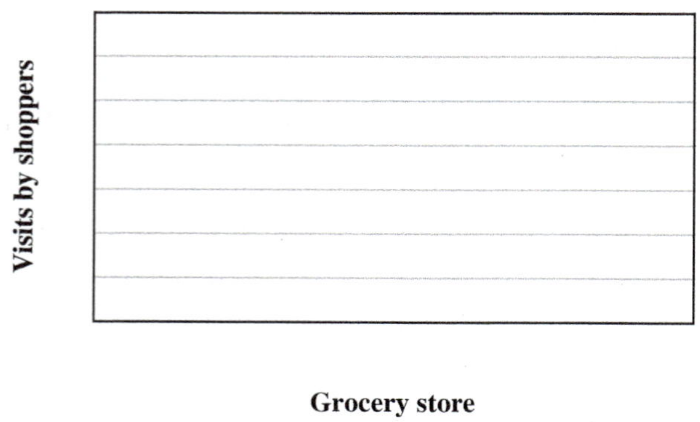

> **Math Note**
>
> When drawing ANY graph on axes, labeling those axes is CRUCIAL. Without a scale on the vertical axis, this bar graph will be completely worthless.

7. What can you observe by looking at a bar graph for these data?

An Important Note on Drawing Bar Graphs

In drawing a vertical bar graph, it's very important that the scale on the vertical axis begins at zero and has consistent spacing. As you'll see in Lesson 4-9 (covering misuses of statistics), an improperly labeled axis can lead to a deceptive graph.

Using Technology: Creating a Bar Graph

To create a bar graph in Excel:

Step 1: Type the category names in one column (or row).
Step 2: Type the category values in the next column (or row).
Step 3: Use the mouse to drag and select all the data in those two columns (or rows).
Step 4: With the appropriate cells selected, click the **Insert** tab, then choose **Chart** and Column or Bar graph. ("Column" gives you vertical bars, "Bar" gives you horizontal.) Again, there are a few different styles you can experiment with, but starting with the simplest is a good idea.

You can add titles, change colors, and adjust other formatting by right-clicking on certain elements or by using the options on the **Charts** menu. Try some options and see what you can learn!

See the Using Tech Video on Creating a Bar Graph in Excel for further information.

Did You Get It

Try this problem to see if you understand the concepts we just studied. The answer can be found at the end of the portfolio section.

1. Two surveys of 100 people were done on calorie information at fast-food restaurants. The first asked how much impact posted calorie information has on buying choices. The second asked which (if any) restaurants the subjects tend to avoid because they think the food is unhealthy. Results are below. Draw a pie chart for one set of data and a bar graph for another. Choose which data set is appropriate for each type of graph.

Impact of Posted Calorie Information

Response	Frequency
Great impact	40
Some impact	42
Not much impact	11
None at all	7

Restaurants Avoided for Health Reasons

Response	Frequency
McDonalds	46
Burger King	38
Wendy's	35
Subway	18

4-2 Class

Histograms and Frequency Polygons

When data are organized into grouped frequency distributions, two types of graphs are commonly used to represent them: histograms and frequency polygons. A histogram is similar to a vertical bar graph in that the heights of the bars correspond to frequencies. The difference is that the boundaries of the classes are placed on the horizontal axis, rather than categories.

Drawing a Histogram

Step 1: Write the scale for the frequencies on the vertical axis and the boundaries of the classes on the horizontal axis. Put labeled tick marks at the lower boundary of every class.

Step 2: Draw vertical bars with heights that correspond to the frequencies for each class.

Note: When drawing a histogram, make sure that there is no horizontal space between the bars (unlike a bar graph) unless two bars are separated by a class with frequency zero. Also make sure that all classes are included, even if the frequency is zero.

In Lesson 4-1, we analyzed and organized data representing the record high temperature in every state. It would therefore seem reasonable, if not expected, for us to illustrate that data graphically. So that's just what we'll do. Below is a grouped frequency distribution for the data. We'll use it to draw a histogram.

Solved Example 1

Drawing a Histogram

For the grouped frequency distribution below, which came from the record high temperature in every state, draw a histogram to illustrate the data.

Class	Frequency
100–104	3
105–109	8
110–114	15
115–119	13
120–124	8
125–129	2
130–134	1

SOLUTION

The key to any good graph, histogram or not, is scaling and labeling the axes appropriately. For a histogram, we'll put tick marks along the horizontal axis at the *lower boundary* of each class, so in this case they'll all be separated by 5 units. I'm going to label the starting point of the horizontal axis as 95, then place and label tick marks at 100, 105, 110, 115, 120, 125, and 130. Then I'll put one final tick mark at 135 to mark the end of the last bar.

As for the vertical axis, I need heights up to 15, so I think a scale of 2 going up to height 18 will look nice. After scaling the axes, my job becomes drawing bars with heights corresponding to the frequencies. The first bar will fill up the horizontal space from the beginning of the first class (100) to the beginning of the next class (105). This is another key difference between a bar graph and a histogram: no horizontal space between the bars. So that first space gets a bar with height 3. I then continue in that manner, drawing bars at heights matching each frequency in the appropriate class range.

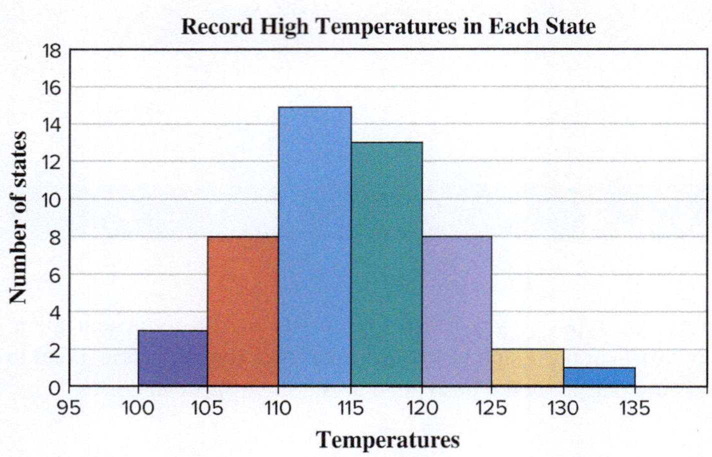

Finally, it's always a good idea to write in a title for your graphs, as well as a verbal description of what the quantities on each axis represent. In this case I decided to clearly note that the numbers on the horizontal axis are temperatures and that the heights of the bars are the number of states whose record high temperature falls in the given range.

Note: Statisticians will sometimes label the horizontal axis of a histogram a bit differently to make it clearer which class a value right on a class limit falls into. For example, if you labeled 100, 105, 110, etc. on the x axis, it wouldn't be clear whether a temperature of 105 lives in the first bar or the second. If it's unclear at some point what values a given bar represents, refer to the frequency distribution for that data.

Using Technology: Drawing a Histogram in Excel

To create a histogram in Excel, we have two choices:

1. We can use the Histogram tool found under the Charts section of the INSERT tab. This option has the advantage of working from the original data set, but it also has some limitations when it comes to the classes.
2. Another method for using Excel to draw a histogram is to first create the frequency distribution where we've already decided on the classes to use. Then we can use the Bar Graph tool to create a bar graph of the frequency distribution. The spacing between the bars can be removed by selecting Format Data Series and then adjusting the Gap Width to 0%.

See the Using Tech Video on Drawing a Histogram in Excel for further information.

1. Write at least three things you can observe from the histogram. Do you feel like these observations would be any different than those you could make by just looking at the grouped frequency distribution? Explain.

2. Our next frequency distribution from Lesson 4-1 organizes the team record for wins in a season by each of the 30 Major League Baseball clubs. Draw a histogram for the data. Don't forget to focus on labeling the axes!

Class	Frequency
92–95	2
96–99	5
100–103	7
104–107	6
108–111	7
112–115	1
116–119	2

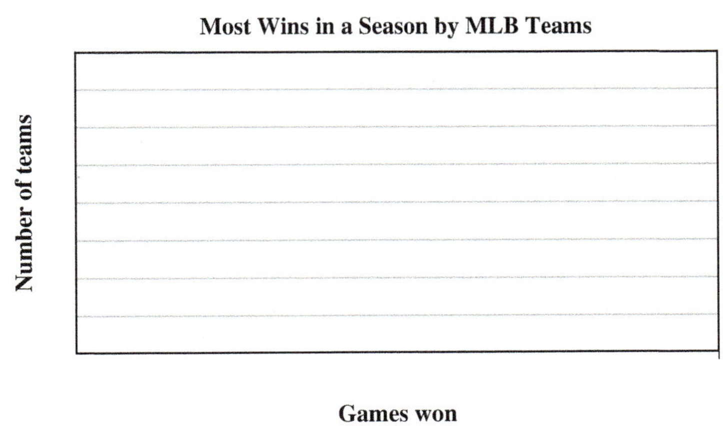

A frequency polygon is similar to a histogram, but instead of bars, a series of line segments is drawn connecting the midpoints of the classes. The heights of those points match the heights of the bars in a histogram.

Drawing a Frequency Polygon

Step 1: Find the midpoint of each class. First, find the midpoint of the first class by adding the upper and lower limits and dividing by 2. You can then find the remaining midpoints by adding the class width repeatedly until you get the midpoint of the last class.

Step 2: Subtract the class width from the lowest midpoint, and add the class width to the highest midpoint.

Step 3: Write a scale for the frequencies on the horizontal axis, and label the vertical axis with all the midpoints from Step 1 and the two extra numbers from Step 2.

Step 4: Plot points at the midpoints you labeled with heights matching the frequencies for each class, then connect the points with straight lines.

Step 5: Draw a line back from the first point on your graph to the horizontal axis, hitting the axis at the first labeled number there. Then draw a line forward from the last point to the horizontal axis at the last labeled number. This represents that the frequencies fall to zero outside the included classes.

Solved Example 2

Drawing a Frequency Polygon

Draw a frequency polygon for the record high-temperature data we looked at in Solved Example 1.

SOLUTION

Step 1: Find the midpoints of the classes. This sounds like we should list out every class and find the midpoint of each, one at a time. But it turns out there's a much smarter approach. The first class is 100–104, and the midpoint of that is 102. (In cases where it's not pretty obvious what the midpoint is, you can find the average of the lower and upper limits of the class.) Now we'll be clever. Since the lower limits of the classes are all 5 units apart, and the upper limits of the classes are all 5 units apart, it makes sense that the midpoints are 5 units apart as well. Awesome! We just need to keep adding 5 to get the rest of the midpoints:

102, 107, 112, 117, 122, 127, 132

Step 2: Find the location on the vertical axis where the frequency polygon will begin and end at height zero. This is accomplished by subtracting the class width (5) from the first midpoint to get 97, and adding the class width to the last midpoint to get 137. The points we put in there at height zero indicate there are no further data values larger or smaller than those temperatures.

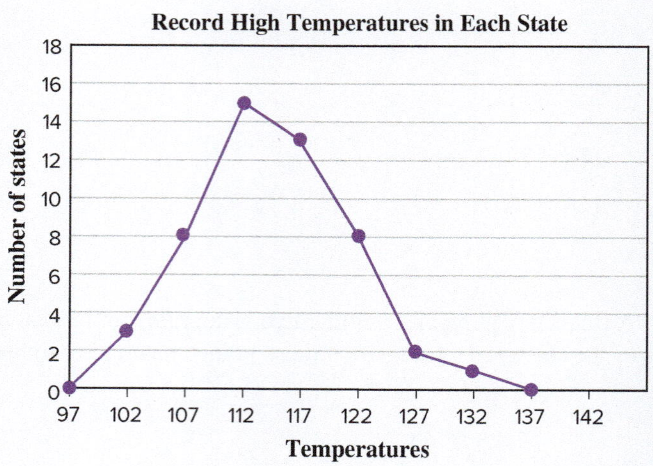

Step 3: Now we'll set up our axes. The highest frequency is still 15, so let's stick with a scale of 2, going up to 18, on the vertical axis. For the horizontal axis, we label all our midpoints, along with 97 and 137. See graph.

Step 4: Once you've got everything set up, drawing the graph is actually the easy part. We plot points at each labeled midpoint, with the height coming from the frequency for that particular class, then play connect the dots with straight line segments.

Step 5: Finally, we extend the graph back down to the horizontal axis in both directions.

Using Technology: Drawing a Frequency Polygon in Excel

To create a frequency polygon in Excel, we will use the scatterplot tool.

1. Start with the frequency distribution.
2. Add an extra column for the midpoints between the classes and the frequencies. Add an extra row above and below the frequency distribution and enter a frequency of 0 for these new rows. These frequencies will serve as the y values for our frequency polygon.
3. Find all the midpoints of the classes. Subtract the class width from the first midpoint and add the class width to the last midpoint to find the values for the added rows. These midpoints will represent the x values for our frequency polygon.
4. To graph the frequency polygon, select the cells containing the midpoints and frequencies and choose the Scatter with Straight Lines and Markers option from the Insert Charts menu. You will need to adjust the minimum and maximum values for the horizontal axis. A good suggestion is to start at the first value in the midpoint column and use the class width as the scale.

Class	Frequency
100–104	3
105–109	8
110–114	15
115–119	13
120–124	8
125–129	2
130–134	1

Class	Midpoint	Frequency
	97	0
100–104	102	3
105–109	107	8
110–114	112	15
115–119	117	13
120–124	122	8
125–129	127	2
130–134	132	1
	137	0

See the Using Tech Video on Drawing a Frequency Polygon in Excel for further information.

3. Draw a frequency polygon for the baseball team data from Question 2.

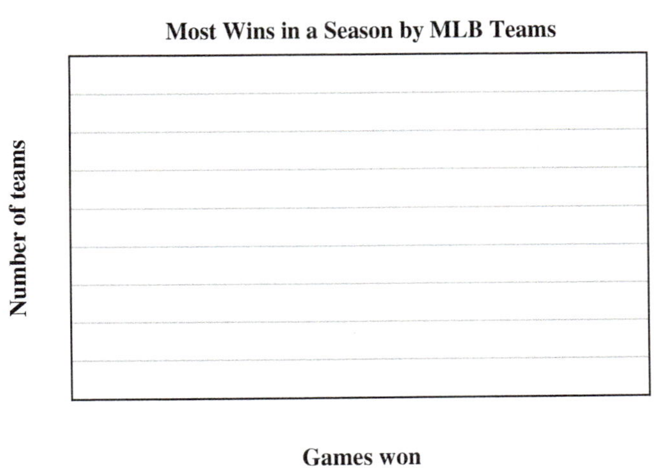

4. Compare your frequency polygon from Question 3 with your histogram from Question 2. Was there any real advantage to the frequency polygon? Feel free to be critical. It won't hurt my feelings one bit. Well, maybe like half a bit … but I'll get over it.

So you may be wondering at this point: Why bother with frequency polygons? Aren't they pretty much the same thing as histograms? Good question. There actually IS a point, though, as we will now see. The two grouped frequency distributions below come from data gathered in two calculus classes that I taught one semester. One is for percentage on the final exam; the other is for overall percentage in the course. Then we see a frequency polygon for each data set.

Final Exam Percentage

Percentage	Frequency
40–49.9	3
50–59.9	2
60–69.9	13
70–79.9	14
80–89.9	14
90–99.9	12

Overall Percentage

Percentage	Frequency
40–49.9	0
50–59.9	1
60–69.9	7
70–79.9	20
80–89.9	19
90–99.9	11

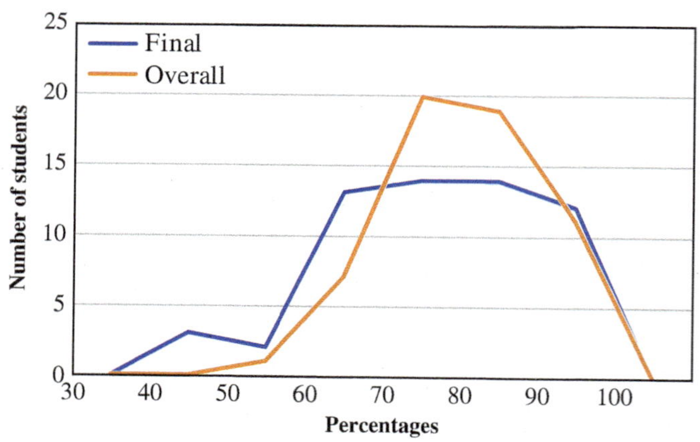

5. What would it look like if we tried to draw a single histogram that compares both data sets?

6. When are frequency polygons particularly useful?

Did You Get It

2. Draw a histogram for the data below, which is the number of losses by the team that won the NCAA men's basketball championship for the years from 1939–2023.

Losses	Frequency
0–1	13
2–3	31
4–5	22
6–7	11
8–9	5
10–11	3

3. Draw a frequency polygon for the same data.

Time Series Graphs

A **time series graph** can be drawn for data collected over a period of time. This type of graph is ideal for recognizing trends, like prices rising or falling, for the time period. We'll close this lesson by drawing and analyzing this type of graph.

Drawing a Time Series Graph

Step 1: Label the horizontal axis with the time units and the vertical axis with a scale (beginning at zero) that's appropriate for the given data.

Step 2: Plot the points from the table and connect them with line segments.

7. The table shows the number of violent crimes committed in the United States per 100,000 citizens for selected years from 2000 to 2024. Draw a time series graph for the data.

Year	Violent Crimes per 100,000 Citizens
2000	506.5
2002	494.4
2004	463.2
2006	479.3
2008	458.6
2010	404.5
2012	387.8
2014	361.6
2016	386.6
2018	368.9
2020	398.0
2022	380.7
2024	326.3*

**Data for first half of 2024*

Comstock/Getty Images

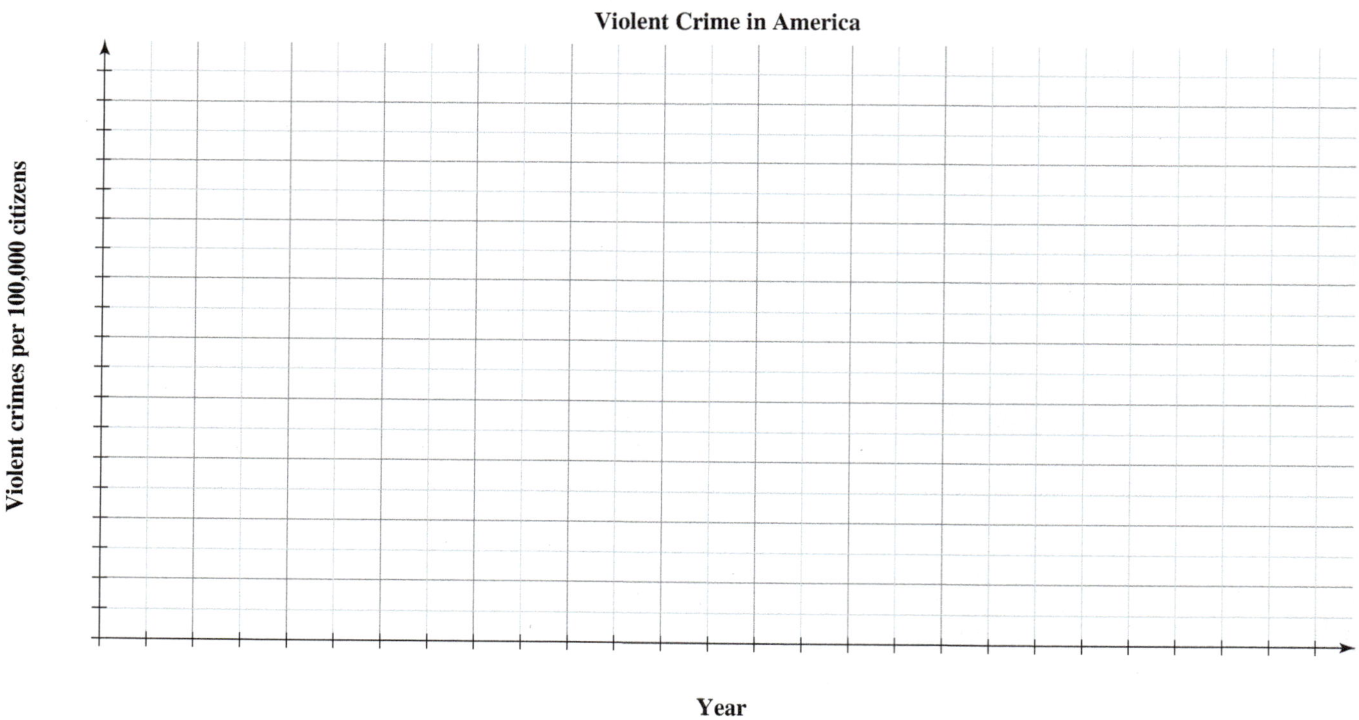

8. Overall, would you say that violent crime is getting worse or better? Explain.

9. When did violent crime increase in the 21st century?

10. In what 2-year span did the biggest decrease occur?

Using Technology: Drawing a Time Series Graph in Excel

To create a time series graph in Excel, we will use the scatterplot tool.

1. Enter the time data in the first column and the data collected over a period of time in the second column.
2. To complete the time series graph, select the two columns of data and choose the Scatter with Straight Lines and Markers option from the Insert Charts menu.

See the Using Tech Video on Drawing a Time Series Graph in Excel for further information.

Did You Get It

4. The number of bankruptcy filings (in millions) in the United States from 2006 to 2022 is shown in the table. Draw a time series graph for the data, then write a description of any trends you notice.

Year	Filings (millions)
2006	0.62
2008	1.12
2010	1.5
2012	1.22
2014	0.94
2016	0.79
2018	0.77
2020	0.54
2022	0.38

4-2 Portfolio

Name ______________________________

Check each box when you've completed the task. Remember that your instructor will want you to turn in the portfolio pages you create.

Technology

1. ☐ In this Excel question, we'll draw a pie chart.
2. ☐ In this Excel question, we'll draw a bar graph.
3. ☐ In this Excel question, we'll draw a histogram.
4. ☐ In this Excel question, we'll draw a frequency polygon.
5. ☐ In this Excel question, we'll draw a time series graph.

ALEKS Practice

1. ☐ Include any written work from the ALEKS assignment along with any notes or questions about this lesson's content.

Applications

1. ☐ Complete the applications problems.

Reflections

Type a short answer to each question.

1. ☐ Discuss the value of illustrating data with a graph after it's been gathered and organized.
2. ☐ When is a pie chart appropriate for illustrating data? When is a pie chart not appropriate?
3. ☐ What type of data is illustrated using a histogram and/or frequency polygon? Why?
4. ☐ Name one thing you learned or discovered in this lesson that you found particularly interesting.
5. ☐ What questions do you have about this lesson?

Looking Ahead

1. ☐ Complete the prep skills for Lesson 4-3.
2. ☐ Read the opening paragraph in Lesson 4-3 carefully and answer Question 0 in preparation for that lesson.

Answers to "Did You Get It?"

Observations will vary.

1.

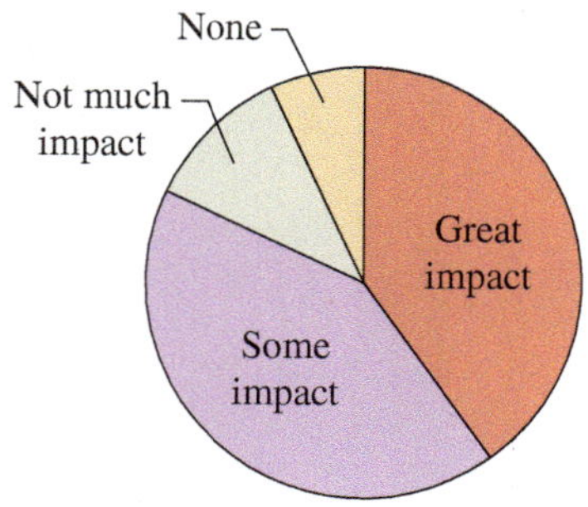

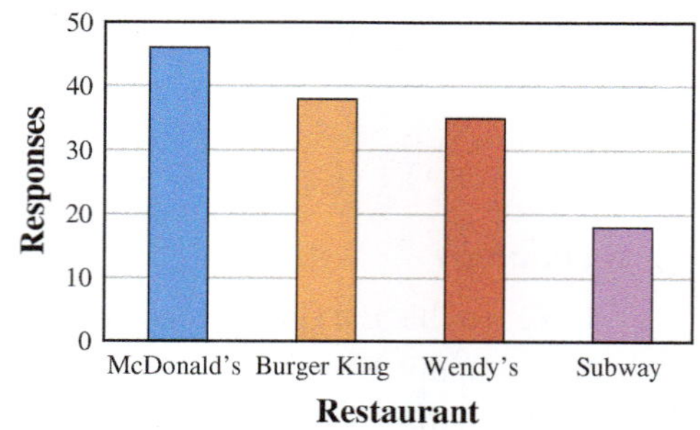

2.

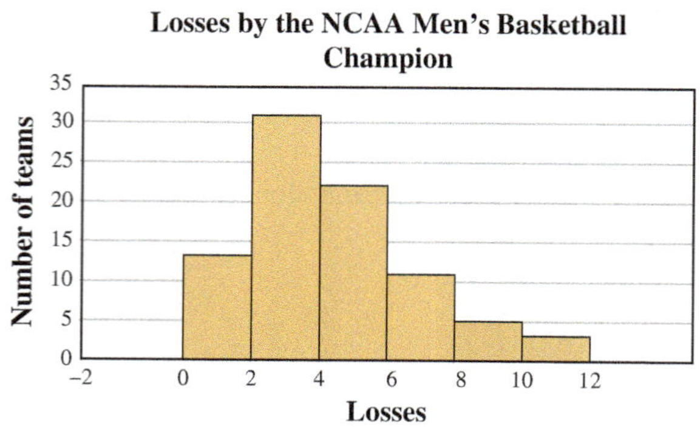

3.

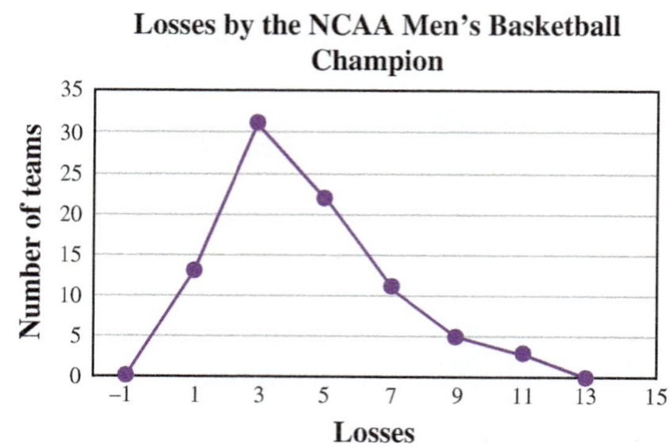

4.

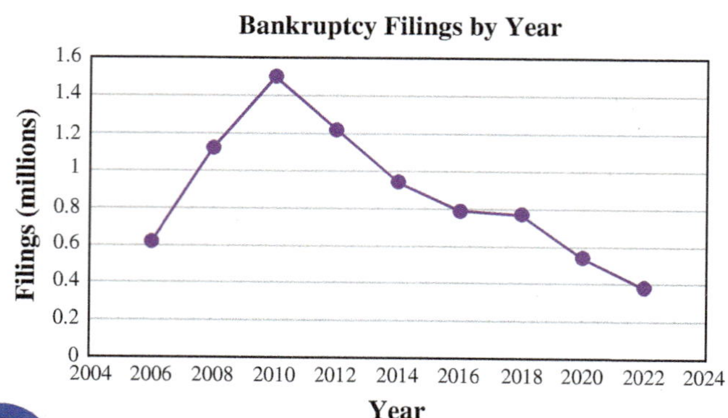

Answers to Prep Skills Questions

1. 97.2
2. $66.\overline{6}\%$
3. 12.5
4. 0.75, 1.25, 1.75, 2.25, 2.75, 3.25, 3.75, 4.25, 4.75, 5.25, 5.75, 6.25, 6.75, 7.25, 7.75
5. 0.25, 0.75, 1.25, 1.75, 2.25, 2.75, 3.25, 3.75, 4.25, 4.75, 5.25, 5.75, 6.25, 6.75, 7.25, 7.75
6.

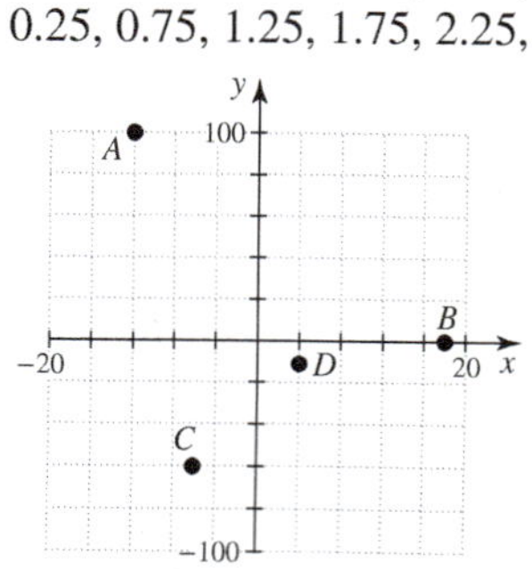

4-2 Applications

Name ______________________________

In Questions 1–6, write if a bar graph, pie chart, or time series graph would be the most appropriate way to represent the given data.

1. The number of students enrolled at a local college each year for the last 5 years

2. The budget for the student activities department at your college

3. The number of students who get to school by automobile, bus, train, or by walking

4. The record high temperatures of a city for the last 30 years

5. The areas of the five lakes in the Great Lakes

6. The amount of each dollar spent for wages, advertising, overhead, and profit by a corporation

In Questions 7–11, we'll use the types of graphs we studied in this lesson to study the issue of climate change.

7. Most climate experts agree that the release of carbon dioxide gas into the atmosphere is a significant factor in rising global temperatures. So where is all that CO_2 coming from? The first table below displays the total CO_2 emissions in kilotons for the top five producers. The second shows per capita emissions for these same producers, which is the number of tons emitted for each resident. Draw a separate bar graph for the data in each table, then write at least three things you can observe from your graphs.

Producer	CO_2 Emissions (kt)
China	12.7 million
United States	4.9 million
European Union	3.2 million
India	2.7 million
Russia	1.9 million

Producer	Tons per Capita
China	8.9
United States	14.2
European Union	7.2
India	1.9
Russia	13.1

4-2 Applications

Name ______________________________

7. (Con't)

8. Americans' attitudes toward climate change have been evolving very rapidly over the last few years. The tables below show responses to the survey question "From what you've read and heard, is there solid evidence that the temperature on earth has been getting warmer over the past four decades?" first in Spring of 2010, then again in Spring of 2024. Use the information to draw two pie charts, then discuss the results.

Spring 2010

Response	%
Yes	52
No	36
Not sure	12

Spring 2024

Response	%
Yes	72
No	16
Not sure	12

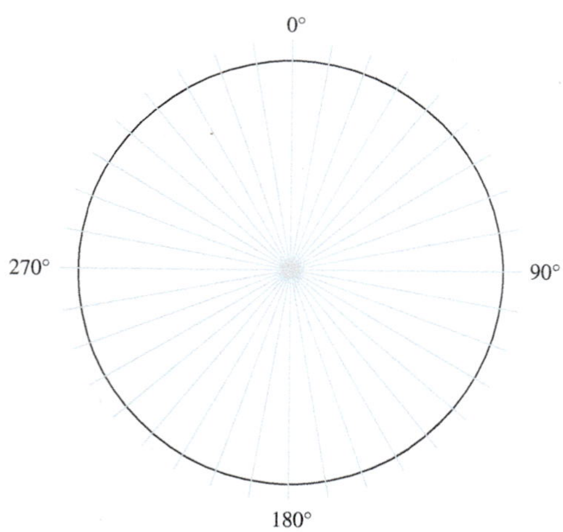

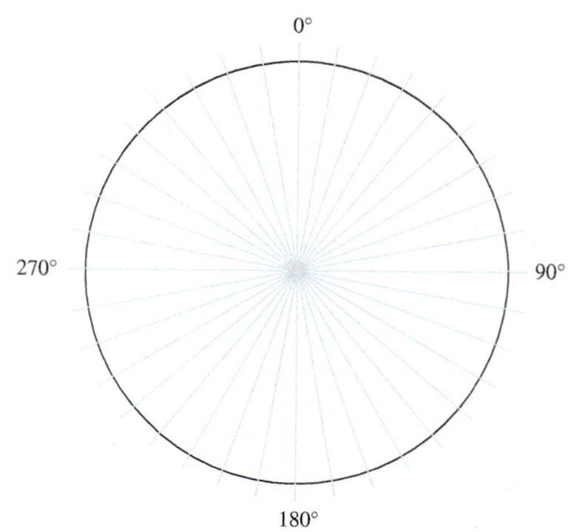

4-2 Applications

Name ______

9. Now let's look at global temperatures. The National Oceanic and Atmospheric Administration tracks the average global temperature every year and compares the result to the average temperature for all years in the 20th century. They report the results in terms of departure from that average. Positive results mean warmer than average, negative cooler. The frequency distribution below summarizes the number of years from 1881 to 2023 that fell into each range of departure from average temperature. Draw a histogram for the data.

Class	Frequency
−0.60 to −0.41	6
−0.40 to −0.21	31
−0.20 to −0.01	42
0 to 0.19	24
0.20 to 0.39	10
0.40 to 0.59	11
0.60 to 0.79	11
0.80 to 0.99	9

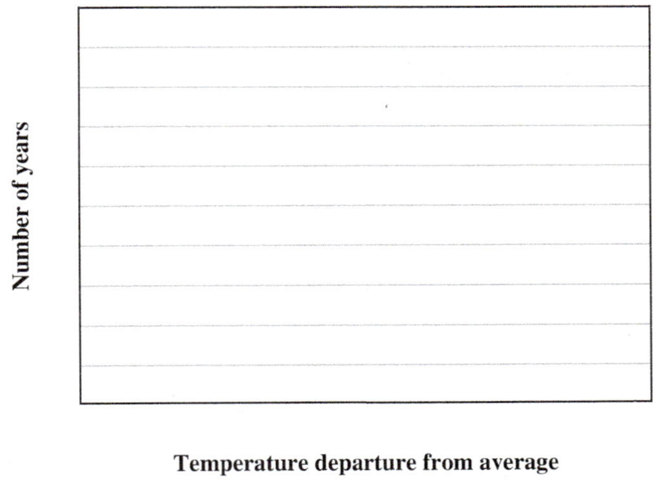

10. The next frequency distribution is drawn from the same data but includes only the years from 1970 to 2023. Draw a frequency polygon, and write any observations you have from comparing the two temperature deviation graphs.

Class	Frequency
−0.60 to −0.41	0
−0.40 to −0.21	0
−0.20 to −0.01	4
0 to 0.19	10
0.20 to 0.39	9
0.40 to 0.59	11
0.60 to 0.79	11
0.80 to 0.99	9

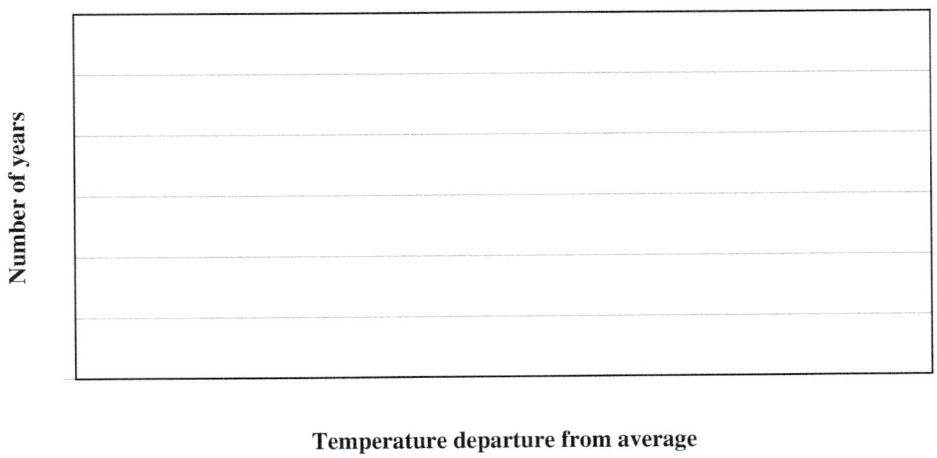

4-2 Applications

Name ____

11. Write about something you learned in this activity that you find interesting or compelling in some way, either about the stats we learned or the data themselves.

Lesson 4-3 Prep Skills

SKILL 1: ORDER A LIST OF NUMBERS

Ordering a list of numbers can be done manually or by using technology. For small lists, manually sorting isn't too bad. To sort a list in increasing order, scan the list and find the smallest value. Then find the second smallest. Then the third smallest, and so on. It is probably a good idea to lightly check off each number as you write them in order to avoid missing a number. To sort a list in decreasing order, start with finding the largest number first and follow a similar process. For larger lists, using the **SORT** feature in a spreadsheet or calculator is the way to go.

SKILL 2: COMPUTE LENGTHY SUMS

To compute a sum of a few numbers by hand is a skill most students should possess, but adding 10 or 15 numbers by hand when a calculator or spreadsheet is nearby seems kind of silly. Entering a list of numbers in a spreadsheet will allow for quickly sorting and analyzing the values. Using the **SUM** command in a spreadsheet is a quick way to add a long list of numbers. Graphing calculators can also perform this task efficiently.

SKILL 3: USE ORDER OF OPERATIONS

The order of operations is a group of agreed-upon rules that allow us to write calculations involving more than one operation without always having to use parentheses to indicate which operation to do first. The order is as follows, with operations higher on the list being performed before those lower on the list.

(1) Grouping symbols	Perform these operations first. If there is more than one set of grouping symbols, work from innermost outward.
(2) Exponents	
(3) Multiplication or division	For operations that live at the same level, work from left to right.
(4) Addition or subtraction	

One good way to avoid common errors is to understand that a fraction bar is considered to have a grouping symbol around the entire numerator and the entire denominator. Below are two different ways of interpreting the expression $\frac{3+4}{5}$.

- $\frac{3+4}{5} = (3+4)/5$ — This calculation would yield the correct result of $\frac{7}{5}$ or 1.4.
- $\frac{3+4}{5} \neq 3 + 4/5$ — This calculation would yield the incorrect result of $3\frac{4}{5}$ or 3.8.

SKILL 4: INTERPRET SUBSCRIPT NOTATION

Subscript notation is a notation that allows you to indicate which number you are talking about from a list of numbers. For example, referring to X_1, X_3, and X_7 from a list of numbers is a way of indicating the first, third, and seventh numbers in that list. X_n is sometimes used to indicate the last number in a list.

- Consider the list of numbers:
  3, 5, 7, 9, 11, 13 — There are 6 numbers in this list, so $n = 6$.

$X_1 = 3$	The 1st number in the list is 3.
$X_2 = 5$	The 2nd number in the list is 5.
$X_3 = 7$	The 3rd number in the list is 7.
$X_4 = 9$	The 4th number in the list is 9.
$X_5 = 11$	The 5th number in the list is 11.
$X_6 = 13$	The 6th number in the list is 13.

PREP SKILLS QUESTIONS

1. Arrange the list of numbers below in order from least to greatest.

   2.4 3.7 1.7 1.7 2.3 3.1 1.2 2.4 1.3 1.7 2.5 2.4 2.8 2.1 1.3 2.5

2. Find the sum of the numbers in the list in Question 1.

For Questions 3 and 4, complete the calculation.

3. $\dfrac{8+11+13+5+6}{5}$

4. $\dfrac{2 \cdot 98 + 4 \cdot 112 + 12 \cdot 13 + 10 \cdot 59 + 6 \cdot 80}{34}$

5. If we use the symbols X_1, X_2, X_3, and so on to represent the numbers in the list in Question 1, what number corresponds to X_7? (In the original list, not the ordered list.)

6. Refer to Question 5. If the entire list is written as $X_1, X_2, X_3, \ldots, X_n$, what number is represented by n?

Lesson 4-3 An Average Joe

LEARNING OBJECTIVES

- ☐ 1. Compute measures of average for given data.
- ☐ 2. Interpret the story told by measures of average.
- ☐ 3. Compute and interpret the mean for grouped data.
- ☐ 4. Compute weighted grades.
- ☐ 5. Use technology to compute measures of average.

Ryan McVay/Photodisc/Getty Images

The average dog is a nicer person than the average person.
—Andy Rooney

Are you an average college student? What does that even mean? According to a variety of sources found on the Web, the average college student is 20 years old, lives on campus, and works off campus during the school year. She is female, talks to her parents every day, and comes from a family with an annual income between \$50,000 and \$100,000. She sleeps 6 hours a night, has \$3,280 in credit card debt, and will leave college with \$29,000 in total debt.

If this doesn't sound exactly like anyone you know, don't feel bad. As we will learn in this lesson, the word "average" is ambiguous—there are a variety of ways to describe an average. So what one source considers average might not match someone else's thoughts, and with so many different facets to a person, it's possible that nobody exactly meets all of the average criteria.

0. Describe what the word "average" means to you right now.

4-3 Group

Our goal in this lesson will be to understand the different measures of average. In casual terms, average means the most typical case, or the center of where the values in a data set are distributed. In this lesson, we'll study four different ways of measuring average: mean, median, mode, and midrange.

Let's suppose that an individual, let's just call him George, is about to finish college and step out into the work force. George finds a small company, Vandelay Industries, which he's considering going to work for. While researching this company, George finds a flier on which that company advertises that its average employee makes almost \$150,000 per year. This sounds pretty darn good to George, so he accepts an entry-level position at a low salary, betting on quick advancement to that alleged average.

Shortly after starting his career, while emptying the recycling bin in the copy room (just one of George's glamorous duties), he stumbles across a list of salaries for all employees at his new company. Let's investigate just how true the company's average salary claim is.

Here's the salary list for Vandelay Industries, which we'll use to study measures of average, which are more technically called **measures of central tendency.**

Employee	Salary
Jerry	\$58,000
Kramer	\$65,000
Newman	\$944,000
George	\$20,000
Elaine	\$52,000
Susan	\$51,000
Tim	\$53,000
Estelle	\$55,000
Frank	\$50,000

The Mean

There's a pretty good chance that when you think of "average" in a math class, what you're thinking of is actually called the **mean.** The mean, also known as the arithmetic average, is found by adding the values of the data and dividing by the total number of values. Here's a good way to think of the mean: If all of the numbers were equal weights put on a number line, the mean is where the number line could be balanced without tipping in one direction or the other.

The Greek letter Σ (sigma) is used to represent the sum of a list of numbers. If we use the letter X to represent data values, then ΣX means to find the sum of all values in a data set. Using this notation, we can write a formula for finding the mean.

The **mean** is the sum of values in a data set divided by the number of values. If $X_1, X_2, X_3, \ldots, X_n$, are the data values (which means there are n of them), we use the symbol $\overline{X}$ to represent the mean, and

$$\overline{X} = \frac{X_1 + X_2 + X_3 + \cdots + X_n}{n} = \frac{\Sigma X}{n}$$

The symbol $\overline{X}$ is read "X bar."

1. Find the mean salary for the company.

2. Newman owns the company, and obviously pays himself a handsome salary. If we ignore Newman's salary, what is the new mean? Based on this, do you feel like the mean is a good measure of what the typical employee makes?

Did You Get It

Try this problem to see if you understand the concepts we just studied. The answer can be found at the end of the portfolio section.

1. The total areas, in thousands of square miles, of the 15 largest states in the U.S. are provided in the list below. Find the mean area for these 15 states.

665	269	164	147	122	114	111	104
98	98	97	87	85	84	82	

The Median

According to the National Organization of Realtors, the mean selling price for existing homes nationally in October 2024 was \$533,400. The **median** selling price, on the other hand, was \$420,400. Since both of these are measures of central tendency, how can they be so far apart? What exactly is the median? In this case, the median tells us that half of all homes sold during that month were priced above \$420,400, and half were priced below that amount. In short, the median of a data set is the value in the middle if all values are arranged in order. The median will either be a specific data value in the set or will fall in between two values. Here's how we find a median:

3. Arrange all of the salaries at Vandelay Industries in order from smallest to largest.

4. Find the salary that's right in the middle of your ordered list. That's the median salary. How does it compare to the mean?

5. If there is an even number of values in a data set, there will be two values in the middle, not one like in Question 4. How do you think you'd find the median in that case?

Finding the Median of a Data Set

Step 1: Arrange the data in order, from smallest to largest. (Actually, largest to smallest is fine, too. Whatever makes you happy.)

Step 2: The way we find the median depends on whether the number of data values is even or odd.

- If there's an odd number of data values, there will be one single value in the middle of that ordered list, with half of the remaining values bigger and half smaller. That value in the middle is the median.
- If there's an even number of data values, there will be two values in the middle of the list. The mean of those two values is the median of the entire list.

6. Find the median of the salaries at Vandelay Industries (without Newman's). How does it compare to the median of the salaries with Newman's included?

7. What does this tell you about median when there are **outliers** (sample data values that are very far away from the majority of values)? This is a VERY key idea, so give it some thought.

8. Find the median of the salaries at Vandelay Industries (without Newman's) if Kramer quits to start his own company (The Peterman Reality Tour). What can you observe?

CASTLE ROCK/WEST-SHAPIRO/ Album/Alamy Stock Photo

Did You Get It

2. Find the median of the land areas provided in Did You Get It 1. Compare to the mean, and discuss the difference. Try to come up with an explanation for the discrepancy.
3. Find the mean and median if you throw out Alaska, which has the largest area by far. What does this tell you?

The Midrange

The **midrange** is another measure of average. The advantage of the midrange is that it's very quick and easy to calculate. The disadvantage is that it totally ignores most of the data values, so it's not a particularly reliable measure. But in a pinch, it can give you a rough idea of average.

9. Find the number halfway between the lowest and highest salaries at Vandelay Industries (including Newman's). That's the midrange.

Finding the Midrange of a Data Set

Step 1: Find the smallest and largest values in the data set.

Step 2: Find the mean of those two values.

10. Find the midrange of the salaries at Vandelay Industries (excluding Newman's). How does it compare to the midrange of all salaries?

11. Explain why the midrange isn't very reliable as a measure of central tendency.

Did You Get It

4. Find the midrange area for the top 15 largest states from Did You Get It 1.

The Mode

The fourth measure of average is called the mode. The **mode** is sometimes said to be the most typical case.

12. These data represent the duration (in days) of the final 20 U.S. space shuttle voyages. Which number of days appears most often on the list? That's the mode.

11 12 13 12 15 12 15 13 15 12 12 15 13 10 13 15 11 12 15 12

fergregory/123RF.com

13. How does the mode compare to the mean? Which do you think describes the typical case best?

The **mode** is the value that occurs most often in a data set.

A data set can have more than one mode (if two or more values are tied for most appearances), or no mode at all (if every value appears only once). While modes aren't extremely useful for small data sets with numerical data, we'll begin there to help get an understanding of what the mode is about.

14. The number of wins in a 16-game season for the Cincinnati Bengals from 1996–2018 is listed below. Find the mode.

8 7 3 4 4 6 2 8 8 11 8 7 4 10 4 9 10 11 10 12 6 7 6

15. How does the mode compare to the mean? Discuss again. Include a description of which measure of central tendency you think is more reliable in general, and why. You might consider the fact that there might not be a single answer—it could depend on what you want to learn from the measure of average.

16. Do the employee salaries at Vandelay Industries have a mode? Why or why not?

The mode is probably most useful when data are classified by groups or categories. In fact, it's the only measure of central tendency that even makes sense when the data values aren't numbers. (For the average college student at the beginning of the lesson, I challenge you to find the mean of {Male, Female}.)

17. A survey of the junior class at Fiesta State University shows the following numbers of students majoring in each field. Find the mode.

Major	Number
Computer Science	632
Liberal Arts	878
Business	1,425
General Studies	95
Education	471

18. Now let's evaluate the "average salary" claim made by Vandelay Industries at the beginning of this lesson. How truthful do you think that claim was? Use the measures of average you found in this lesson to discuss whether or not you think they were being deceptive in any way. I'm not going to say that this question is the main point of the entire lesson, but it is.

Did You Get It

5. The table below lists the average high temperature in degrees Fahrenheit for each month of the year on the island of Antigua (which is delightful, by the way). Find the mode and the median.

Month	Jan	Feb	Mar	Apr	May	Jun	Jul	Aug	Sep	Oct	Nov	Dec
High	81	82	82	83	85	86	87	87	87	86	84	82

Did You Get It

6. Five hundred college graduates were asked how much they donate to their alma mater on an annual basis. Find the mode of their responses, summarized in the table.

Amount	Number
$500 or more	45
Between $0 and $500	150
Nothing	275
Declined to answer	30

4-3 Class

We now return to the mean, this time in a situation that's a bit more complicated: when we have grouped data in a frequency distribution. The procedure for finding the mean for grouped data uses the midpoints and the frequencies of the classes. This procedure will give only an approximate value for the mean, so it's typically used in two situations. One is when the data set is very large and calculating the exact mean using all the data is impractical. The other is when the original raw data are unavailable but have been grouped by someone else. We'll look at an example of finding the mean for grouped data, then summarize what we learned in a cool colored box.

Solved Example 1

Finding the Mean for Grouped Data

Here's another look at the grouped frequency distribution for the record high temperature in each state from earlier in this unit. Find the mean of these high temperatures.

Class	Frequency
100–104	3
105–109	8
110–114	15
115–119	13
120–124	8
125–129	2
130–134	1

SOLUTION

The strategy is based on this idea: If we have a grouped frequency distribution and don't know all of the exact values, it's impossible to add up all of the values. Instead, we'll find the midpoint of each class and estimate that all of the values in that group would likely average to that midpoint. (In this case, by "average" we mean "mean.") So if we multiply the number of values in a class by its midpoint, we're estimating the sum of all values in that class. It might be easier to understand if you think of it this way: there are three values in the 100–104 class, and we don't know what any of them are exactly. But if we multiply three by the midpoint of the 100–104 category, we're essentially replacing all three of the temperatures in that range with 102.

First we'll need to find the midpoint of each class in the grouped data. To make it easier, you can find the midpoint of the first class then just keep adding the class width to find the others. In this case, the first midpoint is 102, and the others are 107, 112, 117, 122, 127, and 132.

Next, multiply the number of values in each class by the midpoint.

$$3 \cdot 102 + 8 \cdot 107 + 15 \cdot 112 + 13 \cdot 117 + 8 \cdot 122 + 2 \cdot 127 + 1 \cdot 132 = 5{,}725$$

This is our estimate of the sum of all data values, so at this point we divide by the total number of values to get the approximate mean.

$$\frac{5{,}725}{3 + 8 + 15 + 13 + 8 + 2 + 1} = \frac{5{,}725}{50} = 114.5°$$

Finding the Mean for Grouped Data

Step 1: Find the midpoint of each class in the grouped data.

Step 2: Multiply the frequency for each class by the midpoint of that class.

Step 3: Add up all of the products from step 2.

Step 4: Divide by the sum of all frequencies (which is the total number of data values).

If you prefer formulas to procedures:

$$\overline{X} = \frac{\Sigma(f \cdot X_m)}{n}$$

where f is the frequency for each class, X_m is the midpoint of each class, and n is the sum of all frequencies.

1. Explain why this formula only gives an approximation to the mean, not the actual mean.

Math Note

Thinking of the mean as a "balance point" is really helpful in understanding why this way of approximating the mean makes sense. Think of stacking a number of coins corresponding to the frequency at the center of each class on a seesaw.

2. The grouped frequency distribution provided illustrates the selling prices of the 95 homes most recently sold in my town. Find the mean selling price.

Class	Frequency
\$101,000–\$200,000	64
\$201,000–\$300,000	26
\$301,000–\$400,000	5

Many college courses have a grading system where different course activities make different contributions to the final grade. The idea is that things like tests typically have a bigger effect on a grade than attendance. Here's an example from a course I taught recently:

Homework average: 50 points

Four tests: 30 points each

Final exam: 50 points

Group project: 20 points

Attendance/Participation: 10 points

The size of the relative contribution for each activity is typically called a **weight,** and the overall grade is called a **weighted mean.** It turns out that we can calculate the weighted mean using the same steps we used to calculate the mean for grouped data. We just have to do some interpretation.

Solved Example 2

Turning Mean for Grouped Data into a Weighted Mean

Tong is a student in a class using the grade structure just described. Her scores for each portion of the course are listed below.

Homework average: 87%

Test scores: 91%, 82%, 86%, 94%

Final exam score: 88%

Group project: 92%

Attendance/Participation: 100%

If we're going to use the procedure for finding the mean for grouped data to calculate her overall grade, what role is played by the weights for each portion, and what role is played by her scores?

SOLUTION

In my view, the key to understanding this is "points." Think of every single point in the course as one occurrence of something being awarded. When you see that homework counts for 50 points, you can think of it as 50 times that a point was awarded, which makes it feel a lot like a frequency. So the fact that Tong has an 87% homework average is like getting 87% of a point 50 different times. If that makes sense, then you get it!

The bottom line: The weights act like frequencies in a grouped frequency distribution, and the scores for each portion act like the midpoints of the classes.

3. Use the results of Solved Example 2 and the mean for grouped data formula to find Tong's average for the course.

Did You Get It

7. The data below are the number of losses by the team that won the NCAA men's basketball championship for the years from 1939–2023. Find the mean number of losses.

Losses	Frequency
0–1	13
2–3	31
4–5	22

Losses	Frequency
6–7	11
8–9	5
10–11	3

4. What are the pros and cons of each of the measures of average we studied?

5. (This is my favorite question in the whole lesson.) Look back at the description of the "average" college student in the opening paragraph. For each characteristic listed, write which measure of average you think was used.

Using Technology to Compute Measures of Central Tendency

Technology is certainly a useful tool in computing measures of average, so let's have a look at using spreadsheets and graphing calculators to do the heavy lifting for us. (For what it's worth, when a list of data has just a few numbers on it, it's probably quicker to find the measures of average by hand. It's the large data sets that make technology incredibly useful.)

To illustrate these features, we'll use the following data set. According to deathpenaltyinfo.org, since 1973, over 200 death row inmates were exonerated and freed (which is absolutely horrifying if you think about it). The table below shows the number freed each year from 2008 to 2023.

Year	'08	'09	'10	'11	'12	'13	'14	'15	'16	'17	'18	'19	'20	'21	'22	'23
Number	4	9	1	1	3	1	7	6	0	5	2	3	6	3	2	3

Excel

There are built-in commands for finding the mean, median, and mode in Excel. There's no midrange command, but we can build one, and ordering the list allows us to find the mode.

First, the data should be entered into a single column, as shown in Figure 1. To calculate the mean, choose an empty cell and enter "= AVERAGE(". Then use the cursor to select all of the cells containing the data, enter a closing parenthesis, and hit enter. This displays the mean, shown in cell B18. To find the median, use the same routine, but enter "MEDIAN" rather than "AVERAGE." This is in cell B19. To find the mode, use the same routine again, but use the MODE function. This is in cell B20.

There's no direct command for midrange, but there are commands to find the highest and lowest numbers on the list; we can embed them in a simple calculation that finds the midrange. The string we need to enter looks like

= (MAX(range)+MIN(range))/2

where "range" represents the range of cells that the data live in. Again, the simplest way to enter that range is to just use the cursor to select it after typing "MAX(" and "MIN(". The resulting midrange is shown in cell B21, and the commands for computing the mean, median, mode, and midrange are shown in Figure 2.

	A	B
1		4
2		9
3		1
4		1
5		3
6		1
7		7
8		6
9		0
10		5
11		2
12		3
13		6
14		3
15		2
16		3
17		
18	Mean	3.5
19	Median	3
20	Mode	3
21	Midrange	4.5

Figure 1

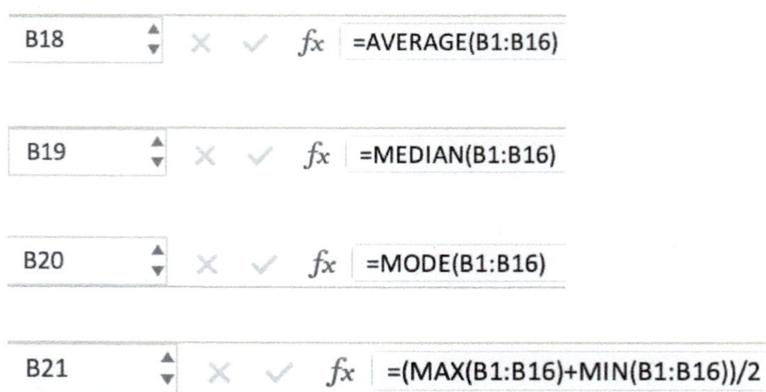

Figure 2

TI-84 Graphing Calculator

In order to find measures of average for a group of numbers, the first step is entering those numbers in a list. To do this, we need to use the list editor: hit the STAT key, then choose 1. Enter the numbers one at a time, hitting ENTER after each. The above data are shown entered in Figure 3 below (a lot of them, anyhow). Then hit STAT again, and use the right arrow to select CALC at the top of the screen, and pick choice 1: 1-Var Stats (Fig. 4). After hitting ENTER twice, you get the 1-Variable Stats menu, shown in Figures 5 and 6.

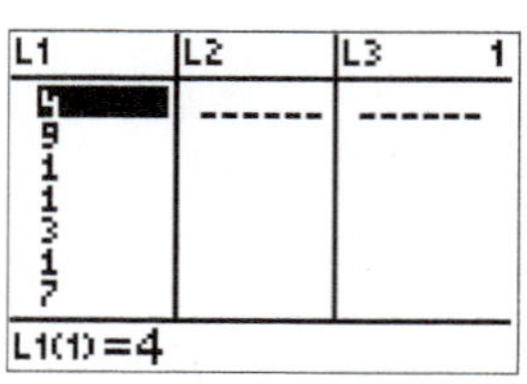

Figure 3

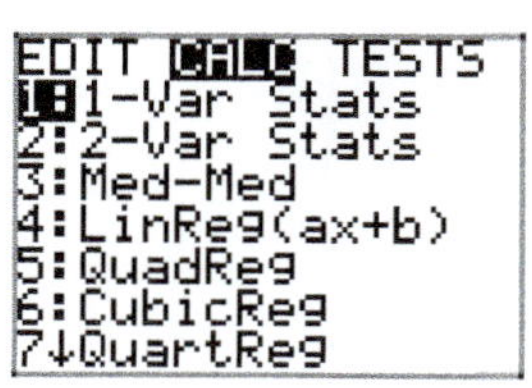

Figure 4

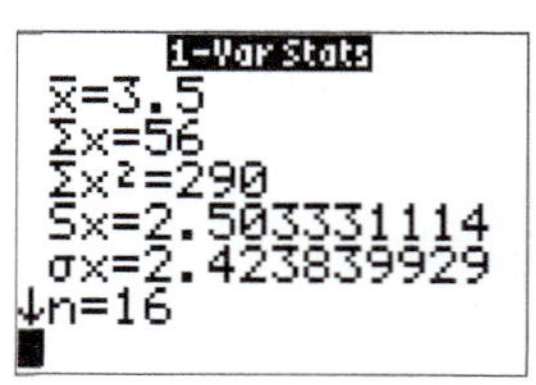

Figure 5

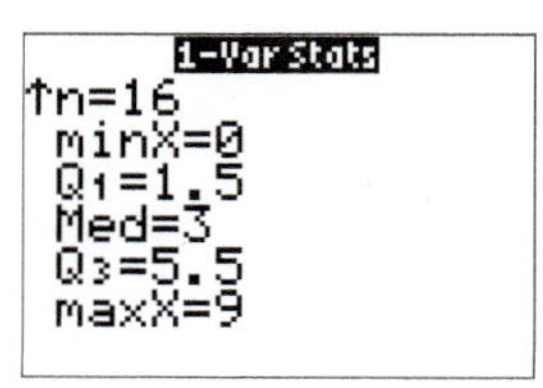

Figure 6

There's actually quite a bit more information on this screen than we're looking for, and some of it will be discussed in later lessons. For now, the mean is the very first thing on the list: $\bar{x} = 3.5$. Using the down arrow to scroll down, we find the median: Med = 3 (Fig. 6). The midrange isn't displayed, but minX = 0 and maxX = 9 provide the lowest and highest values. When you have those, it's simple to add them and divide by 2 to get the midrange of 4.5.

Unlike Excel, there's no specific command for finding the mode, but the calculator will order the data, which makes it a lot easier to find the mode. Hit the STAT key again and choose 2, which is SortA (for "sort ascending"). Then you need to tell it which list to sort. Since we entered the data in list L1, enter that by hitting 2nd 1 (Fig. 7). After you press ENTER, the calculator displays Done, and the list is ordered. When you go back to the list editor, you can easily find which value or values appear most often, giving you the mode (Fig. 8).

Figure 7

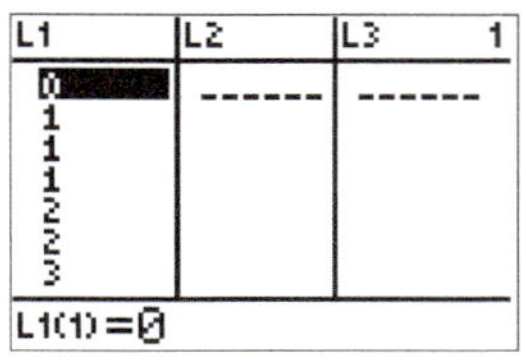

Figure 8

See the Using Tech Videos on Finding Mean, Median, Mode, and Midrange in Excel and on a Graphing Calculator for further information.

Earlier, we asked for your thoughts on the pros and cons of each measure of central tendency. We'll summarize the lesson with our thoughts.

A Comparison of Measures of Average

Measure	Strengths	Weaknesses
Mean	• Unique—there's exactly one mean for any data set • Factors in all values in set • Easy to understand	• Can be adversely affected by a few unusually high or low values • Can be time-consuming to calculate for large data sets
Median	• Divides a data set neatly into two groups • Not affected by one or two extreme values	• Can ignore the effects of large or small values even if they are important to consider
Mode	• Very easy to find • Describes the most typical case • Can be used with categorical data like candidate preference, choice of major, etc.	• May not exist for a data set • May not be unique • Can be very different from mean and median if the most typical case happens to be near the low or high end of the range
Midrange	• Very quick and easy to compute • Provides a simple look at average	• Dramatically affected by extremely high or low values in the data set • Ignores all but two values in the set

4-3 Portfolio

Name ______________________________

Check each box when you've completed the task. Remember that your instructor will want you to turn in the portfolio pages you create.

Technology

1. ☐ In this Excel activity, we'll calculate the mean, median, mode, and midrange of a data set.

ALEKS Practice

1. ☐ Include any written work from the ALEKS assignment along with any notes or questions about this lesson's content.

Applications

1. ☐ Complete the applications problems.

Reflections

Type a short answer to each question.

1. ☐ What did this lesson do to change your view of what the word "average" means?
2. ☐ Which measure of average do you think is the most consistent and the best to use? Why?
3. ☐ Which measures of average are more reliable when you have extreme outliers in the data set?
4. ☐ Name one thing you learned or discovered in this lesson that you found particularly interesting.
5. ☐ What questions do you have about this lesson?

Looking Ahead

1. ☐ Complete the prep skills for Lesson 4-4.
2. ☐ Read the opening paragraph in Lesson 4-4 carefully and answer Question 0 in preparation for that lesson.

Answers to "Did You Get It?"

1. 155,133 sq. mi. **2.** 104,000 sq. mi.; the one extremely large state raises the mean considerably **3.** Mean: 118,714 sq. mi.; median: 101,000 sq. mi.; that one large value did in fact skew the mean upward **4.** 373,500 sq. mi. **5.** Modes: 82° and 87°; median: 84.5° **6.** Nothing **7.** 3.9 losses

Answers to Prep Skills Questions

1. 1.2 1.3 1.3 1.7 1.7 1.7 2.1 2.3 2.4 2.4 2.4 2.5 2.5 2.8 3.1 3.7

2. 35.1

3. $\frac{43}{5} = 8.6$

4. 55

5. 1.2

6. 16

4-3 Applications

Name ____________________

For Questions 1–4, describe which measure of average—mean, median, or mode—was most likely to have been used in each situation. Provide a brief justification.

1. Half of the factory workers make more than \$11.37 per hour and half make less than \$11.37 per hour.

2. The average number of children per family in the Plaza Heights is 1.8.

3. Most people preferred convertibles over any other color.

4. The average age of college professors is 42.3 years.

For Questions 5 and 6, find the mean, median, mode, and midrange. Then decide which you think is the most meaningful measure of average for that data set.

5. A data breach is an incident wherein information is stolen or taken from a system without the knowledge or authorization of the system's owner. Each data breach has the potential to impact millions of citizens. The annual number of data breaches in the United States from 2008 to 2023 is given below:

Year	2008	2009	2010	2011	2012	2013	2014	2015
Data breaches	656	498	662	419	447	614	783	785

Year	2016	2017	2018	2019	2020	2021	2022	2023
Data breaches	1,099	1,506	1,175	1,279	1,108	1,862	1,802	3,205

Source: Statista.com

4-3 Applications

Name ______________________________

6. The 12 states with the highest unemployment rates in October 2024 and their rates of unemployment are given below:

State	Rate
Nevada	5.6
California	5.3
Illinois	5.3
Kentucky	4.9
Washington	4.8
New Jersey	4.7

State	Rate
Alaska	4.6
Rhode Island	4.6
Michigan	4.5
Ohio	4.5
South Carolina	4.5
New York	4.4

Source: U.S. Bureau of Labor Statistics

Forty new automobiles were tested for fuel efficiency (in miles per gallon) by the Environmental Protection Agency. The individual MPG values are displayed below. Use this information for Questions 7–12.

37 27 20 30 24 17 32 26 17 32 25 28 20 18 27 32 36 25 9 26
34 30 25 23 25 24 20 13 31 20 20 25 20 22 37 27 27 25 37 30

7. Find the mean miles per gallon.

8. Complete the frequency distribution.

Class	Frequency
8–12	
13–17	
18–22	
23–27	
28–32	
33–37	

4-3 Applications

Name ______________________________

9. Use the frequency distribution and the procedure for finding the mean of grouped data to approximate the mean. How does your approximation compare to the actual mean?

10. Find the median, mode, and midrange of this data set.

11. Which measure of central tendency would a government official choose to show if she wanted to make the case that gas mileage was getting better for newer vehicles? Why?

12. Which measure of central tendency would an environmental lobbyist choose to show if he wanted to make the case that gas mileage was still not very good for newer vehicles? Why?

4-3 Applications

Name ____

13. Eighty randomly selected compact fluorescent light bulbs were tested to determine their lifetimes (in hours). The grouped frequency distribution was obtained as shown. The full data set for this distribution is in the technology template for this lesson. You'll further analyze these data in the technology assignment. Use the frequency distribution to approximate the mean. Write a sentence explaining what your answer means. Be sure to include units.

Class	Frequency
651–800	6
801–950	12
951–1,100	25
1,101–1,250	18
1,251–1,400	14
1,401–1,550	5

Lesson 4-4 Prep Skills

SKILL 1: FIND THE MEAN

From Lesson 4-3, we learned that the mean is the sum of values in a data set divided by the number of values. If $X_1, X_2, X_3, \ldots, X_n$ are the data values (which means there are n of them), we use the symbol $\overline{X}$ to represent the mean, and $\overline{X} = \dfrac{X_1 + X_2 + X_3 + \cdots + X_n}{n} = \dfrac{\Sigma X}{n}$.

The symbol $\overline{X}$ is read "X bar."

- Find the mean for the list of numbers below.

4 6 12 7 8	There are five numbers in this list.
$\overline{X} = \dfrac{4 + 6 + 12 + 7 + 8}{5}$	Add all five numbers and divide by 5.
$\overline{X} = \dfrac{37}{5} = 7.4$	The mean is 7.4.

SKILL 2: COMPUTE SUMS OF SQUARES

Computing the sum of squares requires remembering the order of operations. The order is as follows, with operations higher on the list being performed before those lower on the list.

(1) Grouping symbols

(2) Exponents

(3) Multiplication or division

(4) Addition or subtraction

Perform these operations first. If there is more than one set of grouping symbols, work from innermost outward. For operations that live at the same level, work from left to right.

Compute each sum of squares.

- $5^2 + (-3)^2 = 25 + 9$ — Since exponents come before addition in the order of operations, compute each square.
  $= 34$ — Then compute the sum.

- $(3 - 4)^2 + (7 - 4)^2 = (-1)^2 + (3)^2$ — First, perform the subtraction inside each pair of parentheses.
  $= 1 + 9$ — Then square each result.
  $= 10$ — Then compute the sum.

SKILL 3: COMPUTE SQUARE ROOTS

A square root of a value is a number that when multiplied by itself gives that specified value. The square root of 25 is 5 because 5 times 5 equals 25. The radical symbol $\sqrt{\ }$ is used to indicate a square root. In symbols, we would write $\sqrt{25} = 5$ because $5 \times 5 = 25$. Unless we are working with a perfect square, computing square roots is typically done with technology. The SQRT function can be used in most spreadsheets. Calculators are also useful for computing square roots.

SKILL 4: USE SUMMATION NOTATION

The Greek letter Σ (sigma) is used to represent the sum of a list of numbers. If we use the letter X to represent data values, then ΣX means to find the sum of all values in a data set. If the data values have been numbered using subscripts, like X_1, X_2, X_3, etc., then the summation notation may be written as ΣX_i.

PREP SKILLS QUESTIONS

1. Find the mean for the list of numbers below.

   26 32 18 65 12 51 162 30

2. Compute the sum:

   $(8 - 13)^2 + (22 - 13)^2 + (13 - 13)^2 + (20 - 13)^2 + (7 - 13)^2$

3. Find the square root of each number. Round to two decimal places.

   112 14,873

4. If $X_1 = 18$, $X_2 = 23$, $X_3 = 50$, and $X_4 = 19$, find ΣX_i.

Lesson 4-4 Your Results May Vary

LEARNING OBJECTIVES

- ☐ 1. Compute measures of variation for a given data set.
- ☐ 2. Interpret standard deviation for a data set.
- ☐ 3. Make meaningful comparisons of standard deviation for two data sets.
- ☐ 4. Analyze the procedure for computing standard deviation.

Tim Davis/Corbis/VCG/Getty Images

Research shows that there is only half as much variation in student achievement between schools as there is among classrooms in the same school. If you want your child to get the best education possible, it is actually more important to get him assigned to a great teacher than to a great school.
—Bill Gates

Now we know quite a lot about measures of average. But just how much information are we getting when we study measures of average? Take a look at the two pictures of dogs on this page. If we look only at measures of average, particularly the mean, we might be fooled into thinking that the two groups are very similar, when clearly they are not. The difference, of course, is that all of the dogs in the first picture are of similar size, while those in the second picture have many different weights. Because there are some small dogs and one very large one, the mean weight in both groups is probably similar. This leads us to the conclusion that there's a lot more to the story told by a data set than simply measures of average. There can be a lot of variation within a group.

Chris Collins/The Image Bank/Getty Images

0. The topic of this lesson is measures of variation. What do you think that means?

4-4 Class

The simplest measure of variation that we'll study is the **range.**

The **range** of a data set is the difference between the largest and smallest values in the set.

Range = Largest value − Smallest value

1. The first list below is the weights of the dogs in the first picture, and the second is the weights of the dogs in the second picture. Find the mean and range for each list.

First picture: 70 73 58 60
Second picture: 30 85 40 125 42 75 60 55

2. Describe what you notice about the range and mean for each group of dogs.

3. Why is the range somewhat limited as a measure of variation for a data set? (Not these particular data sets: ANY data set.)

Did You Get It

Try this problem to see if you understand the concepts we just studied. The answer can be found at the end of the portfolio section.

1. The monthly high temperatures from January to December in Aruba and St. Louis are shown below. Find the median and range for each, then write any observations you make based on this information.

Aruba: 85° 85° 86° 87° 88° 89° 88° 89° 89° 89° 87° 85°

St. Louis: 38° 45° 55° 66° 77° 86° 91° 88° 81° 69° 54° 42°

Standard Deviation

The range is a limited measure of variation because it ignores all the data except the highest and lowest values. If most of the values are similar, but there's just one unusually high value, the range will make it look like there's a lot more variation than there actually is. For this reason, we will next study **standard deviation,** which is a much more reliable (and commonly used) measure of variation.

Standard deviation is a little tricky to compute. Eventually, we'll learn how to use technology to handle the computation, which will allow us to focus on interpretation. But first, we'll develop a procedure for computing standard deviation by hand. This turns out to be useful even if you never actually DO compute it by hand, as it really helps you to understand what exactly standard deviation measures. This is a bit of an oversimplification, but it's still very useful in understanding what standard deviation is designed to measure: *Standard deviation essentially describes the average distance that each data value is away from the mean.*

4. Read the description of standard deviation in italics above carefully again. Why does standard deviation provide a very reasonable way of describing how spread out a data set is?

Interpretation of Standard Deviation

A standard deviation that's large relative to the mean tells us that the data values vary pretty widely from the mean. A relatively small standard deviation means that most data values are clustered fairly close to the mean.

Standard deviation comes in two delightful flavors: population and sample. As we saw in Lesson 4-1, in many cases, data are based not on every single member of a population, but rather a representative sample. When the standard deviation is computed for a data set derived from an entire population under study, it's called the **population standard deviation.** When it's based on data derived from a representative sample of a larger population, it's called the **sample standard deviation.**

As we mentioned, computing standard deviation by hand is pretty complicated, and the truth is that we'll eventually rely on technology to do the computation. In fact, your instructor might have you do that right out of the gate, and that's fine. But we feel like learning and using the formula is useful in that it will help you to understand what exactly standard deviation is measuring, so we're going to start there and learn how to use technology later in the lesson.

In statistics, it's FAR more common to work with a sample than with an entire population, so we will develop a procedure for computing sample standard deviation. As we illustrate the procedure, try to pay special attention to the "key ideas" described in the Solved Example. After that, we'll ask you to think about what those key ideas really tell us.

Solved Example 1

Computing Standard Deviation

Calculate the standard deviation for the weights of the eight dogs pictured in the lesson opener:

30 85 40 125 42 75 60 55

SOLUTION

Step 1: Find the mean. In this case, we've already calculated the mean (Question 1): It's 64 pounds.

Key idea: At its core, standard deviation is a comparison to the mean, so it makes perfect sense to start there.

Step 2: Subtract the mean from each data value. We find it helpful to set up a table to do standard deviation calculations.

X	$X-\overline{X}$
30	$30-64=-34$
85	$85-64=21$
40	$40-64=-24$
125	$125-64=61$
42	$42-64=-22$
75	$75-64=11$
60	$60-64=-4$
55	$55-64=-9$

Key idea: We want standard deviation to be a measure of how far the data values are from the mean, so we should subtract to get the distance from the mean for each value.

Step 3: Square the differences. We'll just add another column to our table, squaring the results from the $X-\overline{X}$ column.

X	$X-\overline{X}$	$(X-\overline{X})^2$
30	$30-64=-34$	$(-34)^2=1{,}156$
85	$85-64=21$	$(21)^2=441$
40	$40-64=-24$	$(-24)^2=576$
125	$125-64=61$	$(61)^2=3{,}721$
42	$42-64=-22$	$(-22)^2=484$
75	$75-64=11$	$(11)^2=121$
60	$60-64=-4$	$(-4)^2=16$
55	$55-64=-9$	$(-9)^2=81$

Key idea: If we just find an average of all the distances from the mean, we'll ALWAYS get zero, which isn't particularly enlightening. (Try it!) By squaring we eliminate the signs to avoid that issue.

Solved Example 1 *Continued*

Step 4: Add up the squares of the distances. This is just plain old addition, and we'll see why in a second.

$$1{,}156 + 441 + 576 + 3{,}721 + 484 + 121 + 16 + 81 = 6{,}596$$

Key idea: To find the mean of the squared distances, we'd add them up and divide by how many, so we start by adding them.

Step 5: Divide the sum by $n - 1$, or one less than the number of data values.

$$\frac{6{,}596}{8 - 1} = 942.29$$

Key idea: To get the actual mean of the squared distances, we would divide by the number of values here. But when you have a sample of data values from a larger population, it's almost certain that there's LESS variation in the sample than in the overall population. So to account for that we divide by $n - 1$ rather than n, making the result a bit larger. Smart!

Step 6: Calculate the square root of the result from Step 5. After this step, we have our standard deviation.

$$\sqrt{942.29} \approx 30.70$$

The sample standard deviation for the dogs in the picture is 30.7 pounds.

Key idea: Remember when we squared all the distances to get rid of the signs? This is where we're basically undoing that square.

Now let's see how you did in understanding the key ideas in the Solved Example.

5. We set out to develop a measurement of how far on average the data values in a sample are from the mean. Based on the key ideas, explain how we were able to accomplish that.

6. Use the procedure from the Solved Example to find the standard deviation for the weights of the four dogs in the first picture at the beginning of the lesson.

70 73 58 60

7. Carefully describe exactly what the standard deviations from the Solved Example and Question 6 tell us about the dogs in the pictures. Include a comparison of the two standard deviations.

The formulas below summarize how to find sample standard deviation, and they also show how to find population standard deviation. For population, the only difference in the procedure is dividing by the total number of data values, rather than one less like we did in finding sample standard deviation.

Formulas for Sample and Population Standard Deviation

Sample: $s = \sqrt{\dfrac{\Sigma(X-\overline{X})^2}{n-1}}$ **Population:** $\sigma = \sqrt{\dfrac{\Sigma(X-\mu)^2}{N}}$

X represents each data value.
$\overline{X}$ is the mean of the sample.
μ is the mean of the population.
n is the number of data values in the sample.
N is the number of data values in the population.

Math Note

The square of the standard deviation is called the *variance* for a data set.

4-4 Group

Now that we have a handle on not only WHAT standard deviation tells us, but WHY it tells us that, we'll look at using technology to do the calculation for us. This allows us to expend more mental energy on interpreting the result. The procedures are summarized in the following box.

Using Technology: Finding Standard Deviation

TI-84 Calculator

The 1-Var Stats screen, which we used to compute the mean and median in Lesson 4-3, also calculates the standard deviation. Below is another look at a screen shot from Lesson 4-3, where the data on death row exonerations were entered in the list editor, and then the 1-Variable Stats command was chosen. The quantities we're interested in are Sx and σx: these are the sample standard deviation (Sx) and the population standard deviation (σx).

```
1-Var Stats
x̄=3.5
Σx=56
Σx²=290
Sx=2.503331114
σx=2.423839929
↓n=16
```

Excel

To use Excel to calculate standard deviation, the data are entered in a single column, just like we did for computing measures of average in Lesson 4-3. Here are sample calculations for the second group of dog weights from Questions 1 and 6 in this lesson, with the formulas used to calculate each statistic illustrated to the right.

	A	B
1		30
2		85
3		40
4		125
5		42
6		75
7		60
8		55
9		
10	Range	95
11	Sample st. dev.	30.696673
12	Population st. dev.	28.714108

Range

B10 *fx* =MAX(B1:B8)-MIN(B1:B8)

Sample Standard Deviation

B11 *fx* =STDEV.S(B1:B8)

Population Standard Deviation

B12 *fx* =STDEV.P(B1:B8)

See the Using Tech Videos on Finding the Sample and Population Standard Deviations in Excel and on a Graphing Calculator for further information.

A word on notation: We use a lowercase s to represent the sample standard deviation for a data set; the symbol Sx is simply what's used on the calculator screen. Similarly, we use a lowercase Greek sigma (σ) to represent population standard deviation, not σx.

1. The table below shows the number of arrests in the U.S. for possession of marijuana, in thousands, for the years from 2008 to 2017. All data was gathered from the annual Uniform Crime Report published by the Federal Bureau of Investigation. Find the mean and sample standard deviation for the number of arrests. Include units.

Year	2010	2011	2012	2013	2014	2015	2016	2017	2018	2019	2020	2021	2022	2023
Arrests (Thousands)	614	541	543	503	516	575	653	599	668	500	318	219	227	200

2. Why was it appropriate to find sample standard deviation, rather than population?

Standard deviation is easiest to interpret when compared to the size of the mean. For example, a standard deviation of 10 in a data set with mean 20 indicates a much wider spread than a standard deviation of 10 in a data set with mean 200. To help make this easier, we can find the **relative standard deviation** (also known as the **coefficient of variation**). This number, usually denoted c_v, is the standard deviation divided by the mean: It's literally what percent of the mean the standard deviation is. In symbols:

The Coefficient of Variation

$c_v = \dfrac{s}{\bar{x}}$ for sample data, and $c_v = \dfrac{\sigma}{\mu}$ for population data.

3. Find the coefficient of variation for the arrest data above. What percent of the mean is the standard deviation?

4. Do you think the standard deviation and coefficient of variation for this data set would be more likely to be used in an argument by someone who is in favor of legalizing marijuana, or against? Explain.

5. Find the coefficient of variation for the eight dogs from the Solved Example in the Class portion. Compare it to your answer to Question 3. What can you conclude?

Did You Get It ?

2. The numbers below are the percentage of all drug possession arrests that were for marijuana for the years from 2010 to 2022. Find the sample mean, standard deviation, and coefficient of variation.

45.8 43.3 42.4 40.6 39.7 38.6 37.4 36.7 36.8 32.1 27.5 16.2 28.5

For Questions 6–8, identify the data set that is likely to have a larger standard deviation. Make sure you EXPLAIN your reasoning. Like with sentences and stuff. No calculations necessary: Just THINK about the data.

6. Data set 1: 12 15 13 10 16 13 12 13

Data set 2: 5 26 31 2 10 25 6 33

7. Data set 1: 40-yard dash times for starting running backs in the National Football League

Data set 2: 40-yard dash times for every student in your math class

8. Data set 1: Average monthly high temperature in Chicago

Data set 2: Average monthly high temperature in Los Angeles

Did You Get It ?

3. Which data set would be likely to have the larger standard deviation? EXPLAIN.

   Data set 1: Weights of all dogs at a golden retriever rescue shelter

   Data set 2: Weights of all dogs for sale in a pet store

9. A professor has two sections of Math 115 this semester. The 8:30 A.M. class has a mean score of 74% with a standard deviation of 3.6%. The 2 P.M. class also has a mean score of 74%, but a standard deviation of 19.2%. What can we conclude about the students' averages in these two sections?

4-4 Portfolio

Name ______________________________

Check each box when you've completed the task. Remember that your instructor will want you to turn in the portfolio pages you create.

Technology

1. ☐ In this Excel activity, we'll calculate the mean and sample standard deviation for a few data sets.

ALEKS Practice

1. ☐ Include any written work from the ALEKS assignment along with any notes or questions about this lesson's content.

Applications

1. ☐ Complete the applications problems.

Reflections

Type a short answer to each question.

1. ☐ If someone asked you informally to describe what standard deviation is, what would you say?
2. ☐ In many ways, standard deviation is most useful when you're comparing the standard deviations of two data sets. Explain why this is reasonable.
3. ☐ What do the mean and standard deviation together tell you about a data set? What do they not tell you?
4. ☐ Name one thing you learned or discovered in this lesson that you found particularly interesting.
5. ☐ What questions do you have about this lesson?

Looking Ahead

1. ☐ Complete the prep skills for Lesson 4-5.
2. ☐ Read the opening paragraph in Lesson 4-5 carefully and answer Question 0 in preparation for that lesson.

Answers to "Did You Get It?"

1. Aruba, median 87.5°, range 4°; St. Louis, median 67.5°, range 53°; Aruba has a much more consistent climate.
2. Mean 35.8%, standard deviation 8.0%, coefficient of variation 0.22.
3. Data set 2: All of the dogs in the first set are the same breed, so you'd expect them to be reasonably similar in size.

Answers to Prep Skills Questions

1. 49.5
2. 191
3. 10.58, 121.95
4. 110

4-4 Applications

Name ______________________________

The table shows the number of class periods missed over the last 10 courses by two students. Use this information for Questions 1–5.

Trey	3	4	3	2	2	4	3	2	2	3
Juanita	0	9	2	1	0	6	0	4	0	8

1. Find the mean number of classes missed for each student.

2. Find the range of the number of classes missed for each student.

3. Without doing the calculations, how do you think the standard deviations compare for the two students? Explain.

4. Now calculate the standard deviation for these two students. Do these values agree with your thoughts from Question 3?

5. Which is a better student in your opinion? What does that say about the importance of measures of average compared to measures of variation in some situations?

4-4 Applications

Name ______________________________

When playing golf, in most cases, the player will want to get the ball as close as possible to the hole when hitting a shot approaching the green. Suppose that two players hit 10 approach shots to a green, and their results are as shown in the two figures. The black dot is the hole, and the white dots are where each shot ended up. Use the diagram for Questions 6–8.

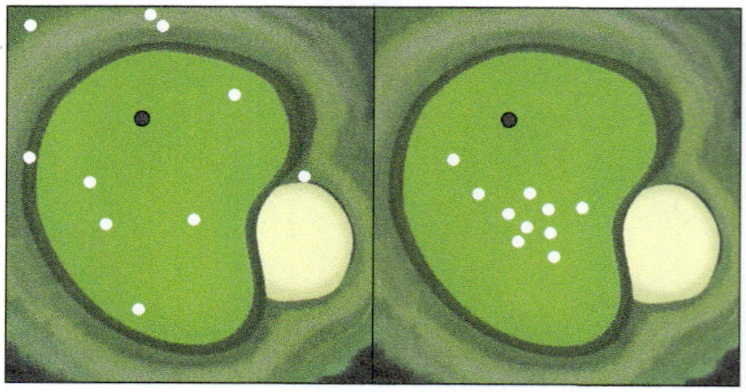

Pat's shots Ron's shots

6. If you were to "average" the locations of all of the results for each player, where would you estimate the average shot would be for each?

7. How would you describe the measures of variation for each of the players?

8. Which is a better player in your opinion? What does that say about the importance of measures of average compared to measures of variation in some situations?

4-4 Applications

Name ____________________

The heights in inches of nine army recruits (first list) and nine randomly selected people walking through a mall (second list) are shown below. Use these data for Questions 9–13.

List 1: 78 72 68 73 75 69 74 73 72
List 2: 32 75 54 48 70 63 67 46 66

9. Compute the mean, range, and standard deviation for the heights of the recruits.

10. Compute the mean, range, and standard deviation for the heights of the mall shoppers.

11. How do the means compare? What does this tell us about the heights for our two groups of subjects?

12. How do the ranges compare? What does this tell us about the heights?

13. How do the standard deviations compare? What does this tell us?

4-4 Applications

Name ____________________

14. The standard deviation for Juanita's number of classes missed (Question 1) was ______________. The standard deviation for the heights of the randomly selected shoppers in a mall was ______________. Explain why comparing these two numbers is meaningless. (Hint: Did you include units?)

15. Find the coefficient of variation for Juanita's number of classes missed and the mall shoppers' heights.

16. The numbers from Question 15 CAN be compared in a meaningful way. Explain what comparing them tells us.

Lesson 4-5 Prep Skills

SKILL 1: COMPUTE A PERCENTAGE OF A WHOLE

As we have seen several times, to compute what percentage of a whole a certain amount is, you perform two steps.

Step 1: Divide the portion by the whole. This gives the percentage in decimal form.

Step 2: Convert that decimal to percent form by moving the decimal point two places to the right. This corresponds to multiplying the decimal form by 100.

- Determine what percent 28 is of 40.

  $\frac{28}{40} = 0.7$, which is 70%

  Divide 28 by 40 to get the decimal form 0.7. Convert this decimal to a percentage by multiplying by 100. 28 is 70% of 40.

SKILL 2: COMPUTE A GIVEN PERCENTAGE OF AN AMOUNT

To compute a number that represents a certain percentage of a given amount, you perform two steps:

Step 1: Convert the percentage to a decimal by moving the decimal point two places to the left.

Step 2: Multiply the decimal form of the percentage by the given amount.

- What is 15% of 80?

  $0.15(80) = 12$

  The decimal form of 15% is 0.15. The calculation tells us that 15% of 80 is 12.

SKILL 3: FIND A MEDIAN

We learned how to find the median in Lesson 4-3.

Step 1: Arrange the data in order, from smallest to largest. (Actually, largest to smallest is fine, too.)

Step 2: The way we find the median depends on whether the number of data values is even or odd.

- If there's an odd number of data values, there will be one single value in the middle of that ordered list, with half of the remaining values bigger and half smaller. That value in the middle is the median.
- If there's an even number of data values, there will be two values in the middle of the list. The mean of those two values is the median of the entire list.

Find the median for each data set.

- 6 7 11 15 16

  The list is sorted from smallest to largest. There are 5 values in this list.

  Median = 11

  The median of this list is 11, which is the middle number in the list.

- 4 6 7 11 15 16

  The list is sorted from smallest to largest.

  There are 6 values in this list. The two middle values are 7 and 11.

  $\text{Median} = \frac{7 + 11}{2} = 9$

  The median of this list is 9, which is the mean of 7 and 11.

PREP SKILLS QUESTIONS

1. If 35 people in a class of 50 score lower than you on a final exam, what percent of people scored lower than you?

2. What percent of people scored higher than you?

3. If 36% of days in March last year were rainy, how many days were rainy? How many were not?

4. A math class recorded the following scores, out of 200, on a final exam. What was the median score?

189	186	183	179	177	175	168	167	166	163	161	158
156	154	153	152	148	146	144	137	122	119	108	87

5. How does the median change if we throw out the lowest score?

Lesson 4-5 Where Do You Rank?

LEARNING OBJECTIVES

- ☐ 1. Compute percentile ranks.
- ☐ 2. Find data corresponding to a given percentile rank.
- ☐ 3. Use percentiles to compare data from different sets.
- ☐ 4. Compute quartiles and use them to analyze spread.
- ☐ 5. Draw and interpret box plots.

Grades are almost completely relative, in effect ranking students relative to others in their class. Thus extra achievement by one student not only raises his position, but in effect lowers the position of others.

—James S. Coleman

PictureQuest/Comstock Images/Getty Images

So you managed to survive your college years and escape with that long-awaited diploma. But maybe the thought of entering the mythical "real world" starts to look a little less appealing, and you realize that a couple more years of college might suit you. Next step: Take the GRE (Graduate Record Examination), which is like the SAT for grad school. When you get your score back, you see that it's 1120. How do you know if that's a good score? The quickest way is to look at the percentile rank on the report, which in this case would probably be 60%. Does this mean that you only got 60% of the questions right? That wouldn't be a very good score, now would it? In fact, that's not at all what it means, and you did just fine.

When raw scores (or other data) are hard to interpret, it's useful to see where certain values rank within the data set, and that's what we'll study in this lesson.

0. Do you have any idea what a percentile rank of 60 means? if so, describe your understanding.

4-5 Group

The term *percentile* is used in statistics to measure the position of a data value in a data set.

A **percentile,** or percentile rank, of a data value indicates the percent of data values in a set that are less than that particular value.

In the case of the GRE score, your percentile rank of 60% means that 60% of all students who took the GRE scored lower than you did. Not bad! Scores on standardized tests are one of the most common uses for percentiles because they help you to put a raw score like 1120 into context.

1. Percentiles are used very commonly in health care, especially pediatrics. To monitor a child's development, doctors compare the child's height, weight, and head size at a certain age to measurements of other children at the same age. What do you think the medical value of doing this is?

> **Math Note**
>
> Some books define a percentile as the percentage of data values less than or equal to a certain value, instead of just "less than."

Percentiles were originally used to analyze data sets with 100 or more values. But when statistical techniques were developed for smaller data sets, the percentile concept came into use for those sets as well. There are several methods that can be used for computing percentiles, and sometimes the answers vary slightly, especially with small data sets. In this lesson, we'll use a basic method that works well for smaller data sets, and rely on technology to handle bigger sets.

Finding a Percentile Rank

The percentile rank for a given data point in a set with n total values can be calculated using this formula:

$$\text{Percentile of a value } x = \frac{\text{Number of data values less than } x}{n} \cdot 100$$

The result should be rounded to the nearest whole number.

2. Suppose you score 77 on a test in a class of 20 people, with the 20 scores listed below. What was your percentile rank? (Hint: It's not necessary to put the data values in order, but it surely doesn't hurt.)

93 82 64 75 98 52 77 88 90 71 78 55 62 99 91 83 89 61 73 80

Did You Get It

Try this problem to see if you understand the concepts we just studied. The answer can be found at the end of the portfolio section.

1. The number of people getting on at one New York City subway stop at various times throughout a Monday are shown below. Find the percentile rank of the time at which 97 people boarded.

   101 120 88 72 75 80 98 91 105 97 78 85

Finding the percentile rank for a certain data value, as we've seen, can give us some useful perspective on how high or low that data value is. But what about the opposite question? Suppose that a researcher is interested in studying the homework habits of students that rank at the 80th percentile in their classes. How do you find the score that corresponds to a certain percentile? Let's see.

Here are the exam scores from your hypothetical class again. The goal is to find the score that corresponds to the 80th percentile.

93 82 64 75 98 52 77 88 90 71 78 55 62 99 91 83 89 61 73 80

3. Arrange the scores in order from smallest to largest.

4. The 80th percentile means that we want 80% of the scores to be less than the one we're looking for. With that in mind, how many scores do we want to have LESS than the score we're trying to identify?

5. Find the score that has 80% of the data below it.

This seems like a reasonable way to find the data value we're looking for, but there's a minor problem. If you're at the 80th percentile, you should have 80% of the scores below you, and 20% above you, right? But in this case . . .

6. How many scores are above the score you found in Question 5? What percentage is that of the total number of scores? What can you conclude?

Now we see the problem. For really large data sets, this isn't much of an issue. For example, if there were 200 scores rather than 20, the 80th percentile would have 80% of them below it (which is 160 scores), and 39 above it. This corresponds to 19.5%, which is pretty close to the 20% we'd want. But for smaller data sets, we don't have "enough" data above the one we're choosing for a certain percentile. The box below describes a slightly modified procedure to address this issue.

Finding the Data Value That Corresponds to a Percentile

Step 1: Arrange the data values in order from smallest to largest.

Step 2: Count the total number of data values, then multiply that number by the percentile rank we're looking for, written as a percentage in decimal form. We'll call this number the *index* and represent it with the letter L.

Step 3: If L is a whole number, then the data value corresponding to our given percentile is halfway between the value in the Lth position from the bottom and the next higher value.

Step 4: If L is not a whole number, round it up to the next whole number. Then the data value at the given percentile is the Lth value from the bottom on the list.

7. For the test score data in Questions 3–6, what would be the data value at the 80th percentile using this procedure? (Note that you already calculated L in Question 4. You just didn't realize it at that point.)

8. The disadvantage of the data value we got in this case is that it's NOT one of the actual data values on the list. The advantage? You tell me by answering this question: How many data values on the list are now below the value at the 80th percentile? How many are above? What percent is each of those of the total number of values?

9. Use the procedure in the box to find the score that's at the 62nd percentile.

Did You Get It

2. The number of wildfires in the United States, in thousands, from 2007 to 2023 is shown in the table. Which year is at the 25th percentile?

'07	'08	'09	'10	'11	'12	'13	'14	'15	'16	'17	'18	'19	'20	'21	'22	'23
91.9	82.4	81.8	72.0	74.1	67.8	47.6	63.6	68.2	67.7	71.5	58.1	50.5	59.0	59.1	69.0	56.6

Percentile ranks are really useful for comparing data that come from two different sets. Here's a great example of that: Two students are competing for one remaining spot in a law school class. Gabriela ranked 51st in a graduating class of 1,400, while Dustin ranked 27th in a class of 540.

10. Would it be fair to award the spot to Dustin because he ranked closer to first in his class? Why or why not? Explain your answer.

11. Which has the better ranking if you compute percentiles within the class? Does that surprise you? Discuss. (That means write more than "yes" or "no.")

Did You Get It

3. In the 2023–2024 school year, the University of Missouri finished the season ranked 14th out of 128 teams in football and 36th out of 363 teams in basketball. Based on percentile rank, which team had the better ranking?

4-5 Class

1. Discuss this question among your classmates and see if you can come up with a good answer: What does the median (which we studied in Lesson 4-3) have to do with percentiles?

Quartiles

Quartiles divide a data set into four equal parts; that is, four intervals that contain the same amount of data. This isn't the same thing as four intervals of the same length relative to the data. For the test score data we've been studying, the highest score is 99 and the lowest is 52, so the range is 47 points. If we divided the data into categories of equal length, each would cover 11.75 points, and the categories would be 52–63.75, 63.75–75.50, 75.50–87.25, and 87.25–99.

2. How many data values would fall into each of those categories? You put the data in order in Question 3 of the Group portion of this lesson, which will help. The data are repeated below if you want them for reference.

93 82 64 75 98 52 77 88 90 71 78 55 62 99 91 83 89 61 73 80

Category	# Values
52–63.75	
63.75–75.50	
75.50–87.25	
87.25–99	

3. So we can see that the data have been divided into four categories with equal class widths, but that's not the same as dividing them into equal groups in terms of number of data values. If, instead, we divide the data so that the same number of data values fall into each of four categories, how many data values will be in each category? (Don't think too hard—this is NOT a hard question.)

Here's a visual that illustrates what quartiles are about:

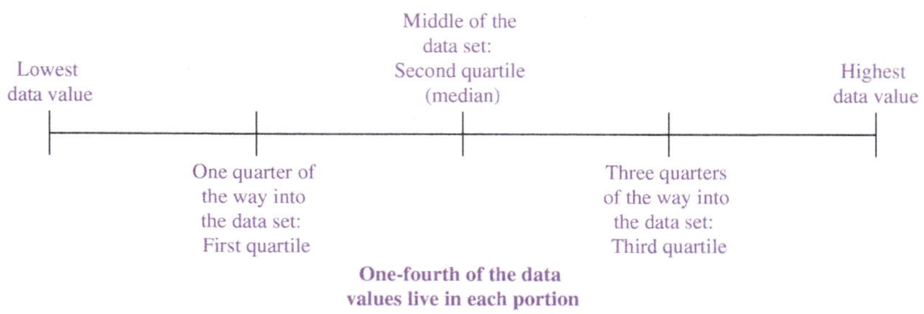

We use the symbol Q_1 to represent the first quartile, Q_2 to represent the second (which is also the median), and Q_3 for the third. This is a nice way to think of what these symbols mean:

Interpretation of Quartiles

Q_1 (first quartile): 25% of data values are less than this, and 75% are greater than it.

Q_2 (second quartile): 50% of data values are less than this, and 50% are greater than it.

Q_3 (third quartile): 75% of data values are less than this, and 25% are greater than it.

4. While quartiles break the data into four equal portions, percentiles break data into 100 equal portions. If we use the symbol P_{60} to represent the 60th percentile, how would we represent the first and third quartiles as percentiles using this notation? What about the median? (Note: There were three questions there.)

Solved Example 1

Finding Quartiles

David Wasserman/Brand X Pictures/Getty Images

Following are the percentages of total electricity generated from nuclear power for the nations with the 12 largest economies in the world, listed by size of economy. Find the quartiles and describe what they mean.

2.2 13.7 4.9 64.8 0 3.1 0 5.6 18.4 31.5 12.5 18.6

SOLUTION

First, we'll find the median, which is also Q_2 (the second quartile). To do so, put the data values in order, from least to greatest.

0 0 2.2 3.1 4.9 5.6 12.5 13.7 18.4 18.6 31.5 64.8

There's an even number of values, so the median is the number halfway in between the two values in the middle, which are 5.6 and 12.5.

$$\frac{5.6 + 12.5}{2} = 9.05$$

So $Q_2 = 9.05\%$.

This divides the distribution into two halves, with six values in the lower half and six in the upper half. Next, we'll find the median of the lower half: This is Q_1 (the first quartile). The two values in the middle of the bottom half are 2.2 and 3.1, and halfway between them is 2.65. So $Q_1 = 2.65\%$.

Finally, we'll find the median of the upper half—that's the third quartile. The two values in the middle of the upper six values are 18.4 and 18.6, and 18.5 is halfway between. So $Q_3 = 18.5\%$.

Now let's summarize. The first quartile is 2.65%, so nations that get less than 2.65% of their power are in the bottom fourth among the world's largest economies. The second quartile is 9.05%, which tells us that nations that get less than 9.05% of their energy from nuclear are in the bottom half. The third quartile is 18.5%, so if a nation gets more than 18.5% of its energy from nuclear, it's in the top fourth in terms of nuclear generation among the world's largest economies.

5. Find the quartiles for the following data, which are the ages as of this writing of the nine justices of the Supreme Court.

74 52 57 54 64 59 69 70 76

Did You Get It ?

4. The data below are the number of cattle on farms in the United States (in millions) for each year that begins a decade from 1920 to 2020. Find the quartiles.

70.4 61.0 68.3 78.0 96.2 112.4 111.2 95.8 98.2 93.9 103.0

The Interquartile Range

The **interquartile range,** or IQR, is a measure of variation that is often used when median is used as your measure of center. IQR measures the spread of the middle 50% of the data.

$$\text{IQR} = Q_3 - Q_1$$

6. Find the IQR for the nuclear power energy data set from Solved Example 1.

When studying measures of average in Lesson 4-3, we learned that median is a more reliable measure of average than mean when there are a few values in the data set that are way higher or lower than the others. In other words, median tends to be resistant to the effects of a few aberrational data values, and for that reason it is called a **resistant measure.**

7. IQR is considered to be a resistant measure of variation. Discuss why you think that is reasonable.

One of the most useful applications of quartiles is using them to draw a **box plot** (sometimes called a box and whisker plot). This is a graphical way to evaluate the spread of a data set. In particular, a box plot makes it easy to identify data points that are *outliers*—those that appear to be aberrational in some way. (We'll learn more about outliers in just a bit.) Following is an example of a box plot for a group of test scores. (The labeling in purple is not typically included in a box plot; it's there to help you understand what the plot is displaying.)

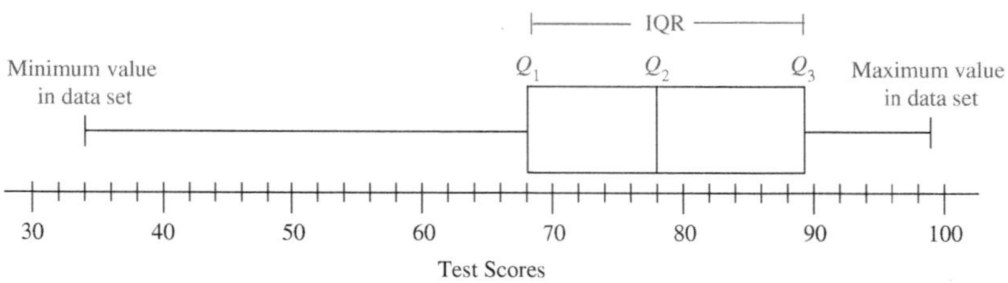

8. Use the box plot to fill in all of the requested information for the test scores it illustrates.

Lowest score ___________ Third quartile ___________

First quartile ___________ Highest score ___________

Median ___________ Interquartile range ___________

The first five numbers you found in Question 8 make up what's known as the **five number summary** for a data set. This is a good way to show how data are distributed in a concise way. In essence, a box plot is a graphical representation of the five number summary.

9. Write the five number summary for the nuclear power data we've been studying. Write it in this format, which is a typical way to write the summary: {Minimum, Q_1, Q_2, Q_3, Maximum}. (You've already found the quartiles.)

10. Draw a box plot for the data.

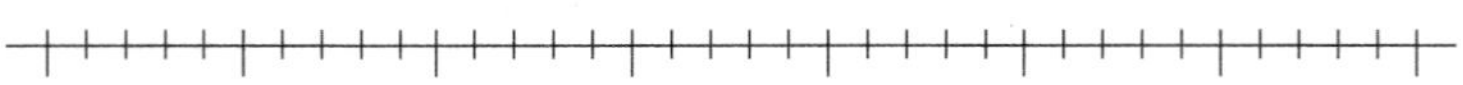

11. What does the box plot tell you about the data set? Write some observations. In particular, what do the numbers inside the box represent? Is the median closer to Q_1 or Q_3? What does that tell you?

Speaking casually, an **outlier** is a data value that's significantly smaller or bigger than the other values in a data set. The idea is that if there are one or two values that are completely different from the others, it might be reasonable to ignore those values when studying the rest of the data.

But in the world of math and stats, casual definitions aren't particularly popular. If we're going to discount some data values when studying a data set, we probably ought to have some actual rules about when it's reasonable to do that, rather than relying on a gut feeling or a subjective opinion. We know that half of the data in any data set live between the first quartile (Q_1) and the third quartile (Q_3). So certainly any outlier would have to be significantly less than Q_1 or significantly greater than Q_3. The question is: How much? That's where the interquartile range comes into play.

Data values are considered to be *outliers* if they are more than 1.5 times the interquartile range below the first quartile, or more than 1.5 times the interquartile range above the third quartile. The numbers $Q_1 - 1.5 \cdot \text{IQR}$ and $Q_3 + 1.5 \cdot \text{IQR}$ are called **fences.** They indicate that any data values outside of them are aberrational somehow. We can draw additional vertical boundaries on our box plot to identify the fences.

Solved Example 2

Finding Outliers in a Data Set

Find the fences for the nuclear power data we've been studying, add them to the box plot from Question 10, and decide if the data set has any outliers.

SOLUTION

The key number here is the distance below Q_1 and above Q_3 we need to go: That distance is 1.5 times the interquartile range. So we have to start by finding that distance.

From Question 6, the IQR is 15.85, and 1.5 times 15.85 is about 23.78. To locate the fences, we need to subtract 23.78 from Q_1 and add 23.78 to Q_3.

$$Q_1 - 1.5 \cdot \text{IQR} = 2.65 - 23.78 = -21.13$$
$$Q_3 + 1.5 \cdot \text{IQR} = 18.5 + 23.78 = 42.28$$

The first number is irrelevant for this data set: No country could possibly get a negative percentage of its energy from nuclear power. So technically, −21.13 is a fence, but not one that we're interested in. The other fence is 42.28, and this one can certainly be added to our box plot:

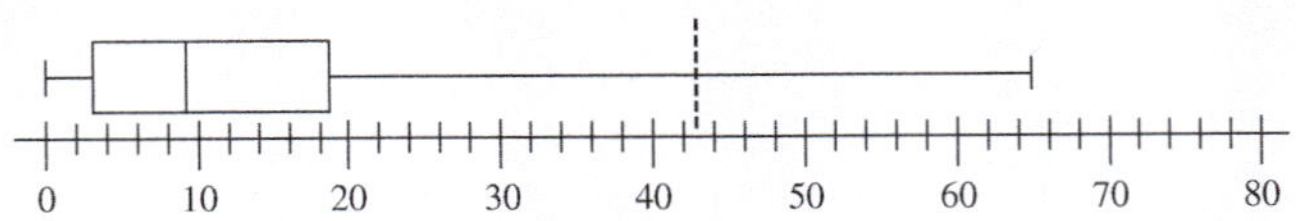

Now we can tell for sure that there is at least one outlier, because the box plot extends past the fence. Looking back at the data set, we can see that the only outlier is the maximum value of 64.8%.

After identifying outliers, typically we would investigate those values. In some cases, they might be mistakes in the data gathering process. In others, they indicate values that are exceptionally high or low compared to the others (and therefore might be likely to skew some measures of average or variation—those that are not resistant.) If you find any outliers, it's worth taking some time to check if they're mistakes in the data. If not, you should consider how they might affect things like measures of average. In many cases, outliers are ignored when evaluating a data set.

12. To identify extreme outliers, we can find **outer fences**. These are values that are three times the IQR below Q_1 and above Q_3. Is the maximum value in our nuclear power data set an extreme outlier? Discuss how you might view that data value in relation to the rest of the set when evaluating how reliant most nations with large economies are on nuclear power.

13. Are there any outliers among the scores that generated the box plot you studied in Question 8? How can you tell?

Did You Get It

5. Draw a box plot for the data in Did You Get It 4. Are there any outliers? If so, what are they?

Using Technology: Finding Percentiles and Quartiles

There are Excel commands that will find all of the information we studied in this lesson. It's all about knowing the syntax. The example below shows the test scores that were used for the box plot sample on page 443. The list was already sorted using the SORT command.

	A	B
1	Test Scores	34
2		62
3		65
4		66
5		66
6		68
7		71
8		73
9		75
10		77
11		78
12		80
13		82
14		83
15		88
16		89
17		90
18		91
19		93
20		98
21		99
22		
23	**Median**	78
24	**1st Quartile**	68
25	**3rd Quartile**	89
26		
27	**65th Percentile**	83
28	**20th Percentile**	66

- We already know the median command.
- To find the first quartile:
  =QUARTILE(cell range,1)
  where cell range is the first and last cell on the list with a colon between them.
- To find the third quartile:
  =QUARTILE(cell range,3)
- To find a given percentile:
  =PERCENTILE(cell range,percentile)
  where the percentile is written as a percentage in decimal form.

fx =MEDIAN(B1:B21)

fx =QUARTILE(B1:B21,1)

fx =QUARTILE(B1:B21,3)

fx =PERCENTILE(B1:B21,0.65)

fx =PERCENTILE(B1:B21,0.2)

See the Using Tech Video on Finding Percentiles and Quartiles in Excel for further information.

4-5 Portfolio

Name ____

Check each box when you've completed the task. Remember that your instructor will want you to turn in the portfolio pages you create.

Technology

1. ☐ In this Excel activity, we'll use what we've learned about percentiles and quartiles to analyze a large data set.

ALEKS Practice

1. ☐ Include any written work from the ALEKS assignment along with any notes or questions about this lesson's content.

Applications

1. ☐ Complete the applications problems.

Reflections

Type a short answer to each question.

1. ☐ Why are percentiles useful in interpreting things like standardized test scores?
2. ☐ Explain what the five number summary tells us about a data set. You should describe each of the numbers and what its significance is and give an overview of what the entire summary is useful for.
3. ☐ If you knew someone whose child scored in the 70th percentile on a standardized test, how would you explain to them what exactly that means?
4. ☐ Name one thing you learned or discovered in this lesson that you found particularly interesting.
5. ☐ What questions do you have about this lesson?

Looking Ahead

1. ☐ Complete the prep skills for Lesson 4-6.
2. ☐ Read the opening paragraph in Lesson 4-6 carefully and answer Question 0 in preparation for that lesson.

Answers to "Did You Get It?"

1. 58th percentile **2.** 2020 **3.** Basketball was a little better; 90th percentile compared to 89th percentile

4. $Q_1 = 70.4$, $Q_2 = 95.8$, $Q_3 = 103.0$ **5.** There are no outliers.

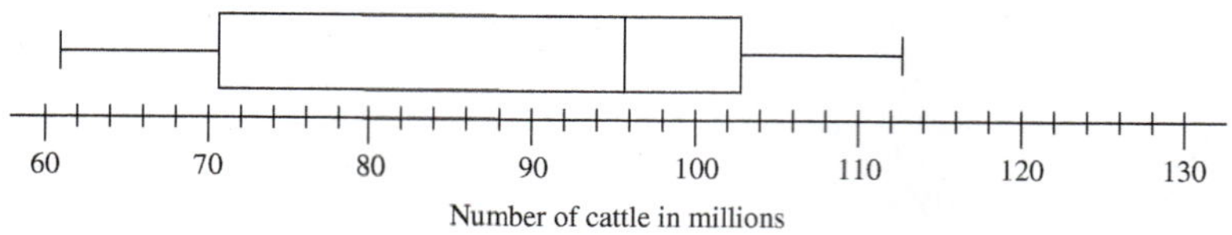

Answers to Prep Skills Questions

1. 70%

2. 28%

3. 11 were rainy, 20 were not

4. 157

5. It goes up by 1.

4-5 Applications

Name ____

It wasn't that long ago that it was reasonably unusual for a college football player to weigh more than 300 pounds, but those days are long gone. Listed below are the average weights for the starting offensive line at 14 of the schools in the Southeastern Conference last season. We'll be studying these weights in this activity.

School	Avg. Weight
Arkansas	321.1
Mississippi	313.4
South Carolina	310.4
Vanderbilt	309.5
Kentucky	308.3
Louisiana State	307.8
Alabama	303.6

School	Avg. Weight
Tennessee	303.3
Mississippi State	302.5
Auburn	300.6
Texas A&M	295.8
Florida	295.3
Georgia	291.1
Missouri	289.5

1. Find the percentile rank for Vanderbilt's offensive line.

2. Which school is at the 15th percentile?

3. After seven weeks of the season, *USA Today* had Kentucky ranked 49th out of 128 teams. Which is higher: their percentile rank for offensive line weight in their conference, or their percentile rank among all 128 teams according to *USA Today?*

4-5 Applications

Name ______________________________

4. Analyze the weight data by drawing a box plot. Feel free to use a spreadsheet to help with the calculations. Make sure you find the interquartile range and fences (and include any that are reasonable to draw on your plot), decide if there are any outliers, then write a paragraph describing any observations you have based on your results.

5. You probably wouldn't be surprised to find that most members of Congress are really rich. But how rich? In the Technology portion of the homework, you'll be analyzing the net worth of almost every member of the House and Senate as of 2024. Senator Dan Sullivan (R-AK) and Representative Suzanne Bonamici (D-CO) both had net worth right around $10 million. Senator Sullivan ranked 25th out of 100 senators, while Congresswoman Bonamici ranked 69th out of 435 members of the House. Based on percentiles, which was richer relative to their congressional body? What do you think that says about the House compared to the Senate?

Lesson 4-6 Prep Skills

SKILL 1: INTERPRET A HISTOGRAM

Interpreting a histogram is pretty straightforward. It's kind of like reading a bar graph. Just be aware that the categories for the bars won't be specifically labeled; you'll have to find them by reading the scale on the horizontal axis. The following histogram represents a distribution of the number of hours per week spent on homework by a group of students.

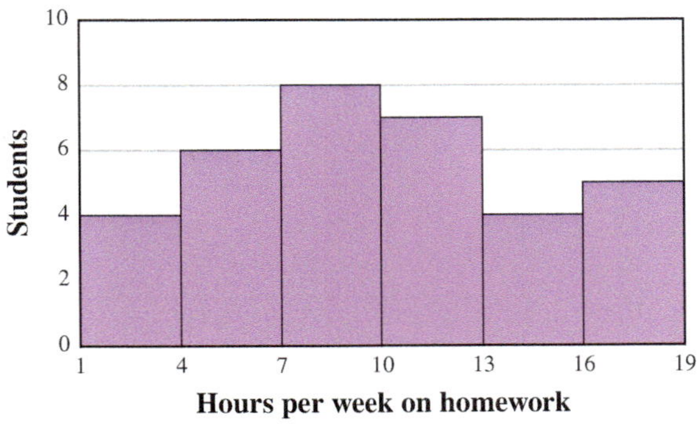

- Which is the most common range?
  7–10

  The most common range is the highest bar.

- How many students spent less than 7 hours per week on homework?
  10

  There are 4 students in the 1–4 range and 6 students in the 4–7 range, making a total of 10 students spending less than 7 hours per week on homework.

SKILL 2: PROBABILITY AND PERCENT CHANCE

The relationship between probability and percent chance is the same as the relationship between any decimal and its percentage form. If the probability of an event occurring is 0.15, that means there is a 15% chance of that event occurring.

SKILL 3: ESTIMATE AN ENCLOSED AREA

Estimating an area of part of a total known area is common in statistics.

- The total area of the circle is 20 square inches.
  Estimate the area of the blue region.

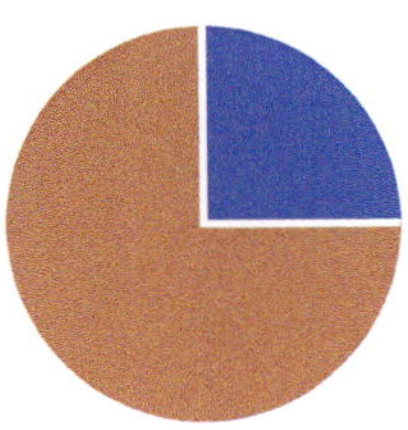

The blue region appears to be about $\frac{1}{4}$ of the circle.

Blue area $= \frac{1}{4}(20) = 5$ sq in.

SKILL 4: COMPUTE A PERCENTAGE OF A WHOLE

As we have seen several times, to compute what percentage of a whole a certain amount is, you perform two steps.

Step 1: Divide the portion by the whole. This gives the percentage in decimal form.

Step 2: Convert that decimal to percent form by moving the decimal point two places to the right. This corresponds to multiplying the decimal form by 100.

- Calculate what percent 21 is of 60.

$\frac{21}{60} = 0.35$, which is 35%

Divide 21 by 60 to get the decimal form 0.35. Convert this decimal to a percentage by multiplying by 100. 21 is 35% of 60.

PREP SKILLS QUESTIONS

Use the histogram to answer Questions 1 and 2. It represents a distribution of the toll required (in cents) at various toll booths on American roads.

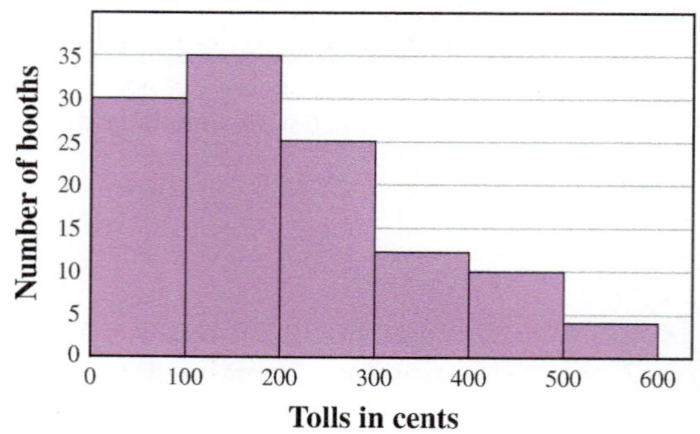

1. What range of tolls is most common?

2. How many toll booths have a toll $4 or more?

3. If there's a 12% chance of a hurricane making landfall in Florida this year, what's the probability of that occurring?

4. If the total area under the following graph is 100, estimate the shaded area.

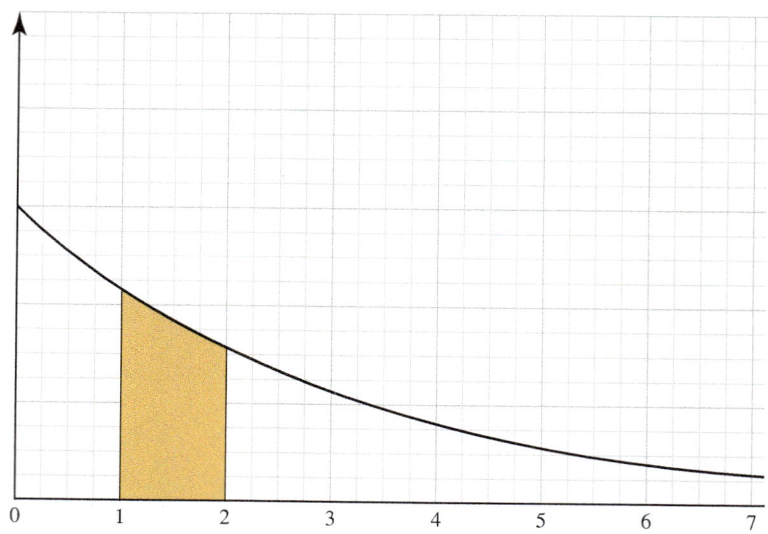

5. There are 712 people watching a play in a theater with 850 seats. What percentage of the seats are filled?

Lesson 4-6 Just a Normal Day

LEARNING OBJECTIVES

- ☐ 1. Recognize characteristics of data that are normally distributed.
- ☐ 2. Understand the connection between area under a normal curve, percentage, and probability.
- ☐ 3. Make an educated guess about the empirical rule, then use the rule to calculate percentages and probabilities.
- ☐ 4. Compare data values from different sets using Z scores.

To study the abnormal is the best way of understanding the normal.
—William James

David Sobecki

If you grew up anywhere near maple trees, you're probably familiar with the seed pods they drop by the thousands in the spring. Like almost all living things, the individual pods vary in size—on any given tree, there's a typical size, with some pods bigger and some smaller.

On a beautiful spring day, I gathered 100 pods from the maple in my back yard, measured each, then grouped them according to their length. The smallest was 42 mm, the largest 59 mm, and in the photo they're grouped in classes 42–43, 44–45, 46–47, etc. An interesting thing happened—we see that the largest number of pods have lengths somewhere in the middle of the range, and the classes further away from the center have fewer pods. A wide variety of quantities in our world tend to exhibit this same phenomenon. In fact, it's so common that frequency distributions of this type came to be known as *normal distributions.* That means that if we get really good at studying normal distributions, we'll be able to analyze tons of interesting things in our world.

0. Why do you think that the lengths of the seed pods in the photo follow that pattern?

4-6 Class

1. List some things that you think might be normally distributed. Try not to focus strictly on physical sizes.

Suppose a researcher selects a random sample of 100 adult women, measures their heights, and constructs a histogram. The researcher would probably get a graph similar to the one shown below.

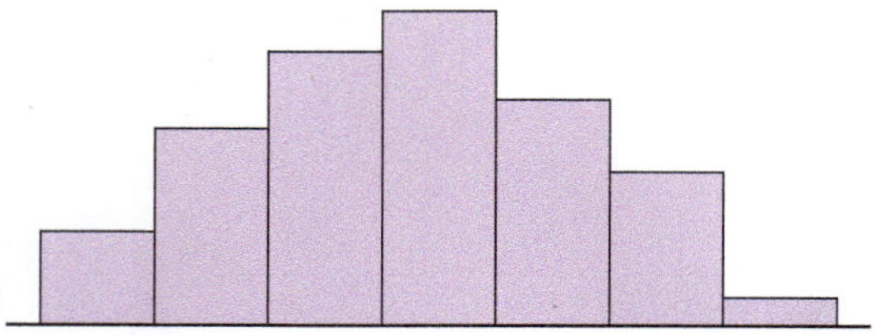

2. Notice that we omitted a scale on the horizontal axis. Put in a scale that you think seems reasonable based on the description of this histogram. What do you think the value right in the middle of the horizontal axis represents?

Next, the researcher decides to go big or go home, and she chooses "big." She dramatically increases the number of women in the sample, and since she has more data to go on, makes the classes narrower as well. Happy with the results, she adds even more women to her sample and makes the classes even narrower. The results would most likely look like the next two graphs.

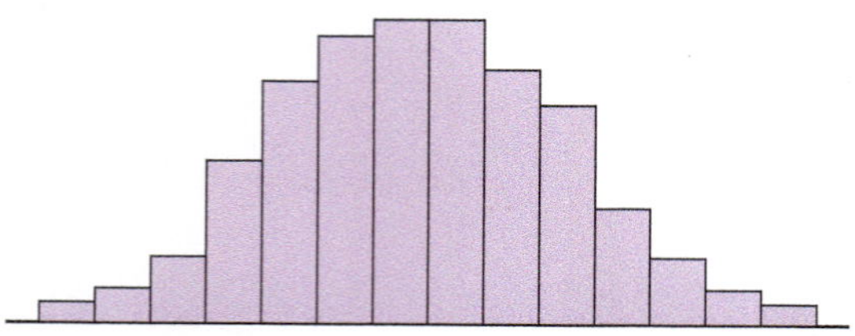

3. Now let's use some imagination. Suppose that our researcher has magical powers, and can continue adding data and improving her histogram until there are infinitely many data points. Wow, that's a lot. Instead of drawing boxes, draw a smooth curve that you think the histogram would become in the long run.

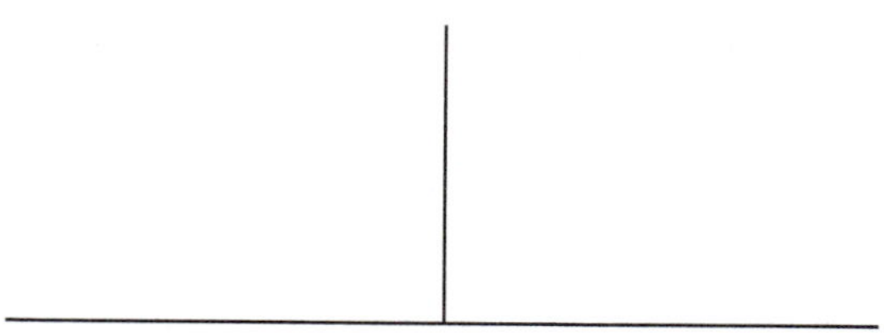

4. If all went well, you've just drawn your first normal probability distribution. (We'll see plenty of examples in the remainder of the lesson if you're unsure how well you did.) How does the drawing relate to the maple seed pod example from the opening paragraphs of this lesson?

To show that normal distributions are not even close to a purely mathematical topic, we're going to borrow our definition of normal distribution from an investment site called "Investopedia":

A probability distribution that plots all of its values in a symmetrical fashion and most of the results are situated around the probability's mean is called a **normal distribution.** Values are equally likely to plot either above or below the mean. Grouping takes place at values that are close to the mean and then tails off symmetrically away from the mean.

Next, we'll highlight some of the important features of a normal distribution (which, in turn, give us the key features of a data set that is normally distributed).

- The value in the middle of the distribution, which appears most often in the sample, is the mean.
- The distribution is symmetric about the mean. This means that the graph has two halves that are mirror images on either side of the mean value.
- (This is the key fact): The area under any portion of the curve is the percentage (in decimal form) of data values that fall between the values that begin and end that region.

Let's explore that last key point, because it's what makes normal distributions so useful in analyzing data. The graph below shows a normal distribution for heights of women in the U.S. The numbers on the horizontal axis are heights in inches, and some areas are labeled for reference.

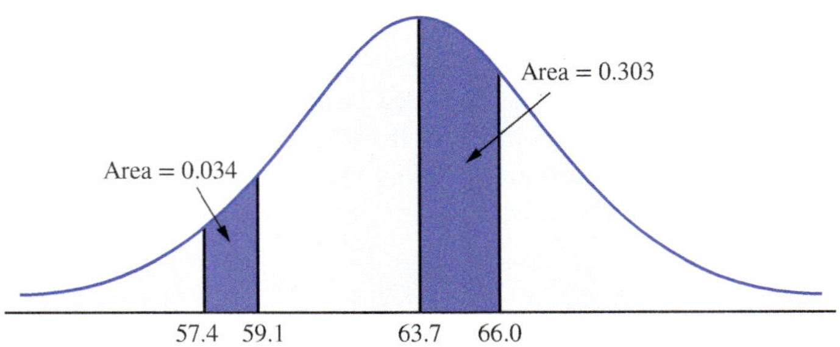

5. What is the mean height for American women?

6. What percentage of American women are between 57.4 inches and 59.1 inches tall?

7. If there are 31,806 women in the stands for a college football game, how many of them would you expect to be between 63.7 and 66.0 inches tall?

Did You Get It

Try this problem to see if you understand the concepts we just studied. The answer can be found at the end of the portfolio section.

1. What percentage of American women are taller than 66 inches? (Hint: Half of all women are taller than 63.7 inches.)

So now we know that if we have data values that are normally distributed, and IF we can find the area under portions of the graph, we can find the percentage of data values that fall in a certain range. That leads to the big question: HOW in the world do you find areas under the graph? In this lesson, we'll rely on a known rule that allows us to quickly find a range in which most of the data values fall. It's known as the *empirical rule.* Before seeing the rule, let's make some educated guesses.

According to recent research, the mean height for women in the U.S. is 63.7 inches (which you already knew, right?), and the standard deviation is 2.7 inches. The normal graph below is marked with the mean and heights that are one and two standard deviations above and below the mean.

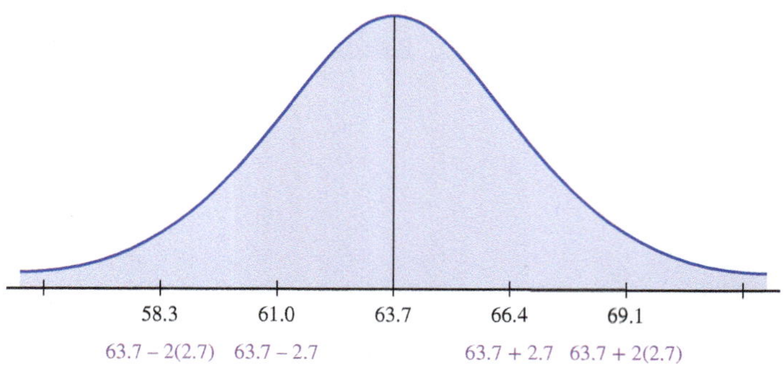

8. The total shaded area under the entire graph is 1, and this is true for ANY normal distribution. Explain why. Think carefully about what we know regarding area under a normal curve.

9. Make an educated guess as to what percentage of the total shaded area is between 61.0 and 66.4 inches. What does this tell us about women's heights?

10. Based on your answer to Question 9, estimate the percentage of area between 63.7 and 66.4 inches.

11. Now make an educated guess on the percentage of total shaded area between 58.3 and 69.1 inches.

Now it's time to see how you did.

The Empirical Rule

When data are normally distributed, approximately 68% of the values are within 1 standard deviation of the mean, approximately 95% are within 2 standard deviations of the mean, and approximately 99.7% are within 3 standard deviations of the mean.

The graph to the right shows an illustration of the empirical rule. Recall that $\overline{X}$ is the mean, and s is the standard deviation.

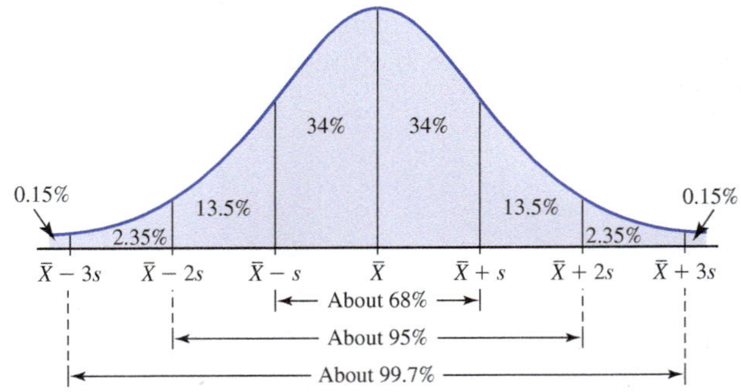

Man, I love M&M's. Can't get enough of them. (An absolutely true fact: I just finished eating a handful of peanut butter M&M's, my new favorite.) When I buy a bag, the LAST thing I want is to get shorted. So how many M&M's is enough? I found a really cool post on joshmadison.com about the number of plain M&M's in a typical bag. Based on the results posted there, the mean number of plain M&M's in a 1 lb. bag is 517, with a standard deviation of 12. Let's see how we can use the empirical rule to study what's likely to happen when you buy a random 1 lb. bag from your friendly neighborhood retailer.

> **Math Note**
>
> In case you're wondering, to get the M&M data, some dude hand-counted almost 3,000 M&M's from close to 50 packages. And I say "Good for him!"

Solved Example 1

Using the Empirical Rule

Assuming that the number of M&M's in a package is normally distributed (which, by the way, it more than likely is), draw an empirical rule diagram using the given mean and standard deviation. Then use your diagram to answer these questions:

a. If you buy 200 1 lb. bags of M&M's (a number I guarantee I've surpassed in my lifetime), how many would you expect to contain between 505 and 529 M&M's?

b. Fill in the blanks: it's likely that 99% of bags contain between ______ and ______ M&M's.

c. What percentage of bags would you expect to have more than 529 M&M's in them?

SOLUTION

The most important part of this example is the first part: In order to use the empirical rule, it's ALWAYS a good idea to draw and label a normal distribution with the values corresponding to the key values from the empirical rule labeled. That means we'll want to start with the mean in the middle, then keep adding and subtracting the standard deviation, since the key values are 1, 2, and 3 standard deviations above and below the mean. The mean in this case is 517 and the standard deviation is 12, so we want to label 517 in the middle. Then, subtracting 12 repeatedly, we get $517 - 12 = 505$, then $505 - 12 = 493$, and $493 - 12 = 481$ for the values to the left of the mean. Next, we add 12 repeatedly, getting $517 + 12 = 529$, $529 + 12 = 541$, and $541 + 12 = 553$ moving out to the right.

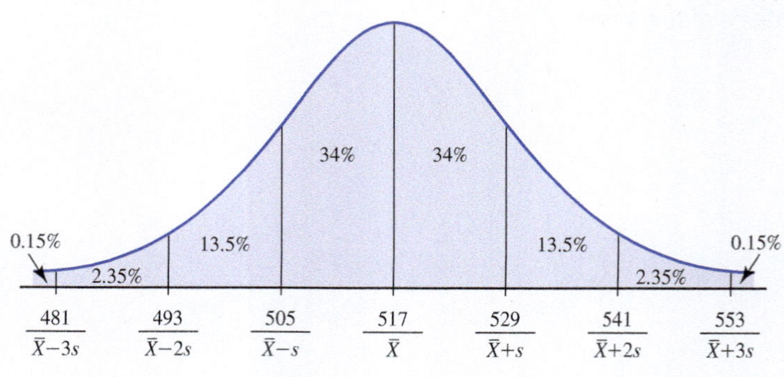

Solved Example 1 *Continued*

a. The thing that makes this an empirical rule problem is that the boundaries we're asked about, 505 and 529, are both labeled on our graph. The first is 1 standard deviation below the mean, and the second is 1 standard deviation above the mean. The diagram shows that this corresponds to 68% of the 200 bags. So we'd expect that $0.68 \times 200 = 136$ bags would contain between 505 and 529 M&M's.

b. According to the empirical rule, over 99% of all data values will fall between the mean minus 3 standard deviations and the mean plus 3 standard deviations. In this case, that's between 481 and 553. So 99% of 1 lb. bags will likely contain somewhere between 481 and 553 M&M's.

c. Bags with more than 529 M&M's correspond to the portion of the graph above 529 on the horizontal axis. Adding up the areas to the right of 529, we get $13.5\% + 2.35\% + 0.15\% = 16\%$. So we'd expect that about 16% of bags will have more than 529 M&M's.

Now let's see if you can use what you learned in Solved Example 1 to answer some similar questions.

12. What percentage of bags have less than 493 M&M's? How bad would that stink if you got one?

13. How unusual would it be to get a bag with more than 553 M&M's, or less than 481? Explain.

Did You Get It

2. Normal distributions are very commonly used for quality control in manufacturing. For example, a manufacturer of light bulbs would surely be interested in studying how long their bulbs last. In one case, the lifetime of one model of light bulb is normally distributed and has a mean $\mu = 1{,}600$ hours and a standard deviation of $\sigma = 300$ hours. What percentage of bulbs would you expect to last longer than 1,900 hours?

4-6 Group

The quantities we can study in relation to the normal distribution—things like sizes of seed pods, heights of human beings, and the number of M&Ms in a bag—are further examples of random variables (which we defined back in Lesson 3-5).

If you're really sharp, you may have thought about the connection between the percentages we've been calculating in this lesson and probability. If you haven't, we're going to ask you to do so now. (There's a good reason that the normal is called a probability distribution, not a percentage distribution.)

1. Using what we know about area under a normal curve and percentage of data values, what does the area under the curve between two data values tell us about probability for a random variable?

2. Based on the work you did earlier, what's the probability that the next 1 lb. bag of M&M's you buy will have between 529 and 541 M&M's in it? Briefly describe how you got your answer.

Did You Get It

3. Refer back to the light bulb manufacturer in Did You Get It 2. On the packaging for this particular bulb, the manufacturer lists the expected life of the bulb as 1,000 hours. If you buy one of those bulbs, find the probability that the manufacturer's claim will be true.

Z Scores

As you probably know, there are two main standardized college entrance exams, ACT and SAT. Since each has a completely different scoring scale, it's really difficult to compare the scores of students who took different exams.

3. Do you think that scores for these exams are normally distributed? Discuss.

One way to compare statistics drawn from different populations is to compare how far above or below the mean they are. The Z score is a statistical measure that was designed to do exactly that.

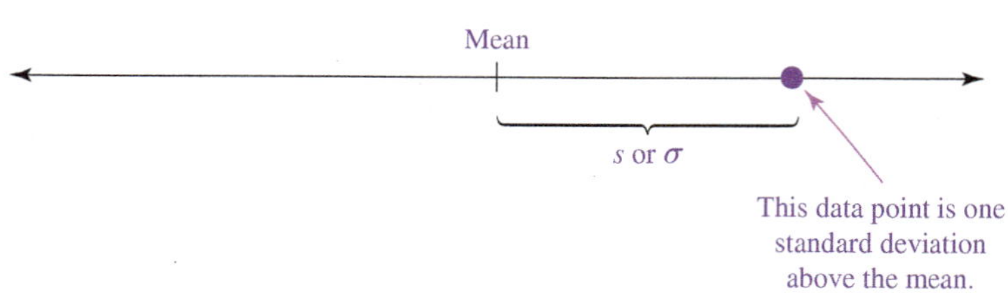

A Z score tells you how many standard deviations a data value is from the mean. The data value on the line above would have a Z score of 1 because it is 1 standard deviation above the mean. Data values that are below the mean would have negative Z scores.

4. Fill in the blanks in the diagram with Z scores, using the description at the bottom of the previous page.

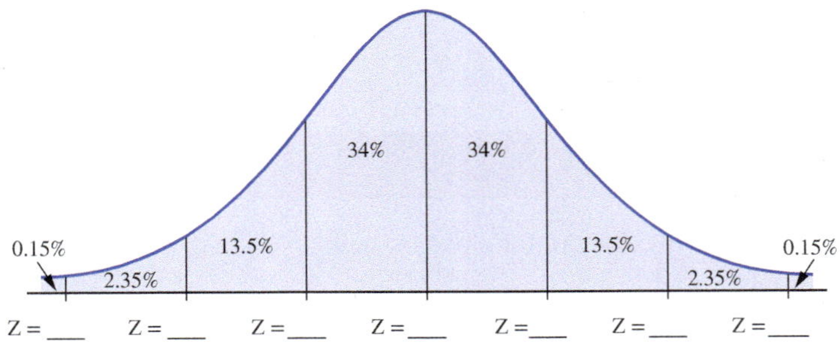

Solved Example 2

Using Z Scores

A standardized test of intelligence is scaled so that the mean IQ is 100, and the standard deviation is 15.

a. Find the Z scores for a person with an IQ of 85 and one with an IQ of 115.

b. What's the probability that a randomly chosen person has an IQ between 85 and 115?

SOLUTION

a. This is all about understanding what Z scores actually mean. Notice that the mean is given as 100 and the standard deviation as 15. That means that 85 is exactly 1 standard deviation below the mean, and 115 is exactly 1 standard deviation above the mean. The Z score for a data value is the number of standard deviations it is above (positive Z score) or below (negative Z score) the mean. So 85 (1 standard deviation below the mean) corresponds to a Z score of -1, and 115 (1 standard deviation above the mean) corresponds to a Z score of 1.

b. Now we're back to interpreting the empirical rule diagram. We can see that 68% of the area lives between $Z = -1$ and $Z = 1$. The total area under the normal curve is 1, so 68% of the area is 0.68, and that's the probability that our chosen individual has an IQ between 85 and 115.

5. If a data value has a Z score of _______, then half of the data are below that value, and half are above.

6. We know that the center of a normal distribution is the mean for a random variable. What does Question 5 tell you is also true about the center of the distribution?

7. A job applicant is told that their IQ score, measured in the screening process, is higher than 84% of the general population. What was their Z score, and their IQ?

Did You Get It

4. On a standardized IQ test, what's the Z score for a person with an IQ of 145?

Computing Z Scores

For a data value from a sample with mean $\overline{X}$ and standard deviation s, or from a population with mean μ and standard deviation σ, the Z score is

$$Z = \frac{X - \overline{X}}{s} = \frac{X - \mu}{\sigma}$$

Some notes:

- A data point is greater than the mean if $Z > 0$ and less than the mean if $Z < 0$.
- Z scores have no units.
- Z scores are typically rounded to two decimal places.

One year the ACT had a mean score of 21.2 and a standard deviation of 5.1. That same year, the SAT had a mean score of 1,060 and a standard deviation of 195. Suppose that a scholarship committee is considering two students, one who scored 26 on the ACT and another who scored 1,230 on the SAT. Both are pretty good scores, but which one is better? That's not so easy to decide because the scales are totally different.

8. First instinct, which score sounds better to you?

Math Note

Z scores are also used to approximate the area under any portion of a normal curve. This would allow you to compute probabilities more accurately than estimating based on the empirical rule.

9. Find the Z score for each test score.

10. Describe exactly what each Z score tells us.

11. Based on the Z scores, which would you say was a better score? Why?

Did You Get It

5. I'm a 6′4″ male, and my wife is 5′10″. We're both pretty tall, but clearly I am taller. I can send you pictures if you don't believe me. Anyhow, based on Z scores, which of us is taller compared to the rest of our sex? (The mean and standard deviation for male heights in the U.S. are 69.1 in. and 2.9 in, and this information for women was provided earlier in the lesson.)

Big sports fans tend to love arguing, and one of their favorite types of argument is "Who was better?" Michael or LeBron? Montana or Manning? Kershaw or Koufax? Blogs and airwaves are filled with arguments of this type. And what makes it so difficult to decide is that players and teams from different eras are hard to compare statistically because sports evolve, and what might have been an amazing statistic 50 years ago could be completely pedestrian today. Fortunately, this is EXACTLY the kind of thing that studying statistics analytically is good for.

The 1942 St. Louis Cardinals were one of the all-time great baseball teams. They won 106 games in an era when teams played only 154 games in a regular season. During the 1942 season, the mean number of games won by a National League team was 75.8 with a standard deviation of 20.5.

Another great Cardinals team was the 2004 squad, which won 105 out of 162 games. In 2004, the mean number of games won by a National League team was 80.9 with a standard deviation of 13.6.

MLB Photos/Stringer/Getty Images

12. Based strictly on winning percentage (the percent of games won, that is), which was the better team?

13. Use information besides the winning percentages in the preceding paragraphs to make a good statistical argument about which team was better. Obviously, showing your work is a must. Write your final answer in the form of a few sentences, showing you know what you are talking about.

14. Look back at your answer to Question 1 on the first page of this lesson. How did you do? Would you be able to come up with more examples now?

4-6 Portfolio

Name ____

Check each box when you've completed the task. Remember that your instructor will want you to turn in the portfolio pages you create.

Technology

1. ☐ In this Excel activity, we'll examine a couple of data sets. We'll calculate the mean and standard deviation for each data set, and then we'll calculate z-scores.

ALEKS Practice

1. ☐ Include any written work from the ALEKS assignment along with any notes or questions about this lesson's content.

Applications

1. ☐ Complete the applications problems.

Reflections

Type a short answer to each question.

1. ☐ Describe a random variable that is normally distributed.
2. ☐ Why are measures of average and variation so important in studying normal distributions?
3. ☐ What's the connection between a normal distribution and probability?
4. ☐ In what setting are Z scores useful? Why?
5. ☐ Name one thing you learned or discovered in this lesson that you found particularly interesting.
6. ☐ What questions do you have about this lesson?

Looking Ahead

1. ☐ Complete the prep skills for Lesson 4-7.
2. ☐ Read the opening paragraph in Lesson 4-7 carefully and answer Question 0 in preparation for that lesson.

Answers to "Did You Get It?"

1. 19.7% **2.** About 16% **3.** About 0.975 **4.** 3 **5.** I am, by just a bit.

Answers to Prep Skills Questions

1. \$1 – \$2
2. 14
3. 0.12
4. About 25 (The exact area is 22.5.)
5. About 83.8%

4-6 Applications

Name ______________________________

According to the National Center for Education Statistics, in 2023 the mean score on the SAT was 1028, with a standard deviation of 221. Use this information for Questions 1–3.

1. Fill in the blanks in the diagram.

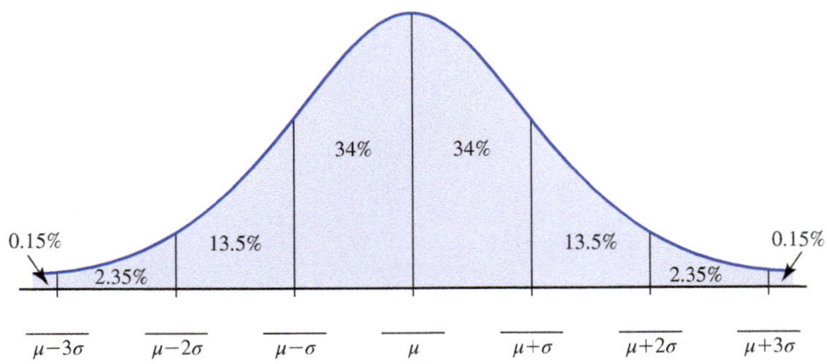

> **Math Note**
>
> Notice that in this case we're using μ and σ, which indicates that the mean and standard deviation are population, not sample, statistics.

2. What percentage of students scored 586 or higher?

3. Out of every 500 students, how many would we expect to score between 807 and 1249?

4. If a student scored 1410 on the SAT and her IQ was tested at 144, use Z scores to decide which test she did better on. Refer to Solved Example 2 for information on IQ scores.

4-6 Applications

Name ______________________________

According to 9monthsafter.com, the lengths of human pregnancies are normally distributed with mean 268 days and standard deviation 15 days. Use this for Questions 5–7.

5. Fill in the blanks in the diagram.

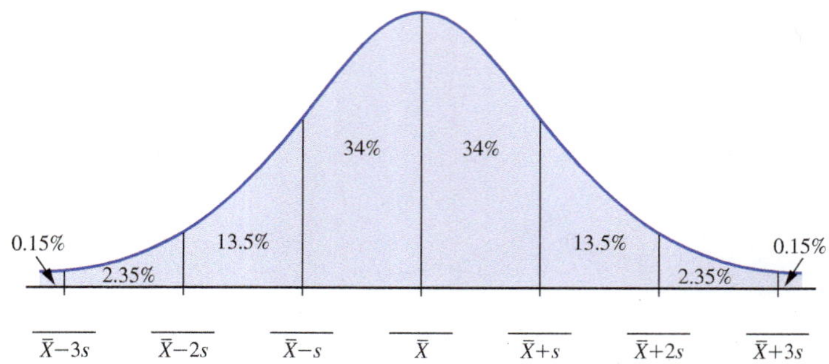

6. What is the probability that you were born more than 298 days after you were conceived?

7. If a child is conceived on April 19, what is the probability that it will be born on or before January 27?

According to the label, a bag of Kettle brand Tias tortilla chips contains 8 ounces of chips. Suppose that the bagging process is normally distributed with a standard deviation of 0.14 ounces.

8. If the machine that does the bagging is set to put 8 ounces of chips in each bag, this would be the mean of the distribution, with standard deviation as mentioned above. What percentage of all bags would have less than the amount stated on the package in this case?

9. Explain why it would be a bad idea for the manufacturer to set the bagging machine at 8 ounces.

4-6 Applications

Name ____

10. If the plant manager wants there to be greater than a 97% chance that any given bag has at least 8 ounces, and the weight they set their filling machine for provides the mean weight, how should they set the machine? Complete the diagram as part of showing your work.

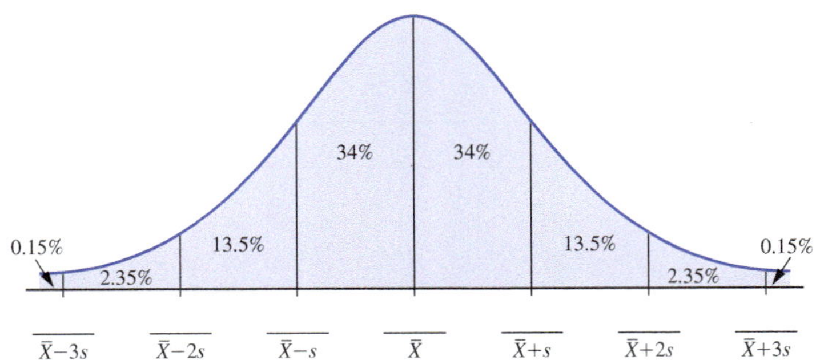

For Questions 11–14, we'll assume that the finishing times for a group of runners in a marathon race are likely to be normally distributed. The mean finish time for male marathon (26.2 miles) runners is 278.42 min with a standard deviation of 39.4 min. The mean finish time for female marathon runners is 294.45 min with a standard deviation of 39.9 min. At the 2024 Boston Marathon, Sisay Lemma won the men's division with a time of 126.28 min and Hellen Obiri won the women's division at 142.62 min.

11. Who finished faster, relative to their sex? Explain why. (Remember that in a race you want to finish in a short time.)

4-6 Applications

Name ______________________________

12. Out of 10,000 runners, where would you expect a woman who finishes in a time of 294.45 min to finish? Complete the diagram to help show your work.

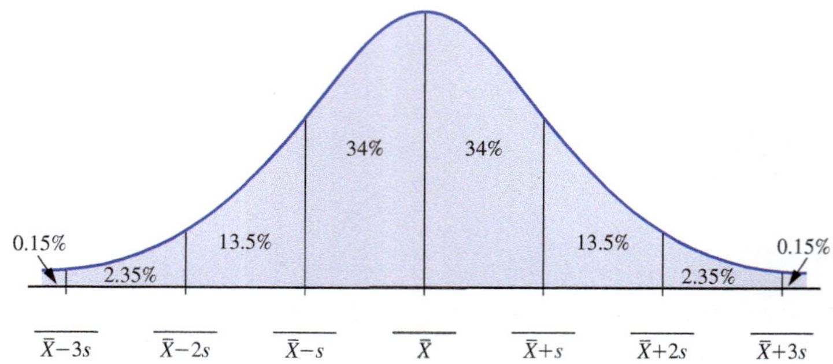

13. A man who finished in a time of 3 1/2 hours would finish in the fastest _______ percent of male racers. Complete the diagram to help you show your work. You'll have to make an estimate of area since that time doesn't correspond to one of the labeled times. (Interesting note after you have the percentage: This guy would finish well over an HOUR after the winner. Wow.) Hint: A time of 160.22 minutes means you're in the fastest 0.15%, so you finished before 99.85% of the runners.

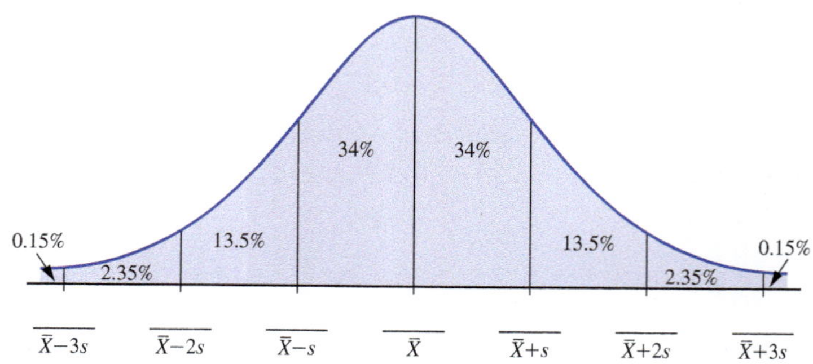

Lesson 4-7 Prep Skills

SKILL 1: USE THE EMPIRICAL RULE

When data are normally distributed, approximately 68% of the values are within 1 standard deviation of the mean, approximately 95% are within 2 standard deviations of the mean, and approximately 99.7% are within 3 standard deviations of the mean.

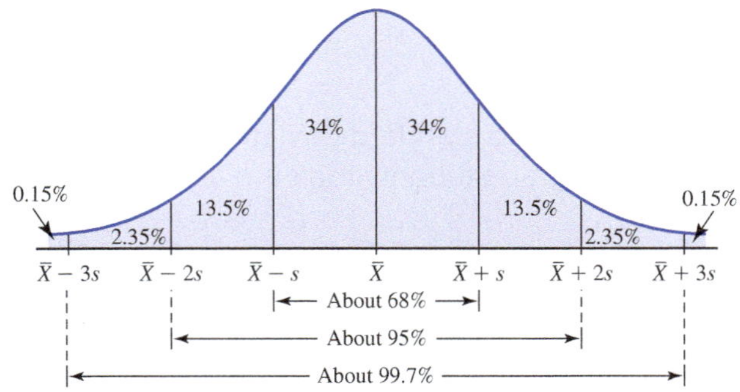

SKILL 2: UNDERSTAND PERCENTILES

A percentile, or percentile rank, of a data value indicates the percent of data values in a set that are less than that particular value.

- A test score at the 70th percentile means that score is higher than 70% of the scores.
- A test score at the 10th percentile means that score is higher than 10% of the scores.

SKILL 3: USE TECHNOLOGY TO FIND MEAN AND STANDARD DEVIATION

On page 406, you'll find the Technology instructions that cover finding the measures of central tendency with a calculator and a spreadsheet. On page 423, you'll find the Using Technology box that covers finding standard deviation with a calculator and a spreadsheet.

SKILL 4: DRAW A HISTOGRAM

In Lesson 4-2, we learned about drawing histograms.

Step 1: Write the scale for the frequencies on the vertical axis and the boundaries of the classes on the horizontal axis.

Step 2: Draw vertical bars with heights that correspond to the frequencies for each class.

Note: When drawing a histogram, make sure that there is no horizontal space between the bars (unlike a bar graph) unless two bars are separated by a class with frequency zero. Also make sure that all classes are included, even if the frequency is zero.

PREP SKILLS QUESTIONS

1. Based on the most recent available data, the Graduate Record Exam (GRE) quantitative portion had a mean score of 151.3 with a standard deviation of 8.7. What percentage of test-takers scored above 160?

2. The mean and standard deviation for the verbal portion were 150.8 and 8.5. In that year, 417,000 prospective grad students took the test. About how many scored between 133.8 and 159.3?

3. For students whose GRE reports indicate that they scored in the 60th percentile, what does that mean?
   a. They got 60% of the questions right.
   b. 60% of everyone that took the test did better than they did.
   c. 60% of everyone that took the test did worse than they did.
   d. Six out of 10 people taking the test earned a score similar to theirs.

4. Use either a graphing calculator or a spreadsheet to find the mean and standard deviation of this list of GRE scores:

   133 133 137 138 139 140 140 142 144 144 145 146
   146 147 151 151 154 156 158 158 160 161 162 164

5. Draw a histogram for the score data in Question 4 using six classes.

Lesson 4-7 The Way the Cookie Crumbles

LEARNING OBJECTIVES

- ☐ 1. Use normal distributions to find probabilities, percentages, and percentiles.
- ☐ 2. Learn how normal distributions are used in manufacturing and packaging.
- ☐ 3. Recognize data that are approximately normally distributed.

John Flournoy/McGraw Hill

I've never been very cookie cutter. If I choose something different from the status quo, it's my responsibility and my choice to live my life that way.
—Sara Ramirez

Surely there must be someone out there who doesn't like Oreos, but I haven't met him or her. A standard package of America's favorite cookie contains 510 grams of sweet temptation. Of course, when you buy a package, you expect every one of those 510 grams. But there's variation in just about everything, so some packages will have more than the intended weight, and some will have less. The folks running Nabisco aren't dummies, though—they know that if someone decides to check the weight and finds it to be less than 510 grams, they won't be a very happy customer. So what to do?

This is exactly the sort of situation where weights tend to be normally distributed. The company would likely design its production and packaging process so that the mean is somewhat larger than 510 grams, with a standard deviation that assures that the vast majority of packages will weigh at least 510 grams. We know a little bit about finding probability for random variables that are normally distributed. In this lesson, we'll learn a whole lot more and be able to carry out in-depth analysis on the related quantities.

0. What percentage of packages do you think Nabisco would want to weigh 510 grams or more? Why?

4-7 Class

In Lesson 4-6, we were able to answer questions about the area under a normal curve because the values we were interested in just happened to be exactly 1, 2, or 3 standard deviations away from the mean. But what if we wanted to know how many men have heights that are between 1.2 and 1.8 standard deviations below the mean height? We'd need to be able to find the area under the graph for that exact situation, which I can guarantee you is NOT easy. Worse still, if we then try to find how many people we'd expect to have an IQ between 1.2 and 1.8 standard deviations below the mean IQ, we're stuck with a similar problem, but the graph (and area calculations) will be different!

The point is that while all normally distributed data have the same general shape for their graph, areas under portions of the graph depend on how tall the curve is and how spread out it is. These characteristics in turn depend on what the mean and standard deviation are for that particular data set. This would require a different set of extremely difficult area calculations for every individual situation involving normally distributed data. Ouch!

Not to worry, though: The picture's not as grim as we've painted it. Statisticians found two ways to work around this problem. (1) Use tables. (2) Use technology. Using tables of Z scores and corresponding areas is a good way to get approximations for probabilities for normally distributed data. The areas have been calculated for a simple normal distribution with mean 0 and standard deviation 1 (if you want to know how, take Calculus 2); then a procedure was developed to apply those areas to other distributions.

Graphing calculators, spreadsheets, and other statistical software can perform these calculations easily and accurately. The user is left with the important part—interpreting the results. We'll develop the process for using a table, then show how to use technology and leave it up to your professor to decide which method to use.

The Standard Normal Distribution

The special distribution used for tables is called the **standard normal distribution,** which is a normal distribution that has mean 0 and standard deviation 1. The standard normal distribution is shown below. The values under the curve, which should look very familiar, indicate the proportion of area in each section. For example, the area between the mean and 1 standard deviation above or below the mean is about 0.341, or 34.1%. Of course, these percentages come from the empirical rule, which we studied in Lesson 4-6.

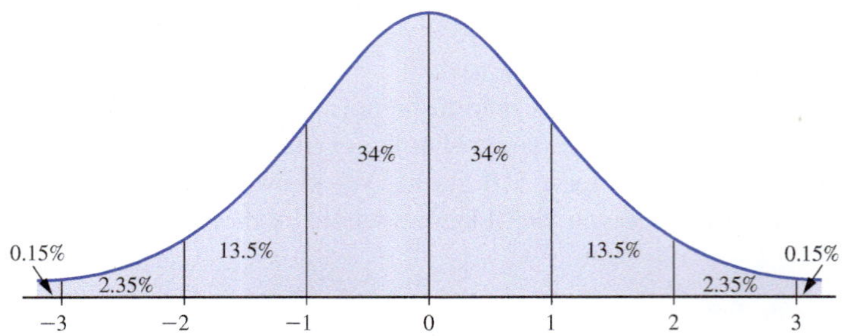

1. Now back to cookies, and not a moment too soon in my opinion. If the weights of Oreos in a package are normally distributed with mean 518 grams and standard deviation 4 grams, what percentage of packages will weigh less than the stated weight of 510 grams?

We were able to answer Question 1 without doing any calculations because we know the empirical rule, and because the weight we're interested in just happens to be exactly 2 standard deviations away from the mean. But what if the plant manager were interested in the probability that a package had a weight between 520 and 527 grams? Those weights don't correspond to whole-number multiples of the standard deviation away from the mean. So at the moment the best we could do is make a rough estimate of area from looking at the graph. I believe that we should be able to do better than a rough guess, so let's make it happen.

2. Based on the mean and standard deviation described in Question 1, label the normal graph on the next page with the appropriate weights.

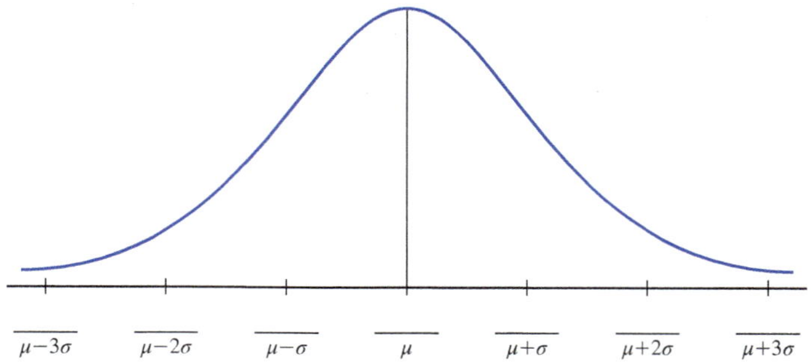

3. Label the two weights we're interested in (520 and 527 grams), and draw vertical bars from the axis along the bottom up to the normal curve. Then shade the area in between. This is the area we want, as it will give us the probability that a package weighs between 520 and 527 grams.

The table in the colored box provides areas under certain portions of a standard normal curve. In Solved Example 1, we'll walk you through how to use the diagram you labeled in Questions 2 and 3 and the table to find a probability.

Area Under a Normal Distribution Curve

The value in the *A* column of the table is the area under the standard normal curve between Z = 0 and each given positive value of Z.

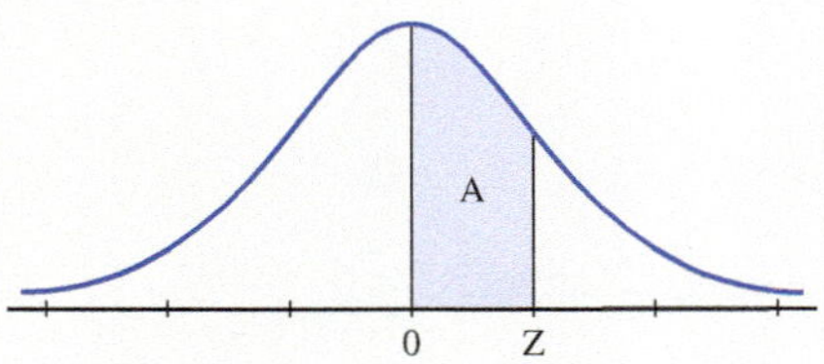

If you need the area for a Z score that isn't listed in this table, you can get an approximate area by choosing an area between areas in the table. In a bit, we'll learn how to use technology to get more accurate areas.

Important notes:

1. The area under any normal curve is divided into two equal halves at the mean. Each of the halves has area 0.500.
2. The area between Z = 0 and a positive Z score is the same as the area between Z = 0 and the negative of that Z score.

These facts are both consequences of the fact that normal distributions are symmetric about the mean.

Z	*A*	Z	*A*	Z	*A*
0.00	0.000	1.10	0.364	2.20	0.486
0.05	0.020	1.15	0.375	2.25	0.488
0.10	0.040	1.20	0.385	2.30	0.489
0.15	0.060	1.25	0.394	2.35	0.491
0.20	0.079	1.30	0.403	2.40	0.492
0.25	0.099	1.35	0.412	2.45	0.493
0.30	0.118	1.40	0.419	2.50	0.494
0.35	0.137	1.45	0.427	2.55	0.495
0.40	0.155	1.50	0.433	2.60	0.495
0.45	0.174	1.55	0.439	2.65	0.496
0.50	0.192	1.60	0.445	2.70	0.497
0.55	0.209	1.65	0.451	2.75	0.497
0.60	0.226	1.70	0.455	2.80	0.497
0.65	0.242	1.75	0.460	2.85	0.498
0.70	0.258	1.80	0.464	2.90	0.498
0.75	0.273	1.85	0.468	2.95	0.498
0.80	0.288	1.90	0.471	3.00	0.499
0.85	0.302	1.95	0.474	3.05	0.499
0.90	0.316	2.00	0.477	3.10	0.499
0.95	0.329	2.05	0.480	3.15	0.499
1.00	0.341	2.10	0.482	3.20	0.499
1.05	0.353	2.15	0.484	3.25*	0.499

*For Z scores greater than 3.25, use *A* = 0.500.

Solved Example

Using the Standard Normal Table

If the mean weight of Oreo packages is 518 grams and the standard deviation is 4 grams, use the standard normal table to find the probability that a randomly selected package weighs between 520 and 527 grams.

SOLUTION

Our goal is to find the area that you shaded in Question 3. The normal table that we have is based on the standard normal curve, with mean zero and standard deviation 1. Obviously that's not the normal curve we're dealing with in our cookie diagram. So our first order of business will be to find Z scores for 520 and 527 grams: This accomplishes turning the problem into one about the standard normal curve, at which point we can use the table.

We'll use the given mean and standard deviation, along with the Z score formula from the previous lesson.

$$\text{For 520 grams: } Z = \frac{520 - 518}{4} = 0.5$$

$$\text{For 527 grams: } Z = \frac{527 - 518}{4} = 2.25$$

Now this is where we need to really understand what the normal table tells us. For any Z score, the table gives us the area between the mean ($Z = 0$) and a given positive Z score. The issue is that we don't want the area of a shaded region bounded by zero: We want the area of the region between $Z = 0.5$ and $Z = 2.25$. So we'll need to be clever.

First, looking for the 0.5 row in the normal table, we see that the area is 0.192. That tells us that this region has area 0.192:

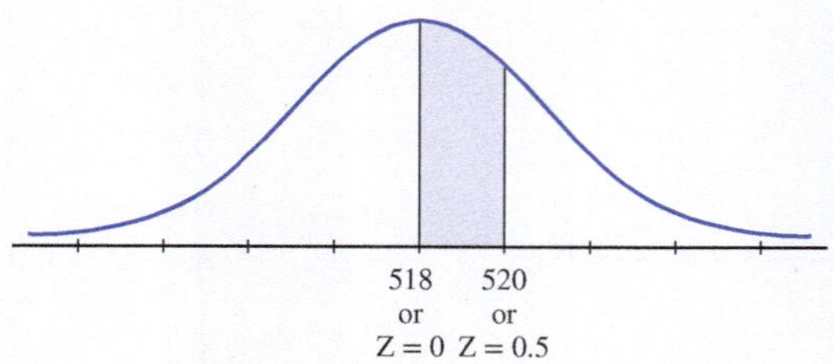

Next, the area corresponding to $Z = 2.25$ in the table is 0.488. That's this area:

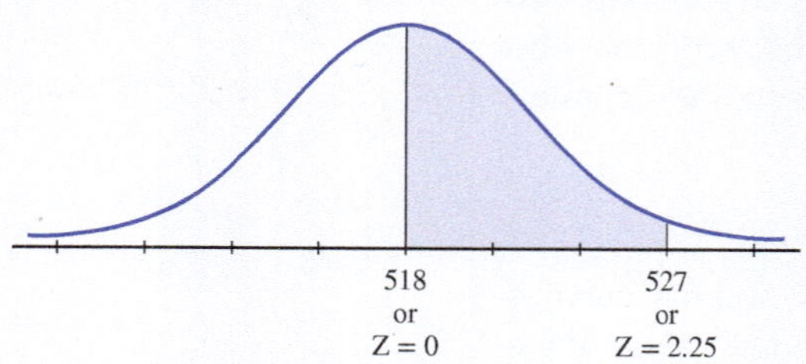

Now the clever part. The area that we want, between 520 and 527 grams, is the second, larger area MINUS the first area. (Think of it in terms of the picture, not the numbers!)

$$0.488 - 0.192 = 0.296$$

The area under the curve between 520 grams (or $Z = 0.5$) and 527 grams (or $Z = 2.25$) is 0.296, so that's the probability that a randomly selected package weighs between 520 and 527 grams.

4. Find the probability that a package weighs between 513 and 518 grams. (Look carefully at comment 2 under "Important Notes" in the colored box that contains the table!)

Did You Get It

Try this problem to see if you understand the concepts we just studied. The answer can be found at the end of the portfolio section.

1. Use the table to find the probability that a package of Oreos weighs less than 512 grams. Report the Z score you used as well.

Using Technology: Computing Probabilities from a Normal Distribution

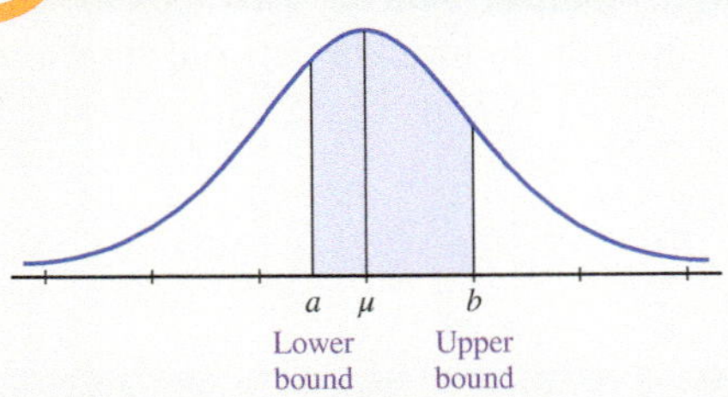

On a TI-84 calculator, the **normalcdf**(a,b,μ,σ) command will find the area bounded by a normal curve, the x axis and vertical lines at a lower bound a and an upper bound b as shown in the figure.

Notice that the inputs in parentheses correspond to the lower bound (a), the upper bound (b), the mean (μ), and the standard deviation (σ). This eliminates the need to calculate Z scores: You just need the mean and standard deviation for the given random variable.

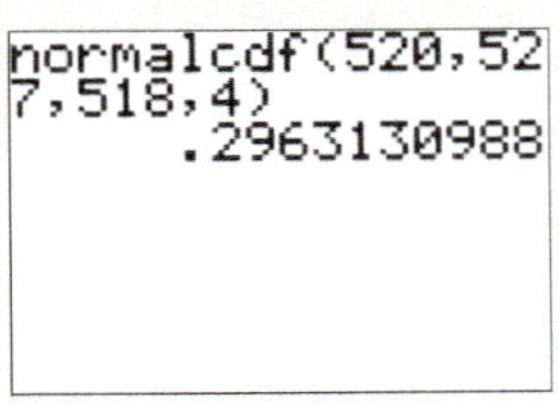

The normalcdf is used to find the probability that a given random variable takes any value between the lower and upper bounds. The screen shot shows a calculation for the probability that a package of Oreos weighs between 520 and 527 grams.

See the Using Tech video on Computing Probabilities from a Normal Distribution with a Graphing Calculator for further information.

In order to use technology effectively, it's important to understand exactly what it's built to do. In this case, the calculator is built to have the user input two values for the random variable in addition to the mean and standard deviation. It calculates the probability that the random variable will be between those two values. If we want to find a probability that doesn't have two boundaries, like the probability that a package weighs more than 525 grams, we'll need to be clever.

5. Shade the area on the graph that corresponds to the probability that a package weighs more than 525 grams.

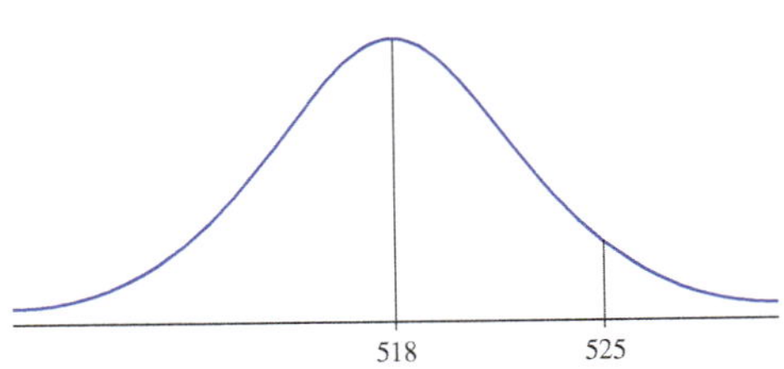

6. Use a calculator to find the probability that a package weighs between 525 and 540 grams. Round to six decimal places.

7. Use a calculator to find the probability that a package weighs between 525 and 1,000,000 grams. Round to six decimal places.

8. What can you conclude about a strategy for finding areas that only have one boundary using the **normalcdf** function?

Excel is set up to calculate normal probabilities differently. Instead of calculating the area between two values, it calculates the area to the LEFT of any value. This works fabulously well if you're interested in the probability of a randomly selected value being LESS than some number. For other types of probability, we'll need to do some arithmetic. First let's look at the syntax. The samples at the bottom show calculations for weights we've studied so far, with the formulas entered shown in the second graphic.

Using Technology: Computing Probabilities from a Normal Distribution

The Excel command is =NORM.DIST. When you start typing =NORM.DIST in Excel, here's what appears, as a guide to help you:

=NORM.DIST(x, mean, standard_dev, cumulative), where

- x is a value of the random variable.
- mean is the mean for the distribution.
- standard_dev is the standard deviation for the distribution.
- cumulative = TRUE; we won't use FALSE in this book.

	A	B
1	**X**	**Probability that a package weighs less than X grams**
2	510	0.022750132
3	520	0.691462461
4	527	0.987775527
5	525	0.959940843

	A	B
1	**X**	**Probability that a package weighs less than X grams**
2	510	=NORM.DIST(A2,518,4,TRUE)
3	520	=NORM.DIST(A3,518,4,TRUE)
4	527	=NORM.DIST(A4,518,4,TRUE)
5	525	=NORM.DIST(A5,518,4,TRUE)

See the Using Tech Video on Computing Probabilities from a Normal Distribution in Excel for further information.

Whether you are using tables or some form of technology, the best way to proceed with any of these problems is to start by drawing a graph. Don't ignore your graph when writing your final answer! As long as you keep in mind that the total area under the distribution is 1, you can visually check if your answer is reasonable.

9. Look back at the drawing you shaded on page 475. The Excel screen grabs on page 478 show the area under the normal curve up to each of 520 grams and 527 grams. Devise a strategy that uses that information to find the probability that a package weighs between 520 and 527 grams. You can use the answer from the Solved Example to check that your strategy worked. (You can round all values to three decimal places.) Show your work, please!

In Questions 10–13, find each area under the standard normal distribution to practice what we've learned so far. Use whatever method you feel most comfortable with (tables, calculator, Excel), unless a particular method is specified by your instructor. Remember that "standard normal" means that the mean is 0 and the standard deviation is 1. We'll shade the figures for the first two: After that, it's up to you.

10. Between Z = −1.75 and Z = 1.50

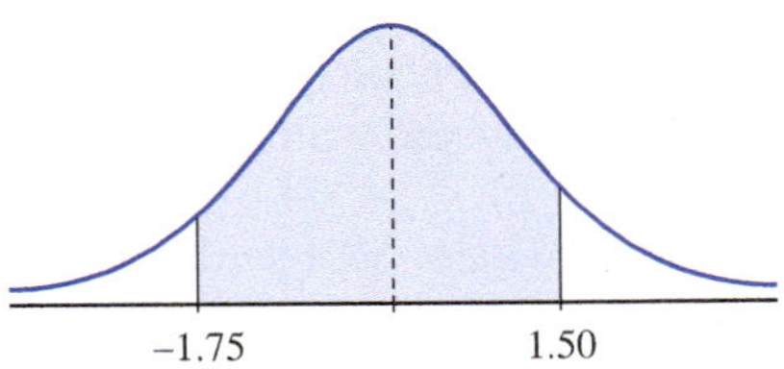

11. To the right of Z = 1.70

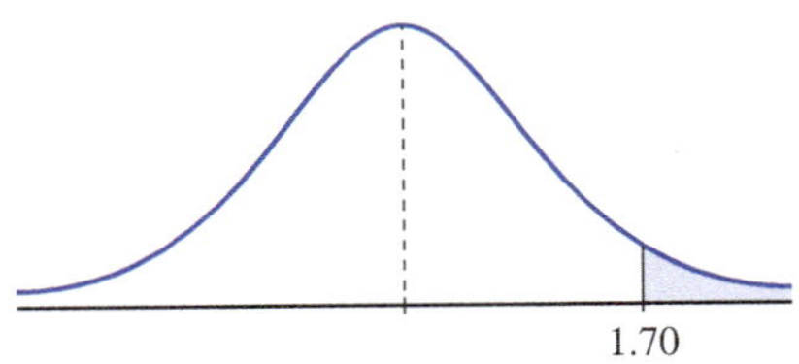

12. To the left of Z = −2.20

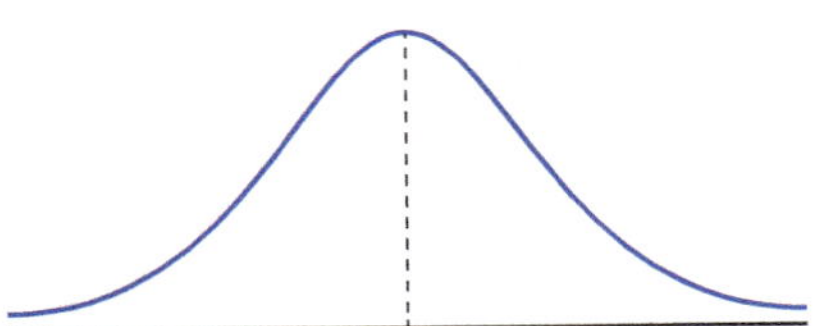

13. To the left of Z = 1.95.

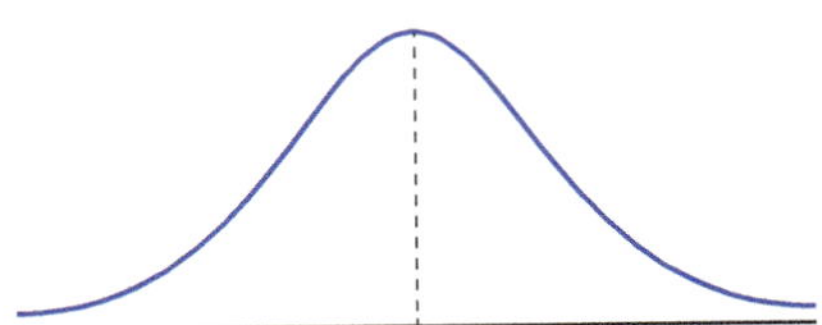

Did You Get It

2. Find the area under the standard normal distribution

   a. Between Z = −3.2 and Z = −2.0.
   b. Between Z = −0.55 and Z = 1.6.
   c. To the right of Z = 0.25.
   d. To the left of Z = 0.1.

4-7 Group

1. How many pounds of trash do you think you personally generate in one year? Discuss with your group and see how your guesses compare.

Based on data in the Statistical Abstract of the United States, the average American generates 1,570 pounds of garbage per year. We'll assume that the number of pounds generated per person is approximately normally distributed with standard deviation 200 pounds.

In Questions 2–4, let's work on finding the probability that a randomly selected person generates between 1,250 and 2,050 pounds of garbage per year.

2. Draw and label a figure to represent the area we need to find.

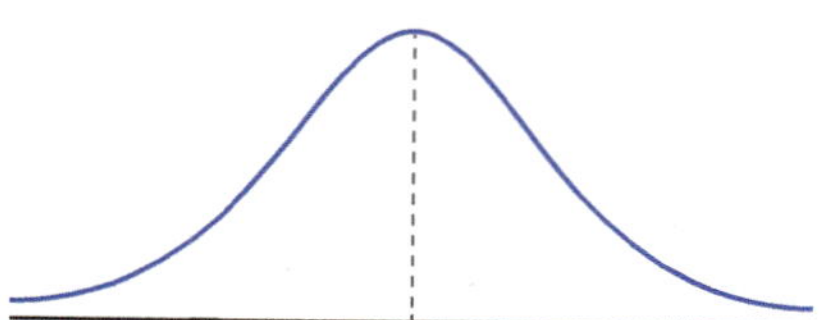

3. Find the area you drew using whatever method makes you happy.

Photodisc/Getty Images

4. Write a sentence or two describing exactly what your answer to Question 3 means.

In Questions 5–7, let's work on finding the probability that a randomly selected person generates more than 2,050 pounds of garbage per year. When you interpret the area you find, think about how likely it is to find someone who generates that many pounds of garbage per year.

5. Draw and label a figure to represent the area we need to find.

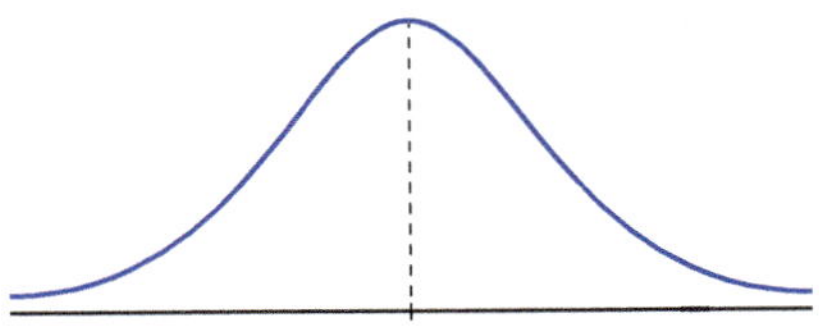

6. Find the area you drew using whatever method makes you happy.

7. Interpret your answer from Question 6.

Did You Get It

3. The Statistical Abstract also indicates that of the 1,570 pounds of garbage generated by the average individual, 872 pounds will end up in a landfill. If these amounts are approximately normally distributed with standard deviation 160 pounds, find the probability that a randomly selected person generates

   a. Less than 600 pounds that end up in a landfill.
   b. Between 600 and 1,000 pounds that end up in a landfill.

The American Automobile Association reports that the average time it takes to respond to an emergency call is 25 minutes. It would be reasonable to assume that the response time is approximately normally distributed. Let's suppose that the standard deviation is 4.5 minutes.

In Questions 8–10, we'll work on finding the approximate number of calls out of 80 randomly selected calls that will have response times less than 15 minutes.

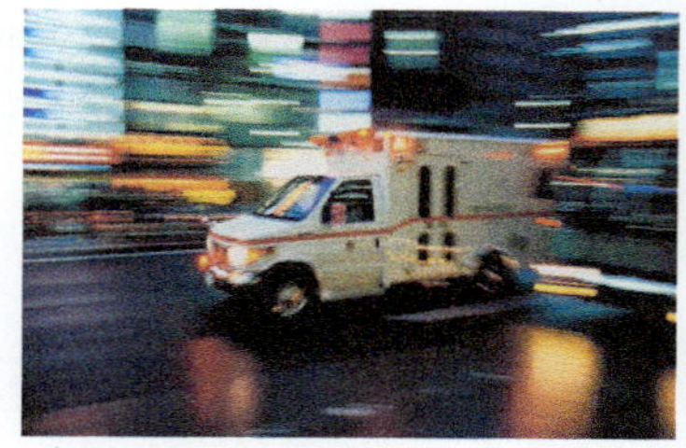

Jeremy Hoare/Life File/Getty Images

8. Draw and label a figure to represent the area we need to find.

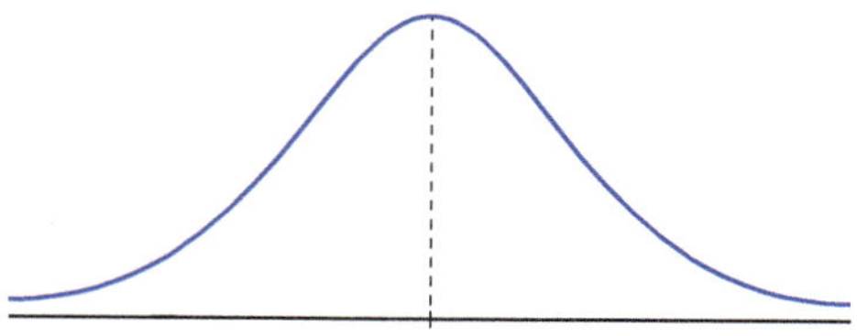

9. Find the area you drew using whatever method makes you happy.

10. Interpret your answer from Question 9. Reread the paragraph above Question 8, and think carefully about how you write your final answer.

Based on what we know so far, if the response time for one emergency call is 30 minutes, is that fast or slow? One way to answer that is to look at the percentile rank for 30 minutes, which we'll do in Questions 11–13.

11. Draw and label a figure to represent the area we need to find.

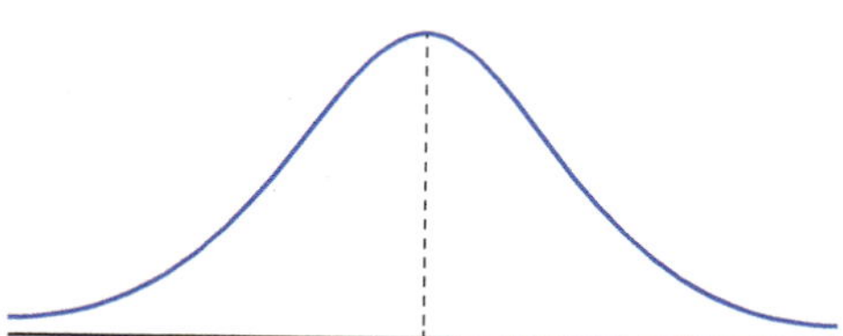

12. Find the area you drew using whatever method makes you happy.

13. Based on your answer to Question 12, in what percentile is a response time of 30 minutes? Explain your answer, and describe what this says about how good or bad that time is.

Did You Get It

4. The mean for a reading test given nationwide is 80, and the standard deviation is 8. The random variable is normally distributed. If 10,000 students take the test, find each.

   a. The number of students who will score above 90

   b. The number of students who will score between 78 and 88

   c. The percentile of a student who scores 94

The random variables we've studied so far in this lesson are described as being normally distributed. But how would you know that if you were gathering raw data? One way is to draw a histogram and see if it's bell-shaped, or close to it.

A random sample of 25 entrées served by a college cafeteria is tested to find the number of calories per serving, with the data in the list below.

845	460	620	752	683	1,088	785	575	580
755	720	650	512	945	672	526	1,050	725
822	740	773	812	880	911	910		

14. Draw a histogram for the data using the classes shown below.

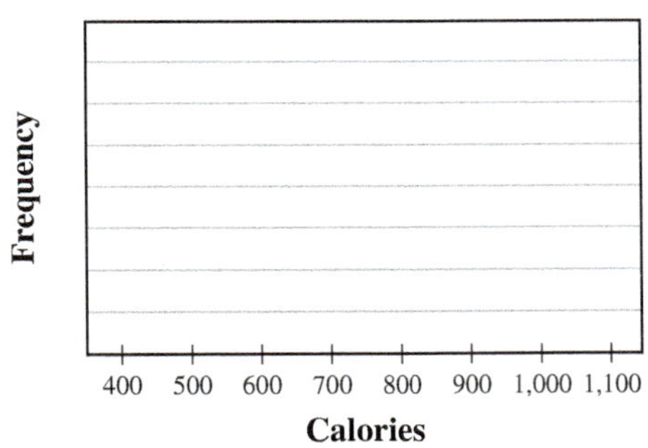

15. Based on your histogram, do the data appear to be normally distributed? Explain.

16. Find the mean and standard deviation for the data. Technology strongly recommended.

17. If the mean and standard deviation from the sample are reasonable values for the mean and standard deviation of all entrées served by that cafeteria, and we assume that the number of calories in all of the entrees are normally distributed, what percentage of their entrées have more than 700 calories? Include a sketch in your answer, please.

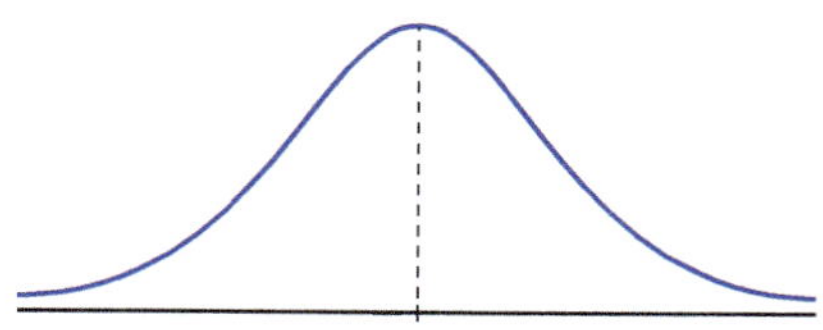

Did You Get It

5. The ages of 33 randomly selected inmates at a county jail are shown in the list.

18 18 20 24 26 26 27 29 30 30 31 33 34 34 36 37 37
38 38 38 39 41 42 42 43 43 46 47 49 49 52 54 55

Draw a histogram to show that the ages are approximately normally distributed, then find the mean and standard deviation and use that information to find how many of the jail's 1,148 inmates are likely to be under the age of 21.

4-7 Portfolio

Name ______________________________

Check each box when you've completed the task. Remember that your instructor will want you to turn in the portfolio pages you create.

ALEKS Practice

1. ☐ Include any written work from the ALEKS assignment along with any notes or questions about this lesson's content.

Applications

1. ☐ Complete the applications problems.

Reflections

Type a short answer to each question.

1. ☐ Given a data set, how could you decide if the distribution of the data was approximately normal?
2. ☐ When you know that a random variable is approximately normally distributed, what do you need to find in order to calculate probabilities of the variable being in some range?
3. ☐ What are some advantages of using technology to find areas under a normal curve?
4. ☐ Name one thing you learned or discovered in this lesson that you found particularly interesting.
5. ☐ What questions do you have about this lesson?

Looking Ahead

1. ☐ Complete the prep skills for Lesson 4-8.
2. ☐ Read the opening paragraph in Lesson 4-8 carefully and answer Question 0 in preparation for that lesson.

Answers to "Did You Get It?"

1. $Z = -1.5$, probability is 0.067 **2. a.** 0.022 **b.** 0.654 **c.** 0.401 **d.** 0.540
3. a. 0.045 **b.** 0.744 **4. a.** 1,056 **b.** 4,400 **c.** 96th percentile
5. Histogram omitted; the mean is 36.5, the standard deviation is 10.0, and about 70 inmates would likely be under 21.

Answers to Prep Skills Questions

1. About 16%
2. 339,885
3. c
4. Mean = 147.875, standard deviation ≈ 9.44
5. 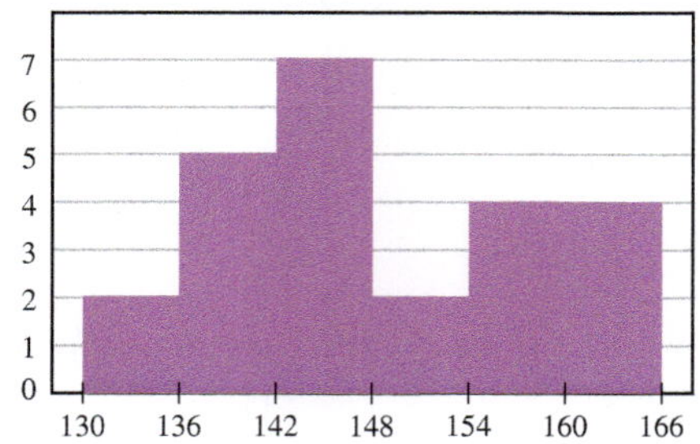

4-7 Applications

Name ______________________________

For Questions 1–5, the average life of one model of automobile tires is 41,000 miles, with a standard deviation of 3,900 miles. Assume the life of this brand of tire is normally distributed. If a tire is selected and tested, find the probability that it will have the given lifetime. Drawings are recommended, but it's up to you.

1. Greater than 41,000 miles

Ingram Publishing/Alamy Stock Photo

2. Greater than 47,000 miles

3. Less than 45,000 miles

4. Between 36,000 and 43,000 miles

5. If you worked at a tire shop, would you expect to see many sets of these tires that lasted 55,000 miles? Why or why not? Make sure your answer is backed up with math! Show the details.

4-7 Applications

Name ______________________________

For Questions 6 and 7, the average waiting time at a local restaurant is 9.2 minutes, with a standard deviation of 2.6 minutes. When a customer arrives at the restaurant, find the probability that they will have to wait the given time. Assume the wait time is normally distributed.

6. Between 5 and 10 minutes

7. Less than 6 minutes or more than 9 minutes

8. For an advertised promotion, the manager of the restaurant decides to give a free appetizer to anyone waiting longer than 12 minutes. Was this a good idea? What percent of customers will be receiving an appetizer?

9. Suppose the manager's team is able to cut the mean wait time down to 7.0 minutes with a standard deviation of 2.1 minutes. If offering the same promotion, how many customers out of every 100 will get a free appetizer?

For Questions 10–13, the average time a person spends in each visit to a certain online social networking service is 62 minutes. The standard deviation is 12 minutes. If a visitor is selected at random, find the probability that he or she will stay on the networking service for the amount of time shown. Assume the times are normally distributed.

10. At least 180 minutes

4-7 Applications

Name ______________________________

11. At least 50 minutes

12. Suppose that your grades have been slipping, and you impose a new rule on yourself: no more than 30 minutes at a time on social networking sites. What percentile would that put you in for this particular site?

13. If you just can't stay away from social media as much as you planned and end up spending 45 minutes at a time, what would be the change in percentile rank from your answer in Question 12?

A researcher is studying reaction times in adult subjects who are alcohol-impaired in an effort to study the effects of impairment on driving. The amount of time in seconds that 25 randomly selected adults took to react to a stimulus when they have a blood alcohol level of 0.08 was recorded and is displayed below. Use this information for Questions 14–18.

1.2	1.5	1.6	1.6	1.7	1.9	2.0	2.1	2.1	2.2	2.2	2.2	2.3
2.4	2.4	2.5	2.5	2.7	2.8	2.8	2.8	3.0	3.1	3.3	3.5	

14. Build a frequency distribution for the data using five classes.

Class	Frequency

4-7 Applications

Name ____________________

15. Draw a histogram for the data, and use it to explain why it's reasonable to assume that reaction times for all impaired adults would be normally distributed.

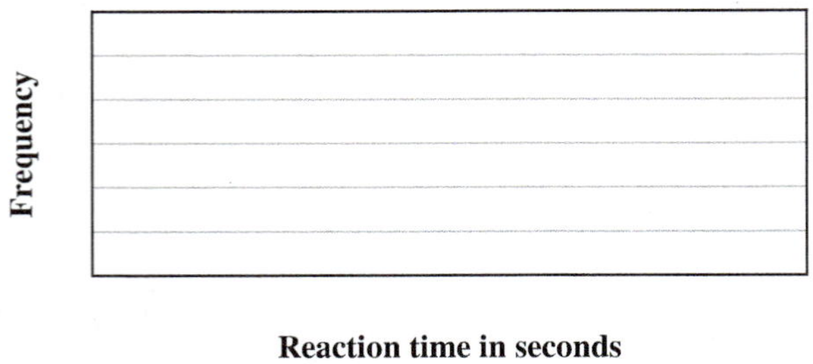

16. Find the mean and standard deviation. Again, technology is pretty much a must.

17. Assuming that reaction times are normally distributed, find the probability that a randomly selected driver with a blood alcohol level of 0.08 will take less than 1.5 seconds to react.

18. Find the probability that a randomly selected driver with a blood alcohol level of 0.08 will take between 2 and 4 seconds to react.

Lesson 4-8 Prep Skills

SKILL 1: EVALUATE A LINEAR EQUATION

To evaluate an equation, replace all occurrences of the input variable or variables with specific numbers. This allows us to find a numeric value for the output variable.

- Evaluate $y = 2x - 5$ for $x = -4$ and for $x = 0$.

Using $x = -4$,
$y = 2(-4) - 5$
$y = -13$

Using $x = 0$,
$y = 2(0) - 5$
$y = -5$

Replace x with the appropriate value and simplify. The value of y when $x = -4$ is -13, and the value of y when $x = 0$ is -5.

SKILL 2: EVALUATE A LINEAR EQUATION IN CONTEXT

It's common to have a linear equation where the input is based on a number of years (or other unit of time) after an initial point in time. An example would be the value of an investment after it was purchased. Rather than use input values like 2021, we assign the starting year to be $x = 0$ so that the variable x represents years after the starting year. Consider an investment made in 2005 with an initial value of \$3,000. If that investment consistently increased in value by \$400 per year, the equation $y = 3{,}000 + 400x$ could be used to model this situation as long as x was defined to be the number of years after 2005. In this case, $x = 0$ (the y intercept) tells us the value of the investment zero years after 2005, which of course is the value in 2005. If we'd instead used the year as input, the value at $x = 0$ would be the value of the investment in the year zero, which of course is silly.

- Find the value of the investment in 2010.

2010 is 5 years after the investment was made.

$y = 3{,}000 + 400(5)$
$y = 5{,}000$

Use an input of 5 to evaluate this equation.

The value of the investment in 2010 was \$5,000.

SKILL 3: GRAPH A LINE

The nice thing about drawing the graph of a line is that it doesn't change direction. This means all we really need are two points on the graph, since you can then draw a straight line that connects them. On the other hand, unless you never, ever make mistakes, it's a pretty good idea to find at least three points when drawing a line: If the three points you found don't line up, then you know at least one of them must be wrong.

Of course, this relies on knowing in advance that the graph is a line. And how do we do that? There are two main ways. First, if the situation described involves a quantity changing at a constant rate, then the graph is going to be a line. Why? Constant rate of change means constant slope, and constant slope means that the graph never changes direction—that's a line. Second, if you're graphing an equation with two variables, and both variables appear only in the numerator and only to the first power, then the graph will be a line.

- To graph $y = 15x - 12$, we first recognize from the form of the equation that the graph is a line. Then we find three points by substituting three values in for x:

x	y
0	−12
−5	−87
5	63

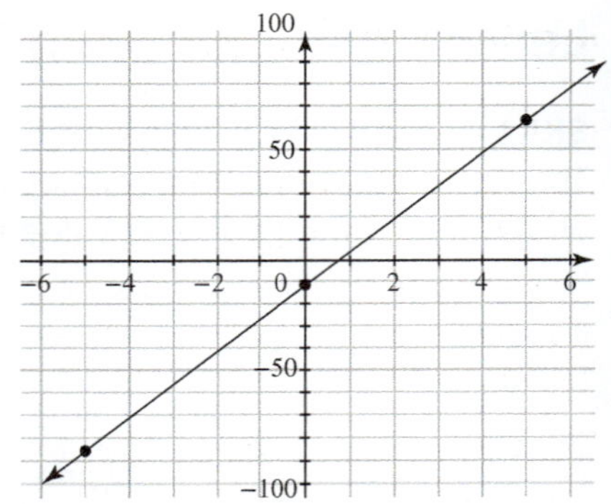

Now we set up a coordinate system, labeling the axes so that all of the points will fit on the graph. Then we plot the points, connect with a line, and include arrows on each end because there's nothing that says the line should have a beginning or an end.

PREP SKILLS QUESTIONS

1. For the equation $y = 4.3x - 7.2$, find the value of y when:

   $x = 10$

   $x = 40$

   $x = 19.2$

2. The equation $y = 179{,}840 + 3{,}250x$ can be used to approximate the value of a certain home in dollars x years after it was built. If the house was built in 1982, approximate the value in 2016.

3. Draw the graph of the line that has equation $y = 126x + 1{,}400$. Make sure you choose a reasonable scale for each axis, and label the scales on your graph.

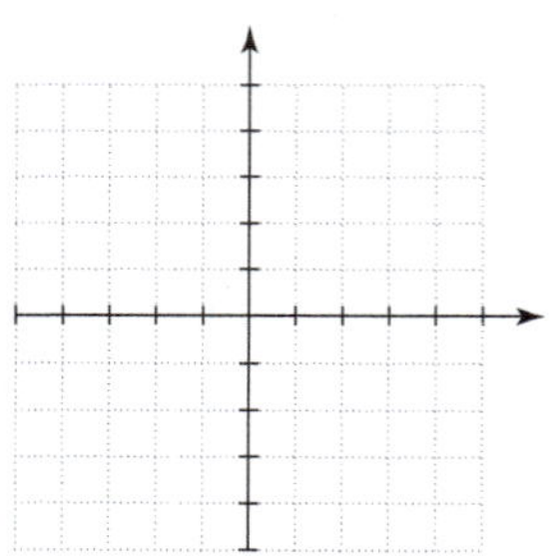

Lesson 4-8 Making Connections

LEARNING OBJECTIVES

- ☐ 1. Draw and analyze scatter plots for two data sets.
- ☐ 2. Define correlation coefficient, and decide if correlation coefficients are significant.
- ☐ 3. Find regression lines and use them to make predictions.
- ☐ 4. Recognize the difference between data sets being related and being linearly related.

David Sobecki

The price of soda in 20 years has gone down 40 percent while the price of whole foods, fruits and vegetables, has gone up 40 percent and obesity goes up right along that curve.
—Tom Colicchio

Unless you're a major shareholder in an oil company, chances are you're not at all fond of high gas prices. Is it possible that maybe prices like the ones shown in the photo aren't all bad? While doing some research for this book, I came across an interesting article online that was based on research a student in economics did for his PhD dissertation. The claim made in the paper is that rising gas prices have a strong connection to a decrease in the number of obese people in the United States. So what exactly does that mean? Do higher gas prices make people thinner? Superficially, those things appear to be totally unrelated. In order to study this claim, we'd want to find some data on both obesity and gas prices in a way that the data can be paired together for comparison. In this case, we could find obesity rates by year, as well as the average price of a gallon of gas in that year. Then we'll see if we can make a connection.

0. So what do you think? Should there be any connection between gas prices and obesity rates?

4-8 Class

Once we've found comparative data, we can then plot a graph designating one set of data as the *x* variable or **independent variable** and the other as the *y* variable or **dependent variable.** This graph is called a *scatter plot.*

A **scatter plot** is a graph of the ordered pairs (*x*, *y*) consisting of data from two data sets.

After a scatter plot is drawn, we can analyze the graph to see if there's a noticeable pattern. If there is, like the points falling in an approximately straight line, we can conclude that the two variables MIGHT be related somehow.

If there appears to be a reasonably clear relationship between the variables, we might study that relationship further by trying to find an equation that relates the variables. This is known as *regression analysis,* which we will learn about later in the lesson. For now, we'll focus on building a scatter plot.

Data on obesity rates in the U.S. were collected from the Center for Disease Control for various years. Data were also collected on the average price of a gallon of gas for those same years from a variety of online sources. These data are organized in the table below.

Year	*x*, Avg. Gas Price per Gallon ($)	*y*, Obesity Rate (% of adult population)
1961	0.27	10.7
1973	0.39	12.1
1978	0.65	12.7
1991	1.18	20.5
1999	1.30	27.7
2001	1.43	28.3
2003	1.56	31.7
2005	2.27	33.8
2007	2.81	32.5
2009	2.35	35.9
2011	3.58	34.9
2013	3.58	37.7
2015	2.51	39.6
2018	2.72	41.9
2023	3.52	40.3

1. Draw a scatter plot for the data by plotting points based on the information in the table. Use the values from the gas prices column as first coordinates and obesity rates as second coordinates.

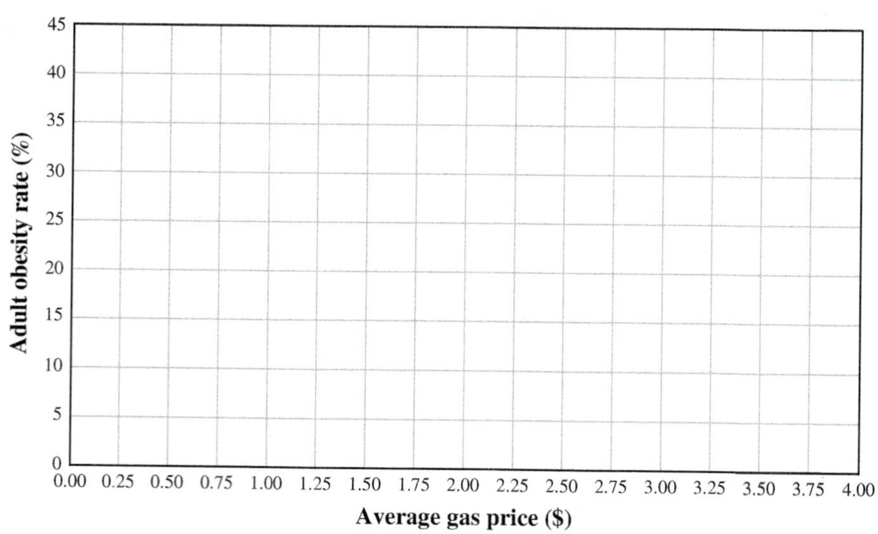

2. Write a general description of trends you see on your scatter plot. Does this appear to support or refute the claim made in the opening paragraph of this lesson?

Did You Get It ?

Try this problem to see if you understand the concepts we just studied. The answer can be found at the end of the portfolio section.

1. The data below represent the heights in feet and the number of stories of the tallest buildings in Pittsburgh. Draw a scatter plot for the data and describe the relationship.

Height (x)	485	511	520	535	582	615	616	635	728	841
Stories (y)	40	37	41	42	38	45	31	40	54	64

Regression

When we draw a scatter plot and it looks like there might be a straight-line relationship between the data sets, we can analyze the data further by trying to find the equation of a line that goes through the data points. This line is known as a **regression line.**

If you look back at the scatter plot in Question 1, you can see a general trend among the points from lower left to upper right. They definitely don't line up exactly in a line, but you could probably put a straightedge down and draw what seems like the closest approximation to a line by "splitting the difference" between points that don't line up exactly. That's what a regression line does. But instead of guessing, we're finding the single line with a special property: The overall distance from each point to the line is a minimum. (For that reason, the regression line is also called the *line of best fit.*)

3. Lightly draw what your eyes tell you is the line of best fit on your scatter plot from Question 1.

We know from our algebra background that the equation of a line in slope-intercept form is $y = mx + b$, where m is the slope and b is the y intercept (that is, the point that has first coordinate zero). In statistics, the equation of the regression line is often written as $y = a + bx$, where a is the y intercept and b is the slope. In this book, we'll write our equations in the form $y = ax + b$ since this is the form we'll get from the technology we use. Since we know what the equation looks like, and since y and x in the general formula are the variables, all we need to find in a given situation is specific values for a and b.

There are formulas that can be used to compute the values of a and b by hand, but they're so cumbersome to work with that it's almost always necessary to use technology to do the final calculations. And if you're going to use technology anyhow, you may as well take full advantage of it, right? In this case, both graphing calculators and Excel have built-in commands for finding the line of best fit. These are described in the technology boxes on the next page. (The formulas and a description are included in an online supplement in case your professor wants you to learn how to use them.)

Using Technology: Finding the Line of Best Fit

To find and graph a line of best fit for two data sets that appear to have an approximate linear relationship:

Graphing Calculator

1. Press STAT ENTER to get to the list editor.
2. Enter the data set you want to use as inputs (x values) under **L1.**
3. Use the right arrow key ▶ to access **L2,** then enter the data set you want to use as outputs (y values). When entering data from a table, make sure you enter the second list in the same order as the first.
4. Turn on **Plot1** by pressing 2nd Y= 1. Set up the screen, as shown below.
5. Press Y= and if the **Y=** screen isn't blank, move the cursor over any entered equations and press CLEAR.
6. Press ZOOM 9 which is the **Zoom Stat** option; this automatically sets a graphing window that displays all of the plotted points.
7. Press STAT followed by the right arrow key to access the **STAT CALC** menu, and choose the **LinReg** (for linear regression) option, which is choice 4. Then press ENTER to calculate the line of best fit. The display will read "y = ax + b," then provide values for a and b. The equation of the line of best fit is written by substituting the given values for a and b into the equation $y = ax + b$.
8. Press Y= and enter the equation of the line of best fit, then press GRAPH to display the scatter plot along with the line of best fit.

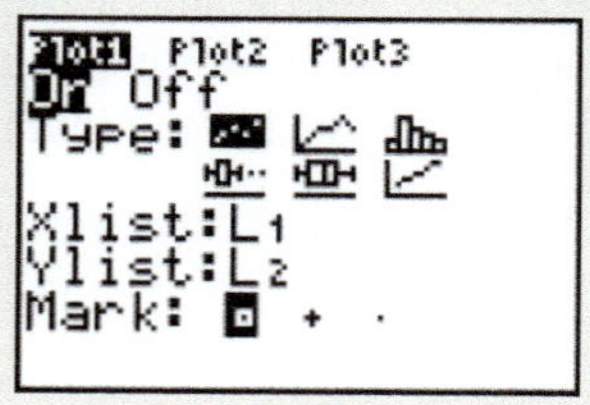

Excel

1. Enter the data in two columns and create a scatter plot.
2. Right-click on one of the points on the scatter plot, then choose "Add Trendline" from the **Chart** menu.
3. In the formatting dialog box that appears, click "Options," then click the checkbox for "Display equation on chart."

See the Using Tech Videos on Finding the Line of Best Fit with a Graphing Calculator and in Excel for further information.

4. Find the equation of the regression line for the data in Question 1. You can use either a graphing calculator or a spreadsheet.

Graphing the Regression Line

After finding the equation of a regression line, we'll want to add it to our scatter plot to further help us study the relationship between the two variables.

Solved Example 1

Adding a Regression Line to a Scatter Plot

Graph the line of best fit on the scatter plot that illustrates a potential relationship between gas prices and obesity rates.

SOLUTION

The simplest way to graph ANY line is to find and plot two points on the line, then draw a straight line connecting those points. In this case, certainly we'll want to find points that live somewhere on our scatter plot, so I decided to pick two inputs for the equation (gas prices) that are between \$0 and \$4. It's also a good idea to pick points that aren't too close together, which makes it easier to draw the line accurately. So I picked \$1 and \$3, which I now plug into the regression equation for *x*:

$$y = 8.2908x + 12.705$$

$$x = 1: y = 8.2908(1) + 12.705 = 20.9958$$

$$x = 3: y = 8.2908(3) + 12.705 = 37.5774$$

Now we'll add the points (1, 21) and (3, 37.5) to the scatter plot, then connect them with a line. (It doesn't hurt to round the *y* values to the nearest 0.5 since we're just using them to locate points on the graph, and visually you can't tell the difference between height 20.9958 and height 21.)

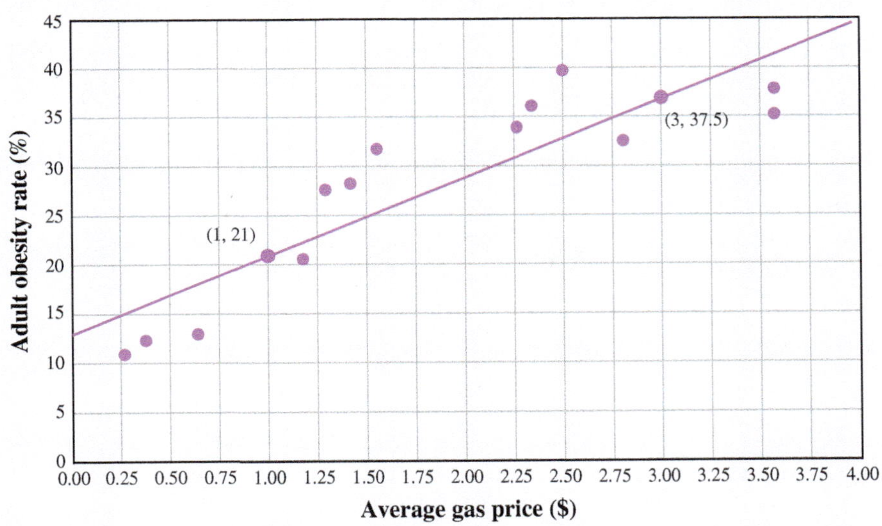

5. Write a sentence or two analyzing how the line of best fit from Solved Example 1 compares to your initial guess from Question 3.

Did You Get It

2. Find the equation of the regression line for the data in Did You Get It 1.

Analyzing a Scatter Plot

There are several types of relationships that can exist between the x values and the y values in a scatter plot. These relationships can be identified by looking at the pattern of the points on the graphs. For the scatter plot that you drew, it was reasonable to guess that there was a linear relationship, but that may not always be the case.

In Questions 6–9, select the graph below that best represents each description and fill in any blanks.

6. A positive linear relationship exists when the points fall approximately in an ascending straight line from left to right. In this case as the x values are increasing, the y values are __________. This relationship is best illustrated by graph ____.
7. A negative linear relationship exists when the points fall approximately in a descending straight line from left to right. In this case as the x values are increasing, the y values are __________. This relationship is best illustrated by graph ____.
8. A nonlinear relationship exists when the points fall in a curved line. The relationship is described by the nature of the curve. This relationship is best illustrated by graph ____.
9. No relationship exists when there is no discernible pattern to the points. This relationship is best illustrated by graph ____.

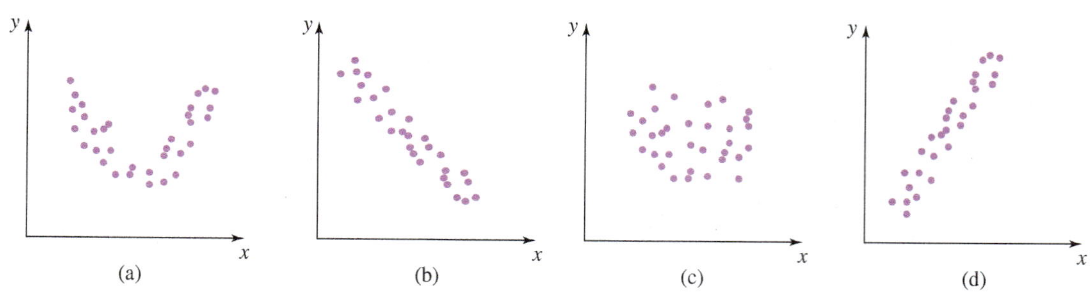

The Correlation Coefficient

Deciding whether or not two data sets are related by simply looking at a scatter plot is a pretty subjective process, so it would be nice to have a way to quantify how strongly connected data sets are. The **correlation coefficient** is a number that describes how close to a linear relationship there is between two data sets. Correlation coefficients range from -1 (perfect negative linear relationship) to $+1$ (perfect positive linear relationship). The closer this number is to one in absolute value, the more likely it is that the data sets are linearly related. A correlation coefficient close to zero indicates that the data sets are most likely not linearly related. This could mean that they're related in some nonlinear way or that they're not related at all.

We use the letter r to represent the correlation coefficient. It doesn't depend on the units for the two data sets, and it also doesn't depend on which set you choose for the x variable.

10. The graphic below describes the scale of correlation coefficients. Make an estimate for r for each of the four graphs in Questions 6–9.

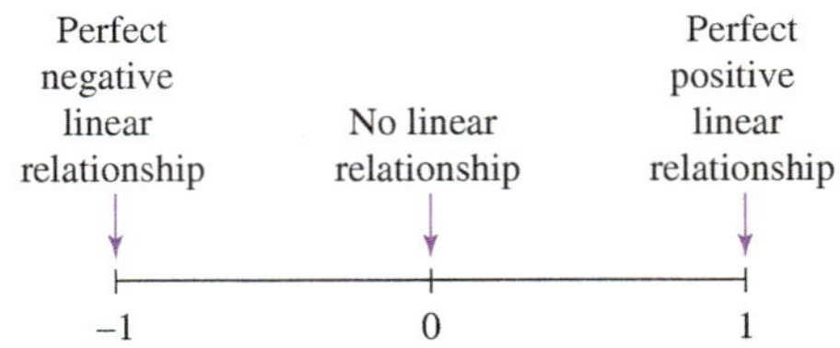

As with the regression line, there's a formula for calculating r, which you can find in the online supplement to this lesson. In terms of technology, you already know how to find the correlation coefficient, since both graphing calculators and spreadsheets automatically calculate r when finding the regression line.

Using Technology: Finding the Correlation Coefficient

Graphing Calculator

The correlation coefficient should display on the screen when you find the regression line. If it doesn't, go to the CATALOG menu (2nd 0), then scroll down and choose DiagnosticOn.

Excel

If you've already found the trendline, on the format trendline menu, click the box that says "Display R-squared value on chart." This finds the square of the correlation coefficient, so you can then use the formula "=SQRT(" to find r. You may need to add a negative sign if the slope of your trendline is negative. To find the correlation coefficient without finding the trendline, use the CORREL command. When you type "=CORREL(" you will be prompted to select the two data sets with a comma between them.

See the Using Tech Video on Finding the Correlation Coefficient for further information.

11. What's the correlation coefficient for the data in Question 1? What does that say about your answer to Question 2?

Did You Get It

3. Find the correlation coefficient for the data in Did You Get It 1, and discuss what you think it indicates.

12. Based on what we've seen so far, would you say that an increase in gas prices led to an increase in obesity rates, or that an increase in obesity rates caused gas prices to rise?

All we know so far is that the two variables appear to be related in a linear way. This says exactly NOTHING about whether one is caused by the other. We'll discuss this very important issue later in the lesson.

The Relationship between r and the Regression Line

Two things should be noted about the relationship between the value of r and the regression line. First, the value of r and the value of the slope (a) always have the same sign. Each indicates whether there's a positive or negative linear relationship between the variables. Second, the closer the value of r is to $+1$ or -1, the better the points will fit the line. In other words, the stronger the relationship, the better the fit.

13. Match each value of r with the appropriate graph.

$r = -1$: ______

$r = -0.9$: ______

$r = -0.5$: ______

$r = 0.5$: ______

$r = 0.9$: ______

$r = 1$: ______

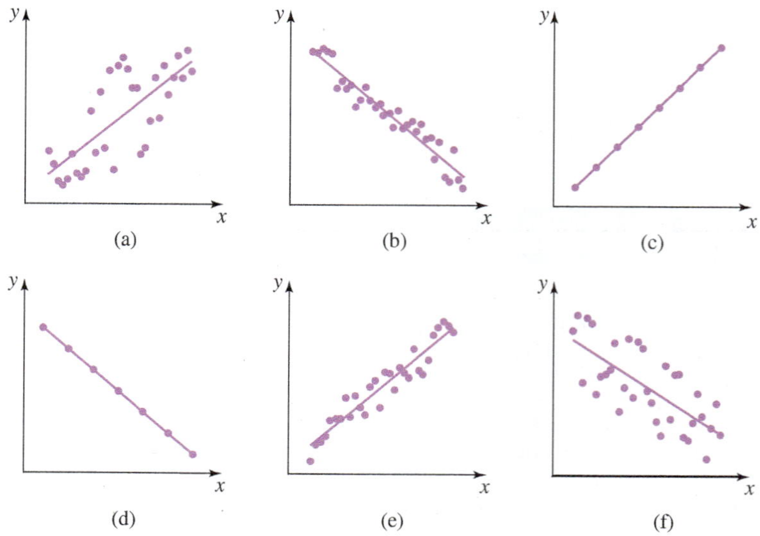

4-8 Group

1. A medical researcher selects a sample of small hospitals in his state and hopes to discover if there is a relationship between the number of beds and the number of personnel employed by the hospital. Draw a scatter plot on the next page for the data shown. Make sure you choose a reasonable scale for each axis, and clearly mark the scale on your graph.

No. of Beds (x)	28	56	34	42	45	78	84	36	74	95
Personnel (y)	72	195	74	211	145	139	184	131	233	366

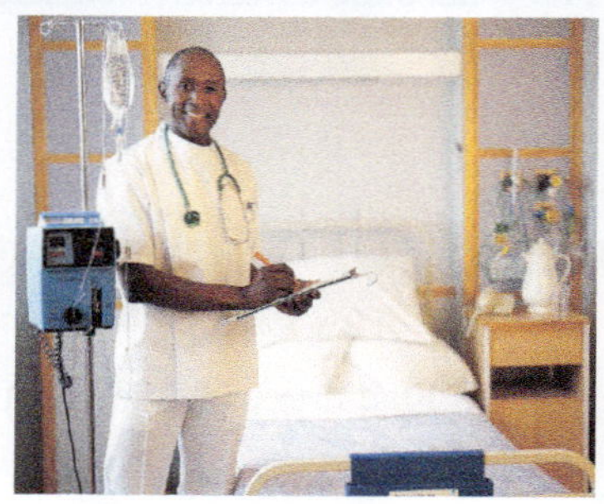

Digital Vision/SuperStock

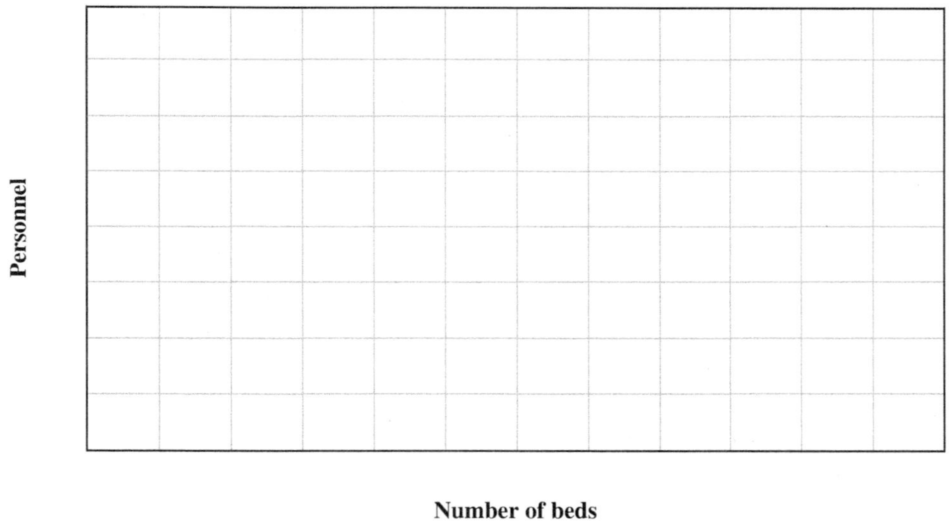

2. From simply looking at your scatter plot, describe the type of relationship (if any) that you think might exist between the number of beds and the number of personnel.

3. Find the equation of the regression line for these data.

4. Graph the line of best fit on the scatter plot from Question 1 by choosing two x values between 0 and 100 (so that the points will appear on the scatter plot).

5. Find the correlation coefficient for these data. Do you think this indicates that the two variables are linearly related? Justify your answer.

Significance Levels

Next we'll tackle the issue of how confident we can be that a relationship exists between two quantities based on the sample size and the correlation coefficient. For the hospital bed data we've been examining, the value of r was based on a sample of data. The population is all hospitals in a certain state with fewer than 100 beds, and a sample of such hospitals was chosen. When data from a sample is used, we can't be positive that it represents the entire population. So conclusions drawn, like the conclusion that there's a connection between numbers of beds and personnel, may not be correct.

Statisticians have traditionally agreed that when we conclude that two data sets have a relationship, we can be satisfied with that conclusion if there is either a 95% or 99% chance that we're correct.

6. A 95% chance that we're correct would correspond to a ______% chance that we're wrong, and a 99% chance that we're correct would correspond to a ______% chance that we're wrong.

These percentages of being wrong are called **significance levels.**

We know that the more data we have, the more likely it is that the data are representative of the population.

7. We would expect that the size of r needed to be reasonably sure that two data sets are correlated is __________ for small sample sizes, and ___________ for large sample sizes.

In the following table, we see the minimum r values needed, for different sample sizes, for us to have a 5% and 1% chance of being wrong when we conclude that two data sets are linearly related. (Actually, these are the minimum values for the absolute value of r, since it could be either negative or positive.)

Sample Size	5%	1%
4	0.950	0.990
5	0.878	0.959
6	0.811	0.917
7	0.754	0.875
8	0.707	0.834
9	0.666	0.798
10	0.632	0.765
11	0.602	0.735
12	0.576	0.708
13	0.553	0.684
14	0.532	0.661
15	0.514	0.641
16	0.497	0.623
17	0.482	0.606
18	0.468	0.590
19	0.456	0.575
20	0.444	0.561
21	0.433	0.549
22	0.423	0.537
23	0.412	0.526
24	0.403	0.515
25	0.396	0.505
26	0.361	0.463
27	0.312	0.402
28	0.254	0.330
29	0.179	0.234

So what exactly do all of these numbers mean? Let's look at an example. Suppose that we've found the correlation coefficient for two data sets with seven values in each, and the correlation coefficient is $r = 0.81$. Look at the row of the table corresponding to a sample size of 7. In the 5% column, the value is 0.754. This means that if the correlation coefficient is 0.754 or higher, we can conclude that the two data sets are linearly related and have less than a 5% chance of being wrong.

Now look at the next column in that row: The value in the 1% column is 0.875. This tells us that if the correlation coefficient is 0.875 or more, we can conclude that the two data sets are linearly related and have less than a 1% chance of being wrong. So in our hypothetical example, the correlation coefficient of 0.81 (which is more than 0.754 but less than 0.875) tells us that we can be 95% sure that there's a linear relationship between the two data sets, but we can't be 99% sure.

Solved Example 2

Using Significant Values for the Correlation Coefficient

Test the correlation coefficient from Question 5 for significance at the 5% level, then describe what the result tells us.

SOLUTION

If all went well, you got $r = 0.748$ in Question 5. The data we were studying were from a sample of 10 hospitals, so we need to focus on the row in the table corresponding to sample size 10. The value in the 5% column of that row is 0.632. Since the correlation coefficient we got is more than 0.632, it IS significant at the 5% level. This tells us that we can be 95% certain that there is a linear relationship between the number of beds and the number of personnel at hospitals in that state.

8. Is the correlation coefficient found in Question 5 significant at the 1% level? What does that mean?

Did You Get It

4. Test the significance of the correlation coefficient obtained from Did You Get It 3. Use 5%, and then 1%.

How should we choose whether to use the 5% or 1% significance level in a given situation? It depends on the seriousness of the situation and the importance of drawing a correct conclusion. Suppose that researchers think that a new medication helps patients with asthma to breathe easier, but that some patients have experienced side effects. A correlation study would probably be done to decide if there is a relationship between the medication and these particular side effects.

9. If the potential side effects are serious, like heart attacks or strokes, which significance level would be used? Why?

10. If the side effects are mild, like headache or nausea, which significance level would be used? Why?

One of the most useful aspects of finding a regression line is that it can be used to make predictions for one of the variables given a value for the other. As part of graphing the line, we've already used the equation to make predictions for two points. **Important note:** If we hadn't already checked whether the correlation is significant, then it would be meaningless to make any predictions.

11. Use the equation of the regression line found in Question 3 to predict the approximate number of personnel for a hospital with 65 beds.

Math Note

Never use a regression line to predict anything unless you've already checked that the correlation is significant! If it isn't significant, the regression line is meaningless.

For any data value from a set, the **residual** is the difference between the actual value and the predicted value that comes from the regression equation. Each point from the data set has one residual. As you'll verify in the Tech assignment, the sum and the mean of the residuals for a data set will be zero. That's a fancy way of saying that the regression line perfectly "splits the difference" between all of the points on a scatter plot.

Notice that the original data set didn't contain the x value of 65 beds, so we can't compute a residual for the value in Question 11.

12. Use the equation of the regression line found in Question 3 to predict the approximate number of personnel for a hospital with 42 beds.

13. What is the actual number of personnel for a hospital with 42 beds?

14. What is the residual for this value? Interpret this result.

15. What is the residual for the number of personnel in a hospital with 78 beds? Pay attention to the sign. Interpret this value.

Did You Get It

5. Use the equation of the regression line found in Did You Get It 2 to predict the number of stories for a 582-foot building. Then compute the residual.

Correlation and Causation

Here's a key phrase to remember when studying correlation and regression analysis: *Correlation is not the same thing as causation.* Knowing that two variables are related doesn't say anything about whether or not one CAUSES changes in the other. Here are some possible interpretations when two variables are significantly correlated.

Possible Relationships Between Data Sets

1. *There is a direct cause-and-effect relationship between two variables.* That is, x causes y. For example, water causes plants to grow, poison causes death, and heat causes ice to melt.
2. *There is a reverse cause-and-effect relationship between the variables.* That is, y causes x. Suppose that we find a correlation between excessive coffee consumption and nervousness. It would be tempting to conclude that drinking too much coffee causes nervousness. But it might actually be the case that nervous people drink a lot of coffee because they think it calms their nerves.
3. *The relationship between the variables may be caused by a third variable.* You could probably find a positive correlation between the amount of ice cream consumed per week and the number of drowning deaths for each week of the year. But this doesn't mean that eating ice cream causes you to drown. Both eating ice cream and swimming are more common during the summer months, and it stands to reason that more people will drown when more people are swimming. In this case, a third variable (seasonal weather) is affecting each of the original two variables.
4. *There may be a variety of complicated interrelationships among many variables.* For example, a researcher may find a significant relationship between students' high school grades and college grades. But there probably are many other variables involved, such as IQ, hours of study, influence of parents, motivation, age, and instructors.
5. *The relationship might simply be coincidental.* For example, historians have noticed that there is a very strong correlation between the party that wins a presidential election and the result of the Washington football team's final game before the election. But good old common sense tells us that this relationship has to be a coincidence.

16. Based on everything you learned in this lesson, write a description of the relationship between gas prices and obesity. Include your thoughts on which of the five relationships above is most likely to hold in this case.

Sidelight **Of Skiing and Killer Bedsheets**

Don't ask how, but someone that's very interested in statistics discovered what appears to be a connection between the total revenue generated by skiing facilities in the United States and the number of Americans who die by becoming tangled in their bedsheets. Don't believe it? Have a look at the graphs. The first shows both of these data sets graphed by year. Note the similarity in their shapes. That's the first indication that they may be related somehow. So the second graph is a scatter plot of the data for matching years: it displays a pretty clear linear trend. How clear? The correlation coefficient, as calculated by Excel, is 0.97—close to a perfect correlation!

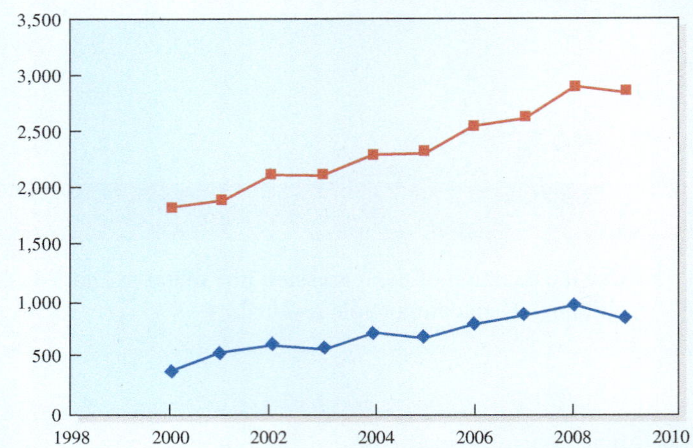

So should you start sleeping on the floor if ski facility revenues are up next year? If someone dies from getting tangled in their bedsheets, does everyone they know cope with their grief by going skiing? Are ski resorts stocked with killer sheets? While that would make a fine plot for a really dreadful movie, in real life it's just plain silly. This is a classic example of the fact that correlation does not in any way equate to causation. Bedsheet deaths don't cause people to go skiing, and visiting ski facilities doesn't make an individual more likely to die in bed. It's also quite a stretch to think that there's some third variable that connects these quantities. More than likely it's just a strange, cosmic coincidence. The moral of the story? Use your head. Don't be a data nerd. Never rely strictly on numbers like correlation coefficients to do your thinking for you. Statistics is a tool to help us draw conclusions, not an excuse to use our heads only to keep our hats from resting on our shoulders.

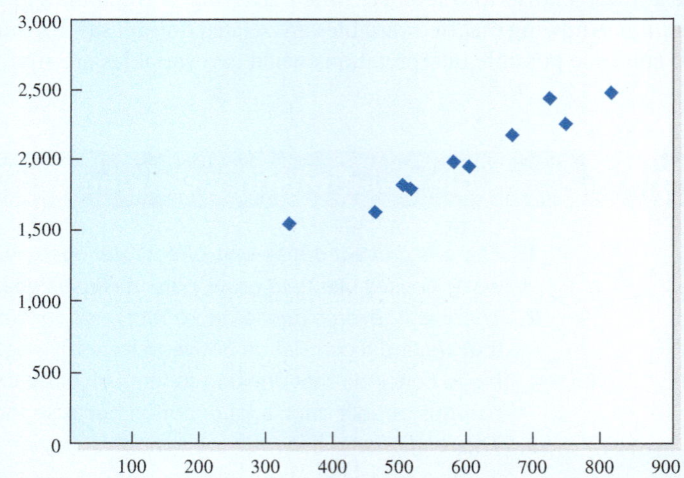

If you find this correlation interesting, do a Google search for the phrase "spurious correlations." It's HOURS of fun.

4-8 Portfolio

Name ______________________________

Check each box when you've completed the task. Remember that your instructor will want you to turn in the portfolio pages you create.

Technology

1. ☐ In this Excel activity, we'll compute the line of best fit and residuals for the hospital bed data from the beginning of the group portion of the lesson.

ALEKS Practice

1. ☐ Include any written work from the ALEKS assignment along with any notes or questions about this lesson's content.

Applications

1. ☐ Complete the applications problems.

Reflections

Type a short answer to each question.

1. ☐ Describe the difference between correlation and causation, and explain why understanding the difference is so important.
2. ☐ If the correlation coefficient for two data sets is close to zero does that mean that they're not related at all? Think carefully about this one, and explain your answer.
3. ☐ Think of an example of two data sets that you'd expect to have a negative linear correlation.
4. ☐ Name one thing you learned or discovered in this lesson that you found particularly interesting.
5. ☐ What questions do you have about this lesson?

Looking Ahead

1. ☐ Complete the prep skills for Lesson 4-9.
2. ☐ Read the opening paragraph in Lesson 4-9 carefully and answer Question 0 in preparation for that lesson.

Answers to "Did You Get It?"

1. 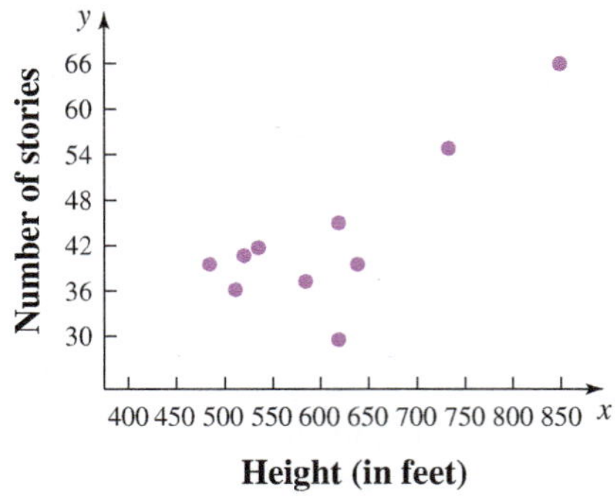

As height goes up, so does the number of stories.

2. $y = 0.0685x + 1.646$
3. $r \approx 0.798$; there is likely a linear relationship between height of a building and number of stories.
4. Significant at both levels.
5. 42 stories; the residual is −4.

Answers to Prep Skills Questions

1. 35.8, 164.8, 75.36
2. $290,340
3. 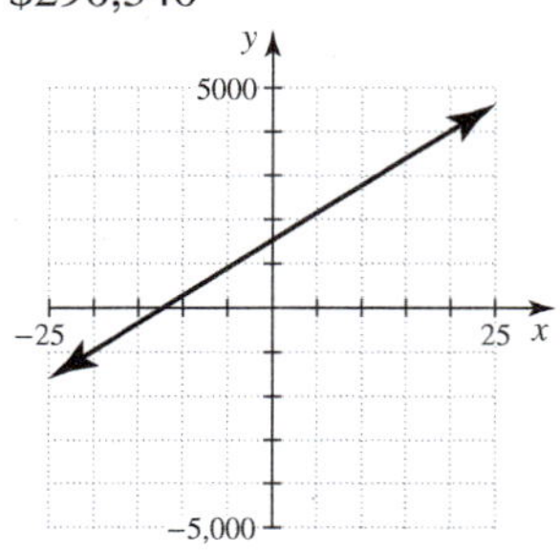

4-8 Applications

Name ____

In Questions 1–4, for each pair of quantities, decide if they are likely to be positively correlated, negatively correlated, or uncorrelated. Explain your reasoning.

1. Total number of years of school completed and average annual salary

2. Number of class meetings missed by college students and their average scores in the course

3. The age of a person and the probability that he or she wears glasses or contact lenses

4. The height of adults and the amount of money they earn relative to their coworkers

For the data sets in Questions 5–7,

- **a.** Draw a scatter plot.
- **b.** Find the value for r.
- **c.** Test the significance of r at the 5% level and at the 1% level. If you conclude that r is not significant at either level, be sure to explain why.
- **d.** Find the equation of the regression line and draw the line on the scatter plot, but only if r is significant.
- **e.** Describe the nature of the relationship if one exists.
- **f.** If possible, make the requested prediction using your regression line.
- **g.** If possible, compute the requested residual and explain the meaning of the result.

4-8 Applications

Name ________________________________

5. A researcher hopes to determine if there is a relationship between the number of days an employee missed a year and the person's age. The following table displays information for 10 randomly selected employees at a large company. Predict the number of days missed by a 56-year-old employee. Compute the residual for the point (50, 7).

Age, x	22	30	25	35	65	50	27	53	42	58
Days Missed, y	0	4	1	2	14	7	3	8	6	4

4-8 Applications

Name ______________________________

6. A researcher believes there is a connection between violent crime and the average high temperature in a city. He randomly selects nine cities from a list of U.S. cities over 250,000 in population and records the average high temperature in April and the violent crime rate (per 100,000 residents). Predict the crime rate for a city with an average high temperature in April of 75°. Compute the residual for the point (63, 473).

Avg. Apr. High, *x*	62	70	80	64	65	63	60	84	67
Crime Rate	458	1,122	841	795	1,185	473	1,334	1,060	1,685

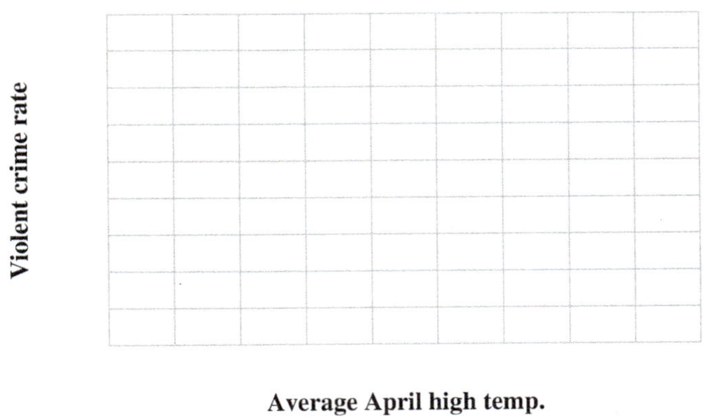

4-8 Applications

Name ______________________________

7. Is there a connection between the unemployment rate in a state and the average household income there? The next table shows 2023 data for nine randomly selected states. Predict the average household income for a state with an unemployment rate of 4%. Predict the unemployment rate for a state with average household income of $62,500. (Careful about which variable represents which quantity.) Compute the residual for the point (2.2, 85.2).

Unemployment %, *x*	2.9%	4.3%	4.9%	2.9%	2.7%	4.5%	2.2%	4.2%	2.0%
Household Income (thousands $), *y*	97.4	76.9	62.0	102.0	89.2	73.8	85.2	61.0	81.7

Lesson 4-9 Prep Skills

SKILL 1: COMPUTE MEASURES OF AVERAGE

We looked at the four measures of average in Lesson 4-3. A reminder of the process used to find each measure is given below.

Mean If $X_1, X_2, X_3, \ldots, X_n$ are the data values (which means there are n of them), we use the symbol $\overline{X}$ to represent the mean, and $\overline{X} = \dfrac{X_1 + X_2 + X_3 + \cdots + X_n}{n} = \dfrac{\Sigma X}{n}$.

The symbol $\overline{X}$ is read "X bar."

Median **Step 1:** Arrange the data in order, from smallest to largest. (Actually, largest to smallest is fine, too.)
Step 2: The way we find the mean depends on whether the number of data values is even or odd.

- If there's an odd number of data values, there will be one single value in the middle of that ordered list, with half of the remaining values bigger and half smaller. That value in the middle is the median.
- If there's an even number of data values, there will be two values in the middle of the list. The mean of those two values is the median.

Midrange **Step 1:** Find the smallest and largest values in the data set.
Step 2: Find the mean of those two values.

Mode The **mode** is the value that occurs most often in a data set.

SKILL 2: COMPUTE RELATIVE CHANGE

The relative change between two quantities measures the change as a fraction of the original value.

$$\text{Change} = \text{New value} - \text{Original value}$$

$$\text{Relative change} = \frac{\text{Change}}{\text{Original value}} = \frac{\text{New value} - \text{Original value}}{\text{Original value}}$$

- Compute the change and relative change if the cost of a gallon of gas jumps from \$2.80 to \$3.50.

$\text{Change} = \$3.50 - \$2.80 = \$0.70$ The cost of a gallon of gas increased by \$0.70.

$\text{Relative change} = \dfrac{\$0.70}{\$2.80} = 0.25$ The relative change from \$2.80 to \$3.50 is 0.25. This indicates a 25% increase.

SKILL 3: DRAW A BAR GRAPH

In Lesson 4-2, we saw that a bar graph is a visual way to compare the sizes of different values. Pie charts are great for comparing parts to a whole; bar graphs are effective for comparing parts to other parts. In a bar graph, the length of each bar gives an indication of how the size of a quantity compares to other related quantities. Using a spreadsheet is a great tool for quickly creating a bar graph, but you can use the following procedure when creating a bar graph by hand.

Step 1: Label the name of each category in your data, equally spaced along the bottom of your graph (or along the left if you want to draw horizontal bars).

Step 2: Label an appropriate scale along the left of your graph (or along the bottom for horizontal bars). Make sure your scale starts at zero and extends out a little bit further than the largest number in your data set.

Step 3: Draw in rectangular bars whose heights or lengths correspond to the number representing each category in your data set.

SKILL 4: GRAPH PAIRED DATA

When graphing pairs of data sets, the key things to remember are (1) write the given information as ordered pairs; (2) put the information corresponding to the first coordinates on the horizontal axis and the information corresponding to the second coordinates on the vertical axis; and (3) put an appropriate scale on those axes so that all of the points on your plot fit in the space you have, and don't leave a ton of empty space with no points. In almost every case, you'll want the scale on the vertical axis to begin at zero so that the heights of points don't give a misleading view of differences in the data.

- This table displays the number of viewers for a TV show over the first 5 weeks of its (presumably short) life. The data are illustrated in graphic form next to the table.

Weeks after Premiere	Viewers
0	12,500,000
2	11,100,000
4	9,700,000
6	8,300,000
8	6,900,000
10	5,500,00

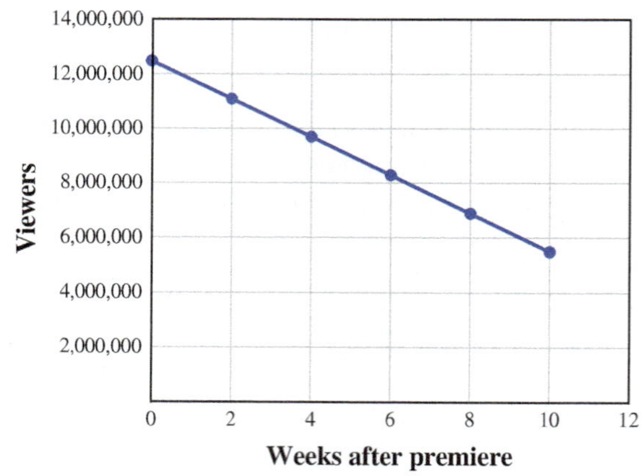

PREP SKILLS QUESTIONS

1. Find all four measures of average for this data set:

   12 18 21 18 9 33 97 20 19 25 13

2. After a promising earnings report, the price of a tech stock jumps from $7.45 to $11.30 in 1 day. Find the change in price and the relative change in price.

3. Draw a bar graph that illustrates the data in the table, which is the maximum depth in feet of each of the Great Lakes. The scale on the vertical axis should go from zero to 1,500 feet.

Lake	Superior	Erie	Michigan	Ontario	Huron
Max. Depth	1,333	210	923	802	750

4. The data in the table are the percentage of total power in the U.S. generated by wind for selected years from 2002 to 2023. Graph the data on a rectangular coordinate system with the year on the x axis.

Year	2002	2005	2008	2011	2014	2017	2023
Percent	0.26%	0.46%	1.25%	3.20%	3.95%	6.30%	10.2%

Lesson 4-9 Trust No One

LEARNING OBJECTIVES

- ☐ 1. Identify misuses of sampling and evaluate their effect on statistical results.
- ☐ 2. Recognize and describe common misuses of compiling and reporting statistics that make them meaningless or deceiving.
- ☐ 3. Study ways that graphs can be manipulated to tell a desired story.

FOX Image Collection via Getty Images

There are two ways of lying. One, not telling the truth and the other, making up statistics.

—Josefina Vazquez Mota

"Trust No One" was the unofficial motto of the classic TV show *The X Files,* in which intrepid agent Fox Mulder and his faithful assistant Dr. Scully work to find the truth behind incredible events amid vast conspiracies. In the world of that show, things were rarely what they seemed, and the best course of action for the investigators (and viewers) was to assume that anything that anyone said may have been at best a stretching of the truth. In our modern world, there's a lot of wisdom in that approach. We are constantly bombarded with information and statistics, and it would be quite a leap to expect that most of the people disseminating that information have your best interests in mind. An educated adult knows that it's a good idea to view information suspiciously, and also knows how to recognize when statistics are being manipulated to tell the story that someone wants you to hear, not the truth. But, as Mulder was fond of pointing out, the truth is out there. You just have to want to find it.

0. What are some circumstances in which someone might want to mislead you with a misuse of statistics?

4-9 Group

Ideally, statistical research is used to provide knowledge and information to help us make intelligent decisions about our health and welfare, or maybe just to give us information about things we find interesting. But the world is a less-than-ideal place, and there are people who will misuse statistics to sell us products that don't work; to attempt to prove something that isn't true; or to get our attention by using fear, shock, or outrage supported by bad statistics.

Just because we read or hear the results of a research study or an opinion poll in the media, this doesn't mean that these results are reliable or that they can be applied to any and all situations. For example, reporters sometimes leave out critical details like the size of the sample used or how the research subjects were selected. Without this information, you can't properly evaluate research and properly interpret the conclusions of a study or survey.

The purpose of this lesson is to show you some ways that statistics can be misused. The point isn't to make you reject any statistical study or result; it's to help you to recognize when information or conclusions you're being given might be presented in a biased or dishonest way. Throughout this unit, we've sprinkled in some information on ways that statistics can be misused; in this lesson, we'll compile and expand on those misuses.

Suspect Samples

Since there are over 300 million people in the U.S. (and over 8 billion in the world), the vast majority of statistical studies you run across involve sampling: This is where any study is going to begin. That's why we started our study of statistics with sampling, and it's also why that's where we'll begin our look at statistical mischief.

Unreported Sample Sizes

In the first part of the 21st century, there was a chewing gum brand that ran a large campaign claiming that "Four out of five dentists surveyed" would recommend their product. That sounds like a ringing endorsement. But what does it really mean?

1. By making this statement, what is the advertiser trying to imply? Calculate and use a percentage in your answer.

2. How many dentists do you think there are in the United States? Make a guess.

3. How many dentists do you feel like they would have to survey in order to make this statistic meaningful?

The bottom line is that the advertiser (of course) provided no information on how many dentists were actually surveyed. If they only surveyed five, their claim could still be technically true, but they may have stumbled across four that by coincidence would recommend their product, even if the vast majority of dentists would rather see their patients chewing razor blades. Worse still, they could easily have chosen to survey five that would be predisposed to recommend their product for one reason or another (like say a check with multiple zeroes on it).

If samples in a study are too small, the results could be completely meaningless. And if the size of a sample is conveniently left out, there's a pretty good chance that the person reporting the statistics would just as soon have you not know what that sample size was.

4. Why would an advertiser be likely to leave out the size of a sample on which a statistic is based?

Did You Get It

Try this problem to see if you understand the concepts we just studied. The answer can be found at the end of the portfolio section.

1. Evaluate this claim by an advertiser: 60% of people surveyed prefer our brand of cola over the leading national brand.

Samples Relying on Volunteer Participation

5. Suppose that a national poll is conducted by telephone to gauge the favorability rating for a particular candidate for the presidency. What types of people will not be represented in this survey? List as many as you can think of.

People who are annoyed by unsolicited phone calls won't appear in a sample of people contacted by phone.

Plush Studios/Blend Images

Studies using volunteers sometimes have a built-in bias because volunteers generally don't represent the population at large. Sometimes they are recruited from a particular socioeconomic background. For example, unemployed people are far more likely to volunteer for research studies in order to get paid, and because they're likely to have more free time available to participate.

6. What groups of people would be available for studies that require the subjects to spend several days or weeks in an environment other than their home or workplace?

7. In the past, many studies have used only men but have tried to generalize the results to both men and women. Do you think this is a good idea? Why? Does it depend on what is being studied?

8. To conduct market research on their products, some companies will offer coupons, samples, or even cash to participate in a study. Comment on how you think this is likely to affect the outcome.

Samples Relying on Convenience

It can be difficult and time-consuming to conduct a study involving a truly random sample. It's not uncommon for people to cut corners and work with a sample that's easy to get to for one reason or another.

9. A sports reporter in Chicago is asked to write an article about the popularity of certain baseball teams in the Midwest. She decides that conducting a poll to gather some data is a fine idea, and while she's at it, she may as well enjoy a beverage and an appetizer. So she hops on the L (public transportation) and goes to a very popular sports bar on the north side of the city, then asks everyone that will pay attention to her which is their favorite baseball team. Over the course of 4 hours, she manages to gather information from over 300 people. Do you think this was a reasonable approach to gathering information? Explain.

Frank Romeo/Shutterstock

A **convenience sample** is one where members of a preexisting group are selected. This can lead to bias because there may be something that connects all the members of that group.

10. An educational researcher wants some data on the number of hours per week that average college students study outside of class. He sends around an email asking colleagues to administer a short survey in their classes, and ends up surveying two 300-level chemistry courses and students in a senior-level physics course. List several different reasons why this is likely to be a nonrepresentative sample.

If you're trying to interpret the results of a study done using a small sample, a volunteer sample, or a convenience sample, you have to give a lot of thought to whether the characteristics of that sample make it representative of an entire population.

11. There are many sites on the Internet where college students can rate their professors in terms of teaching effectiveness. Suppose that one instructor has seven reviews at a site, with six of them describing the teacher as being a wonderful educator. Discuss how reliable you think that would be for a student deciding whether or not to take that instructor. Back up your conclusions with evidence based on the ideas in this lesson.

Did You Get It ?

2. In a study to determine how long it takes to fall asleep, would college students be a good sample of the overall population? Why?

Ambiguous Averages

There was a time when we all thought that "average" meant one thing: the result of adding all data values and dividing by how many there are. But we know better now, don't we? When someone uses the term "average" in making some sort of claim instead of letting us know which measure of average (mean, median, mode, and midrange) that they're using, it's time to trust no one and be suspicious.

We know that the values of these different measures of average sometimes differ quite a lot. People who know that there are several types of average can, without lying, select the one for the data that most lends evidence to support their position. For example, suppose a storeowner employs four salespeople. The number of years each has been employed by the store is 22, 10, 2, and 2.

12. What is the mean number of years of service for the four salespeople?

13. What is the median?

14. What is the mode?

15. What about the midrange?

16. If the storeowner wanted to advertise the fact that their employees have many years of experience, which average do you think they would use? Why?

Faulty Survey Questions

Surveys and opinion polls obtain information by using questionnaires. There are two types of studies: interviews and self-administered questionnaires. The interview survey requires a person to ask the questions either in person or by telephone. Self-administered questionnaire surveys require the participant to answer the questions by mail, computer, or in a group setting, such as a classroom. When analyzing the results of a survey using questionnaires, you should be sure that the questions are properly written, since the way questions are phrased can often influence the way people answer them.

Photographer's Choice/SuperStock

17. A group of people was asked two questions relating to gun control:

- Do you favor a waiting period before guns are sold?
- Do you favor a national gun-registration program costing about 20% of all dollars spent on crime control?

Both questions ask people if they're in favor of stricter gun restrictions. One question was answered "Yes" by 91% of respondents, and 6% said "No." The other was answered "Yes" by just 37% and "No" by 61%. Which is which? Explain why.

When reading and interpreting the results obtained from questionnaire surveys, watch out for some of these common mistakes made in the writing of the survey questions.

- *Asking Biased Questions.* By asking a question in a certain way, the researcher can lead the respondents to answer the question the way the researcher wants them to.
- *Asking Double-Barreled Questions.* Sometimes two ideas are contained in one question, and the respondent may answer one or the other in his or her response.
- *Using Double Negatives.* Survey questions containing double negatives often confuse the respondent.
- *Using Confusing Words.* Using words in a survey question that are not well-defined or understood can invalidate the responses. Most people wouldn't admit that they don't understand a question, so they just answer it in some random way. Clearly, this is not optimal.

In Questions 18–21, consider each survey question. Why do you think each question is a poor one? Think about the common mistakes described above.

18. "Are you going to vote for Candidate Jones, even though the latest survey shows he will lose the election?"

19. "Do you think people would live longer if they were on a diet?"

20. "Are you in favor of a national health program, and do you think it should be subsidized by a special tax as opposed to other ways to finance it, such as a national lottery?"

21. "Do you feel that it is not appropriate to have areas where people cannot smoke?"

Other factors that could bias a survey would include anonymity of the participant, the time and place of the survey, and whether the questions were open-ended or closed-ended.

22. Participants will, in some cases, respond differently to questions based on whether or not their identity is known. Researchers try to ensure confidentiality rather than anonymity; however, many people will be suspicious in either case. List some types of issues that people might be surveyed on where lack of anonymity might affect their responses.

23. The time and place where a survey is taken can influence the results. For example, if a survey on airline safety is taken right after a major airline crash, the results may differ from those obtained in a year with no major airline disasters. Think of at least two other similar examples where time and/or place could bias results.

Did You Get It ?

3. Two "researchers" (by which we mean "spin doctors") conduct surveys on people's attitudes toward health care. One survey contains the question "Do you feel that the high cost of Obamacare is justified by what it accomplishes?" The other survey has the question "Are you in favor of expanding the Affordable Health Care Act so that even more families can get quality medical care?" Describe how each "researcher" wants those surveyed to respond.

Convenient Use of Percentages

Pool/Getty Images

Another type of statistical distortion can occur when different values are used to represent the same data. For example, one political candidate who is running for reelection might say, "During my administration, expenditures increased a mere 3%." His opponent, who is trying to unseat him, might say, "During my opponent's administration, expenditures have increased a whopping \$6,000,000." Both figures could be correct; but expressing \$6,000,000 as "a mere 3%" makes it seem like a very small increase, while expressing the same increase as "a whopping \$6,000,000" makes it sound like a very large increase. If you want to get to the truth, ask yourself, "Which measure best represents the data?"

24. In Kentucky, the individual income tax rate was 5% in 2018. In 2024, a plan was announced to lower that rate to 3.5% in 2026. What is the difference in the two tax rates? Answer as a percent.

25. What is the relative difference in the two tax rates?

26. If someone earned \$70,000 in income, what is the difference this tax cut will have on that person's income?

27. Which of the three differences would be used by someone wanting to argue that this tax cut won't hurt the state's budget much?

28. Which of the three differences would be used by someone wanting to argue that this tax cut will have a devastating effect on the state's budget?

Did You Get It

4. An upcoming school levy in my town would raise the property taxes on a $150,000 home from $248 per month to $280 per month. Calculate the annual increase and the percent increase, then pick which one of these numbers would most likely be publicized by an advocacy group that opposes the raise.

Implied Connections

This one is a classic—if you look for it, you'll find it in advertising pretty much every day. Many claims imply that there's a connection between two variables without having any compelling evidence (or maybe any evidence at all) to back up that connection. To cover themselves legally, advertisers will use words like "may," "suggest," "some," etc. so that if the connection actually doesn't exist, they're not lying.

29. Explain why each of the following claims, found in online advertisements, are essentially meaningless.

a. "Eating fish may help to reduce your cholesterol."

b. "Studies suggest that using our exercise machine will reduce your weight."

c. "Taking calcium will lower blood pressure in some people."

Misleading Graphs

Graphs are great. We love them. In general, they're a fantastic way to illustrate information, making it easier to interpret. But there's a dark side to graphs. They can often be found hanging out with shady characters and committing crimes. Well, that's an exaggeration, but an inappropriately drawn graph can (either intentionally or not) misrepresent data and lead to false or misleading conclusions.

30. Draw a bar graph illustrating the change in Kentucky's tax rate on each of the axes below (see Question 24). Note that you'll be using the same information on each graph.

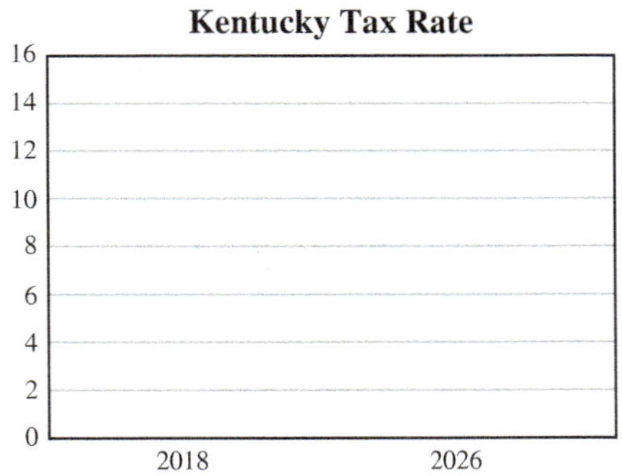

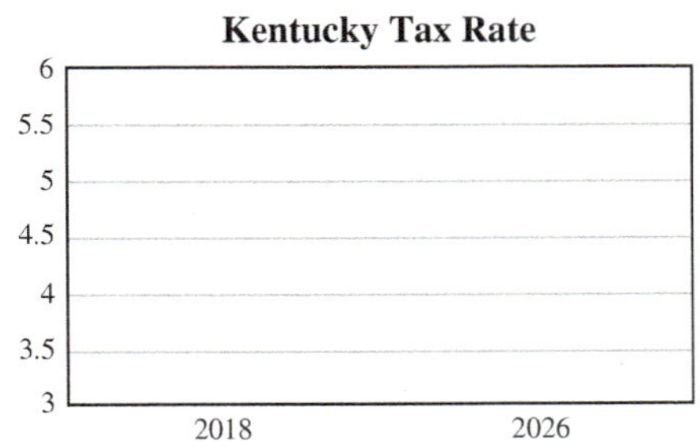

Math Note

It isn't necessarily *wrong* to truncate an axis (like the vertical axis here) when drawing a graph. Sometimes it may even be necessary. But the reader should be aware of this fact and interpret the graph accordingly. Don't be misled if an inappropriate explanation is given. Trust no one!

31. Which of the two bar graphs would be used by someone wanting to argue that this tax cut won't hurt the state's budget much? Why?

32. Which of the two bar graphs would be used by someone wanting to argue that this tax cut will have a devastating effect on the state's budget? Why?

The average lifespan for Americans has been increasing for decades (although, to be fair, we still rank 34th in the world). The data below display the average life expectancy for Americans according to the National Center for Health Statistics.

Year	1985	1990	1995	2000	2005	2010	2015	2020
Life Expectancy	74.7	75.4	75.8	76.8	77.4	78.3	78.8	78.9

33. Draw a graph for the pairs of data on each graph below, with the year on the horizontal axis.

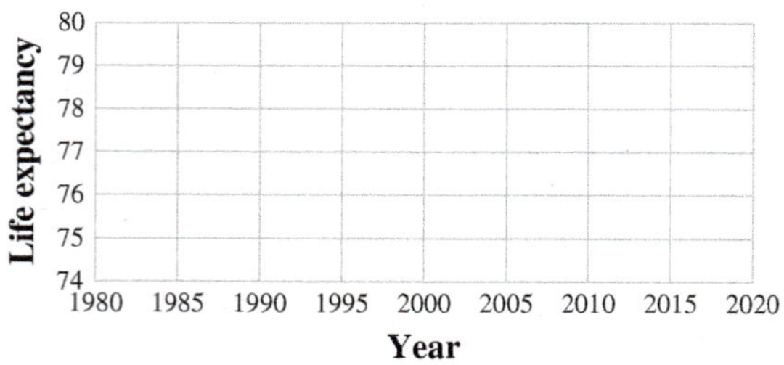

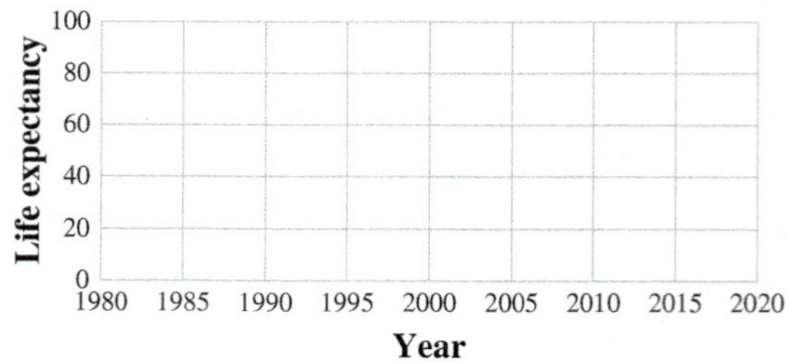

34. Describe how the life expectancy appears to be changing based on what you observe in each graph. (Two separate descriptions, please.)

35. Which graph would you show if you wanted to convince someone that the advances of modern medicine will soon have us living to be 150?

36. Which graph would you show if you were arguing that we're really not living much longer than we were way, way back in 1985?

4-9 Portfolio

Name ______________________________

Check each box when you've completed the task. Remember that your instructor will want you to turn in the portfolio pages you create.

ALEKS Practice

1. ☐ Include any written work from the ALEKS assignment along with any notes or questions about this lesson's content.

Applications

1. ☐ Complete the applications problems.

Reflections

Type a short answer to each question.

1. ☐ Why is a truly representative sample so important in gathering data?
2. ☐ Write about an example you've seen or heard recently where someone made a fishy claim backed either by dubious evidence, misleading evidence, or maybe no evidence at all.
3. ☐ How can you use the ideas in this lesson throughout your lifetime?
4. ☐ Name one thing you learned or discovered in this lesson that you found particularly interesting.
5. ☐ What questions do you have about this lesson?

Looking Ahead

1. ☐ Complete the prep skills for Lesson 5-1.
2. ☐ Read the opening paragraph in Lesson 5-1 carefully and answer Question 0 in preparation for that lesson.

Answers to "Did You Get It?"

1–3. Answers can vary. Talk to your group mates and instructor to make sure you understand the key ideas! **4.** Increase: \$384; percent increase: 13%. Which the group might choose is an opinion. Justify yours!

Answers to Prep Skills Questions

1. Mean: about 25.9; median: 19; mode: 18; midrange: 53
2. $3.85, about 51.7%
3.

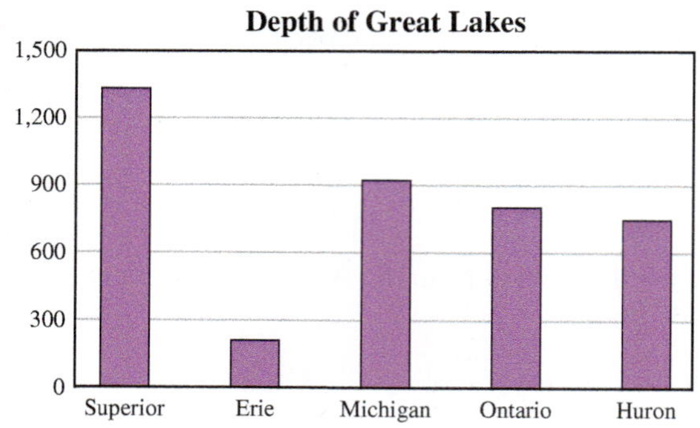

4.

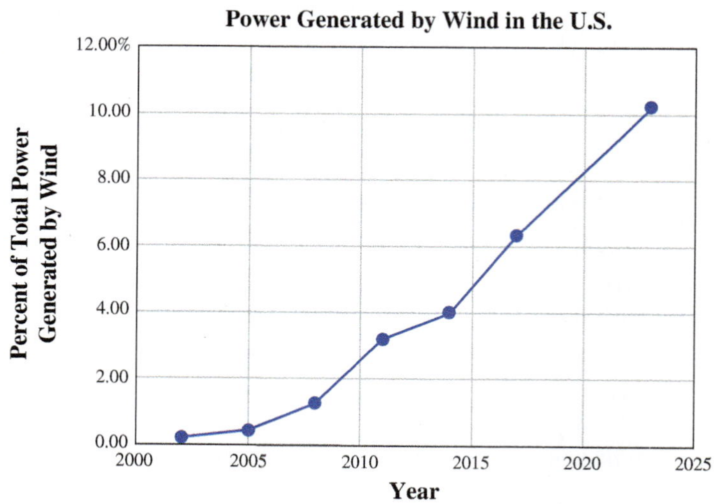

4-9 Applications

Name ______________________________

1. According to a pilot study of 20 people conducted at the University of Minnesota, daily doses of a compound called arabinogalactan over a period of 6 months resulted in a significant increase in the beneficial lactobacillus species of bacteria. Why can't it be concluded that the compound is beneficial for the majority of people?

2. Comment on this statement taken from a magazine advertisement: "In a recent clinical study, Cell-Tech was proved to be 1,950 percent better than creatine!"

3. In one advertisement, the following statement was made: "For every hundred women, 91 have taken the road less traveled." Comment on this statistic.

4. In many ads for weight-loss products, under the product claims and in small print, the following statement is made: "These results are not typical." What does this say about the product being advertised?

5. An article in a leading magazine stated that "When 18 people with chronic, daily, whiplash-related headaches received steroid injections in a specific neck joint, 11% had no more headaches." Think of a possible reason why the figure 11% was used.

4-9 Applications

Name ____________________

6. "Vitamin E is a proven antioxidant and may help in fighting cancer and heart disease." Is there anything ambiguous about this claim?

7. The results of a survey reported in *USA Weekend* stated that "9% would drive through a toll booth without paying," and "13% would steal cable television or inflate their resumes." Explain why these figures may not be representative of the population in general.

8. In a recent year, there were 6,067 male fatalities in the workplace and 521 female deaths. A government official made this statement: "Over 90 percent of the fatal injuries the past year were men, although men accounted for only 54 percent of the nation's employment." Can we conclude that women are more careful on the job? Explain.

9. At one point in the voting process for the 2015 baseball All-Star team, the Kansas City Royals had the leading vote getter for seven of the nine starting positions. Fans were allowed to vote online as many times as they wanted. Do you think it was likely that the vote was representative of the population as a whole? Why?

10. A stats instructor has recorded the scores of a recent quiz worth 20 possible points below.

20 18 18 15 15 18 15 15 15 20 19 20 15 20 19 0 0 18 19 15

a. If he wanted to be a jerk and tell his students the average on the quiz was terrible, which measure of average is he likely to use? Give that value in your explanation.

4-9 Applications

Name ______________________________

b. If he wanted to be a great guy and tell them the average on the quiz was fantastic, which measure of average is he likely to use? Give that value in your explanation.

For Questions 11–14, explain why each survey question might lead to an erroneous conclusion. Also suggest a better question.

11. "How often do you run red lights?"

12. "Do you think gun manufacturers should put safety locks on all guns sold even though it would increase the cost of the gun by 20%?"

13. "Do you think that it is not important to give extra tutoring to students who are not failing?"

14. "Which restaurant in town has the best food and the greatest service?"

4-9 Applications

Name ______________________________

15. The two graphs below were found in online discussions of calorie content in fast food. Which graph is misleading and why?

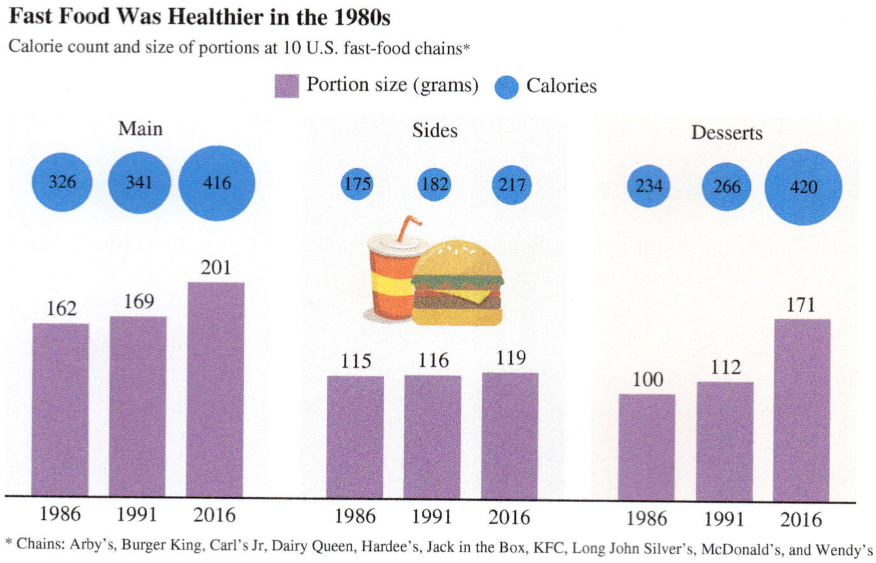

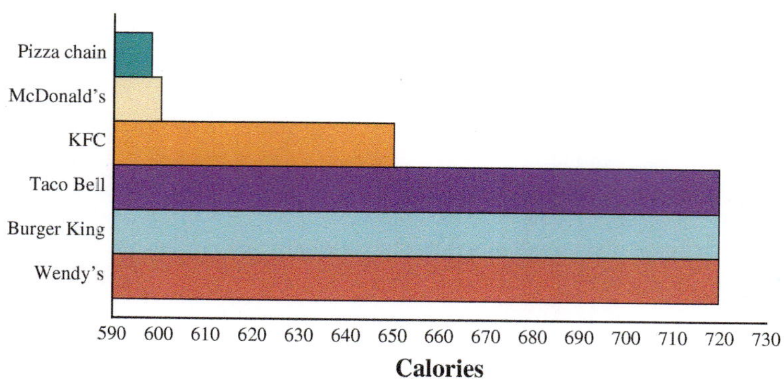

16. The graph to the right is one of the most intentionally deceiving graphs I've ever found. What about it makes that true?

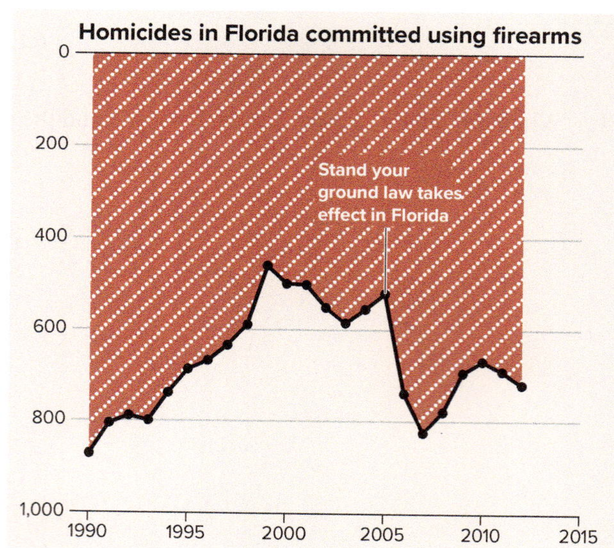

Source: Florida Department of Law Enforcement

Math in Sociology REVISITED

James W. Hall

1. The bar graph is below. It does in fact appear that homicides go up generally through spring and into summer, then decrease as summer ends.

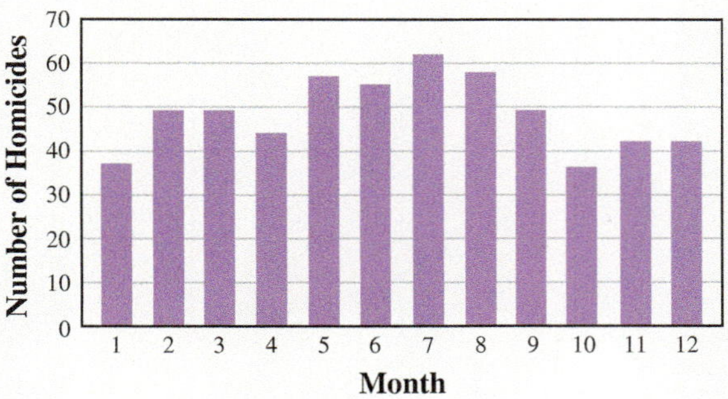

2. Mean: 48.3; median: 49.0; midrange: 49.0; mode: 49. Interestingly, for this data set all of the measures of average are amazingly similar. Because of the lack of variation here, you could choose any of them.
3. The scatter plot follows. There does appear to be a general trend from lower left to upper right, which would indicate the possibility that homicides tend to go up along with temperature.

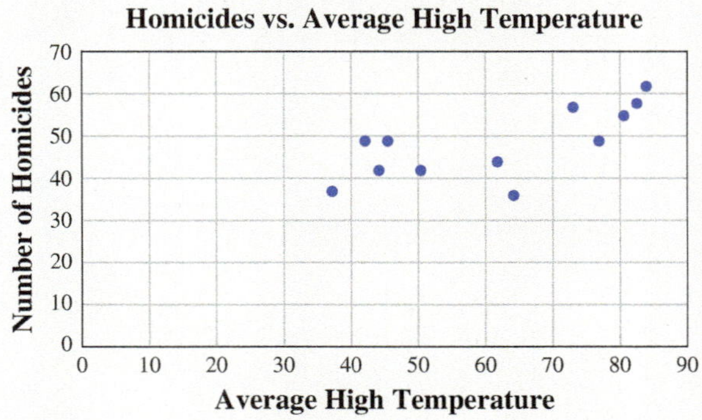

4. The correlation coefficient is $r = 0.710$. Looking at the table of significance values on page 502, for $n = 12$, the 5% significance value is 0.576 and the 1% value is 0.708. Since r in this case is bigger than both, we can be 99% sure that there is in fact a linear relationship between average high temperatures and number of homicides in Chicago.
5. Absolutely not. In fact, no matter how strongly the data sets are related, we can't draw any conclusions about whether one causes the other. All we know is that for some reason, they appear to be related. There are a wide variety of theories to account for this phenomenon, but none have been proven or even widely accepted.

Unit 5
Building Models

SofikoS/Shutterstock

Outline

Math in Drug Administration

Have you ever looked at the dosage information on a bottle of aspirin and thought "It just doesn't seem reasonable to recommend the same dosage for all adults"? People come in all shapes and sizes, and the effect of a certain dosage is in large part dependent on the size of the individual. If a 105-pound woman and her 230-pound husband both take two aspirin the morning after their wedding reception, she is in effect getting more than twice as much medicine as he is.

Because there's a lot of variation in the world, the study of algebra was created based on a brilliantly simple idea: using a symbol, rather than a number, to represent a quantity that can change. Since the word "vary" is a synonym for change, we call such a symbol a variable. The use of variables is the thing that distinguishes algebra from arithmetic and makes it extremely useful to describe phenomena in a world in which very few things stay the same for very long. An understanding of the basics of algebra is, in a very real sense, the gateway to higher mathematics and its applications. Simply put, you can only go so far with just arithmetic. To model real situations, expressions involving variables are almost always required.

In this unit, we'll review some key ideas from algebra, but one thing we won't do is find x just for the heck of it. A lot of students, maybe even most, believe that algebra was conceived as an abstract study, with early nerds sitting around simplifying expressions and solving equations on scrolls. Then a few centuries later, some genius said "Holy cow, we can use this stuff to find when two trains left San Francisco!"

Nothing could be farther from the truth. Algebra, and really all of classical math, was developed 100% as a tool for solving very real problems. So our study of algebra will focus on modeling, which is using algebra to represent situations in our world, like dosage calculations. You're accustomed to losing points when you make a mistake in a calculation, but outside the classroom, the stakes can be a LOT higher. The wrong dose of aspirin might upset your stomach, but more serious drugs carry with them more serious consequences. In many cases, an incorrect dosage could lead to death. Using the skills you learn in the chapter, you should be able to answer the following questions.

Suppose that a new pain-killing drug is being tested for safety and effectiveness in a variety of people. The recommended dosage for an average 5′ 10″, 170-pound individual is 400 mg. The manufacturer claims that the minimum effective dose for a person of that size is 250 mg, and that anything over 1,500 mg could be lethal. Two of the patients in the test are a 6′ 5″, 275-pound college football player recovering from a major injury, and a 4′ 2″, 65-pound girl being treated for sickle-cell disease. According to the manufacturer, dosages for this drug are proportional to body weight.

1. What would the recommended dosage be for each patient?
2. If the medications are mixed up and the recommended dosage for the football player is given to the child, is she in danger of dying?
3. Would the child's recommended dosage be at all effective for the football player?

For many medications, the effective dosage depends not just on body weight but also on body surface area (BSA), which accounts for the overall size of the patient. One model for body surface area is

$$BSA = \frac{\sqrt{1.15\,wh}}{60}$$

where w is weight in kilograms and h is height in centimeters.

4. Find the ratio of body surface area for the football player compared to the girl. How does it compare to the ratio of their weights?
5. Find the body surface area for an average 5′ 10″, 170-pound man.
6. If the dosages for the medication in the drug trial are based on body surface area rather than weight, rework Questions 2 and 3 to see if the child would be in danger, or if the medication would be at all effective for the football player.
7. Write any conclusions about using body surface area rather than weight when doing dosage calculations.

For answers, see Math in Drug Administration Revisited on page 651

Lesson 5-1 Prep Skills

SKILL 1: REDUCE FRACTIONS

The process of writing fractions in lowest terms is called reducing fractions. It's useful to do because it's easier to work with and interpret fractions when they're in lowest terms. For example, it's a lot easier to see that $\frac{5}{2}$ is two and a half than it is to see that $\frac{180}{72}$ is two and a half. To reduce a fraction, divide BOTH the numerator and the denominator by any common factors.

- $\frac{10}{15}=\frac{10\div 5}{15\div 5}=\frac{2}{3}$ The greatest factor common to 10 and 15 is 5.
- $\frac{40}{48}=\frac{40\div 8}{48\div 8}=\frac{5}{6}$ The greatest factor common to 40 and 48 is 8.

SKILL 2: MULTIPLY A WHOLE NUMBER BY A FRACTION

Multiplying a whole number by a fraction is an easily misunderstood problem. The multiplication $\frac{1}{2}\cdot 8$ should be thought of as one-half of eight. This expression is often misunderstood as $\frac{1}{2}\cdot\frac{8}{8}$ when, in fact, it is equivalent to $\frac{1}{2}\cdot\frac{8}{1}$. Once you've rewritten as the product of two fractions, multiply the numerators and denominators separately.

- $\frac{4}{7}\cdot 11=\frac{4}{7}\cdot\frac{11}{1}=\frac{44}{7}$
- $$\begin{aligned}\frac{12}{15}\cdot 3&=\frac{12}{15}\cdot\frac{3}{1}\\&=\frac{12}{3\cdot 5}\cdot\frac{3}{1}\\&=\frac{12}{5}\end{aligned}$$ Instead of multiplying 12 and 3 in the numerator, notice that 15 can be written as $3\cdot 5$. Now it's easier to see that the result can be reduced. Divide numerator and denominator by the common factor of 3, and the result is in lowest terms.
- $\frac{3}{2x}\cdot x=\frac{3}{2x}\cdot\frac{x}{1}=\frac{3}{2}$ Divide away the common factor of x from the numerator and the denominator.

SKILL 3: SOLVE A LINEAR EQUATION

Solving equations involves performing a series of algebraic steps that produce simplified but equivalent equations. Remember, equivalent equations are equations that have the same solution(s).

Here is a general strategy that will work for solving any linear equation. You may not need all of these steps, but working through the steps you do need, in order, will lead to a solution if one exists.

Step 1: Simplify each side of the equation.

a. If the equation contains fractions, simplify by multiplying both sides of the equation by the least common denominator (LCD) of all the fractions.

b. Look at the two sides of the equation to see if they can be simplified. If so, simplify by using the distributive property to remove any grouping symbols and then combine like terms.

Step 2: By adding or subtracting the same quantities from both sides of the equation, isolate the variable terms on one side of the equation and the constant terms on the other side.

Step 3: By multiplying or dividing both sides of the equation by the same quantity, make sure that the variable is all by itself on one side. This solves the equation produced in step 2.

The following equation requires only step 2.

- Solve $x + 3 = 8$.

$$x + 3 - 3 = 8 - 3$$
$$x = 5$$

Isolate x by subtracting 3 from both sides of the equation. $x + 3 = 8$ and $x = 5$ are considered equivalent equations. The solution is $x = 5$. This means both sides of the original equation will have the same value when $x = 5$, making the equation a true statement.

The following equation requires only step 3.

- Solve $\frac{x}{2} = 5$.

$$\frac{x}{2} \cdot 2 = 5 \cdot 2$$
$$x = 10$$

Isolate x by multiplying both sides of the equation by 2.

The following equation requires all three steps.

- Solve $4(3x - 2) = 52$.

$$4 \cdot 3x - 4 \cdot 2 = 52$$

Use the distributive property.

$$12x - 8 = 52$$
$$12x - 8 + 8 = 52 + 8$$

Add 8 to both sides.

$$12x = 60$$
$$\frac{12x}{12} = \frac{60}{12}$$

Divide both sides by 12.

$$x = 5$$

PREP SKILLS QUESTIONS

Write each fraction in lowest terms:

1. $\frac{70}{100}$
2. $\frac{6}{10}$
3. $\frac{18}{48}$

Perform each multiplication.

4. $\frac{9}{14} \cdot 7$
5. $y \cdot \frac{11}{y}$

Find all solutions to each equation.

6. $8x = 30$
7. $-4(2x - 3) = 10$

Lesson 5-1 Keeping Things in Proportion

LEARNING OBJECTIVES

- ☐ 1. Compare two quantities using ratios.
- ☐ 2. Describe the value of using ratios to compare quantities rather than differences.
- ☐ 3. Solve proportions.
- ☐ 4. Solve problems that involve proportional reasoning.

Digital Vision/SuperStock

One of the tricks of life is to have sense and money in roughly equal proportions.
—Craig Brown

The most obvious way to compare the sizes of two numbers is to subtract them. But is that the best way? Suppose you're comparing the cost of an item at two different stores, and you find that the item is a dollar more at Target than at Wal-Mart. If that item is a bottle of Coke, and the prices are \$1 and \$2, that dollar difference is significant. You'd most likely feel pretty strongly about buying from the cheaper store.

But if the item is a state-of-the-art laptop and the prices are \$1,201 and \$1,200, would you really care? Those two prices are essentially the same. It's still a dollar difference, but that difference just feels inconsequential. If you divide the prices rather than subtract them, however, something interesting happens:

$$\text{Coke: } \frac{\$2}{\$1} = 2 \quad \text{Laptop: } \frac{\$1{,}201}{\$1{,}200} = 1.0008$$

Do you see the point? For the Coke, a dollar more is twice as much—kind of a big deal. For the laptop, when you divide the prices, you basically get 1, meaning the two prices are pretty much the same. Conclusion? The most meaningful way to compare the sizes of two numbers is to divide them, forming what we call a *ratio*. The study of ratios as a means of comparing numbers will lead us into a type of equation that can be used in a lot of modeling situations, the *proportion*.

0. Think of another situation where dividing two numbers might provide a more useful comparison than subtracting them.

5-1 Class

Ratios

A **ratio** is a comparison of two quantities using division. For example, according to a survey conducted by the Pew Research Center, for every 71 Americans who reported some religious affiliation, there were 29 who did not. We would say that the ratio of those with an affiliation to those without is 71 to 29.

Ratios are typically written in two different ways: using a colon between the quantities, or in fractional form. So we could write the ratio in the previous paragraph as either

$$71{:}29 \text{ or } \frac{71}{29}$$

Notice that when we're writing the ratio of those with a religious affiliation to those without, the number associated with the first group (those with) comes before the colon and is the numerator of the fraction.

In this course, we'll mostly use fractional form so that we can do arithmetic and algebra with ratios. Notice that if you perform the division in the ratio 71/29, you get about 2.4 in decimal form (rounded to one decimal place). This shows us that there are about two and a half times as many Americans with a religious affiliation as without. And this brings us full circle on what the point of ratios is: comparing sizes of quantities. Nice.

Solved Example 1

Writing and Interpreting Ratios

From 1969 through 1977, 4 of the teams in Major League Baseball made the playoffs each year, while 20 did not. Now, 12 make the playoffs and 18 do not.

a. Write the ratio of teams making the playoffs to the total number of teams in 1969 and today. Write each ratio in fractional form, then reduce the fraction if you can.

b. Describe what those ratios tell us about how likely a given team is to make the playoffs now compared to the mid-1970s.

SOLUTION

a. There are two keys to answering this question correctly. The first is understanding how to write ratios, the second is reading the question carefully. Because the first sentence of the statement gives us the numbers 4 and 20, a lot of students want to write the ratio 4/20. But that's not the ratio that was asked for, because 20 is the number of teams NOT making the playoffs. The ratio we're asked for here is the number of teams making the playoffs to the TOTAL number of teams. Since 4 made the playoffs and 20 did not, there were 24 teams total, and the ratio of teams that made the playoffs to the total number of teams from 1969 to 1977 was 4/24, which reduces to 1/6.

Now there are 12 teams that make the playoffs and 18 that do not, which means there are 30 teams total. So the modern ratio of teams making the playoffs to teams not making the playoffs is 12/30, which reduces to 2/5.

b. Here's a simple interpretation of what those two ratios tell us: From 1969 to 1977, one out of every six teams made the playoffs. Now, two out of every five teams make the playoffs. The decimal forms of these ratios are 0.17 (to two decimal places) and 0.4. This makes it easier to see that a team today is more than twice as likely to make the playoffs as a team from 1969 to 1977.

1. In October 2024, 25 of the 100 U.S. senators were women. Write and reduce the ratio of female senators to male senators.

2. Why was it useful to reduce the ratios in Solved Example 1 and Question 1?

3. A sub shop advertises a new 18-inch sandwich. Big! Not to be outdone, a rival chain comes out with a 2-foot-long sandwich. Explain why writing the ratio below to compare the size of the sandwiches is deceiving:

$$\frac{2}{18}$$

4. How should you write the ratio in order to make it meaningful as a size comparison? What does this tell us about the importance of units when writing ratios?

Did You Get It

Try this problem to see if you understand the concepts we just studied. The answer can be found at the end of the portfolio section.

1. Write each ratio as a fraction and reduce if possible. Think about consistency of units before writing!
   a. The U.S. Department of Health and Human Services reported in 2023 that 8 out of 100 Americans did not have health insurance. Write the ratio of those that have health insurance to those that do not.
   b. Write the ratio of 40 ounces to 2 pounds.

Proportions

A **proportion** is an equation stating that two ratios are equal. In some cases involving only numbers, a proportion can be obviously true. For example, you most likely know that the ratios 1/2 and 4/8 are identical: if something occurs four out of every eight times, that means the same as saying that it occurs one out of every two times. In other words,

$$\frac{1}{2} = \frac{4}{8}$$

This is a simple example of a proportion that happens to be true.

Proportions are a bit more interesting when they contain variable quantities. We know that when you flip a coin, on average it will land on heads half the time. So if you flip a coin 80 times, how many times would you expect it to land on tails? If you said 40, that's great. Here's what that question would look like in proportion form:

$$\frac{1}{2} = \frac{t}{80}$$

where t represents the number of times that tails would come up. The solution to this equation (which is a value of the variable quantity t that makes the equation true) is the number of times we'd expect to get tails.

It turns out that there's a simple method for solving proportions that contain a variable quantity, and this method will come in really handy when we get to the key idea in this lesson, which is using proportions to solve applied problems. So before we get to the applied part, we'll practice the algebra part.

The goal will be to solve the proportion below, and we'll develop a simple method for doing so along the way.

$$\frac{12}{60} = \frac{3}{y}$$

Solved Example 2

Building a Method for Solving Proportions

a. Multiply both sides of the proportion $\frac{12}{60} = \frac{3}{y}$ by 60, then by y.

b. Explain why this was useful in solving the equation.

c. Find the solution.

SOLUTION

a. First, we're instructed to multiply both sides by 60:

$$60 \cdot \frac{12}{60} = \frac{3}{y} \cdot 60$$

60 divided by 60 is 1; 3 times 60 is 180.

$$12 = \frac{180}{y}$$

Next we were asked to multiply both sides by y:

$$y \cdot 12 = \frac{180}{y} \cdot y$$

y divided by y is 1.

$$12y = 180$$

Solved Example 2 *Continued*

b. If it's not reasonably clear why this was useful, compare the equation we now have with the one we started with, keeping in mind that the goal is to solve the equation. In short, it's a LOT easier now. In particular, no more fractions! That's the good part. By multiplying both sides first by the denominator on the left, then by the denominator on the right, we cleared away all the fractions, leaving a much simpler equation to solve.

c. Now all we need to do is divide both sides by 12 to isolate y, which makes the solution obvious:

$$\frac{12y}{12} = \frac{180}{12} \Rightarrow y = 15$$

The solution to the proportion is $y = 15$. Remember: That means 15 is the only value you can plug in for y to make the proportion a true statement.

So there's a perfectly good, reliable method for solving a proportion: Multiply BOTH sides first by the denominator on the left, then by the denominator on the right. After that all the fractions will be gone (if you multiplied correctly, that is) and you'll be left with an easier equation, which you of course still have to solve.

Now we can make an interesting observation that you might find helpful in solving proportions. Look carefully at the original proportion and the resulting equation after we multiplied to clear fractions:

$$\frac{12}{60} = \frac{3}{y} \qquad 12y = 180$$

Notice that the product on the left, $12y$, is the product of the first numerator and the second denominator. Notice also that the number on the right, 180, is the product of the second numerator and the first denominator. This diagram might help you to visualize the result:

$$\frac{12}{60} = \frac{3}{y}$$

Think of this as multiplying along those diagonal arrows. It turns out that this will always result in the same equation as multiplying both sides by the two denominators, like we did in Solved Example 2. This method goes by the name **cross-multiplying,** and if it makes sense to you, then it provides a shortcut for solving proportions. But keep in mind that cross-multiplying is not a real math procedure! It's just a trick that happens to work because it accomplishes the same thing as multiplying both sides by the two denominators one at a time.

5. Use cross-multiplying to solve the proportion from Solved Example 2.

6. Solve this proportion: $\frac{x-5}{10} = \frac{x+2}{20}$. You can use either cross-multiplying, or multiply by each of the denominators one at a time. Try to decide which method works best for you.

7. How can you check to see if you solved the proportion in Question 6 correctly? Explain, then do it.

Did You Get It

2. Use cross-multiplying to solve the proportion:

$$\frac{x+6}{15} = \frac{x-2}{5}$$

5-1 Group

If it doesn't make you uncomfortable, exchange the following information with the classmates in your Unit 5 group. This will be your small group for the fifth unit. It would be a good idea to schedule a time for the group to meet to go over homework, ask/answer questions, or prepare for exams. You can use this table to help schedule a mutually agreeable time.

Name	Phone Number	Email	Available Times

Applications of Proportions

The reason that we practiced solving proportions, and in fact the reason this lesson is even in the course, is that there are TONS of applied problems that can be solved by setting up and solving a proportion. In fact, proportions have been around for a really long time in one form or another—a written record of their use goes back at least to 400 BCE or so, but most math historians feel that the idea is almost as old as formal numeric thought. That doesn't happen unless a topic is legitimately useful.

Here's the first problem we'll study: While on a spring break trip, a group of friends burns 12 gallons of gas in the first 228 miles, then stops to refuel. If they have 380 miles yet to drive, and their SUV has a 21-gallon tank, can they make it without refueling again?

Keep in mind that a proportion is an equation stating that two ratios are equal. Keep in mind also that ratios are used to compare the sizes of quantities. So a good way to recognize when proportions might help to solve a problem is to identify situations that are comparing the sizes of related quantities. That's certainly the case here: There's a comparison between gas used and miles traveled. There's also a comparison between the distance traveled in the first segment of the trip, and the distance that can be traveled after refueling. If we start writing some comparisons, maybe we'll be able to write an equation.

1. We were given a comparison of gallons of gas burned and miles driven for the first part of the trip. Write a ratio in fraction form describing this comparison, with gallons of gas in the numerator.

NithidPhoto/iStock/Getty Images

2. Next we'd like to write a similar ratio that compares the amount of gas to the number of miles driven for the second part of the trip. We know the amount of gas that they have to burn—it's the capacity of the tank. What we don't know is how many miles they can drive with that much gas. This is exactly the kind of situation where using a letter to represent an unknown quantity is perfect! Use m for the number of miles, and write a ratio of gas used to miles driven for the second part of the trip.

3. Explain why the two ratios that you wrote in Questions 1 and 2 should have the same value. Then set the two ratios equal to each other and solve the resulting proportion.

4. Use your result to decide if our friends can make the rest of the trip without refueling. Write a sentence explaining your reasoning.

The procedure outlined in Questions 1–4 is a pretty darn good model for how to solve proportion problems: It's usually about writing ratios by finding information that compares the sizes of quantities, and recognizing that there's a connection between ratios allowing you to set them equal to each other.

In order to decide that a certain species is endangered, biologists have to know how many individuals are in a population. But how do they do that? It's actually an interesting application of proportions illustrated in our next example problem.

As part of a research project, a biology class plans to estimate the number of fish living in a lake thought to be polluted. They catch a sample of 37 fish, tag them, and release them back into the lake. A week later, they catch 80 fish and find that 6 of them are tagged. About how many fish live in the lake?

5. Using the letter f to represent the total number of fish in the lake, write a ratio comparing the total number of fish in the lake to the number that are tagged.

Photo by Scott Bauer, USDA-ARS

6. Write a ratio that compares the number of fish that were caught a week later to the number that were tagged.

7. Explain why the two ratios that you wrote in Questions 5 and 6 should have the same value. Then set the two ratios equal to each other and solve the resulting proportion.

8. Use your result to answer the original question posed in this problem. Make sure your answer makes sense in the problem!

Math Note

One good way to make sure that your answer to an application problem makes sense is to write your answer in the form of a sentence by rephrasing the original question. Do that here!

Did You Get It

3. In 2024, roughly 14 of every 100 people in the United States were African-American. A marketing company wants to select a group of 250 people that accurately reflects the racial makeup of the country. How many African-Americans should be included?

There are 39 grams of sugar in one 12-oz can of Coca Cola. If you're like most Americans, that doesn't come as much of a surprise. Not because you have a good perspective on how much sugar they pack in there—because you have no idea how much sugar that is. Let's check it out.

9. There are just about 21 grams of sugar in 4 teaspoons of regular granulated sugar. How many teaspoons of sugar are in one can of Coke? To find out, set up and solve a proportion.

10. How many grams of sugar would you ingest if you drank exactly one can of Coke every day for an entire year?

11. One hundred grams is just about 0.22 pounds. How many 5-pound bags of sugar would you ingest in a year if you drank that single can of Coke each day? Show all work, and make sure you're using a proportion, seeing as that's like, the point.

12. There are about 12.5 grams of sugar in one standard-sized Reese's peanut butter cup. How many of those would you have to eat in a year to get the same amount of sugar that you get in a daily can of Coke?

Did You Get It

4. According to a study reported by Wallethub in 2024, the metropolitan areas with the highest and lowest proportions of adults suffering from obesity were McAllen, TX, and Asheville, NC, respectively. In McAllen, for every 56 people with obesity, there were 69 without. In Asheville, for every 37 people with obesity, there were 163 without. At that time, the estimated populations were 146,593 for McAllen and 95,056 for Asheville. How many more people were suffering from obesity in McAllen than in Asheville?

Understanding ratios can be really helpful in making sure that you've set up a proportion problem correctly. We'll illustrate the idea with one final example.

Based on a combination of worldwide surveys conducted from 2021 to 2023 and reported at Gallup.com, for every 84 adults that would be happy to stay in the country they live in, 16 would prefer to emigrate to a different country. At the time, the adult population of the world was very close to 6 billion. Based on these results, how many of those people were happy where they lived?

First, we write a ratio based on the survey:

$$\frac{\text{Want to stay}}{\text{Want to leave}} \quad \frac{84}{16}$$

Math Note

When writing a ratio, writing it first in words is a really great way to keep track of what it represents.

If we use the letter h to represent the variable quantity representing the number of adults worldwide that are happy in the country where they live, it would seem reasonable to write the ratio

$$\frac{h}{6 \text{ billion}}$$

13. This is a fine ratio. It's delightful. But it won't help answer the question. Write the ratio in word form, like the one above, and explain why.

To avoid inconsistency in setting up proportions, we can use a snazzy rule:

The Rule of Matching Parts

When setting up a proportion, read the numbers straight across the numerators. There should be a connection between the pieces of information they represent. Repeat for the denominators. Then read up and down on each side of the equation—the same should be true.

If we set up a proportion using the two ratios above and apply the rule of matching parts, we can see that we've made a mistake.

	Numbers from survey		Overall population	
Want to stay →	84		h	← Want to stay
Want to leave →	16		6 billion	← Total adults

$$\frac{84}{16} = \frac{h}{6 \text{ billion}}$$

Reading up and down, we're good: Both parts on the left are numbers from the survey, and both parts on the right are numbers from the overall population. Reading across the numerators, we're also good: Both numbers represent folks who want to stay in their country. It's reading across denominators that we find our mistake: On the left, we have people who want to leave, on the right, we have total adults. The easiest fix? Change the 16 in the left-side denominator to 100, which represents a total from the survey ($16 + 84 = 100$). We'll ask you to make this change then answer the question in the Applications.

5-1 Portfolio

Name ______________________________

Check each box when you've completed the task. Remember that your instructor will want you to turn in the portfolio pages you create.

Technology

1. ☐ In this Excel activity, we'll calculate the unit price of several different items to determine the better buy for each product.

ALEKS Practice

1. ☐ Include any written work from the ALEKS assignment along with any notes or questions about this lesson's content.

Applications

1. ☐ Complete the applications problems.

Reflections

Type a short answer to each question.

1. ☐ Why is it more useful to compare the sizes of quantities using division than subtraction?
2. ☐ What kinds of problems are likely to be solvable using proportions?
3. ☐ Describe why units are so important in writing ratios.
4. ☐ Name one thing you learned or discovered in this lesson that you found particularly interesting.
5. ☐ What questions do you have about this lesson?

Looking Ahead

1. ☐ Complete the prep skills for Lesson 5-2.
2. ☐ Read the opening paragraph in Lesson 5-2 carefully and answer Question 0 in preparation for that lesson.

Answers to "Did You Get It?"

1. a. $\frac{23}{2}$ **b.** $\frac{5}{4}$ **2.** $x = 6$ **3.** 35 **4.** 48,088 people

Answers to Prep Skills Questions

1. $\frac{7}{10}$
2. $\frac{3}{5}$
3. $\frac{3}{8}$
4. $\frac{9}{2}$
5. 11
6. $x = \frac{15}{4}$
7. $x = \frac{1}{4}$

5-1 Applications

Name ______________________________

In Questions 1–3, a comparison of two quantities is given. Write a ratio and reduce if possible. Make sure the comparisons are consistent in terms of units.

1. 3 pounds to 12 ounces

2. 12 years to 2 decades

3. 5 feet to 30 inches

Near the end of this lesson, we began to study this question: For every 84 adults who would be happy to stay in the country they live in, 16 would prefer to emigrate to a different country. At the time, the adult population of the world was very close to 6 billion. Based on these results, how many of those people were happy where they lived?

4. Review the work we did on page 550 and use it to solve this problem. Then find the number of people who are interested in moving to another country.

5. Among the people in the survey who wanted to move, 9 out of every 50 listed the U.S. as the country they'd most like to emigrate to. Based on this, and on what you learned in Question 4, how many people worldwide would like to move to the U.S.?

5-1 Applications

Name ____

6. A true story about how I recently used a proportion: My car has one of those touch-screen thingies for maps, music, and so forth, and you can upload your own picture to be the background screen. The catch is that the size of the picture has to be 800 pixels wide by 384 pixels high, or it gets all stretched out and looks goofy. Actually, the picture doesn't have to be EXACTLY those dimensions: It just has to have the same ratio of width to height. The most recent picture I chose to use as background screen had a native width of 1,164 pixels. To what height did I need to crop it?

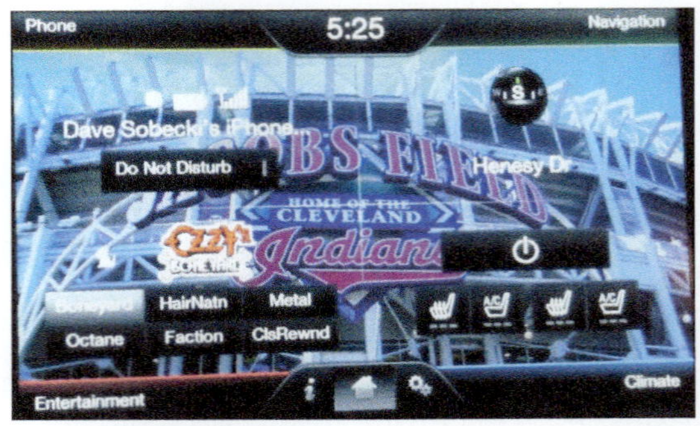

7. A small college has 1,200 students and 80 professors. The college is planning to increase enrollment to 1,500 students next year. How many new professors should be hired to keep the ratio of students to professors the same?

Lesson 5-2 Prep Skills

SKILL 1: EVALUATE ALGEBRAIC EXPRESSIONS

To evaluate an expression (or a formula), replace all occurrences of the variable or variables with specific numbers. This allows us to find a numeric value.

- Evaluate $-4x + 5$ for $x = 3$. — This means to find the value of $-4x + 5$ when $x = 3$.

  $-4(3) + 5 = -12 + 5 = -7$ — Replace x with 3 and simplify the result.

- Evaluate $3x^2 - x + 4$ for $x = -2$.

  $3(-2)^2 - (-2) + 4$ — Replace x with -2 and simplify the result.
  $= 3 \cdot 4 + 2 + 4$
  $= 12 + 2 + 4$
  $= 18$

SKILL 2: SOLVE A LINEAR EQUATION

As we have seen in previous lessons, solving equations involves performing a series of algebraic steps that produce simplified but equivalent equations. Remember, equivalent equations are equations that have the same solution(s).

Here is a general strategy that will work for solving any linear equation. You may not need all of these steps, but working through the steps you do need, in order, will lead to a solution if one exists.

Step 1: Simplify each side of the equation.

a. If the equation contains fractions, simplify by multiplying both sides of the equation by the least common denominator (LCD) of all the fractions.

b. Look at the two sides of the equation to see if they can be simplified. If so, simplify by using the distributive property to remove any grouping symbols and then combine like terms.

Step 2: By adding or subtracting the same quantities from both sides of the equation, isolate the variable terms on one side of the equation and the constant terms on the other side.

Step 3: By multiplying or dividing both sides of the equation by the same quantity, make sure that the variable is all by itself on one side. This solves the equation produced in step 2.

The following equation requires steps 2 and 3.

- Solve $1 = 2x + 9$.

  $1 - 9 = 2x + 9 - 9$ — Subtract 9 from both sides.
  $-8 = 2x$
  $\frac{-8}{2} = \frac{2x}{2}$ — Divide both sides by 2.
  $-4 = x$ — The solution is $x = -4$.

SKILL 3: WRITE INEQUALITIES

Writing inequalities involves correctly interpreting words that indicate inequalities, which can be a little confusing. Reading them carefully and thinking about the meaning is the best way to interpret these phrases accurately. ***Less than*** ($<$), ***greater than*** ($>$), ***less than or equal to*** ($\le$), and ***greater than or equal to*** ($\ge$) are straightforward because they mean exactly what they say. Other phrases like **at most** ($\leq$) and **at least** ($\geq$) can be used to indicate inequalities.

Write each statement as an inequality.

- A is a number greater than 20.

  $A > 20$ — The symbol $>$ is used to indicate greater than.

- B is a number less than or equal to 30.

  $B \leq 30$ — The symbol $\leq$ is used to indicate less than or equal to.

- C is a number that is at most 100.

  $C \leq 100$ — "At most" indicates less than or equal to.

Between is a way of indicating two inequalities at the same time. This is called a compound inequality or a three-part inequality. In this case, always write the smaller number first, the less than symbol, the given variable, the less than symbol, and then the larger number. If you stick to only using less than symbols you can't go wrong.

Write the statement as an inequality.

- X is a number between 25 and 40.

  $25 < x < 40$ — If you cover up the "< 40" part, this says that x is greater than 25. If you cover up the "$25 <$" part, this says that x is less than 40. So overall, this expression is equivalent to saying that x is both greater than 25 and less than 40.

PREP SKILLS QUESTIONS

1. Evaluate each expression for the given values of x.

   $2x + 12; x = -10, 0, 10$

   $700 - 0.3x; x = 0, 350, 1{,}100$

   $x^2 + 7x - 3; x = -3, 0, 5$

   $-2x^2 + 400; x = 0, 10, 20$

2. Find all solutions to each equation.

   $2x + 12 = 0$

   $0 = 700 - 0.3x$

3. Write each statement as an inequality.

   x is a number less than 140.

   y is a number greater than or equal to -10.

   z is a number between 4 and 12.

Lesson 5-2 Making Some Extra Cash

LEARNING OBJECTIVES

- ☐ 1. Demonstrate an understanding of the significance of a rectangular coordinate system.
- ☐ 2. Describe what the graph of an equation is.
- ☐ 3. Use and interpret function notation.
- ☐ 4. Graph and interpret linear functions.
- ☐ 5. Graph and interpret quadratic functions.

Ingram Publishing/SuperStock

A dead tree, cut into planks and read from one end to the other, is a kind of line graph, with dates down one side and height along the other, as if trees, like mathematicians, had found a way of turning time into form.
—Alice Oswald

Pretty much everyone I know has been in a situation at least once in their lives where they really need some quick cash. When you find yourself in that position, what to do? Picking up a temporary job would certainly be one way to raise some funds. In a lot of locations, business owners pay people to stand near a busy intersection to advertise the services offered by their company. Sometimes workers hand out flyers or coupons that are designed to encourage customers to stop in. Of course, the boss would want to know that the time he or she is paying for is being well-spent, so it would be reasonable to keep track of how many customers they get from their investment. In turn, this might affect how much money you could make from the job. If you're handing out coupons for $2 off at a local restaurant, your pay would most likely depend on how many of those coupons get used by actual customers.

This is the type of situation that the field of math modeling was designed to study. What does the relationship between the number of coupons used and the amount of money you earn look like? In this lesson, after reviewing some basic graphing skills, we'll dive into the study of math modeling by studying questions about how the size of one quantity affects another.

0. What do you think the term "math modeling" means?

5-2 Class

We've already talked about scatter plots in this course, but before we dive in on graphing equations, let's review some basic facts about the rectangular coordinate system.

The Rectangular Coordinate System

The foundation of graphing in math is a system for locating data points using a pair of perpendicular number lines. We call each one an **axis.**

1. The horizontal number line is called the __________, the vertical number line is called the __________, and the point where the two intersect is called the __________.

Collectively, they form what is known as a **rectangular coordinate system,** sometimes called the **Cartesian plane.** The two axes divide the plane into four regions called **quadrants,** which we number using Roman numerals I, II, III, and IV as shown below.

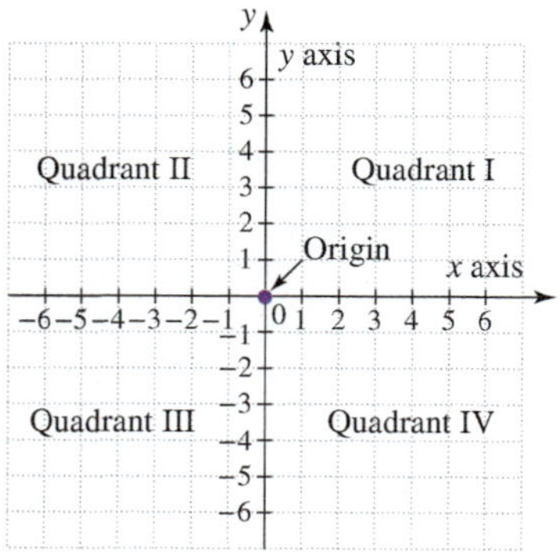

When visualizing data, it's often natural to have two quantities paired together (like the number of coupons that get used and the amount of money you make). We can associate pairs of numbers in a rectangular coordinate system with points by locating each number on one of the two number lines that make up the coordinate system. We call the two numbers **coordinates** of a point. Together they form an **ordered pair,** and we write them as (x, y), where the first numeral describes a number on the x axis and the second describes a number on the y axis.

2. The coordinates of the origin are ________.

3. A point P whose x coordinate is 2 and whose y coordinate is 5 is written as $P =$ (___, ___).

4. The point in Question 3 is plotted by starting at the origin and moving two units ________ and five units ________.

5. Negative coordinates correspond to negative numbers on the axes, so a point like $(-5, -4)$ is plotted by starting at the origin, moving five units ________ and four units ________.

6. Plot and label the points from Questions 3 and 5 on the figure to the right.

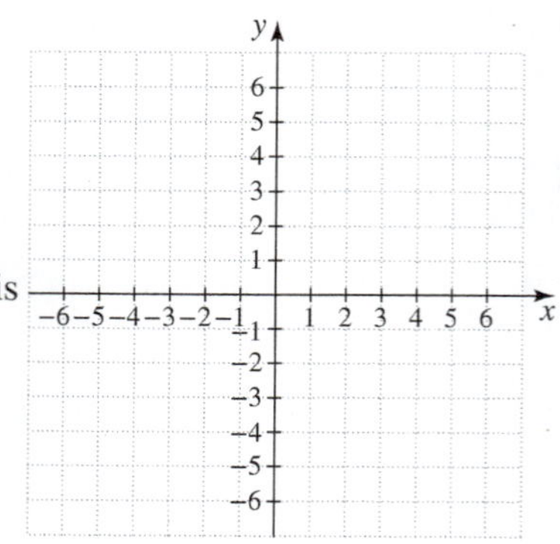

7. The four quadrants in a rectangular coordinate system can be described by noting the signs of the coordinates of points in those quadrants. For example, points in the first quadrant live to the right of the origin, meaning the first coordinate is positive. Use this idea to fill in the table below.

Quadrant	Sign of *x* Coordinate	Sign of *y* Coordinate
I	Positive	
II		
III		
IV		

8. Now look at points that live on the x axis. They all have one thing in common. What is it?

9. What do all the points on the y axis have in common?

Solved Example

Modeling with a Linear Equation

Suppose you get offered six bucks to stand on a street corner and hand out flyers for a restaurant, plus an extra \$2 for every customer that brings your flyer into the restaurant and orders a meal.

a. Write an equation that describes the amount of money you'll earn if x of your flyers get used. Use the letter y to represent the amount made.

b. Substitute $x = 4$ and $y = 14$ into your equation. Is the resulting statement true or false?

c. Substitute $x = 14$ and $y = 4$ into your equation. Is the resulting statement true or false?

SOLUTION

a. Let's start with the easy part: You're getting paid \$6 regardless of how many flyers are used, and you'll get extra if any actually are used. So the amount you make better look like $y = 6 +$ something. We just have to figure out what the "something" is. Think of it this way: Since you get \$2 for each flyer used, how much extra (beyond the \$6) would you make if three were used? Six dollars, right? Three times \$2. What if 10 were used? Ten times \$2, or \$20. See the pattern? The extra amount you make is \$2 times the number of flyers turned in. We were told to use x to represent that number of flyers turned in, so the extra amount earned is 2 times x, which we write as $2x$.

Now we know how to model the amount we'll earn: Add the initial \$6 to the $2x$ dollars earned from x flyers turned in.

$$y = 6 + 2x$$

b. Replacing y with 14 and x with 4, we get

$$14 = 6 + 2(4) \quad \text{or} \quad 14 = 6 + 8$$

This is a true statement.

c. Replacing y with 4 and x with 14, we get

$$4 = 6 + 2(14) \quad \text{or} \quad 4 = 6 + 28$$

This is a false statement.

Linear Functions

Two really important things happened in parts (b) and (c) of the Solved Example. First, we found that if we substitute $x = 4$ and $y = 14$ into the equation, the result is a true statement. This tells us the following:

- The pair (4, 14) is called a **solution** to the equation.
- We say that (4, 14) **satisfies** the equation.
- The point (4, 14) is on the graph of the equation $y = 2x + 6$.

Second, we found that substituting $x = 14$ and $y = 4$ into the equation makes it false. This tells us that

- The pair (14, 4) is not a solution to the equation.
- The pair (14, 4) does not satisfy the equation.
- The point (14, 4) is not on the graph of the equation $y = 2x + 6$.

Here's a list of some pairs that satisfy the equation $y = 2x + 6$:

(0, 6), (1, 8), (2, 10), (3, 12), (4, 14)

10. Plot the points corresponding to these pairs. What do you notice?

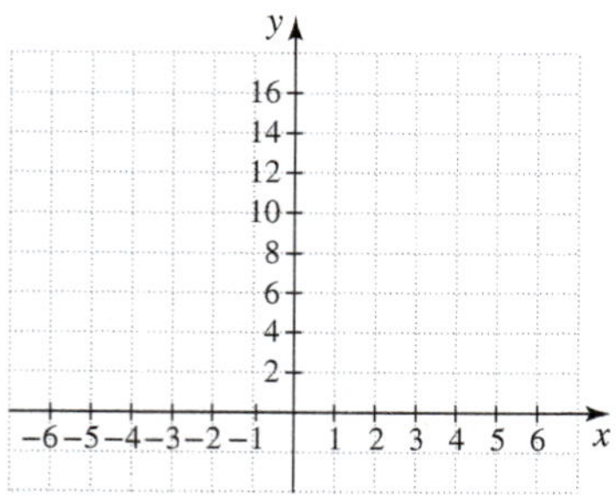

Hopefully, you noticed that all of the points appear to line up in a straight-line pattern. This is not a coincidence. In fact, for this reason, equations like $y = 2x + 6$ are called **linear equations.** (More on that later.)

11. Connect the points you plotted above with a line.

12. Explain why in this case it's appropriate to include an arrow on the right end of the line you drew, but NOT the left end.

The result of plotting some sample points and then making an educated guess on how to fill in the gaps is called the *graph of the equation.* This is one of the most important ideas in all of math, so it deserves a cool definition box:

The **graph of an equation** is a way to geometrically represent every pair of numbers that is a solution to the equation. Each of those pairs corresponds to a point on the graph.

We may not have realized it, but we've been working with a really important concept in math for quite some time now. Finally, we'll give it a name and symbols.

A **function** is a relationship between two quantities where each input produces a unique output.

In the case of that job handing out flyers, for every possible number of those bad boys that can be redeemed at the restaurant, there corresponds a specific amount of money that you'll make. In this case, the number of flyers redeemed is the input, and the resulting output is your pay. In order to make it clear that a given equation actually defines a function, we use *function notation.*

Function Notation

For a given input x, the output of a function named f is described by the symbol $f(x)$, which is read "f OF x," NEVER "f TIMES x."

The use of the word "of" is to indicate that every output for a function is a result **OF** applying that function to a given input. The symbol $f(x)$ does NOT indicate multiplication; it's a way to indicate that the function f determines the output for a given input x.

13. For example, the function describing your pay from handing out x flyers could be written as $f(x) =$ __________.

14. In that case, $f(20)$ represents the output of the function (amount of money you make) when the input is 20 (meaning 20 flyers get handed out and turned in). Compute the value of $f(20)$. This is called **evaluating a function for a particular value of the input.** Then describe what your result means in context.

Another advantage of function notation is efficiency. By writing $f(8) = 22$, we can communicate a bunch of information in one clean notation.

15. Write a sentence explaining what is meant by $f(8) = 22$ in the context of this problem.

Did You Get It

Try this problem to see if you understand the concepts we just studied. The answer can be found at the end of the portfolio section.

1. The tuition at a community college is based on a flat fee of \$200, plus an additional \$120 for each credit hour you register for. The cost of registering for x hours can then be modeled by $T(x) = 200 + 120x$. Find $T(3)$ and $T(15)$, and describe what each result means.

Using Technology: Finding Ordered Pairs

Graphing Calculator

The table feature on a graphing calculator can be used to quickly find points on a graph.

1. Enter the equation next to **Y1** on the **Y=** screen.
2. Access the Table Setup by pressing **2nd, WINDOW.** Decide on the lowest input value you want an output for, and enter it next to **TblStart.** Then decide on the distance you want between input values, and enter it next to **ΔTbl** for your input values
3. View the table by pressing **2nd GRAPH.**
4. Selecting **Ask** next to **Indpnt** in the **TABLE SETUP** screen will allow you to input any x values and have the calculator calculate the corresponding y values. Hey, that's probably why it's called a calculator!

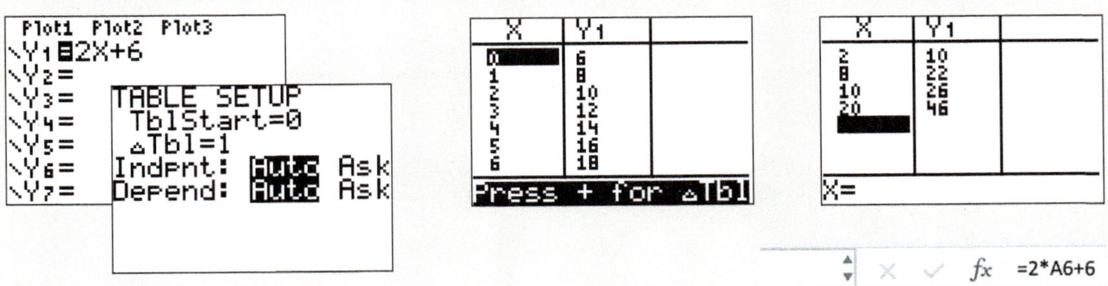

Excel

Step 1: Enter x values you'd like to input in column A.

Step 2: In cell B2, enter "=" followed by the formula for the function, with A2 in place of x. For the ordered pairs we used in Question 10, we would enter "=2*A2+6". This will put the output in cell B2.

Step 3: Copy the formula in B2 down to as many rows as you have x values for.

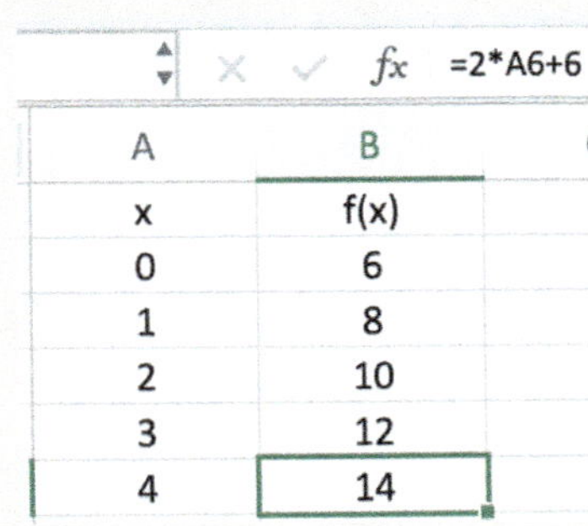

See the Using Tech video on Finding Ordered Pairs for further information.

Intercepts

We've seen that, for the function describing your pay in terms of flyers turned in, $f(8) = 22$. This in turn tells us that the point (8, 22) is on the graph of that function. When setting up the axes to graph that function, would you feel good about having to draw in 22 little tick marks on the y axis? If you make each tick mark correspond to 1 unit, that's what you'd be stuck with. The smarter thing to do is make a wise choice of what number each tick mark corresponds to on each axis. This is called choosing a **scale** for each axis, and it's ALWAYS important when drawing a graph.

In fact, it's literally the first thing you should think about when starting to draw a graph. One good way to get started on choosing a scale is to find the key points where the graph crosses each of the axes. These are called the **intercepts** for the graph. More specifically, the point where a graph crosses the x axis is called the ***x* intercept,** and the point where a graph crosses the y axis is called the ***y* intercept.** When we asked you to think about the coordinates of points on each axis earlier, we were setting up finding intercepts. If you keep in mind that points on the x axis all have y coordinate zero, and that points on the y axis all have x coordinate zero, then you'll UNDERSTAND the following rules and won't have to memorize them.

Finding Intercepts

To find the x intercept, substitute zero for y and solve the equation for x.

To find the y intercept, substitute zero for x and solve the equation for y.

16. Find the x intercept for the equation $y = 2x + 6$. Explain the significance of this point in the context of handing out flyers, or explain why it doesn't have one.

17. Find the y intercept for the equation $y = 2x + 6$. Explain the significance of this point in the context of handing out flyers, or explain why it doesn't have one.

Did You Get It

2. Find both intercepts for the function describing tuition in Did You Get It 1, then explain the significance of each in the context of the problem, or explain why it doesn't have one.

5-2 Group

So far, we've looked at an example of a linear function. Linear functions are particularly convenient to work with because they're easy to recognize, both from the graph and from the equation.

- All linear functions can be written in the form $f(x) = mx + b$ where m and b are real number constants.
- The graph of every linear function is a straight line. (Which explains why they're called linear functions.)

1. In the general form $f(x) = mx + b$, what letter represents the input of the function? What letter represents the output?

In using function notation, math folks often use letters other than f for the name of the function. Sometimes we use letters to help us remember what the function stands for. In the next function we study, we'll use the capital letter C to represent the cost of a mobile phone plan.

The function $C(x) = 12x + 40$ represents the cost of a data-restricted cell phone plan, where x represents the number of gigabytes the user goes over the 10 gigabyte data allowance.

2. Complete the table of ordered pairs, and use it to draw the graph of $C(x)$.

x	$C(x)$
0	
1	
2	
3	
4	
5	

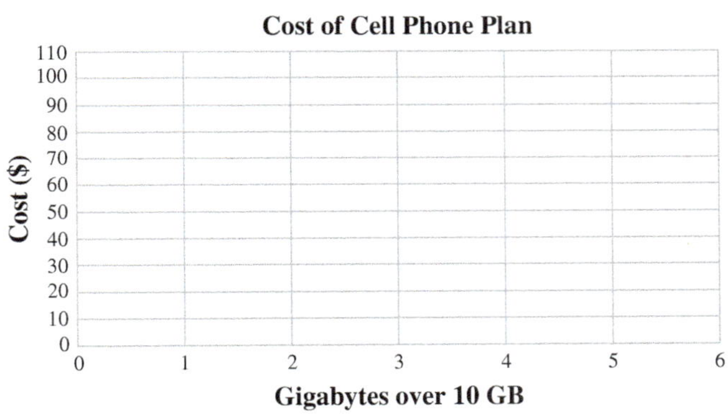

3. Fill in the blank. $C(3) =$ ______. Explain what both numbers in this statement mean.

Math Note

According to the mobile phone company Ericsson, the average mobile phone user around the world consumes 17 GB of data per month.

4. Complete the blank. $C(\_\_\_\_) = 100$. Explain what both numbers in this statement mean.

5. Explain the conditions for which the cost of this phone plan would be more than $88.

6. Write the question and the answer to Question 5 in symbols.

7. Find the x intercept for $C(x)$ and describe if it has meaning in the context of the problem.

8. Find the y intercept for $C(x)$ and describe if it has meaning in the context of the problem.

9. As the x (input) values increase from 0 to 1, the y (output) values change from ____ to ____, which is an increase of ____.

10. As the x (input) values increase from 3 to 4, the y (output) values change from ____ to ____, which is an increase of ____.

11. Summarize what you observe about how the y values change compared to the x values. A good answer will include units.

Did You Get It

3. The function $C(x) = 2.40x + 3.60$ represents the cost of a taxi ride of length x miles.

   a. Complete the table and draw the graph of the function.

x	C(x)
0	
2	
4	
6	
8	
10	

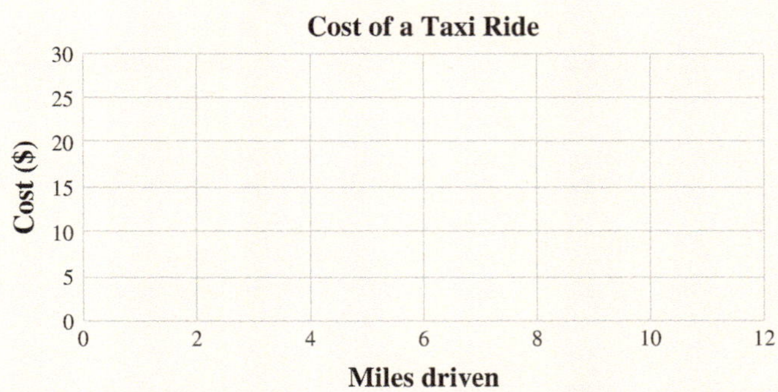

   b. Find the intercepts and explain the meaning of each, if there is one.
   c. Fill in the blanks and explain the meaning of each resulting ordered pair.

$C(2) =$ _____ $\qquad C($_____$) = 22.80$

   d. Explain the conditions for which the cost of a ride in this taxi would be less than \$13.20 and write this question and answer in symbols.

Quadratic Functions

Linear functions are useful in a lot of different settings, but they're not the only game in town. Next, we'll turn our attention to the graph of a *quadratic function.* A function of the form $f(x) = ax^2 + bx + c$, where a, b, and c are real numbers and $a \neq 0$, is called a **quadratic function.** The $a \neq 0$ part provides the distinguishing feature of quadratic functions: The highest power of the variable that appears is two. The graph of a quadratic function is called a **parabola.** If you toss a ball across a room and think of the path it would take if viewed from the side, then you're thinking of a parabola.

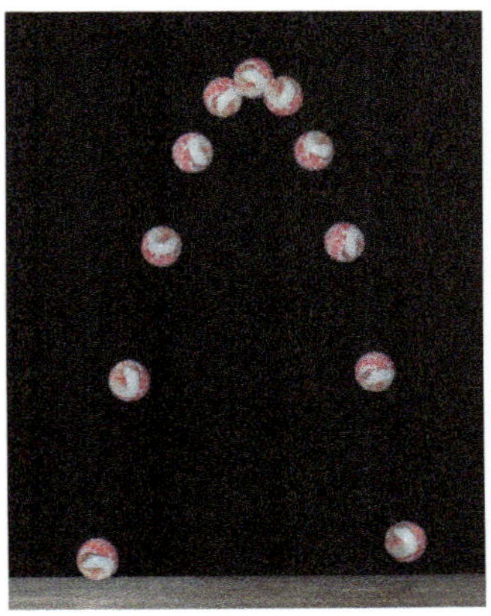

Custom Life Science Images/ Alamy Stock Photo

After hiring an economic consultant, the owner of a small business is informed that the profit from producing x model toys per week can be approximated by the function $P(x) = -x^2 + 70x - 600$.

12. Complete the table of ordered pairs. Using technology will be a big help here. You can use a graphing calculator or spreadsheet (See tech box on page 561). Doing all the calculations by hand will take forever. Then plot the points and complete the graph by connecting the points with a smooth line. If you found and plotted all of the points correctly, you should get a shape similar to the time-lapse photo on page 565—a parabola. Think about whether it is or is not appropriate to put arrows on the ends of your graph.

x	$P(x)$
0	
5	
10	
15	
20	
25	
30	

x	$P(x)$
35	
40	
45	
50	
55	
60	

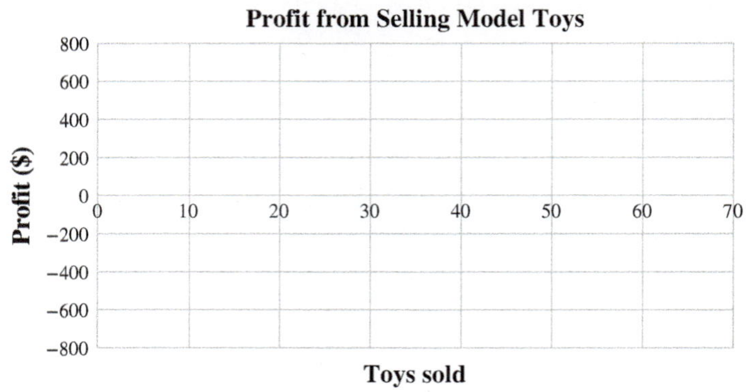

Math Note

It might strike you as odd that the profit starts to go down at some point when selling more toys. But think of it this way: When too much of a product is available, the price you can get for it goes down, and profits can be adversely affected.

13. Fill in the blank. $P(5) =$ __________. Explain what both numbers in this statement mean.

14. Complete the blank. $P($__________$) = 400$. If there's more than one correct answer for this blank, give them all. Explain what all numbers in this statement mean.

15. What has to happen for the weekly profit for this business to be more than $600?

16. Write the question and the answer to Question 15 in symbols.

17. Do you see any x intercepts on your graph? If so, what are they and what do they mean?

18. Do you see a y intercept on your graph? If so, what is it and what does it mean?

Summary of Linear and Quadratic Functions

Linear Function

Algebraically: $f(x) = mx + b$

Graphically: Straight line

Example: $f(x) = -2x + 3$

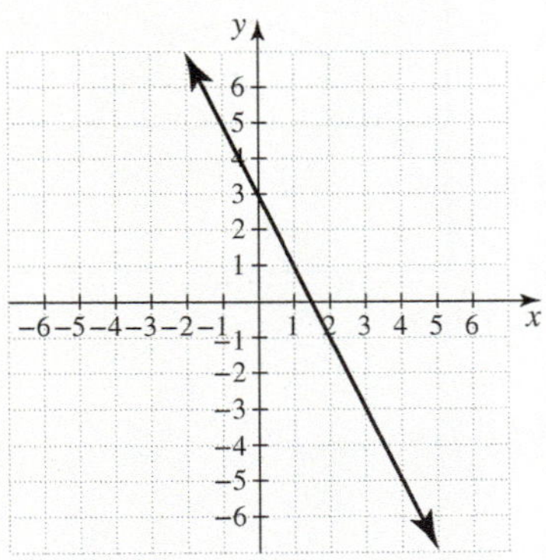

Quadratic Function

Algebraically: $f(x) = ax^2 + bx + c$

Graphically: Parabola

Example: $f(x) = x^2 - 4$

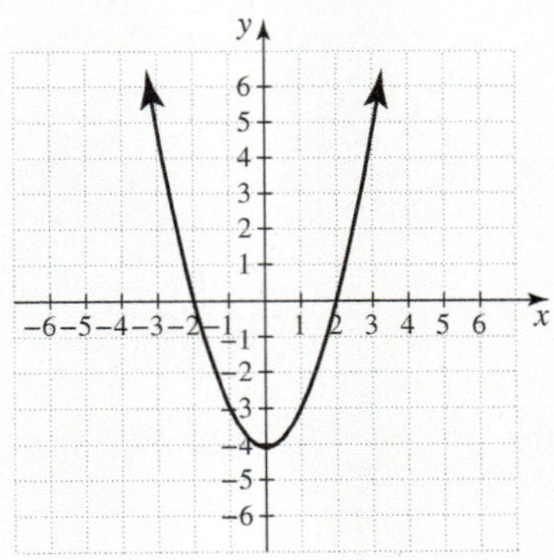

Using Technology: Graphing Lines and Parabolas

The thing that makes a graphing calculator a graphing calculator is the ability to draw graphs. Here's a quick look at how.

1. Press **Y=** and enter the equation to be graphed.
2. Press **WINDOW** and adjust the viewing window to display the portion of the graph that you're interested in seeing. The two screenshots show settings and the result for the cell phone problem and the profit problem from this lesson. Notice how the **Xmin** and **Xmax** settings match the least and greatest *x* values displayed on the graphs you drew earlier, and the **Xscl** setting matches the width of each box on the grid in the horizontal direction. Likewise, the **Ymin, Ymax,** and **Yscl** settings reflect the lowest and highest values on the *y* axis as well as the height of each box on the grid.
3. Press **GRAPH.**

Note: If you don't see anything when you press **GRAPH**, the most likely issue is that you have bad settings on the **WINDOW** screen.

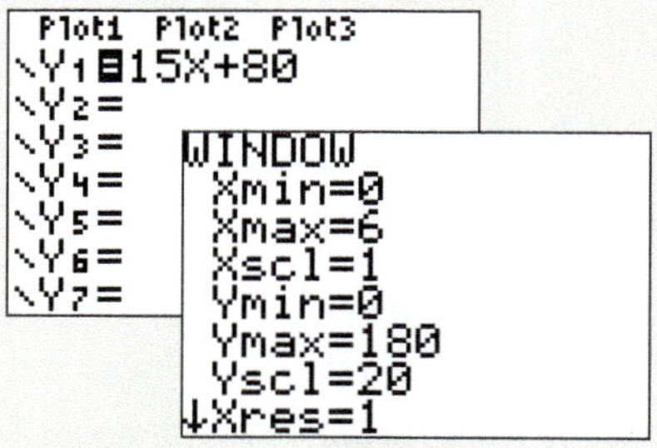

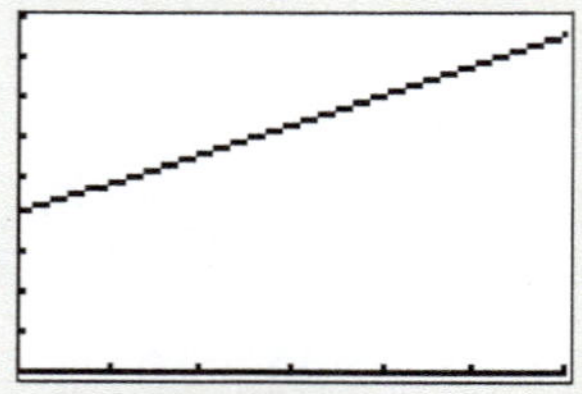

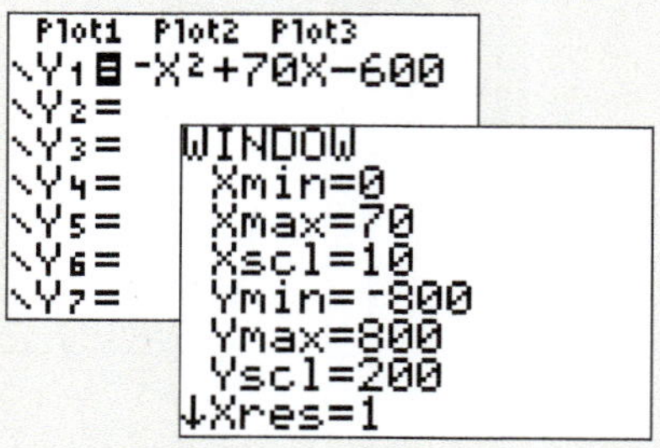

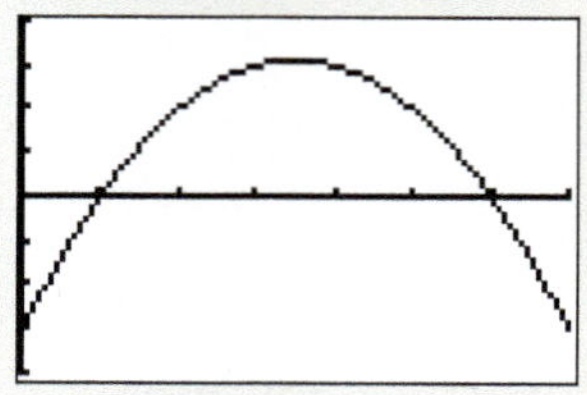

See the Using Tech video on Graphing Lines and Parabolas for further information.

In the Technology activity for this lesson, you'll learn how to use the scatter plot feature of a spreadsheet to create graphs of functions.

5-2 Portfolio

Name ______________________________

Check each box when you've completed the task. Remember that your instructor will want you to turn in the portfolio pages you create.

Technology

1. ☐ In this Excel activity, we'll use an equation to graph a line.
2. ☐ In this Excel activity, we'll use an equation to graph a parabola.

ALEKS Practice

1. ☐ Include any written work from the ALEKS assignment along with any notes or questions about this lesson's content.

Applications

1. ☐ Complete the applications problems.

Reflections

Type a short answer to each question.

1. ☐ What do you think the term "math modeling" means?
2. ☐ Explain what the point of drawing the graph of an equation is in a modeling situation.
3. ☐ Name one thing you learned or discovered in this lesson that you found particularly interesting.
4. ☐ What questions do you have about this lesson?

Looking Ahead

1. ☐ Complete the prep skills for Lesson 5-3.
2. ☐ Read the opening paragraph in Lesson 5-3 carefully and answer Question 0 in preparation for that lesson.

Answers to "Did You Get It?"

1. $T(3) = 560$, $T(15) = 2{,}000$. It would cost \$560 to register for 3 hours, and \$2,000 for 15 hours.
2. The x intercept is $(-5/3, 0)$, which has no meaning in context, and the y intercept is $(0, 200)$, which means that you'll pay \$200 to be a student without taking any hours.

3. a.

x	0	2	4	6	8	10
$C(x)$	3.60	8.40	13.20	18.00	22.80	27.60

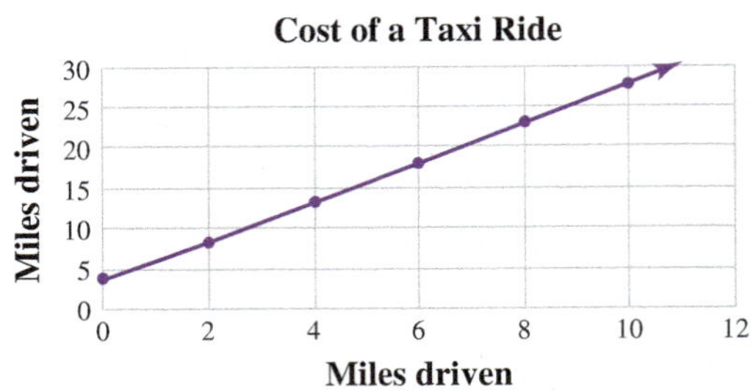

b. The x intercept is $(-1.5, 0)$ which has no meaning. The y intercept is $(0, 3.6)$ which means that it costs \$3.60 just to get into the cab. **c.** $C(2) = 8.4$ and $C(8) = 22.8$. A cab ride of 2 miles costs \$8.40, while riding for 8 miles costs \$22.80. **d.** Any ride less than 4 miles will cost less than \$13.20; Question: $C < 13.2$; Answer: $x < 4$.

Answers to Prep Skills Questions

1. −8, 12, 32; 700, 595, 370; −15, −3, 57; 400, 200, −400
2. $x = -6$; $x = 2,333.\overline{3}$
3. $x < 140$; $y \geq -10$; $4 < z < 12$

5-2 Applications

Name ______________________________

For skydivers, it is recommended that the parachute is opened by the time they reach an altitude of 2,500 ft. Once a parachute opens, skydivers fall at approximately 22 ft/sec. Assuming the chute opens at 2,500 ft, the height in feet of a skydiver can be modeled by the function $H(x) = 2{,}500 - 22x$, where x is the number of seconds after the chute opens.

1. Would you describe this as a linear function or a quadratic function? Explain.

Purestock/SuperStock

2. Complete the table of ordered pairs. Remember that you can use the table feature on a graphing calculator, or a spreadsheet. Either is hugely helpful for this type of repeated calculation.

x	$H(x)$
0	
15	
30	
45	
60	
75	
90	
105	
120	

3. Do all of the ordered pairs make sense in the context of this problem?

4. Find the y intercept and explain what it means in the context of this problem.

5. Find the x intercept and explain what it means in the context of this problem.

5-2 Applications

Name ____

6. Using the intercepts and a few of the ordered pairs from the table, draw the graph of $H(x)$.

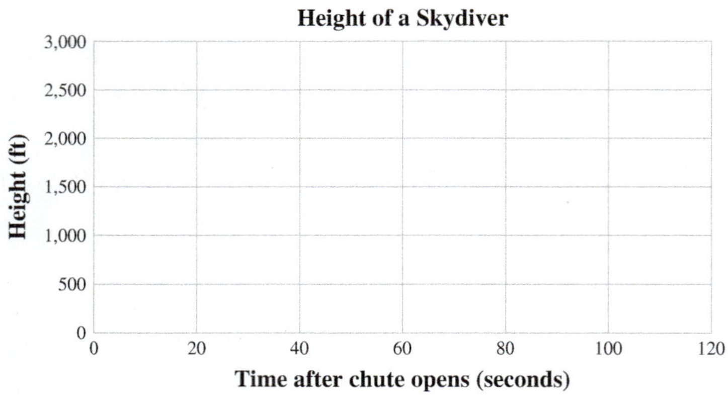

7. Should we put arrows on either end of the graph? Explain.

8. Fill in the blank. $H(12) =$ ________. Explain what both numbers in this statement mean.

9. Fill in the blank. $H($________$) = 520$. Explain what both numbers in this statement mean.

5-2 Applications

Name ______________________________

10. Explain when the height of the skydiver would be less than 850 ft. Write the question and the answer in symbols.

The profit for a company that produces discount sneakers can be modeled by the function $P(x) = 4x^2 - 32x - 210$, where x is the number of pairs of sneakers made.

11. Would you describe this as a linear function or a quadratic function? Explain.

12. Complete the table of ordered pairs and use them to draw a graph of $P(x)$. Using the table feature on a calculator (or making a table in a spreadsheet) will make this a lot easier.

x	P(x)
0	
2	
4	
6	
8	
10	
12	
14	
16	

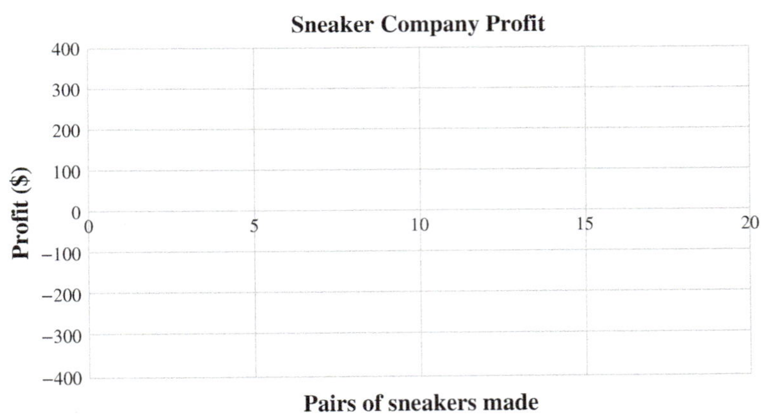

13. Do all of the ordered pairs make sense?

14. Find the y intercept and explain what it means in the context of this problem.

5-2 Applications

Name ______________________________

15. Write the equation you would need to solve to find the x intercept. Then USE THE GRAPH to approximate this x intercept and explain what it means in the context of this problem.

16. Should arrows be placed at either end of this graph? Explain.

17. Fill in the blank. $P(11) =$ _____. Explain what both numbers in this statement mean.

18. Fill in the blank. $P($_____$) = 302$. Explain what both numbers in this statement mean.

19. Look over your graph. There should appear to be a point on the graph that is lower than all other points. Write a statement containing both coordinates of this point using function notation. Explain the meaning of that statement in the context of this problem.

Lesson 5-3 Prep Skills

SKILL 1: REDUCE FRACTIONS

As we've seen several times, the process of writing fractions in lowest terms is called reducing fractions. It's useful to do because it's easier to work with and interpret fractions when they're in lowest terms. For example, it's a lot easier to see that $\frac{5}{2}$ is two and a half than it is to see that $\frac{180}{72}$ is two and a half. To reduce a fraction, divide BOTH the numerator and the denominator by any common factors.

- $\frac{10}{30} = \frac{10 \div 10}{30 \div 10} = \frac{1}{3}$ The greatest factor common to 10 and 30 is 10.
- $\frac{24}{16} = \frac{24 \div 8}{16 \div 8} = \frac{3}{2}$ The greatest factor common to 24 and 16 is 8.

SKILL 2: USE FUNCTION NOTATION

Remember from Lesson 5-2 that a function is a relationship between two quantities where each input produces a unique output. For a given input *x*, the output of a function named *f* is described by the symbol $f(x)$, which is read "*f* OF *x*," NEVER "*f* TIMES *x*."

Use the function $f(x) = -3x + 7$ to find $f(5)$, $f(0)$, and $f(-4)$.

- $f(5) = -3(5) + 7$
  $f(5) = -15 + 7$
  $f(5) = -8$
  Finding $f(5)$ requires substituting 5 for *x* and simplifying the resulting expression.
- $f(0) = -3(0) + 7$
  $f(0) = 0 + 7$
  $f(0) = 7$
  Finding $f(0)$ requires substituting 0 for *x* and simplifying the resulting expression.
- $f(-4) = -3(-4) + 7$
  $f(-4) = 12 + 7$
  $f(-4) = 19$
  Finding $f(-4)$ requires substituting -4 for *x* and simplifying the resulting expression.

SKILL 3: FIND THE INTERCEPTS OF A LINE

When finding the intercepts of a line, it helps to remember that an intercept is a point on a graph. **Finding the *x* intercept:** The *x* intercept is the point where a graph intersects the *x* axis. By looking at any graph, you can see that all points on the *x* axis have a *y* coordinate (or height) of zero. To find the *x* intercept, we substitute 0 for *y* or $f(x)$ and find the associated *x* value.

Finding the *y* intercept: The *y* intercept is the point where a graph intersects the *y* axis. By looking at any graph, you can see that all points on the *y* axis have an *x* coordinate of zero. To find the *y* intercept, we substitute 0 for *x* and find the associated *y* value or output.

- Find the intercepts of $f(x) = 20x - 250$.

  x intercept:

  $$0 = 20x - 250$$
  $$250 = 20x$$
  $$\frac{250}{20} = x$$
  $$x = 12.5$$

  To find the x intercept, substitute 0 for $f(x)$, which is another name for y, and solve the resulting equation to find the x value associated with $y = 0$.

  The x intercept is (12.5, 0).

  Sometimes the x intercept will be reported as the x value, but it is important to remember that it is a point.

  y intercept:

  $$f(0) = 20(0) - 250$$
  $$f(0) = -250$$

  The y intercept is (0, −250).

  To find the y intercept, substitute 0 for x and find the y value or output.

PREP SKILLS QUESTIONS

Write each fraction in lowest terms:

1. $\frac{40}{32}$

2. $\frac{39}{52}$

In Questions 3–5, use the function $P(x) = 50x - 200$ to find each value:

3. $P(0)$

4. $P(4)$

5. $P(10)$

6. Find the intercepts of $f(x) = -30x + 240$.

7. Find the intercepts of $f(x) = 15x + 450$.

Lesson 5-3 A Slippery Slope

Grant V Faint/Photodisc/Getty Images

LEARNING OBJECTIVES

- ☐ 1. Calculate slope and interpret as rate of change.
- ☐ 2. Solve problems using linear modeling, both algebraically and using technology.

When I see a slippery slope, my instinct is to build a terrace.
—John McCarthy

In the movie *The Shawshank Redemption,* there's a character who was released from prison after being locked away for a very long time. The first thing that jumped out at him after going back out into the world was how quickly everything and everyone seemed to move. In a letter back to his friends in prison, he wrote "The world went and got itself in a big damn hurry." Things do change very quickly in our modern world, and those who can't adapt to change get left behind. So far, we've studied math modeling from kind of a static viewpoint: The information we get from points is like a snapshot of what happened at a particular point. In order to study our changing world, we'll be interested in studying relationships between different points in a math model. One of the most common, and most useful, ways of doing so is studying the rate at which a quantity changes, which happens to be the best reason for studying the concept of slope.

0. Describe your understanding of what the slope of a line is.

5-3 Class

Now that we've reviewed graphing lines and used them to model situations, we'll study these graphs and the functions that generate them in greater depth to increase our understanding of the usefulness of math modeling.

More About Linear Functions

Slope

Once you know a single point on any line, all you need to get the rest of the line is what direction it heads in. (That's what makes a line a line: The direction never changes.) This makes the steepness of a line one of its key characteristics geometrically. Eventually we'll see that this provides much more than just geometric information, but for now we'll focus on graphs. Loosely defined, *slope* is a number that indicates how steep a line is. Understanding how slope is calculated, and why that calculation makes perfect sense, is key to understanding the most important stuff we'll be using slope for in modeling, so let's start there.

The slope of a line on a graph is very much like the slope (or steepness) of a road. Look at the two roads in the figure.

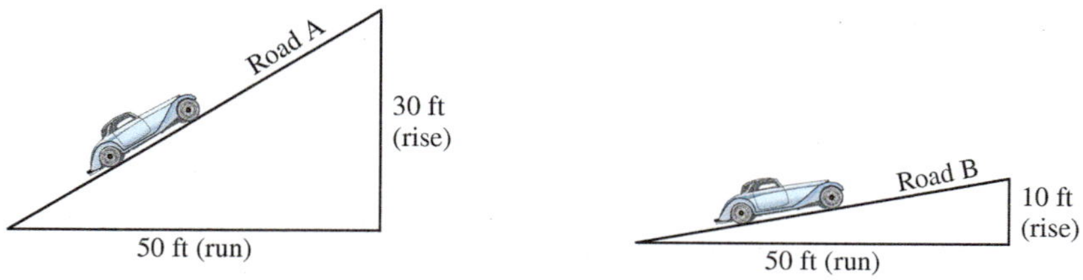

1. If we want slope to measure the steepness of a line, which should have a bigger slope? Why?

2. A good way to define steepness of each road is to compare the change in height (sometimes called the *rise*) and the horizontal distance covered (the *run*). And the best way to compare the sizes of two numbers is with a ratio. The question is: Which distance should be put in the numerator? Use your answer from Question 1 to decide whether we want to divide change in height by horizontal distance, or horizontal distance by change in height.

3. Summary: To find the slope of a line, we divide ________________ by ____________________.

4. Find the slope of each road above; make sure that your results make sense based on your answer to Question 1.

Now that we have some perspective on what slope means physically, let's talk about how this connects to the graphs we studied in the last lesson. Consider the two labeled points on this graph:

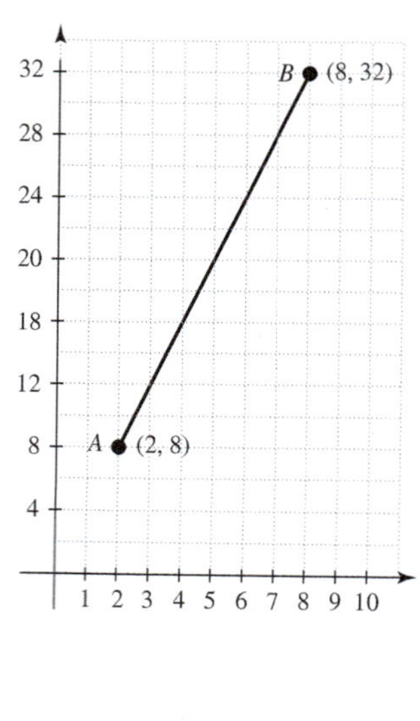

5. How much higher is point B than point A? Give a very specific description of how you got your answer.

6. How much further to the right is point B than point A? Do that detailed description thing again.

7. Use your answers from Questions 5 and 6 to find the slope of the line connecting points A and B.

You might not realize it, but you just discovered the connection between the geometric idea of slope as rise over run, and a way of describing it on a graph. To find the rise, you subtracted the y coordinates of two points on the line, and to find the run you subtracted the x coordinates. This turns out to be SUPER important in terms of understanding what we'll use slope for: slope is a comparison between the change in y and the change in x. More on that in a bit. For now, here's a summary of what we've learned.

Formula for the Slope of a Line

The **slope** of a line (designated by m) is

$$m = \frac{y_2 - y_1}{x_2 - x_1} \quad \frac{\text{Change in } y \text{ coordinate}}{\text{Change in } x \text{ coordinate}}$$

where (x_1, y_1) and (x_2, y_2) are two points on the line. In words, the slope of a line can be calculated by subtracting the y coordinates (the vertical height) of two points and dividing that difference by the difference obtained from subtracting the x coordinates (the horizontal distance) of the same two points.

So now we know the slope of the line connecting the points (2, 8) and (8, 32) is 4. Look back at the graph to answer this important question.

8. If you start at point A and move along the line one unit to the right, what happens to the y coordinate? What's the significance of that answer?

And there it is! The key to understanding what slope measures on a line: It tells us how much the y coordinate changes when the x coordinate increases by 1. That observation will come in handy throughout our study of slope in context, which starts . . . NOW!

Let's think back to the cell phone problem in Lesson 5-2. The completed graph and table are reproduced here for convenience.

x	C(x)
0	40
1	52
2	64
3	76
4	88
5	100

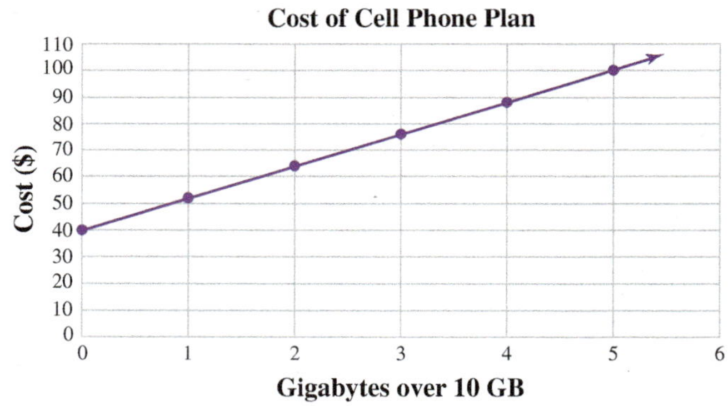

9. Choose any two points on this line and perform the calculation to find the slope.

10. Choose two completely different points and calculate the slope again.

11. Explain why it makes no difference which two points you choose.

Did You Get It

Try this problem to see if you understand the concepts we just studied. The answer can be found at the end of the portfolio section.

1. In Did You Get It 3 of Lesson 5-2, you completed the table below, which describes the cost of a cab ride in terms of miles driven. Find the slope of the line that describes the cab fare.

x	0	2	4	6	8	10
C(x)	3.60	8.40	13.20	18.00	22.80	27.60

12. Remember the key observation earlier? The slope of a line measures the amount of change in the y coordinate when the x coordinate increases by 1. What quantities in this problem are represented by the x and y coordinates? Make sure you think about (and write) which is which.

13. Big moment: Use the result of Question 12 to describe exactly what the slope you calculated tells us about the cell phone plan. Be specific, and use units.

A comparison like the one you described in Question 13 is called a **rate of change.** In this case, we would say that the slope represents the rate at which the monthly cost is changing as the number of gigabytes of data used changes.

14. In the case of the cost of the cell phone plan, we can say that the cost is increasing by ________ for each additional GB over the 10 GB limit.

15. You already answered this question in the previous lesson, but what is the y intercept of this line?

Slope-Intercept Form

The $y = mx + b$ form we've been using is called the **slope-intercept** form of a linear equation with a slope m and y intercept $(0, b)$. This is very useful information because the slope and y intercept can be observed directly from the equation.

16. The linear function $C(x) = 12x + 40$ that models the phone plan problem in the graph on the previous page is also in slope-intercept form. Does this make sense with what you've written in Questions 14 and 15? Explain.

Now here is the key point: When we have a linear equation in the form $y = mx + b$, we know the slope (the coefficient of x) and the y intercept (the constant term). But it works the other way around, too: If we know the slope and the y intercept for a linear equation, then we can write the equation!

17. A physical therapist made \$73,000 per year at the time she was hired, and her salary has been increasing by \$2,700 each year. Write a linear function $P(x)$ that describes her salary, where x represents the number of years since she was hired.

18. Use the function you wrote in Question 17 to calculate the therapist's salary 11 years after she was hired.

Let's conclude the Class portion with an example that ties together some of the key ideas in our study of linear modeling.

Solved Example

Modeling with a Linear Function

The following table gives the cost of a shipment of hard drives from a supplier to a big box store.

x Drives	*C*(*x*) Cost ($)
70	4,250
140	8,450
250	15,050
400	24,050

a. Assuming that the relationship between the number of drives and the cost is linear, find the slope of the line that models the data.

b. Interpret the meaning of the slope.

c. Find a function in the form $C(x) = mx + b$ that models the data.

SOLUTION

a. If the relationship is in fact linear, then we can choose any pair of points from the table to find the slope, then divide the change in outputs by the change in inputs to get the slope. Using the points (70, 4,250) and (140, 8,450):

$$m = \frac{8{,}450 - 4{,}250}{140 - 70} = \frac{4{,}200}{70} = 60$$

b. The numerator is a change in price, while the denominator is a change in number of hard drives supplied. So the slope should tell us how the cost changes when the number of hard drives increases by 1. A slope of 60 means that when the number of hard drives supplied goes up by 1, the cost goes up by $60. In simpler terms, this means that each additional hard drive is costing the store $60.

c. Since we know the slope, if we also knew the y intercept, we could use $y = mx + b$ to write the equation. But we don't know the y intercept based on the given data. So we'll have to be clever. Forget about the y intercept for a second: The slope is 60, so the equation has to look like $C(x) = 60x + b$; we just don't know what b is at the moment.

Now look at the top row of the table: 70 hard drives cost the store $4,250, so when x is 70, $C(x)$ has to be 4,250. In other words, $C(70) = 4{,}250$. But we just decided that the function has to look like $C(x) = 60x + b$, in which case $C(70) = 60(70) + b$. Do you see what happened there? $C(70)$ has to be both 4,250 AND $60(70) + b$. That means that $60(70) + b$ has to equal 4,250, and when you write that in symbols, you get an equation you can solve for b.

$$60(70) + b = 4{,}250 \Rightarrow 4{,}200 + b = 4{,}250 \Rightarrow b = 50$$

Solved Example *Continued*

Now we know both the slope (60) and the y intercept (50), and we can write the function:

$$C(x) = 60x + 50$$

A final note before moving on: In theory, we've found a linear model for the data in the table. A really good way to check our answer would be to make a table of inputs and outputs for the function we found using a calculator or spreadsheet and make sure it matches the original data table. To the right is the result from Excel: notice that it does indeed match the original table, so we know we got the right equation.

fx =60*A1+50

A	B	C
70	4,250	
140	8,450	
250	15,050	
400	24,050	

Here's a summary of the procedure that we used in the Solved Example to find a linear equation when we aren't given the y intercept.

Finding the Equation of a Line Through Known Points

Step 1: Pick two of the points and find the slope of the line connecting them using the usual procedure: Divide the change in y coordinates by the change in x coordinates.

Step 2: In the slope-intercept form $y = mx + b$, substitute the slope you found for m, and the x and y coordinates from one of the points for x and y. This results in an equation that you can solve for b.

Step 3: Once you know m and b, you can write the equation using $y = mx + b$.

Important note: NEVER use this procedure unless you know that all of the points you know live on a line.

19. Use the procedure from the Solved Example to solve this problem: Herman makes some spending money in his spare time by selling hot dogs at the local ballpark. The table shows how much he makes depending on how many hot dogs he sells. Find a linear function that models his pay $P(x)$ in terms of the number of hot dogs he sells x.

Hot Dogs Sold (x)	20	40	60	80
Pay ($P(x)$)	\$42	\$72	\$102	\$132

Did You Get It

2. What are the slope and y intercept of the line that has equation $y = -43x - 210$?
3. Write the equation of a line that has slope $\frac{11}{2}$ and y intercept (0, 15).

5-3 Group

The standard fare for a taxi in one city is \$3.25, plus \$1.80 per mile. Our goal is to write a linear function that describes the cost of a cab ride in terms of the length of the ride in miles.

Steve Mason/Photodisc/Getty Images

1. Identify the quantity in that description that is a rate of change (slope).

2. Identify the quantity that is a fixed or starting value (y intercept).

3. Use your answers to Questions 1 and 2 to write the linear function that describes the cost of the ride. You can use whatever letters you want to represent the cost and the number of miles, but make sure you identify what each letter stands for.

4. Show work and use your function to find the cost of a 6-mile ride, an 8.5-mile ride, and a 12-mile ride.

Math Note

When writing a function to model a situation, it's ALWAYS important to clearly identify what quantity each variable represents.

5. Use the table in a spreadsheet or a calculator to check your work from Question 4.

6. Show work and use your function to find the length of a ride that costs \$10.90. This will involve setting up and solving an equation.

Did You Get It

4. The cost of a medium cheese pizza at Mario's Campus Pizzeria is \$6.75, and each additional topping costs \$0.35.
   a. Write a linear function that describes the cost of a pizza in terms of the number of toppings.
   b. Use your function to find the cost of a pizza with three toppings and one with five toppings.
   c. Use your function to find the number of ingredients on a pizza that costs \$8.85.

Using Technology: Solving an Equation Using the Intersect Feature

We've already covered how to use the table feature to create ordered pairs. Now, we'll show how to use the intersect feature to solve equations.

1. Enter the left side of the equation next to **Y1.**
2. Enter the right side of the equation next to **Y2.**
3. Adjust the window until you can see where the two graphs cross.
4. Press **2nd TRACE** and select option 5. Then press **ENTER** twice to select the two curves. Before pressing **ENTER** a third time, it might be necessary to move the cursor close to the intersection you are trying to find.
5. The *x* value displayed is a solution to the equation. If the graphs cross more than once, repeat to find additional solutions.

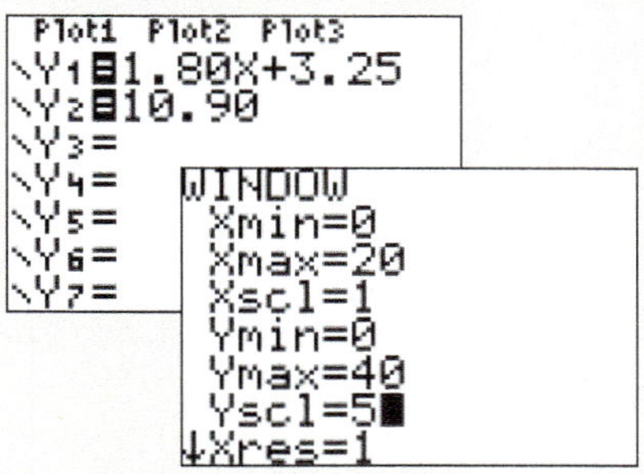

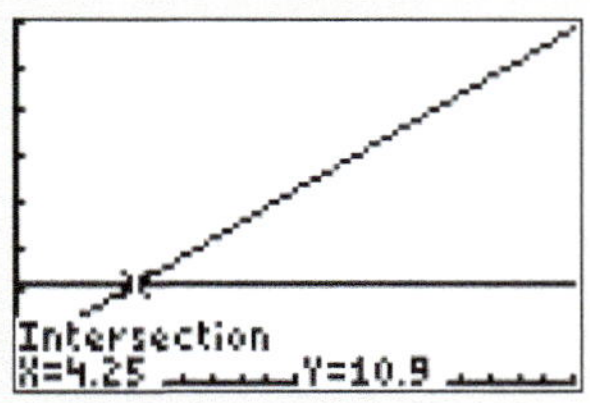

See the Using Tech video on Solving an Equation Using the Intersect Feature for further information.

At this point, we've looked at linear modeling in two different ways: thinking of slope strictly as a rate of change, and from more of an algebraic perspective, writing and using equations. Let's conclude this lesson by taking a look at how the same problem can be solved using either approach. Hopefully, this will better equip you to work the applications homework, since in some cases you can decide which approach makes the most sense to you.

Here's the problem we'll study: Suppose that the altitude (in feet) of an airplane is a linear function of time (in minutes) after it starts its descent for landing. The altitude of the plane was 4,100 feet 18 minutes after starting its descent. When graphed, a function describing the altitude gives a line with a slope of −400. What was the altitude 6 minutes after that?

First, we'll use a rate of change approach that requires some reasoning skills but no algebra.

7. Based on what we've learned about interpreting slope, write a careful description of what the fact that the slope is –400 tells us about the altitude of the plane.

8. Use your answer to Question 7 to decide how far the plane will descend in the next 6 minutes.

9. What will the altitude be at that time?

Next we'll walk you through solving the problem by writing and using a linear function.

10. If we're going to use a linear model, it will be of the form $y = mx + b$. Use the slope we were given to write a preliminary equation with variables x, y, and b.

11. The statement of the problem provides a point on the graph. What is it?

12. Substitute the coordinates of the point you wrote into your equation from Question 10. This should give you an equation with b as the only variable.

13. Solve the equation for b, then write the linear equation describing the altitude in terms of minutes after descent started.

14. What was the altitude 6 minutes after it was 4,100 feet? (Careful! THINK about what value you need to use for x.)

5-3 Portfolio

Name ______________________________

Check each box when you've completed the task. Remember that your instructor will want you to turn in the portfolio pages you create.

ALEKS Practice

1. ☐ Include any written work from the ALEKS assignment along with any notes or questions about this lesson's content.

Applications

1. ☐ Complete the applications problems.

Reflections

Type a short answer to each question.

1. ☐ What is meant by the term "rate of change," and what does that have to do with slope?
2. ☐ Describe two different ways that we can use slope-intercept form in studying data that can be modeled with a linear function.
3. ☐ Discuss the value of finding the vertex for a quadratic model.
4. ☐ Name one thing you learned or discovered in this lesson that you found particularly interesting.
5. ☐ What questions do you have about this lesson?

Looking Ahead

1. ☐ Complete the prep skills for Lesson 5-4.
2. ☐ Read the opening paragraph in Lesson 5-4 carefully and answer Question 0 in preparation for that lesson.

Answers to "Did You Get It?"

1. $m = 2.4$ **2.** slope: -43; y intercept: $(0, -210)$ **3.** $y = \frac{11}{2}x + 15$

4. a. $C(x) = 0.35x + 6.75$, where C is the cost and x is the number of toppings.
b. $C(3) = \$7.80$, $C(5) = \$8.50$ **c.** 6

Answers to Prep Skills Questions

1. $\frac{5}{4}$ **2.** $\frac{3}{4}$ **3.** $P(0) = -200$
4. $P(4) = 0$ **5.** $P(10) = 300$ **6.** x int: $(8, 0)$, y int: $(0, 240)$
7. x int: $(-30, 0)$, y int: $(0, 450)$

5-3 Applications

Name ______________________________

The U.S. Census Bureau uses demographic information to set a poverty threshold that is used to determine how many Americans are living in poverty based on annual income. For a single individual, the poverty threshold was \$3,790 in 1980, and it has increased by about \$226 per year since then.

1. Which piece of information in the problem is a rate of change? What would that represent in a linear function modeling the poverty threshold?

2. When modeling information that changes with time, we almost never use the actual time—whether it's clock time or year—as input. Instead, we choose a beginning time for the problem and call that $x = 0$. In this case, we would decide that $x = 0$ corresponds to 1980, since that's the earliest time we have information for. In that case, what is the y intercept for our function?

3. Write a linear function that describes the poverty threshold in dollars in terms of years after 1980. Then use your function to estimate the poverty threshold in 2020, and the year that it will pass \$18,000 per year.

4. Use the Internet to find the most recent poverty threshold as set by the Census Bureau, and discuss how accurately your model predicted that value.

The next problem is similar to the one you just solved, but this time instead of stepping you through it, we'll just state the problem and you're on your own. If you need to, using Questions 1–3 as a guide will be helpful. Make sure that you clearly identify what the letters you use in your model represent.

5-3 Applications

Name ______________________________

5. Two days after yet another hurricane passed by my area, the nearby Myakka River reached 8.6 feet above flood stage, then began to recede at 1.2 feet per day. Write a linear function that describes the river height in terms of days, then use it to calculate how long it should have taken for the river to return to a non-flood height.

The value of many items you buy decreases steadily after you make the purchase, making these values good candidates to be modeled by linear functions. Maureen has a used pickup truck that was worth $11,400 6 months after she bought it and $9,800 14 months after she bought it.

6. If we were to model the value with a linear function V in terms of months after Maureen bought the truck (x), what would the slope be?

7. Describe exactly what the slope tells us about the value of the truck. Be specific!

8. How much is the truck going to be worth 3 years after Maureen bought it? You can either use a rate of change approach or write and use a linear function. Make sure you show your work to make it clear which approach you chose.

Lesson 5-4 Prep Skills

SKILL 1: USE ORDER OF OPERATIONS

The order of operations is a group of agreed-upon rules that allow us to write calculations involving more than one operation without always having to use parentheses to indicate which operation to do first. The order is as follows, with operations higher on the list being performed before those lower on the list.

1. Grouping symbols — Perform these operations first. If there is more than one set of grouping symbols, work from innermost outward.
2. Exponents
3. Multiplication or division — For operations that live at the same level, work from left to right.
4. Addition or subtraction

One good way to avoid common errors is by understanding that a fraction bar is considered to have a grouping symbol around the entire numerator and the entire denominator. Following are two different ways of interpreting the expression $\frac{3+\sqrt{4}}{10}$.

- $\frac{3+\sqrt{4}}{10} = (3+\sqrt{4})/10$ — This calculation would yield the correct result of $5/10$ or 0.5.
- $\frac{3+\sqrt{4}}{10} \neq 3+\sqrt{4}/10$ — This calculation would yield the incorrect result of $3\frac{1}{5}$ or 3.2.

SKILL 2: EVALUATE QUADRATIC FUNCTIONS

To evaluate a quadratic function, replace all occurrences of the input variable or variables with specific numbers. This allows us to find a numeric value for the output variable.

Use the function $f(x) = -2x^2 + 3x - 10$ to find $f(0)$, $f(-2)$, and $f(12)$.

- $f(0) = -2(0)^2 + 3(0) - 10$
$f(0) = -10$ — Replace x with 0 and simplify. The value of the function $f(x)$ when $x = 0$ is -10.
- $f(-2) = -2(-2)^2 + 3(-2) - 10$
$f(-2) = -24$ — Replace x with -2 and simplify. The value of function $f(x)$ when $x = -2$ is -24.
- $f(12) = -2(12)^2 + 3(12) - 10$
$f(12) = -262$ — Replace x with 12 and simplify. The value of the function $f(x)$ when $x = 12$ is -262.

SKILL 3: USE THE DISTRIBUTIVE PROPERTY

The "distributive property" as most people call it is actually the distributive property of multiplication over addition. The operation of multiplication is distributed over the operation of addition (or subtraction).

$$a(b + c) = ab + ac$$

The expression $a(b + c)$ is equivalent to $ab + ac$. The fact that they are equivalent allows us to replace one with the other depending on what we are trying to do.

- $5(3x + 2) = 5 \cdot 3x + 5 \cdot 2$
  $= 15x + 10$

  The distributive property 5 times the quantity $3x + 2$ is equivalent to $5 \cdot 3x + 5 \cdot 2$, which can be simplified to $15x + 10$.

- $-3x(4x - 9) = -3x(4x) - (-3x)(9)$
  $= -12x^2 + 27x$

  In distributing $-3x$ to each term of the expression inside parentheses, be careful to use exponent rules correctly and to change the sign of each term in the parentheses.

PREP SKILLS QUESTIONS

Evaluate each expression.

1. $\dfrac{2 - \sqrt{9}}{4}$

2. $\dfrac{1 + \sqrt{25}}{9}$

For each quadratic function, find the requested outputs.

3. $K(x) = 3x^2 + 14x - 12$; find $K(0)$, $K(-3)$, and $K(10)$

4. $f(x) = -0.2x^2 - 1.4x + 5.4$; find $f(4.125)$ and $f(11.9)$

Simplify each expression using the distributive property.

5. $3x(x + 15)$

6. $-16x(4x - 10)$

Lesson 5-4 Ahead of the Curve

LEARNING OBJECTIVES

☐ 1. Identify quantities that are and are not good candidates to be modeled with quadratic functions.

☐ 2. Solve problems using quadratic modeling, both algebraically and using technology.

Juan Camilo Bernal/Shutterstock

Curve: The loveliest distance between two points.
—Mae West

The reason that linear functions are called linear functions is that their graphs are straight lines. And the reason that their graphs are straight lines is that the slope never changes. From a modeling perspective, that tells us that quantities can be modeled by linear functions when their rate of change is constant. But there are tons of quantities out there that have varying rates of change. So in the remainder of this unit we'll think about whether or not the quantities we're studying are a good fit for linear modeling, and when they're not we'll have to look at different types of functions.

Have you ever wondered how much your driving speed affects fuel efficiency? I know I have, and since it's 2025, I decided to do some online research to see what I could learn. As you might expect, going either really slow or really fast isn't great for gas mileage: Somewhere in between is the sweet spot. I found the numbers in the following table, which come from an experiment involving over 2,000 driving hours for a group of 2009–2013 Subaru Outbacks, reported on a site called Automatic.com.

Speed	MPG
30	26.6
35	30.2
40	33.1
45	34.2
50	33.9

Speed	MPG
55	33.0
60	32.3
65	31.5
70	28.6
75	27.3

A quick glance shows that fuel efficiency increases until you get to about 45 miles per hour, then it starts to decrease at higher speeds. That tells us for sure that the rate of change isn't constant. As usual, if we look at a graph of the data, the pattern is even more clear:

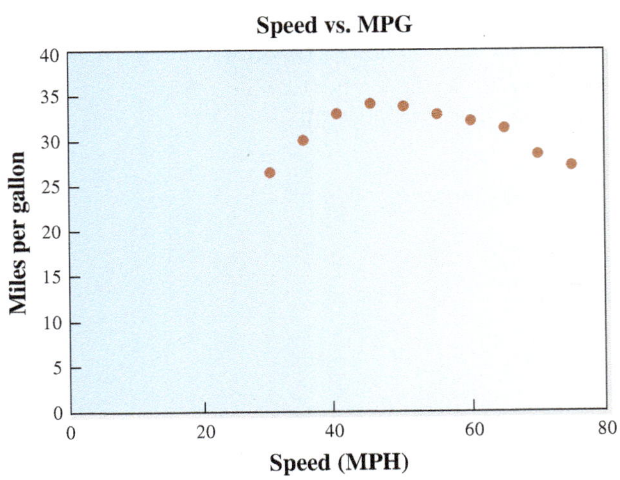

If you completed Lesson 5-2, I surely hope this scatter plot will bring quadratic functions to mind. This is just a small taste of the wide range of real quantities that can be modeled by functions that aren't linear.

0. Think of a real quantity that would most likely not have a constant rate of change. Well, don't just think of it. Write it down, too.

5-4 Class

When we studied quadratic functions in Lesson 5-2, we got the general shape but left out two details that can be useful in analyzing quadratic models—finding the vertex and finding the x intercepts.

1. Recall that a function of the form $f(x) = ax^2 + bx + c$, where a, b, and c are real numbers and $a \neq 0$, is called a

____________ ____________, and the graph of a quadratic function is called a ____________.

When a parabola decreases in height, reaches a low point, and then increases, we say that it *opens upward.* (See the first graph below.) This occurs when a, the coefficient of x^2, is positive. When a parabola increases in height, reaches a high point, then decreases, as in the second graph below, we say that it *opens downward.* This occurs when a is negative.

One of the key features that makes a parabola a parabola is the fact that it changes direction at some point. The point where this occurs is crucial in studying quantities modeled by quadratic function. It's called the **vertex** of the parabola, and it represents either the lowest or highest point on the entire graph, depending on whether the parabola opens up or down.

Every parabola has two distinct halves, with the divider being a vertical line through the vertex. We call this line the **axis of symmetry.** It's not part of the parabola, but a guide to help us draw one. We say that a parabola is **symmetric** about its axis of symmetry, meaning that the parts of the parabola on either side are mirror images of each other. Two representative parabolas are shown in the following figure.

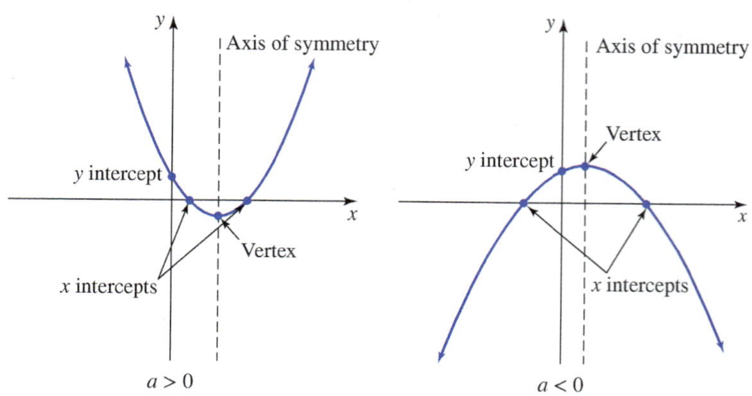

2. How many x intercepts does a parabola have? (This is a harder question than you might think: Give it some thought, and maybe draw a few sample parabolas.)

If a parabola has x intercepts, they are places where the output is zero. So we can find them by substituting zero in for the output, then solving the equation for x. (This is how we found x intercepts for linear functions in Lesson 5-2.) In order to do that efficiently, we'll need to be able to solve equations of the form $ax^2 + bx + c = 0$. There are different methods for solving quadratic equations, but there's one that we'll focus on because it works for EVERY equation of that form: the *quadratic formula.* This is a formula that provides all solutions to any quadratic equation, and you don't have to do any algebra—just plug numbers into the formula. The only catch is that the equation HAS TO BE in the form $ax^2 + bx + c = 0$.

The Quadratic Formula

To solve an equation in the form $ax^2 + bx + c = 0$, where a is not zero, identify the coefficients a, b, and c, then substitute them into the formula

$$x = \frac{-b \pm \sqrt{b^2 - 4ac}}{2a}$$

The $\pm$ symbol indicates that there are two possible solutions: one where you add the value of the radical, and one where you subtract it. If the number inside the radical is negative then there are no solutions, and the graph of the quadratic has no x intercepts.

Important note: Remember that in the standard form of a quadratic equation, $ax^2 + bx + c = 0$, a, b, and c are NUMBERS. They're the coefficients of the three terms, so never include the variable of the equation when using the quadratic formula. Also, note that if one or two of the terms are missing from the equation, that means their coefficient is zero.

Before using the quadratic formula, let's get some practice on identifying the coefficients.

3. Identify the values of a, b, and c for each quadratic equation.

$-2x^2 + 5x - 11 = 0$

$a =$ ________ $b =$ ______ $c =$ ________

$4y^2 - 12 = 0$

$a =$ ______ $b =$ ______ $c =$ ________

$6 - 2t + 10t^2 = 0$

$a =$ ______ $b =$ ________ $c =$ ______

$3x = \frac{1}{2}x^2$ (Careful! You'll need to rearrange to get zero on one side before deciding.)

$a =$ ______________ $b =$ ___________ $c =$ ______

Solved Example

Solving a Quadratic Equation in a Model

The photo back on page 565 shows that a ball flying through the air travels in a parabolic arc. One particular ball is tossed, and its path can be modeled by the function $H(x) = -0.1x^2 + x + 2$, where H is the height of the ball when its horizontal position is x feet from where it was thrown. Use the quadratic formula to find how far the ball travels horizontally before hitting the ground.

SOLUTION

Like most problems that involve using an equation or function to model some quantity, the key is recognizing what the input and output represent. In this case, the input x is the horizontal distance traveled, which is what we're looking for. So if our answer doesn't end up looking like "$x =$ something," it's wrong. The output $H(x)$ represents the height, and we're interested in when the ball hits the ground. That happens when $H(x)$ is zero, so we plug in zero for $H(x)$:

$$0 = -0.1x^2 + x + 2$$

This gives us a quadratic equation, which we can solve using the quadratic formula. Better still, we'll be finding values for x, which is exactly what the problem asked for. Sometimes things just work out.

In this equation, we have $a = -0.1$, $b = 1$, and $c = 2$.

$$x = \frac{-b \pm \sqrt{b^2 - 4ac}}{2a}$$ Plug in $a = -0.1$, $b = 1$, $c = 2$.

$$= \frac{-1 \pm \sqrt{(1)^2 - 4(-0.1)(2)}}{2(-0.1)}$$ Do the arithmetic under the radical first.

$$= \frac{-1 \pm \sqrt{1 + 0.8}}{-0.2} = \frac{-1 \pm \sqrt{1.8}}{-0.2}$$

At this point, the key is to recognize what the symbol "±" means. This is actually providing two distinct solutions: one for adding the radical, and one for subtracting it. So the two solutions are

$$x = \frac{-1 + \sqrt{1.8}}{-0.2} \quad \text{and} \quad x = \frac{-1 - \sqrt{1.8}}{-0.2}$$

Now it's just final arithmetic. Make sure to observe order of operations and do the calculation in the numerator first, then divide the result by -0.2. The results are $x \approx -1.7$ and $x \approx 11.7$. Since the ball couldn't be in two places at once, we need to decide which of these provides the answer to the question. The variable x represents a distance, so the negative solution doesn't make sense in this case, and we conclude that the ball covered 11.7 feet horizontally before hitting the ground.

Now let's talk about technology. When a formula is complicated (and clearly the quadratic formula is), using technology is almost always a stellar idea. It takes some practice, but with a graphing calculator you can enter the entire calculation all at once. Well, twice actually—once for plus in the numerator and once for minus. The following screenshots show the results for each in this problem. Note the parentheses around the entire numerator! They're crucial.

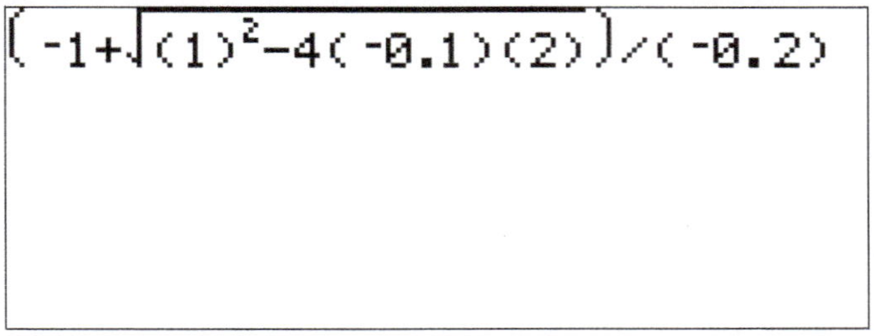

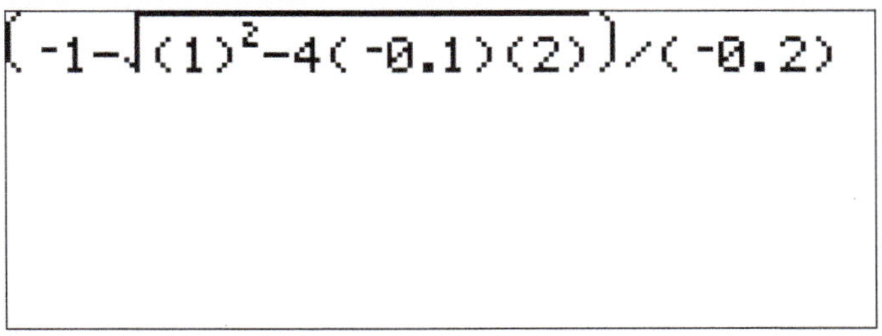

In case you're wondering, we don't have a mutant calculator with a ludicrously wide screen. This is two screens cobbled together through the magic of technology so you can see the entire calculation.

Did You Get It

Try this problem to see if you understand the concepts we just studied. The answer can be found at the end of the portfolio section.

1. The daily profit in dollars made by a cupcake shop can be modeled by the function $P(x) = -0.5x^2 + 34x - 200$, where x is the number of dozen cupcakes they bake that day. How many cupcakes do they need to make in order to break even?

Let's study the function that represents the ball flight more carefully by looking at a table of values and a graph.

x (Horizontal distance)	y (Height)
−2	−0.4
−1	0.9
0	2
1	2.9
2	3.6
3	4.1
4	4.4
5	4.5
6	4.4
7	4.1
8	3.6
9	2.9
10	2
11	0.9
12	−0.4

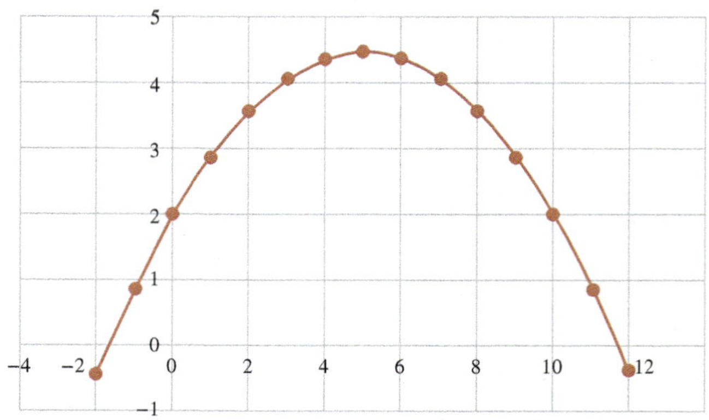

4. Do the table and graph match the result of the Solved Example? Explain.

5. What is the vertex of the parabola in the graph? What does it tell us?

6. For the ball flight function, $H(x) = -0.1x^2 + x + 2$, calculate $\frac{-b}{2a}$. What do you observe?

The result of Questions 5 and 6 is not a coincidence. When a parabola has two x intercepts, they're provided by the solutions to the quadratic formula, and if you look at it the right way, you can see that $x = \frac{-b}{2a}$ is always exactly halfway between the two solutions. Those two solutions are at the same height (on the x axis), and parabolas are symmetric, so the vertex has to be halfway in between. This puts it at $x = \frac{-b}{2a}$.

In any case, this provides us with a simple formula for locating the vertex of a parabola.

Finding the Vertex of a Parabola

To find the vertex of a parabola:

Step 1: Make sure that the equation is written in the form $y = ax^2 + bx + c$.

Step 2: Use the formula $x = \frac{-b}{2a}$ to find the first coordinate of the vertex.

Step 3: Substitute that input into the equation to find the second coordinate.

Did You Get It

2. For the quadratic function $f(x) = 2x^2 + 12x - 1$, find the x intercepts and the vertex.
3. How many cupcakes should the bakery in Did You Get It 1 bake to make the highest possible profit? What will the daily profit be?

In Lesson 5-2, we studied the profit for a toy company. The function $P(x) = -x^2 + 70x - 600$ modeled the profit, where x is the number of models they make each week. Let's see about finding the intercepts and the vertex for this function without first making a table or a graph.

7. The y intercept is pretty simple. Use the function and show your work to find it.

8. The x-intercept is a little more complicated. Use the quadratic formula (or factoring if you're good at it) to find the two x intercepts. Do your answers agree with your observations from the graph and table?

9. Use the formula to find the vertex. Then describe what exactly it tells us about the toy company. Make sure you talk about how the profit CHANGES.

5-4 Group

A family buys a new puppy because puppies are awesome. They plan to fence in a rectangular portion of their backyard for an exercise area. They buy 60 feet of fencing and plan to put the fenced area against the house so that fencing is needed on only three sides. The goal will be to write a quadratic function that describes the area enclosed and use it to find the dimensions that will enclose the largest area.

1. Using the diagram below as a guide, if the sides perpendicular to the house are 10 feet long, how long will the side parallel to the house be? SHOW YOUR WORK. That's important.

Nikaletto/Shutterstock

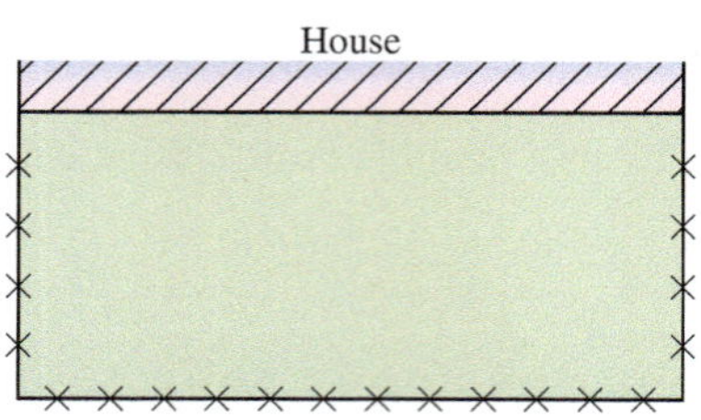

2. Let's use the letter x to represent the length of the two sides perpendicular to the house. Label the diagram using that variable. Then use the same line of reasoning you did in Question 1 to label the length of the side parallel to the house.

3. Write a function $A(x)$ that describes the area enclosed by the fence. Use the dimensions in the diagram.

4. Multiply out the parentheses in your function so that it's in the form $A(x) = ax^2 + bx + c$.

5. Use our vertex formula to find the dimensions that will provide the largest area for the fuzzy puppy to roam around in.

Did You Get It

4. Suppose the family from Questions 1–5 decides to move the dog pen away from the house, so that fence is needed on all four sides. What dimensions will provide the largest area?

Back in the lesson opener, we looked at a relationship between driving speed and gas mileage. We can model these data with the function $y = -0.0137x^2 + 1.4193x - 2.7323$, where x is the speed and y is the gas mileage. Let's take a look at just how well this function models the data.

6. Use the table feature on a calculator or spreadsheet and the modeling function provided in the above paragraph to complete the last column in this table, which repeats the data from page 593.

Speed	Miles per Gallon (MPG)	MPG from Model
30	26.6	
35	30.2	
40	33.1	
45	34.2	
50	33.9	
55	33	
60	32.3	
65	31.5	
70	28.6	
75	27.3	

7. Examine and discuss the accuracy of the model. Do you think it's a good model for the data?

8. Use the vertex formula to find the speed that provides the best fuel efficiency. How does the result compare to the actual data?

9. Use the quadratic formula to find the x intercepts of the modeling function.

10. What is the significance of the x intercepts in this context? Do the intercepts you found make sense in context? Discuss.

Math Note

The fact that we've seen some very unrealistic results from models illustrates why statisticians caution against using a model to make predictions outside of the range of a data set. Use caution and THINK about reasonableness!

11. What have you learned about modeling real data with functions from Questions 6–10?

One of the goals of this course is to turn you into a mathematician. Not a PROFESSIONAL mathematician, mind you—that's my job and you can't have it. By "mathematician" I mean someone who thinks about how math can help us to understand and interpret the world around us, and someone who's interested in knowing how formulas work and where they came from.

If you're successfully on the road to being a mathematician, you may have wondered this: Where in the world did I get the function that models the gas mileage data? Am I some kind of savant who can just look at a table and write a function from it? The answer, sadly, is no. But if you are wondering where modeling functions like that one come from, then as a budding mathematician you'll really enjoy Lesson 5-5.

5-4 Portfolio

Name ______________________________

Check each box when you've completed the task. Remember that your instructor will want you to turn in the portfolio pages you create.

ALEKS Practice

1. ☐ Include any written work from the ALEKS assignment along with any notes or questions about this lesson's content.

Applications

1. ☐ Complete the applications problems.

Reflections

Type a short answer to each question.

1. ☐ What types of quantities might be modeled by a quadratic function?
2. ☐ What is the quadratic formula used for?
3. ☐ Discuss the value of finding the vertex for a quadratic model.
4. ☐ Name one thing you learned or discovered in this lesson that you found particularly interesting.
5. ☐ What questions do you have about this lesson?

Looking Ahead

1. ☐ Complete the prep skills for Lesson 5-5.
2. ☐ Read the opening paragraph in Lesson 5-5 carefully and answer Question 0 in preparation for that lesson.

Answers to "Did You Get It?"

1. $6\frac{1}{2}$ or $61\frac{1}{2}$ dozen
2. x intercepts: $(-6.08, 0)$ and $(0.08, 0)$; vertex: $(-3, -19)$
3. 34 dozen, \$378
4. 15 feet on all sides

Answers to Prep Skills Questions

1. $-1/4$ **2.** $2/3$ **3.** $K(0) = -12$, $K(-3) = -27$, $K(10) = 428$
4. $f(4.125) \approx -3.78$, $f(11.9) \approx -39.58$ **5.** $3x^2 + 45x$ **6.** $-64x^2 + 160x$

5-4 Applications

Name ____

When an object is launched or thrown from some height h_0 (in feet) with an initial upward velocity v_0 (in feet per second), the height in feet of that object t seconds later is given by the function $h(t) = -16t^2 + v_0 t + h_0$. (This ignores the effects of friction and air resistance, which can be negligible for an object like a ball bearing and significant for an object like a paper airplane.) Suppose that a bowling ball is launched upward from the top of a 230-foot building with initial velocity 60 ft/sec. (Legal disclaimer: If you try this, you're a jerk.)

1. In this case, what numbers would correspond to v_0 and h_0 in the function? Use that information to write the specific function describing the height of the bowling ball.

2. Find the vertex of the parabola, then describe what it tells us physically about the situation.

3. Find any x intercepts, and describe what they tell us physically about the situation. (You can call them t intercepts if you like, since we're using t as the input.)

5-4 Applications

Name ______________________

4. In a calculus class, you would learn that if the height of the ball is given by $h(t) = -16t^2 + 60t + 230$, then its velocity at any time t is modeled by $v(t) = -32t + 60$. How fast would the ball be going when it hit the ground?

5. Why can't the information we've used in this problem tell us how far away from the building the bowling ball will land?

6. A decent major league pitcher can throw a ball at 90 miles per hour, which is 132 feet per second. How high could one of these guys throw a ball straight up, and how long would he have to get out of the way to avoid getting hit in the melon when it comes back down? (Let's say the pitcher is 6 feet tall, and he releases the ball from a point a foot above his head.) Again, you're on your own this time, but you can use Questions 1–3 as a guide.

5-4 Applications

Name ______________________________

You might think that my curiosity about gas mileage would have been satisfied by the data I found online. And you might be wrong. Did I mention that I'm a mathematician? I decided to perform my own experiment.

The slope of a road has a huge effect on gas mileage, which will come as no surprise to you if you've ever walked up a hill. So I needed a flat road for my experiment, and U.S. 1 in Florida might be the flattest road in the eastern U.S. Perfect! It was also lightly traveled on this particular day, which allowed me to drive for exactly 2 minutes at speeds from 30 up to 70, resetting the gas mileage computer once I had cruise set at each speed. The vehicle was a 2015 Ford Explorer. My results are shown in the following table.

Speed	Miles per Gallon (MPG)
30	31.2
35	32.6
40	36.3
45	34.9
50	31.7
55	29.2
60	28.1
65	25.9
70	25.4

7. Once again, we'll want to use speed as input and miles per gallon as output. Why?

5-4 Applications

Name ______________________________

8. Draw a scatter plot for the data, then discuss how well you think a quadratic function might model the data.

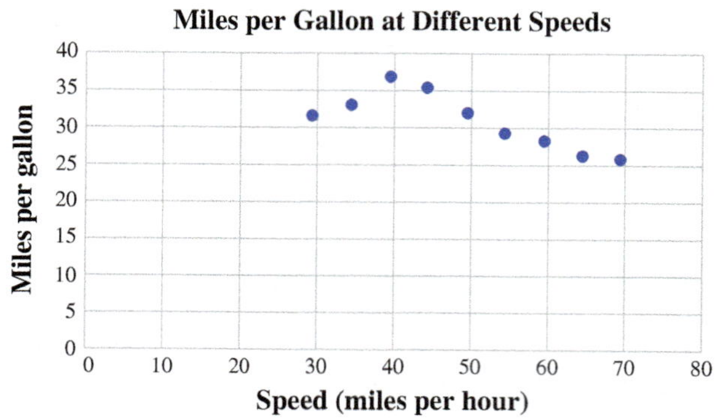

9. My data can be modeled by the function $y = -0.0106x^2 + 0.8409x + 16.782$ where y is the miles per gallon at x miles per hour. Use the table feature on a calculator or spreadsheet and the modeling function provided to complete the last column in this table.

Speed	Miles per Gallon (MPG)	MPG from Model
30	31.2	
35	32.6	
40	36.3	
45	34.9	
50	31.7	
55	29.2	
60	28.1	
65	25.9	
70	25.4	

5-4 Applications

Name ______________________________

10. Examine and discuss the accuracy of the model. Do you think it's a good model for the data?

11. Use the vertex formula to find the speed that provides the best fuel efficiency. How does the result compare to the actual data?

12. Use the quadratic formula to find the x intercepts of the modeling function, then discuss if they provide realistic results.

13. Compare everything you learned about this model, and the one for Subaru Outbacks within the lesson. Which model do you think is better? Justify your choice.

Lesson 5-5 Prep Skills

SKILL 1: EVALUATE AN EQUATION

To evaluate an equation, replace all occurrences of the input variable or variables with specific numbers. This allows us to find a numeric value for the output variable.

- Evaluate $y = 3.2x + 7.5$ for $x = 0$ and for $x = 4$.

Using $x = 0$,
$y = 3.2(0) + 7.5$
$y = 7.5$

Replace x with 0 and simplify. The value of y when $x = 0$ is 7.5.

Using $x = 4$,
$y = 3.2(4) + 7.5$
$y = 20.3$

Replace x with 4 and simplify. The value of y when $x = 4$ is 20.3.

- Evaluate $y = 1.6x^2 - 9.2x + 3.4$ for $x = 0$ and for $x = 5$.

Using $x = 0$,
$y = 1.6(0)^2 - 9.2(0) + 3.4$
$y = 3.4$

Replace x with 0 and simplify. The value of y when $x = 0$ is 3.4.

Using $x = 5$,
$y = 1.6(5)^2 - 9.2(5) + 3.4$
$y = -2.6$

Replace x with 5 and simplify. The value of y when $x = 5$ is -2.6.

SKILL 2: FIND THE VERTEX OF A PARABOLA

As we saw in Lesson 5-4, follow these steps to find the vertex of a parabola:

Step 1: Make sure that the equation is written in the form $y = ax^2 + bx + c$.

Step 2: Use the formula $x = -\frac{b}{2a}$ to find the first coordinate of the vertex.

Step 3: Substitute that input into the equation to find the second coordinate.

Find the vertex of the parabola $y = 2x^2 + 5x - 1$.

- Note that: $a = 2$, $b = 5$, and $c = -1$.

Find the x coordinate:

$$x = -\frac{b}{2a}$$

$$x = -\frac{5}{2(2)}$$

$$x = -\frac{5}{4} = -1.25$$

Substitute $b = 5$ and $a = 2$ in the formula for the x coordinate of the vertex and simplify.

Find the y coordinate:

$y = 2x^2 + 5x - 1$

$y = 2(-1.25)^2 + 5(-1.25) - 1$

$y = -4.125$

Then substitute the value for the x coordinate into the equation to find the corresponding y coordinate.

Vertex: $(-1.25, -4.125)$

Put the two coordinates together to form the vertex.

PREP SKILLS QUESTIONS

1. Evaluate $y = -1.5x + 8.4$ for $x = 2$ and $x = 8$.

2. Evaluate $y = -2.8x^2 + 1.7x - 5.6$ for $x = -2$ and for $x = 0$.

3. Find the vertex of the parabola $y = -2.8x^2 + 1.7x - 5.6$. Round the x coordinate to two decimal places before finding the y coordinate.

Lesson 5-5 Progressing Regressively

Digital Vision/Getty Images

LEARNING OBJECTIVES

☐ 1. Decide whether a linear or a quadratic model is most appropriate for a data set.

☐ 2. Create a line or parabola of best fit from data.

Without continual growth and progress, such words as improvement, achievement, and success have no meaning.

—Benjamin Franklin

The story of humanity is one of progress—socially, physically, technologically, and intellectually. Some folks rue progress and long for a romanticized version of "the good old days" that probably has only a loose connection to reality. But progress is kind of like a speeding train: You can hop on for the ride, get passed by, or get flattened on the tracks.

Statistics make it clear that human athletic achievement has progressed at a remarkable rate for the last hundred years or so. For an example, have a look at the length of the gold medal–winning jump in the Olympic men's long jump for selected years from 1900 to 2000. Rather than using the year as input, we're again setting a time scale, calling 1900 year zero and making the input years after 1900. A scatter plot for the data is included as well. We've started the y axis at 7 to emphasize the change in length.

Year	x	Length (meters)
1900	0	7.18
1912	12	7.60
1920	20	7.15
1932	32	7.64
1948	48	7.82
1960	60	8.12
1972	72	8.24
1980	80	8.54
1996	96	8.50
2000	100	8.55

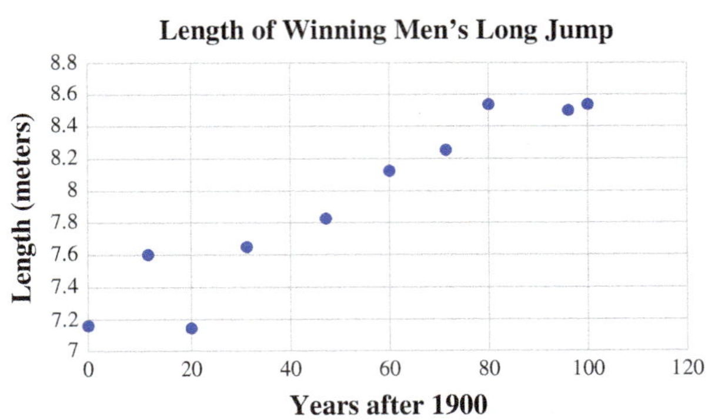

Two things are pretty clear to me when looking at the table and scatter plot. First, the distance men can jump has increased a LOT. Second, a line will not model these data perfectly. But would a line model them approximately? Or maybe some type of curve? Deciding is a pretty subjective process. And even if we make a good decision on what type of model we'd like, that doesn't help us to actually FIND that model. These aren't easy questions because none of us is a computer. But your calculator IS a computer (and Excel, as you know, runs on computers), and it turns out that both are really good at answering these types of questions. In this lesson we'll learn how to find models for raw data using technology, and we'll study the strengths and limitations of the process.

0. What type of model do you think would be reasonable for the long jump data? Why do you feel that way?

5-5 Class

For Questions 1–3, think about the data in each scatter plot. From what you know about the shape of linear and quadratic models, which type of model do you think is more appropriate for each data set (linear, quadratic, or something else)?

1. 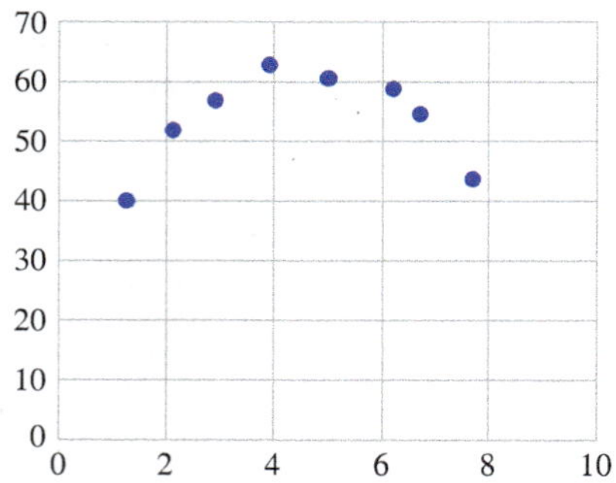

2. 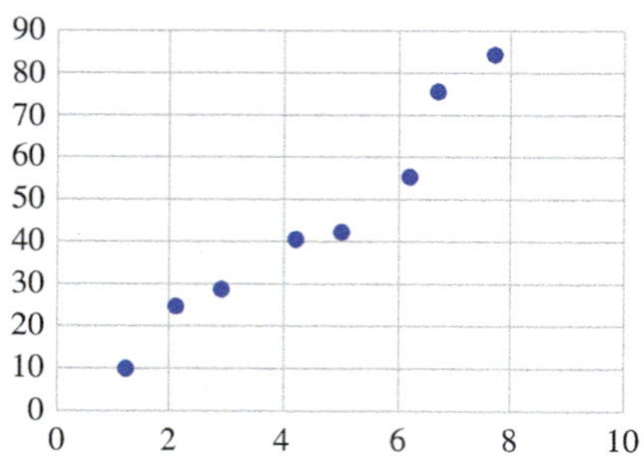

3. 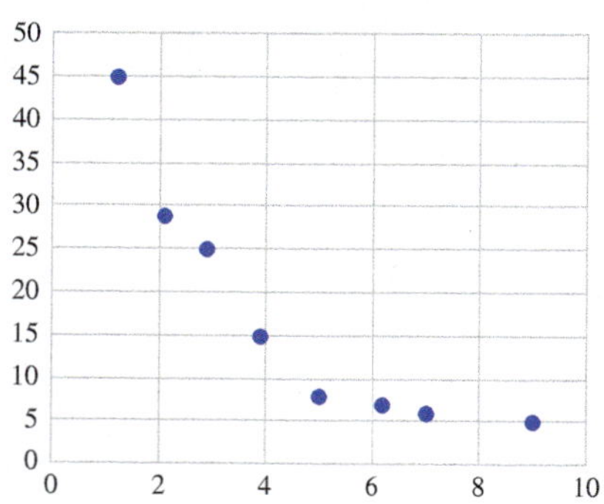

If you covered Lesson 4-8 in your course, then you're already familiar with the idea of finding lines of best fit. In that case, think of the next couple of pages as review. If, on the other hand, you haven't covered Lesson 4-8, then this will be your introduction to using technology to find modeling functions.

The process of finding a function that best models a given data set is called **regression.** In short, you will enter data into the calculator or spreadsheet as inputs and outputs, and built-in programs will run statistical routines that identify the equation of a certain type that comes closest to fitting the data. It's like magic!

Using Technology: Finding the Line of Best Fit

To find and graph a line of best fit for two data sets that appear to have an approximate linear relationship:

Graphing Calculator

1. Press STAT ENTER to get to the list editor.
2. Enter the data set you want to use as inputs (*x* values) under **L1.**
3. Use the right arrow key > to access **L2,** then enter the data set you want to use as outputs (*y* values). When entering data from a table, make sure you enter the second list in the same order as the first.
4. Turn on **Plot1** by pressing 2nd Y= 1. Set up the screen, as shown below.
5. Press Y= and if the **Y=** screen isn't blank, move the cursor over any entered equations and press CLEAR.
6. Press ZOOM 9 which is the **Zoom Stat** option; this automatically sets a graphing window that displays all of the plotted points.
7. Press STAT followed by the right arrow key to access the **STAT CALC** menu, and choose the **LinReg** (for linear regression) option, which is choice 4. Then press ENTER to calculate the line of best fit. The display will read "y = ax + b", then provide values for *a* and *b*. The equation of the line of best fit is written by substituting the given values for *a* and *b* into the equation $y = ax + b$.
8. Press Y= and enter the equation of the line of best fit, then press GRAPH to display the scatter plot along with the line of best fit.

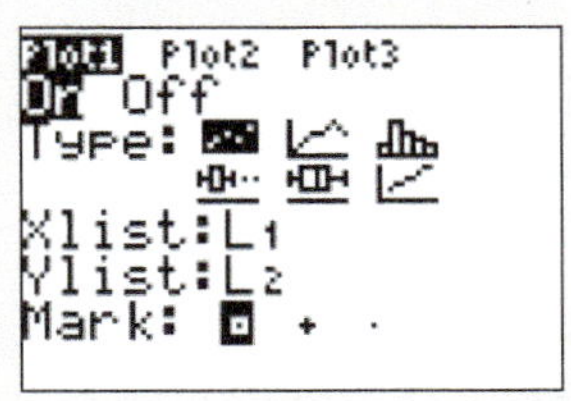

Excel

1. Enter the data in two columns and create a scatter plot.
2. Right-click on one of the points on the scatter plot, then choose "Add Trendline" from the **Chart** menu.
3. In the formatting dialog box that appears, click "Options," then click the checkbox for "Display equation on chart."

See the Using Tech videos on Finding the Line of Best Fit with a Calculator and Excel for further information.

Let's walk through an example of this process on a graphing calculator, using the long jump data from the lesson opener.

Solved Example

Finding a Line of Best Fit

Find a linear equation to model the length of the Olympic-winning men's long jump for the years from 1900 to 2000. Use years after 1900 as input and length in meters as output.

SOLUTION

First, we'll look at the process on a graphing calculator. We begin by hitting STAT-ENTER and entering the x values from the table in list L1 and the accompanying lengths in list L2. Note that only part of the data are shown in the screenshot.

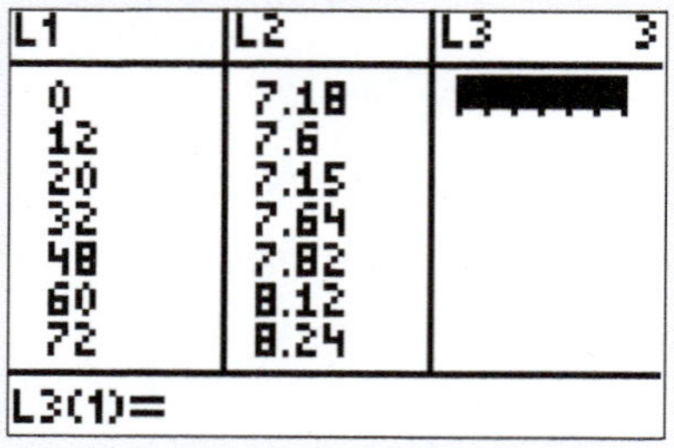

Next, we hit $2^{nd} - Y = -1$ to turn on the scatter plot, choosing the setting as shown. We follow this with ZOOM-9 which produces the scatter plot that should look familiar from the lesson opener.

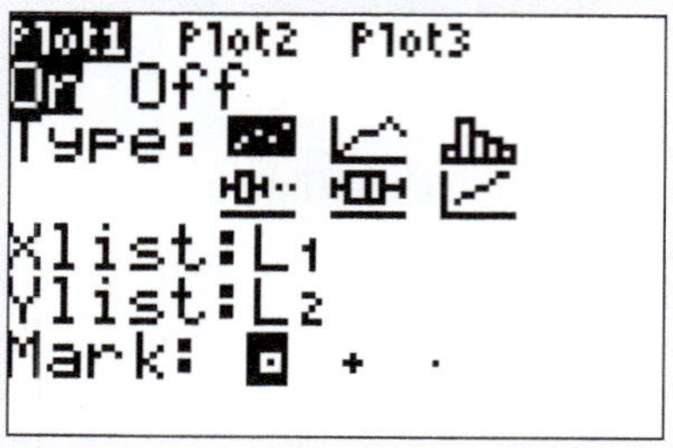

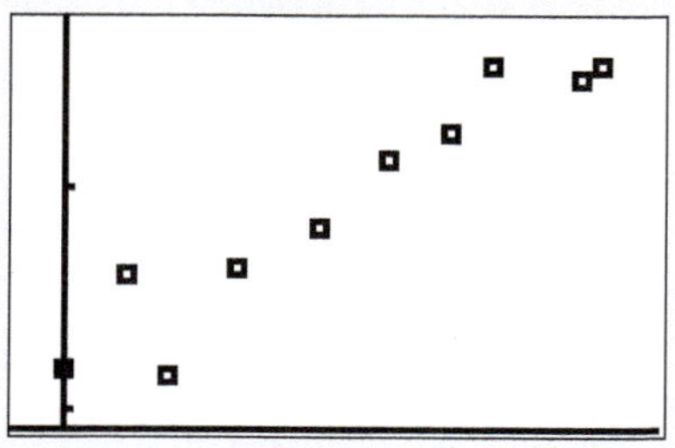

Now here's where the real magic happens. We hit STAT and right arrow to get the STAT-CALC menu and then hit 4 to choose LinReg(ax+b). We follow that up with a few ENTERS until the display shows the equation of the line that comes closest to fitting the data that we entered. The way it displays, though, requires a bit of interpretation.

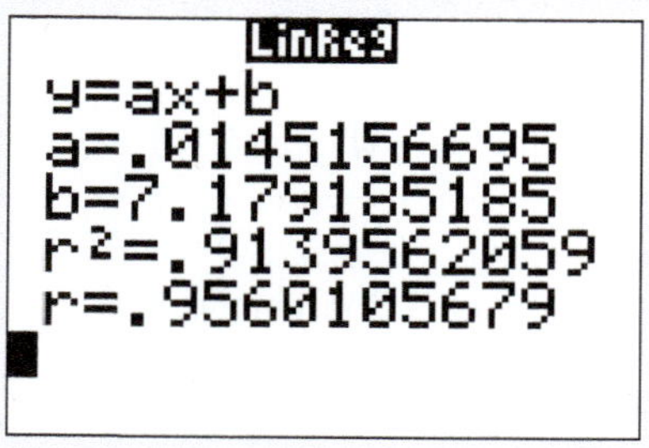

The key is the top line: The calculator is telling us that it found an equation of the form $y = ax + b$ (which of course is why it's a linear equation). It then tells us what numbers to use for the coefficients a and b. In this case, it says that a is about 0.015 and that b is about 7.179. That tells us that the line of best fit is

$$y = 0.015x + 7.179$$

Solved Example *Continued*

To check our work, we then entered this equation in the Y= screen and hit the GRAPH key. Looking at the result gives us some perspective on what the line of best fit is. We already know that no line can go through all of the data points exactly. The line of best fit is designed to minimize the total distance from the data points to the line. In essence, it "splits the difference" as efficiently as possible. Hopefully a look at the graph will allow you to think about what that means.

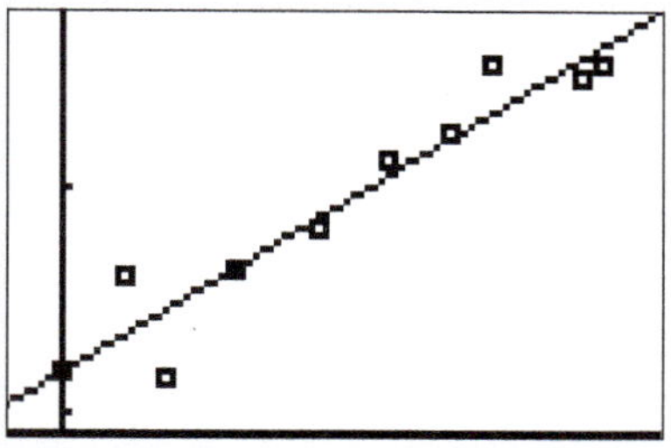

But is this a reasonable model? Are we supposed to just decide by eyeballing the graph? Yes on the reasonable, no on the eyeballing. Looking back at the screen that gave us the equation of the line, notice that there are two pieces of information we haven't described yet: r^2 and r. The number r (0.956 in this case) is the calculator's way of describing how well a linear equation models the data. There's quite a bit more information on this number, called the **correlation coefficient,** in Lesson 4-8. For our purposes, we'll just need to know that $r = 1$ indicates a perfect linear relationship, and the closer r is to 1, the more effective the model is. The fact that the correlation coefficient is almost 0.96 in this case indicates that a linear equation should do a very effective job of modeling the data. If your calculator isn't displaying r^2 and r, hit $2^{nd} - 0$, then scroll down to select "DiagnosticOn" and hit ENTER.

Now we'll look at finding a line of best fit using Excel. We enter the x values and the lengths from the table in two columns, then select the data and choose scatter plot from the Insert pane or menu.

x	Length
0	7.18
12	7.6
20	7.15
32	7.64
48	7.82
60	8.12
72	8.24
80	8.54
96	8.5
100	8.55

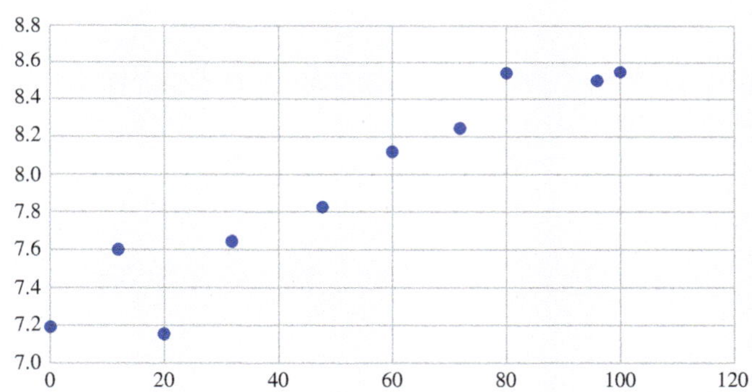

On the scatter plot, right clicking any of the points will bring up a contextual menu, on which you'll choose Add Trendline. This adds the line to the graph and brings up a formatting pane, with these choices:

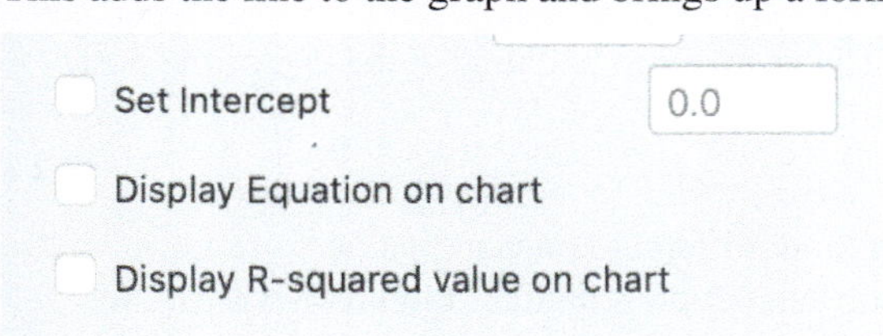

Clicking the "Display Equation on chart" and "Display R-squared value on chart" boxes will, as you might expect, show the desired information. Since the value we're getting is r^2, you can enter =SQRT(followed by the given r^2 into an empty cell to get the correlation coefficient. You will need to include a negative sign if the slope of your trendline is negative.

After checking those boxes, Excel gives the equation $y = 0.0145x + 7.1792$, and an r^2 value of 0.914. The square root of that number rounds to 0.956, so both answers agree with the graphing calculator results.

Now that we've found a model for the long jump length, let's do some stuff with it.

4. The next table lists the length of the winning jumps for some years not in the table. Use the model from the Solved Example to find the predicted length of each winning jump.

Year	x	Length (meters)
2004	104	8.59
2008	108	8.34
2012	112	8.31
2016	116	8.38

Math Note

If one quantity goes down while the other goes up, the correlation coefficient will be negative. In this case, if it's close to −1, that indicates that the data are linearly related.

5. How do the predictions compare to the actual lengths? What can you conclude?

6. Do an Internet search for "1968 men's long jump." After reading about the event that year, explain why we purposely left that year out of the table.

Did You Get It

Try this problem to see if you understand the concepts we just studied. The answer can be found at the end of the portfolio section.

1. Forensic scientists occasionally have the unpleasant but important duty of trying to identify the remains of a human being. They can estimate a person's height based on the length of different bones in that person's body. The table shows some experimental data comparing the length of certain people's femurs (thigh bones if you're not into anatomy) to those people's heights. Find the line of best fit for the data, and use it to predict the height of a person with a 41.5 cm femur.

Length of Femur x (cm)	43.0	43.5	44.0	44.5	45.0	45.5	46.0	46.5	47.0	47.5
Height y (cm)	164	166	166	168	170	169	172	173	175	175

Recall that when we navigated to the STAT-CALC menu, linear regression was just one of the choices. That's a good thing because data come in many different flavors, and regression is not a one-size-fits-all process.

In architecture and engineering, it's very important to know how much weight support structures are able to bear. The following experimental data relate the maximum load in kilograms that a beam can support to its depth in centimeters. In Questions 7–12, you'll use regression to study this relationship.

Depth, *x* cm	Load, *y* kg
10	4,900
12	7,200
14	9,900
16	12,500
18	16,000
20	20,000
22	24,000
24	28,500

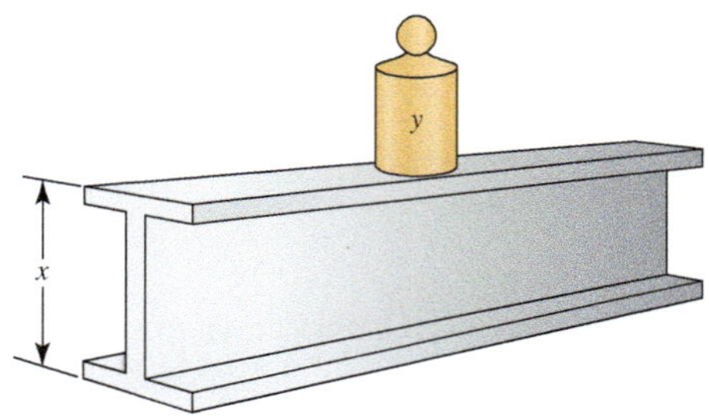

7. Use a calculator or spreadsheet to create a scatter plot of these data.

8. Based on your scatter plot, do you think a linear or quadratic model seems more appropriate?

9. Use linear regression to find the line of best fit, and record the correlation coefficient. Round a and b to the nearest whole number.

10. Now find the quadratic regression model. The command you want is QuadReg, which is choice 5 on the STAT-CALC menu. Round coefficients to one decimal place if necessary. (Note: In Excel, to find the quadratic function of best fit, after choosing "Insert trendline," go to "format trendline" and select "Polynomial," then choose 2 on the "Order" menu.)

11. Your calculator may have displayed R^2 rather than r. This is called the coefficient of determination and is the square of the correlation coefficient. If R^2 is displayed on a calculator or a spreadsheet and you want r, the correlation coefficient, then calculate the square root of R^2 to find the correlation coefficient. Write the correlation coefficient, and discuss which model seems like the best fit by comparing the values of r.

12. Both models provide a good fit, but it's a known principle of physics that the relationship between depth and load actually is quadratic. Using your quadratic model, predict the load that a 15-inch beam can support.

Did You Get It

2. My speed vs. miles per gallon data from Lesson 5-4 are repeated here. Use the QuadReg command to find the quadratic function of best fit and compare to the model from that lesson. If I went completely bonkers and drove 100 miles per hour, how many miles per gallon would I expect to get? What can you conclude? (Legal disclaimer: If you actually drive 100 miles per hour and you're not a race car driver, you're a jerk.)

Speed (mph)	30	35	40	45	50	55	60	65	70
MPG	31.2	32.6	36.3	34.9	31.7	29.2	28.1	25.9	25.4

5-5 Group

The amount of power an engine produces depends on the speed at which it is running in revolutions per minute (rpm). The graph relating speed to power is called a power curve. For some engines, this power curve is approximately parabolic. The point of maximum power is important because engines usually perform best when running faster than the maximum power.

Suppose the engine on a boat is temporarily put under extra load due to a wave or debris on the propeller. For an engine running faster than the maximum power level, this will briefly slow down the engine, increase the power, and smoothly overcome the extra load. We can use data collected on an engine to determine its power curve and its maximum power level. This is illustrated in Questions 1–7.

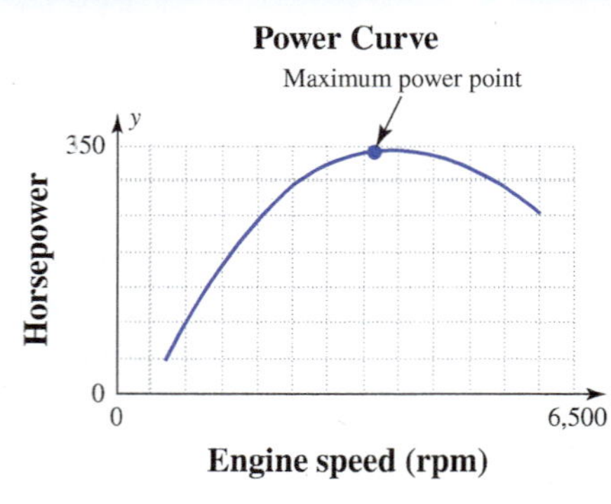

1. The data points shown in the table give the horsepower of one engine running at some tested revolutions per minute (rpm). Draw a scatter plot by hand for these data, then create one on a calculator or spreadsheet and compare.

Revolutions per Minute, x	Horsepower, $f(x)$
0	0
500	130
1,000	240
2,000	340
3,000	280
3,500	180
4,000	50

2. What kind of model do you think would be appropriate for these data?

3. Create a quadratic power curve that best fits these data. Round the coefficients of this quadratic function to three significant digits. (If you're unfamiliar with significant digits, round each coefficient so that it has three digits not counting any zeros that immediately follow a decimal point.)

4. Use your power curve to find the point of maximum power for this engine. Round to the nearest whole number.

5. Write a verbal description of exactly what the maximum power point tells us about the engine. Make sure you mention both coordinates.

6. Compare the horsepower at 3,000 rpm from the provided data with the horsepower at 3,000 rpm predicted by your model. What can you conclude?

7. What horsepower does the model predict for 4,500 rpm? What can you conclude? Think about the physical application of the math!

The following table shows the median price for all homes sold in the U.S. in January for each year from 2003 to 2018. Then we see two scatter plots: The first is for home prices from 2003 to 2009, and the second from 2010 to 2018.

Year	Median Price
2003	$186,000
2004	$212,700
2005	$232,500
2006	$247,700
2007	$257,400
2008	$233,900
2009	$208,400
2010	$222,900
2011	$226,900
2012	$238,400
2013	$258,400
2014	$275,200
2015	$289,200
2016	$299,800
2017	$313,100
2018	$331,800

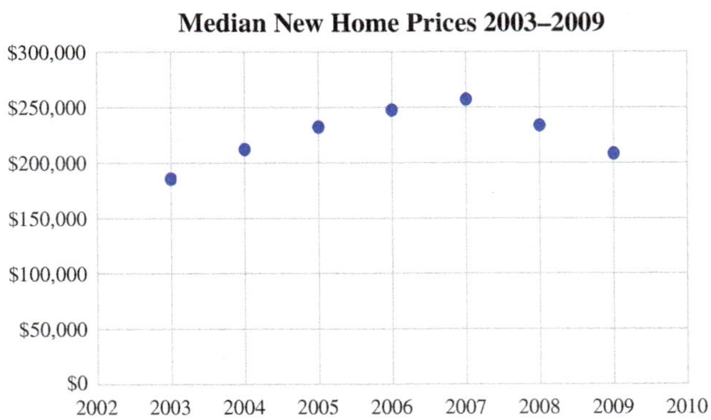

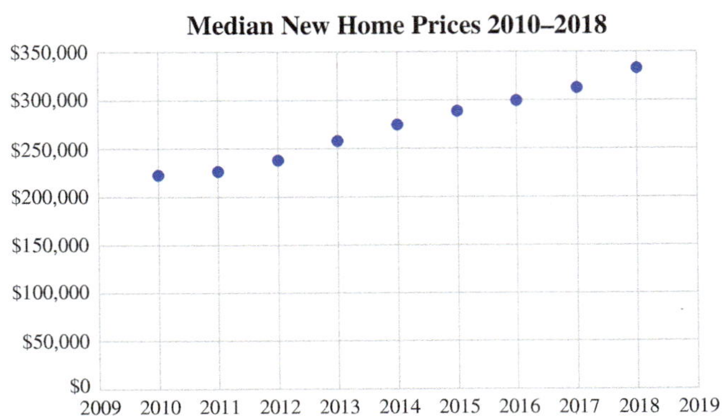

8. Which type of model appears to be a good fit for the 2003–2009 data? What about for the 2010–2018 data?

9. Use a graphing calculator or a spreadsheet to find the models you decided on in Question 8. For best results, don't use years as input; use years after 2003. Round coefficients to the nearest whole number.

10. What's the value of the correlation coefficient for each model that you found? Explain if these results back up the choices you made in Question 8.

11. Using past data to predict future behavior is a dicey proposition because trends change all the time. Do an Internet search for the median price for new home sales in the U.S. in January of the current year, then see if either of the models you found in Question 9 provides an accurate prediction.

5-5 Portfolio

Name ______________________________

Check each box when you've completed the task. Remember that your instructor will want you to turn in the portfolio pages you create.

ALEKS Practice

1. ☐ Include any written work from the ALEKS assignment along with any notes or questions about this lesson's content.

Applications

1. ☐ Complete the applications problems.

Reflections

Type a short answer to each question.

1. ☐ Describe what the line of best fit is for two data sets that appear to be linearly related.
2. ☐ What is the correlation coefficient for a regression function? How can you interpret the correlation coefficient?
3. ☐ After finding a function of best fit, would you expect the function outputs to match data values exactly? Discuss.
4. ☐ Name one thing you learned or discovered in this lesson that you found particularly interesting.
5. ☐ What questions do you have about this lesson?

Looking Ahead

1. ☐ Complete the prep skills for Lesson 5-6.
2. ☐ Read the opening paragraph in Lesson 5-6 carefully and answer Question 0 in preparation for that lesson.

Answers to "Did You Get It?"

1. $y = 2.52x + 55.72$; 160.3 cm **2.** $y = -0.0106x^2 + 0.8409x + 16.782$; this matches the model from Lesson 5-4 (because that's how I found it!). For 100 mph, the model predicts mileage of −5.2 mpg. This indicates that the model is not useful at speeds that high.

Answers to Prep Skills Questions

1. For $x = 2$, $y = 5.4$; for $x = 8$, $y = -3.6$.
2. For $x = -2$, $y = -20.2$; for $x = 0$, $y = -5.6$.
3. (0.30, −5.34)

5-5 Applications

Name ______________________________

It's common knowledge that smoking during pregnancy is a tremendously bad idea. In particular, smoking has a known connection to lowered birth weights. Some experimental data relating the number of cigarettes smoked daily to birth weight are shown in the table.

Number of Cigarettes *x*	Birth Weight *y* (kg)
5	3.20
10	3.15
15	3.10
20	3.06
25	3.09
30	3.05
35	3.01
40	3.00

1. Draw a scatter plot for the data, and write what type of function would be a good choice to model the relationship. Explain why you made the choice you did.

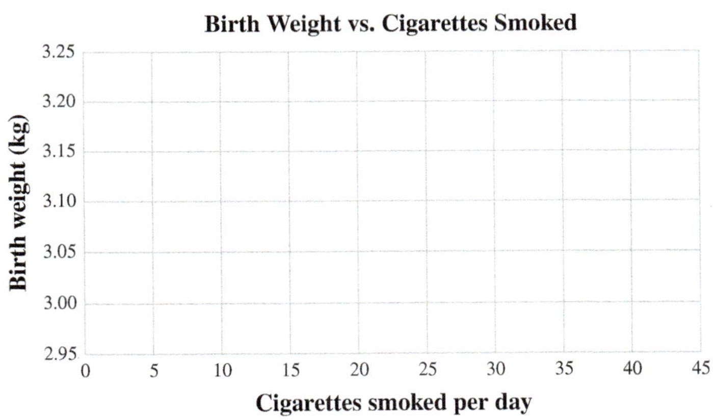

2. Find the line of best fit for these data, including the correlation coefficient. (Note: You should expect a negative value of r since the weights go down as the number of cigarettes goes up.)

5-5 Applications

Name ____________________

3. Based on the correlation coefficient, was a linear function a good choice? Explain.

4. What does the slope of the line tell us about birth weights? Be specific.

5. Use your model to predict the birth weight for a mother who smoked 18 cigarettes per day.

Next we'll study the number of women serving in the U.S. House of Representatives starting in 1979. The first table and scatter plot show the data for every 4 years from 1979 to 2015.

Year	x	Women in the House
1979	79	16
1983	83	22
1987	87	23
1991	91	28
1995	95	48
1999	99	56
2003	103	60
2007	107	72
2011	111	73
2015	115	84

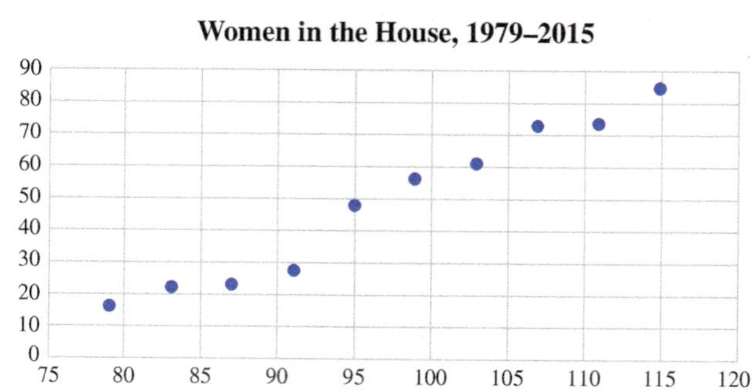

6. I think that a linear function could model these data pretty accurately. Find the line of best fit and the correlation coefficient, then comment on whether you agree with my assessment that a linear model is a good choice. Make sure that you use years after 1900, shown in the x column, rather than actual years as input.

5-5 Applications

Name ______________________________

Something interesting happened starting in 2019: The number of women in the House increased dramatically, continuing a relatively recent trend. Here are the data and scatter plot for every 2 years from 2007 to 2023.

Year	x	Women in the House
2007	107	72
2009	109	73
2011	111	73
2013	113	80
2015	115	84
2017	117	87
2019	119	102
2021	121	120
2023	123	124

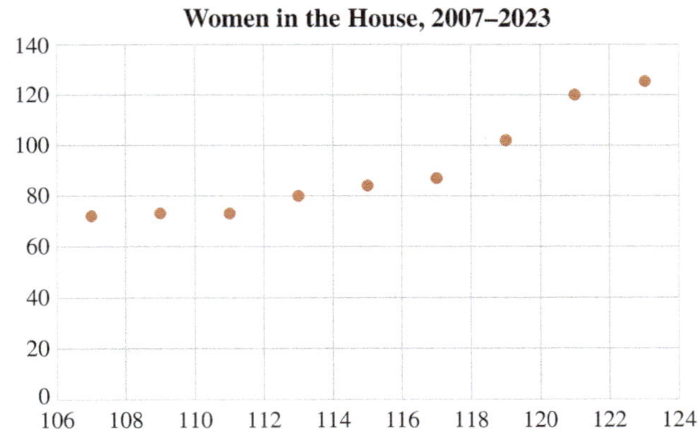

7. If you find the line of best fit for these data, the correlation coefficient drops to about 0.94. Still pretty high, but maybe a quadratic function will model these data better. Find a quadratic model along with the correlation coefficient.

8. Which is the better choice for this second data set, linear or quadratic? Why?

9. Now let's study the models you found. Which one gives a more accurate prediction of the number of women in the House for 2003? What about for 2019?

5-5 Applications

Name ______________________________

10. In 1971, there were 13 women in the House. What does each model predict for that year? Comment on the accuracy.

11. How many women does each model predict will be in the House in 2050? Do you find either prediction to be reasonable?

12. Based on our entire study of women in the House of Representatives, discuss the effectiveness of each model. Consider as many aspects as you can—how good the fit is, short-and long-term predictions, etc.

Lesson 5-6 Prep Skills

SKILL 1: EVALUATE EXPRESSIONS WITH EXPONENTS

Evaluating expressions with exponents requires remembering the order of operations, which is a group of agreed-upon rules that allow us to write calculations involving more than one operation without always having to use parentheses to indicate which operation to do first. The order is as follows, with operations higher on the list being performed before those lower on the list.

1. Grouping symbols — Perform these operations first. If there is more than one set of grouping symbols, work from innermost outward.
2. Exponents
3. Multiplication or division — For operations that live at the same level, work from left to right.
4. Addition or subtraction

Often expressions with exponents will require a calculator, spreadsheet, or other computing device. To evaluate these expressions correctly, it's important to pay attention to the intended meaning of the expression. That means looking carefully at exponents and grouping symbols (including fraction bars) that group the entire numerator and denominator separately.

- Use a calculator to evaluate $500(1 + 0.075)^{24}$.
  Enter as
  500*(1 + 0.075)^24 ≈ 2,836.44.

  Calculators will usually interpret implied multiplication (* would not be needed), but spreadsheets require * for multiplication.

SKILL 2: FIND INTERCEPTS FOR A FUNCTION

When finding the intercepts of a function, it helps to remember that an intercept is a point on a graph.
Finding the *x* intercept: The *x* intercept is the point where a graph intersects the *x* axis. Remember that all points on the *x* axis have a *y* coordinate of zero. To find the *x* intercept, we substitute 0 for *y* and find the associated *x* value(s).
Finding the *y* intercept: The *y* intercept is the point where a graph intersects the *y* axis. Remember that all points on the *y* axis have an *x* coordinate of zero. To find the *y* intercept, we substitute 0 for *x* and find the associated *y* value.

It is possible for a function to have more than one *x* intercept but not more than one *y* intercept.

- Find the intercepts of $f(x) = 5x - 3$.

  x intercept:

  $0 = 5x - 3$

  $3 = 5x$

  $\frac{3}{5} = x$

  The *x* intercept is $\left(\frac{3}{5}, 0\right)$.

  To find the *x* intercept, substitute 0 for $f(x)$, which is another name for *y*, and solve the resulting equation to find any *x* values associated with $y = 0$.

  y intercept:

  $f(0) = 5(0) - 3$

  $f(0) = -3$

  The *y* intercept is $(0, -3)$.

  To find the *y* intercept, substitute 0 for *x* and find the *y* value.

SKILL 3: SOLVE EQUATIONS USING RADICALS

Nearly every tool we use for solving equations involves the idea of "undoing" the action of a number or symbol. For example, to solve $x + 2 = 5$ we undo the action of the "+2" by subtracting 2 from both sides of the equation. For certain types of equations involving exponents, we use radicals to undo the exponents. This idea of a given radical acting as the opposite of the matching power is important to understand for Lesson 5-6.

- Solve $x^2 = 25$.

$\sqrt{x^2} = \sqrt{25}$	Square root (second root) acts as the opposite of second power. Take the square root of both sides to undo the power of 2 on x.
$x = 5$ or $x = -5$	Any solution to this equation has to involve 5, the square root of 25. Notice that both 5 and −5 will make the equation true. $5^2 = 25$ and $(-5)^2 = 25$, so both are solutions.

- Solve $x^2 = 7$.

$\sqrt{x^2} = \sqrt{7}$	Take the square root of both sides to undo the power of 2 on x.
$x = \sqrt{7}$ or $x = -\sqrt{7}$	Any solution to this equation has to involve the square root of 7. Notice that both $\sqrt{7}$ and $-\sqrt{7}$ will make the equation true. $(\sqrt{7})^2 = 7$ and $(-\sqrt{7})^2 = 7$
	Sometimes a decimal solution is needed. In decimals, $\sqrt{7} \approx 2.65$, so $x = 2.65$ and $x = -2.65$ would be decimal approximations. Check that $(2.65)^2 \approx 7$ and $(-2.65)^2 \approx 7$.

- Solve $x^3 = 27$.

$\sqrt[3]{x^3} = \sqrt[3]{27}$	Cube root (third root) acts as the opposite of third power. Take the cube root of both sides to undo the power of 3 on x.
$x = 3$	Check: $(3)^3 = 27$

SKILL 4: SOLVE A LINEAR EQUATION

As we have seen in previous lessons, solving equations involves performing a series of algebraic steps that produce simplified but equivalent equations. Remember, equivalent equations are equations that have the same solution(s).

Here is a general strategy that will work for solving any linear equation. You may not need all of these steps, but working through the steps you do need, in order, will lead to a solution if one exists.

Step 1: Simplify each side of the equation.

a. If the equation contains fractions, simplify by multiplying both sides of the equation by the least common denominator (LCD) of all the fractions.

b. If the equation contains grouping symbols, simplify by using the distributive property to remove the grouping symbols and then combine like terms.

Step 2: By adding or subtracting the same quantities from both sides of the equation, isolate the variable terms on one side of the equation and the constant terms on the other side.

Step 3: By multiplying or dividing both sides of the equation by the same quantity, solve the equation produced in step 2.

The following equation requires steps 2 and 3.

- Solve $3x + 7 = -23$.

$$3x + 7 - 7 = -23 - 7$$ Subtract 7 from both sides.

$$3x = -30$$

$$\frac{3x}{3} = \frac{-30}{3}$$ Divide both sides by 3.

$$x = -10$$ The solution is $x = -10$.

PREP SKILLS QUESTIONS

1. Find the value of each expression.

   $100(1.15)^{15}$ $\quad$ $31{,}000\left(1 + \frac{0.04}{12}\right)^{24}$

2. Find all intercepts for $f(x) = -23x + 70$.

3. Find all solutions to the equation $y^2 = 144$.

4. Find all solutions to the equation $x^3 = 90$.

5. Solve the equation: $12 + 3x = 40$

6. Solve the equation: $-7.3 - 2.3f = 11.1$

Lesson 5-6 Going Viral

LEARNING OBJECTIVES

- ☐ 1. Identify quantities that are and are not good candidates to be modeled with exponential equations.
- ☐ 2. Solve problems by exponential modeling, both algebraically and using technology.
- ☐ 3. Define logarithms as inverses of exponentials.
- ☐ 4. Solve problems by logarithmic modeling, both algebraically and using technology.

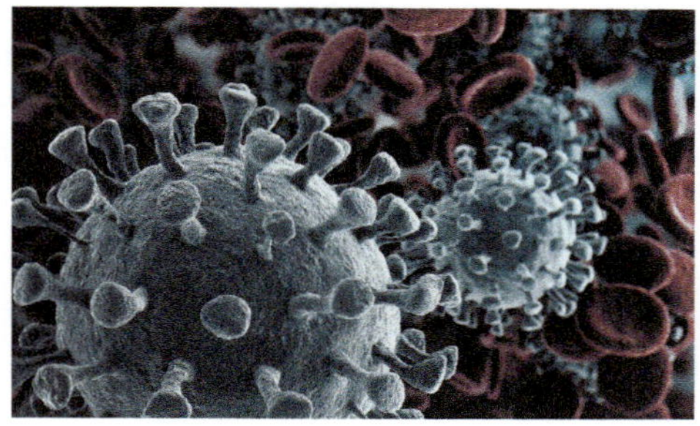
Creativeneko/Shutterstock

Every time one person gets a piece of information, the likelihood of that information being exposed grows exponentially. It's no longer two people. It's two people squared.

—Jon Cryer

In late spring 2020, health officials worldwide started to warn the public about a new version of a type of pathogen known as coronavirus. On March 1 of that year, there had been a total of 29 confirmed cases of what would eventually come to be called COVID-19 in the United States. Some very prominent people proclaimed that the threat was being overstated, that it was a media creation, that the number of cases would be down to zero in short order. As we know, they were tragically wrong. Within two weeks the number of cases in the United States was being measured in the thousands, and of course that was just the beginning of a long health nightmare. In those early days, in fact, the number of cases increased so rapidly that the types of functions we've modeled with so far aren't very effective: We'll need to develop a different type of function that allows for a rapidly increasing growth rate. Does that ring a bell? Think about ways to grow your fortune . . .

0. How can you tell that the growth shown in the graph below couldn't be modeled well by a linear function?

5-6 Class

If we look at a bar graph of COVID-19 cases in the first twelve days of March 2020, there's one striking feature: It starts low, then grows in a BIG hurry.

COVID-19 Cases in March 2020

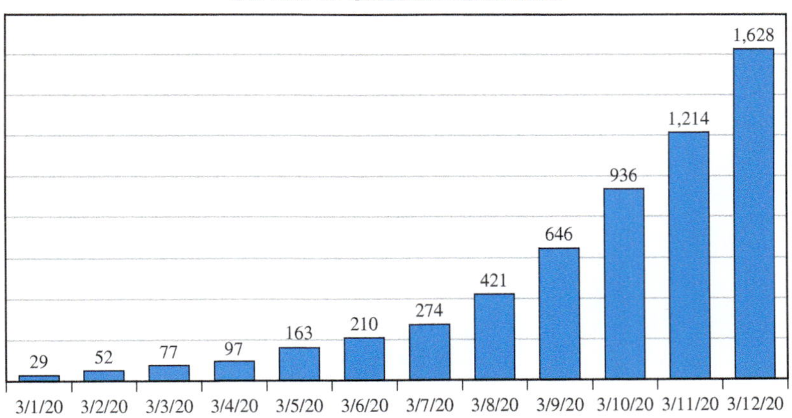

Every one of the functions we've studied so far in this book has one thing in common: All the exponents are numbers. Functions that model rapid growth, like we're seeing in the COVID graph, often have the variable in an exponent. This changes almost everything in terms of the behavior of the function, so the functions known as *exponential functions* merit some special attention. These functions are very useful in modeling situations we've examined in previous units like investments and loans. They are also useful for modeling depreciation, growth and decay, and population changes.

To understand these functions, we'll look at some of the same key characteristics we've studied about linear and quadratic functions. We'll also look at some new ones.

An **exponential function** has the form $f(x) = c \cdot b^x$, where b is a positive real number, but not 1, and c is any nonzero real number.

1. Which of the following are examples of exponential functions?

   a. $f(x) = 2^x$ **b.** $g(x) = 200 \cdot 10^x$ **c.** $h(x) = 5\left(\frac{1}{3}\right)^x$ **d.** $y(x) = (0.5)^x$

Math Note

Why the restrictions on the base of the exponent b in the definition of exponential function? If $b = 0$, then $f(x) = c \cdot b^x$ always has output zero. If $b = 1$, the output is always c. If b is negative, we get a MESS: The graph has infinitely many holes in it!

Let's start by looking back at a situation we've already studied—compound interest.

The formula $A = P(1 + r)^t$ tells us how much an account that starts at P dollars is worth after t years at an interest rate r, where interest is compounded annually.

2. For an account that currently has $5,000 in it and grows at a rate of 3% compounded annually, write a function $A(t)$ that describes the value of this account after t years.

One of the things we typically find when graphing a function is the y intercept. For exponential functions, this is pretty simple, as we will see shortly.

3. For the function you wrote in Question 2, to find the y intercept, we substitute zero for ______ and find the value of ______.

4. Perform that calculation now and write the y intercept for this function, then plot your y intercept on the graph at the top of the next page.

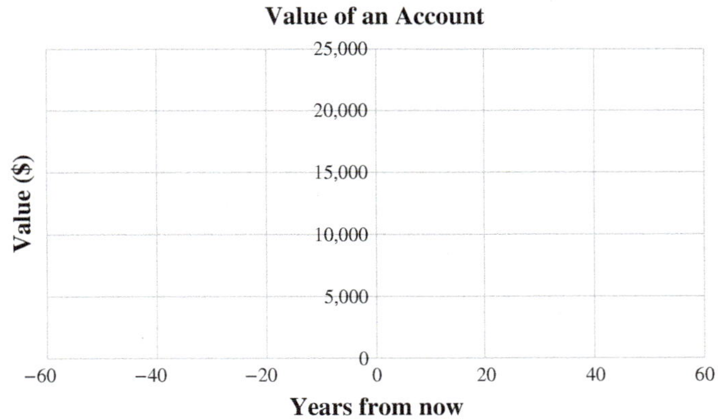

Another key fact we'd like to know is whether there are any x intercepts.

5. In the context of the problem, what would it mean if there was an x intercept?

6. Use a calculator or a spreadsheet to complete the table of values. Then plot these points on the graph.

t	$A(t)$
−40	
−20	
0	
20	
40	

7. What does a time of −40 mean in the context of this problem?

8. Now insert −500 at the top of the t column and 500 at the bottom. Evaluate the function using these two new values. What do you observe?

9. Based on your answers to Questions 7 and 8, do you think this function has an x intercept? Describe what you think happens if you trace the graph further and further in each direction.

Summary of Exponential Function Graph Features

- For any acceptable value of b, the expression $b^0 = 1$, so every exponential function of the form $f(x) = c \cdot b^x$ has y intercept (0, c).
- The graph approaches the x axis in one direction but never touches it. When this happens, we say that the x axis is a **horizontal asymptote** of the graph.
- In the other direction, the height of the graph increases without bound.
- The graph can have two shapes, depending on whether the base b is greater or less than 1.

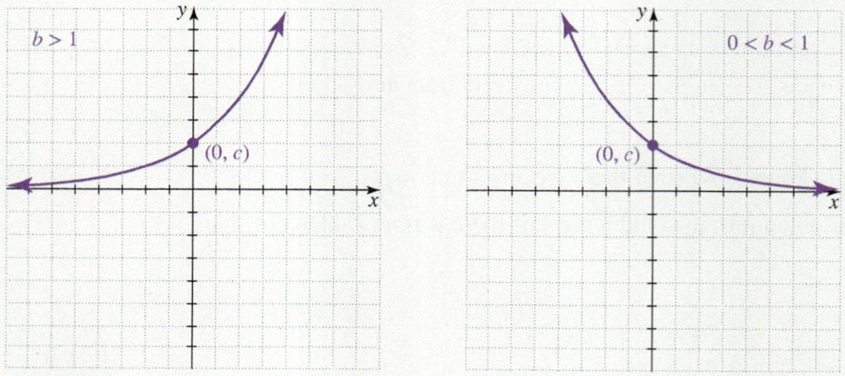

Since all of the exponential functions have the same basic shape, we can draw the graph by plotting a handful of points and drawing a curve similar to the ones in the purple box.

10. Use this knowledge to complete your graph of the exponential function on Page 637.

11. Following is a table of values for the function representing the account that started at $5,000. Recall that the percent change in some quantity is the actual change (new value – old value) divided by the old value. Find the percent change for each 1-year period until you observe a clear pattern. What can you conclude?

t	$A(t)$
0	5,000
1	5,150
2	5,305
3	5,464
4	5,628

Question 11 is really important to understanding what exponential functions are all about. We've seen an algebraic definition—a function of the form $f(x) = c \cdot b^x$—but from an applied perspective, the result tells us something important. The thing that makes a function exponential is the fact that it changes by the same percentage for each 1-unit change in the input. In this case, the new amount is always 3% more than the previous.

So what's the key difference between linear change and exponential change? Linear change happens when the AMOUNT of the change is constant. Exponential change happens when the PERCENT CHANGE is constant. This also explains why there's no x intercept: If the value is ever zero, any percentage of zero is still zero, and the amount would stay at zero forever, which we can all agree is not particularly interesting.

When investing a certain amount of money, like say $5,000 at 3% interest compounded annually, it certainly would be reasonable to wonder how long it would take the investment to grow to a certain amount. Let's see what that would look like.

12. If we wanted to know how many years it took until the account was worth \$10,000, what equation would we need to solve? Write it down. Don't try solving it yet. We'll get there.

13. Use the graph you drew on Page 637 to estimate the solution to this equation.

14. Use the **Intersect** feature on your calculator to get a really good approximation for the solution to this equation. (This command was introduced in Lesson 5-3.)

15. We're ready to start solving the equation. First, divide both sides of the equation by 5,000. What's left?

Now we have a problem. In order to solve the equation, we'll need a way to get the variable out of the exponent, and none of our usual algebraic tools can do that. A lesser scholar might quit at this point, but not you! In much the same way that we use roots to "undo" powers when solving equations, we can use a group of functions called **logarithms** to "undo" exponentials.

Roots are actually a really good analogy for understanding logarithms, or logs for short. There are many different powers: square, cube, fourth power, etc., so there are many different roots: square root, cube root, fourth root, etc. Exponential functions have many potential bases, so there are many different logarithms, depending on the base. For example, the logarithm function with base 3, written $\log_3 x$, is the function that "undoes" an exponential with base 3. That is,

$$\log_3(3^x) = x \qquad \text{Compare this to } \sqrt[3]{x^3} = x.$$

For any base *b* that is positive but not 1, the **logarithm function with base *b*,** denoted $\log_b x$, is a function with the following property: For any real number *x*, $\log_b(b^x) = x$.

Solved Example

Solving an Exponential Equation

We've already seen that the solution to the equation $10{,}000 = 5{,}000(1.03)^t$ provides the length of time needed for an investment of \$5,000 at 3% interest to grow to \$10,000. Use the definition of logarithms to solve the equation and find the time needed.

SOLUTION

As we suggested, the first step in solving an equation where the variable lives in an exponent is to isolate the base to a power. In this case, dividing both sides by 5,000 left us just $(1.03)^t$ on one side: That's what we're shooting for.

$$\frac{10{,}000}{5{,}000} = \frac{5{,}000(1.03)^t}{5{,}000} \Rightarrow 2 = 1.03^t$$

Now here's the issue: The variable that we'd like to isolate is in the exponent. It's common for students to want to divide by 1.03, but that won't work. The right side isn't 1.03 TIMES *t*, it's 1.03 raised to the *t* power. That's where logarithms come in handy: their entire job is to undo exponentials and leave behind the exponent. Perfect.

Solved Example *Continued*

So our job is to apply the correct logarithm function to both sides. Since the base of the exponential is 1.03, we want the base of the log to be 1.03 also. In other words, we apply log base 1.03 to both sides of the equation.

$$\log_{1.03} 2 = \log_{1.03}(1.03)^t$$

Now we use the definition of logarithm: Applying log base 1.03 to an exponential with base 1.03 simply leaves behind the exponent, which is exactly what we want here: $\log_{1.03}(1.03)^t = t$ for the same reason that $\sqrt[3]{6^3} = 6$. And we have a solution!

$$\log_{1.03} 2 = t$$

So how long did it take the money to double? No idea. It's $\log_{1.03}2$ years, whatever that is. That's a pretty unsatisfying answer. Our next order of business is to come up with a way to evaluate expressions like $\log_{1.03}2$.

Now we have an answer, but we can't do much to interpret it until we go to a calculator. The fact that there are many different logarithms (infinitely many, in fact) presents a problem when programming a calculator. You can't put infinitely many log buttons on a calculator: You can try, but you would not be successful. Most calculators have a button for log base 10, which is simply denoted log (if there's no base listed, it means the base is 10). Fortunately, there's a formula that allows us to change the base of a logarithm. Shockingly, it's called the **change of base** formula.

The Change of Base Formula

For any acceptable base b and any positive real number x,

$$\log_b x = \frac{\log x}{\log b}$$

16. Use the change of base formula to put the answer to the Solved Example in terms of log base 10. Then use the log button on a calculator to find a decimal equivalent.

Math Note

In words, the change of base formula says that to calculate the log to any base of some number, you find log base 10 of that number, then divide by log base 10 of the original base.

Finally, we have an answer to the question about how long it will take until our \$5,000 is worth \$10,000.

17. How does your answer compare to the visual estimate and the calculator approximation?

18. Using this same situation, find how long it will take the \$5,000 to be worth \$16,000. Check your answer using the graph and the intersect feature on your calculator.

19. Now find when the \$5,000 was worth \$2,000. Check your answer using the graph and the intersect feature on your calculator.

Did You Get It

Try these problems to see if you understand the concepts we just studied. The answers can be found at the end of the portfolio section.

1. For an account that currently has \$8,000 and grows at a rate of 2.5% compounded annually, write a function $A(t)$ that describes the value of this account after t years.
2. How much will the account be worth in 6 years?
3. How long will it take for this account to grow to \$19,000?

5-6 Group

The town I live in (North Port, Florida) had a little over 6,000 residents in 1980, and by 2024 it had over 90,000. Most of that growth was before I moved here, so my popularity is likely only partly responsible for the growth. The scatter plot shows the city's population from 1980 to 2024.

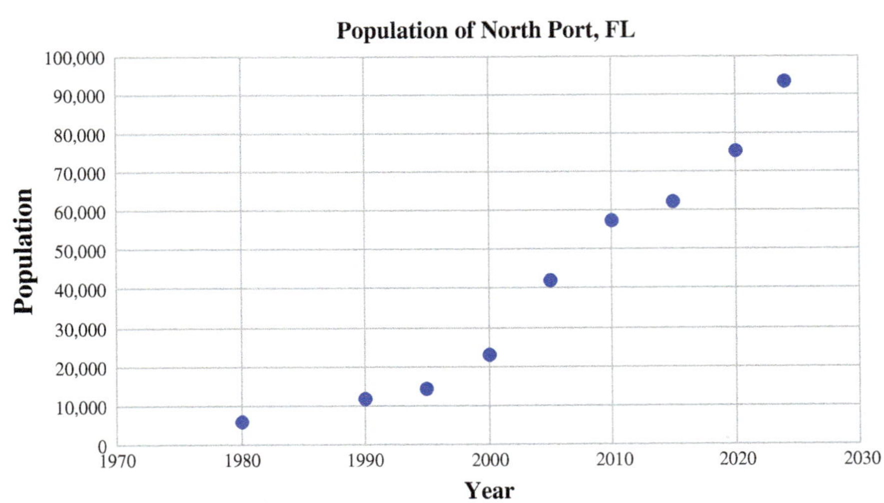

1. What type of model do you think would be best to model this data? Give as many reasons as you can think of.

There was a housing boom and bust in Florida between 2000 and 2010, and the city's population growth was somewhat erratic in that decade. But after reaching a population of 57,392 in 2010, the growth has been more consistent since then. The function $P(t) = A_0(1.037)^t$ can be used to model the population since 2010, where A_0 corresponds to the population at time $t = 0$ and t is in years.

2. Use information in the previous paragraph to write a specific function that describes the population of North Port in terms of years after 2010.

3. Use your model to predict the population in 2030.

4. When does the model predict the population will reach 150,000?

Did You Get It

4. Riverside County in California has been one of the fastest-growing counties in the U.S. for the last decade. One model for the population is $P(t) = A_0(1.02)^t$, where A_0 is the population at time $t = 0$, and t is time in years. The county had grown to 2,189,000 by 2012. What does this model predict the population will be in 2027? Use the model to predict when the population will reach 3,000,000.

As we saw in Unit 2, interest on a savings account can be compounded annually, semiannually, quarterly, or daily. The formula for compound interest is given by the formula

$$A = P\left(1 + \tfrac{r}{n}\right)^{nt}$$

where A is the amount of money, which includes the principal plus the earned interest, P is the principal (amount initially invested), r is the yearly interest rate in decimal form, n is the number of times the interest is compounded each year, and t is the time in years that the principal has been invested. This is another excellent example of a quantity that is modeled with an exponential function.

After being injured in an accident with a driver who was texting while driving, Maureen is awarded an \$80,000 settlement. She wants to invest the money so that she has \$100,000 to put toward a home. If the investment draws 6% interest compounded monthly, how long will it take for her to reach that goal?

5. Use the formula above to write a function describing the amount of Maureen's investment t years from now.

6. How much interest will the investment earn in the first year? Use that to make an educated guess on how long it will take to grow to \$100,000.

> **Math Note**
>
> Statistics show that if you text while driving, you're more than TWENTY times more likely to get into an accident. The National Safety Council reports that cell phone use is a factor in 1.6 million crashes every year.

7. Set up and solve an equation to find how long it will take for the investment to reach \$100,000. Be careful after applying a log to get the variable out of the exponent: You won't quite be done solving at that point.

Did You Get It

5. If Maureen finds an investment that offers 7.5% interest compounded daily, how much sooner will she have the \$100,000 she's hoping for?

Carbon-14 is a radioactive isotope found in all living things. It begins to decay when an organism dies, and scientists can use the proportion remaining to estimate the age of objects derived from living matter, like bones, wooden tools, or textiles. The function

$$f(x) = A_0 2^{-0.0175x}$$

describes the amount of carbon-14 in a sample, where A_0 is the original amount and x is the number of centuries (100 years) since the carbon-14 began decaying. A wooden bowl was unearthed at an archaeological dig in Peru, and when analyzed, it was found to have 62% of the amount of carbon-14 found in a living sample of that type of wood.

8. If A_0 is the amount of carbon-14 that was originally in the sample, write an expression for the amount left in the bowl (which is 62% of the original amount).

age fotostock/Alamy Stock Photo

9. Using your answer to Question 7, set up an equation whose solution will tell you the age of the wood in the bowl.

10. Solve the equation. How old is the bowl?

Did You Get It

6. It's very common for archaeological digs to have several layers of artifacts from different eras. An ax handle found above the wooden bowl in Questions 7–9 has 72% of the original amount of carbon-14 remaining. How much later was it made than the bowl?

We've studied logarithmic functions only from the standpoint of using them to solve exponential equations, but they can be used to model things in our world as well. One example comes from the world of chemistry, where the pH scale is used to measure the hydrogen ion concentration in a solution. Distilled water has a pH of 7, acids have a pH of less than 7, and bases have a pH of more than 7. The formula for the pH of a solution is

$$\text{pH} = -\log(H^+)$$

where H^+ measures the concentration of hydrogen ions in moles per liter (mol/L).

11. The hydrogen ion concentration of a well-known brand of light beer, which may or may not get hauled around on a cart pulled by very large horses with fuzzy white feet, is measured at 1.58×10^{-5} mol/L. What is its pH?

12. The best-selling soft drink in the world, on the other hand, which might come in a red can with a white stripe, has a hydrogen ion concentration of 0.00316 mol/L. Find the pH.

5-6 Portfolio

Name ______________________________

Check each box when you've completed the task. Remember that your instructor will want you to turn in the portfolio pages you create.

Technology

1. ☐ In this Excel activity, we'll compare the exponential model from a calculator with the model generated by Excel.

ALEKS Practice

1. ☐ Include any written work from the ALEKS assignment along with any notes or questions about this lesson's content.

Applications

1. ☐ Complete the applications problems.

Reflections

Type a short answer to each question.

1. ☐ What types of quantities are likely to be modeled well with exponential functions? Why?
2. ☐ What is a logarithm? Why are there many different logarithms? Why are logarithms useful in this lesson?
3. ☐ Name one thing you learned or discovered in this lesson that you found particularly interesting.
4. ☐ What questions do you have about this lesson?

Looking Ahead

1. ☐ Complete the prep skills for Lesson 6-1.
2. ☐ Read the opening paragraph in Lesson 6-1 carefully and answer Question 0 in preparation for that lesson.

Answers to "Did You Get It?"

1. $A(t) = 8{,}000(1.025)^t$ **2.** $A(6) = \$9{,}277.55$ **3.** About 35 years
4. 2,946,105; near the end of 2028 **5.** About 0.7 years, or a bit less than 8 1/2 months
6. About 1,230 years

Answers to Prep Skills Questions

1. About 813.71, about 33,577.43
2. (0, 70) and $\left(\frac{70}{23}, 0\right)$
3. $y = -12, 12$
4. $x = \sqrt[3]{90}$
5. $x = \frac{28}{3}$
6. $f = -8$

5-6 Applications

Name ______________________________

When a drug is administered to a patient, doctors and nurses must be aware of the minimal effective concentration, which is the concentration below which the drug doesn't have the desired effect. One way health professionals talk about this concentration is the half-life of elimination, which is the amount of time it takes for the concentration to be cut in half. The following data are based on theophylline, an antiasthma drug, where the time is the number of hours after the medication peaks and starts to decline in the bloodstream.

Time (hr)	Concentration (mg/L)
0	10
8	5
16	2.5
24	1.25

Terry Vine/Blend Images LLC

1. Many drugs are known to leave the body exponentially. Find the exponential regression equation of best fit using **ExpReg** on a graphing calculator. The process for doing this is identical to that for linear and quadratic regression from earlier in this unit. If you're using Excel, after selecting Add Trendline, you'll choose Exponential in trendline options. In that case, the exponential function provided will have base e; you can replace that with a decimal equivalent of 2.718.

2. Graph the points from the table and the curve of best fit. In this case it should be a perfect fit.

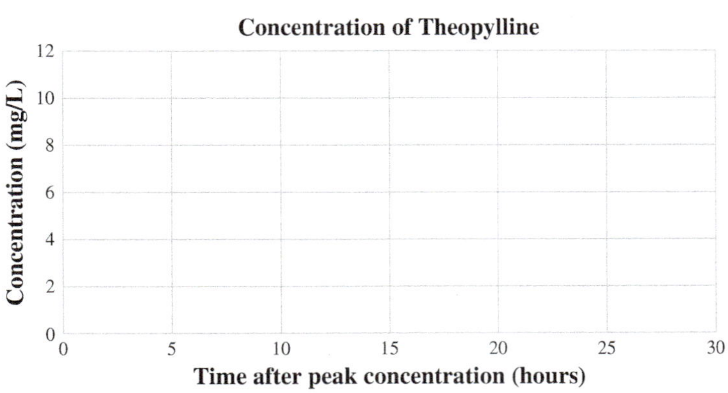

3. Use your equation to find the concentration of this asthma medicine after 4 hours. Does this agree with your graph?

5-6 Applications

Name ______________________________

4. What is the concentration after 32 hours? If you're clever, you won't need your equation.

5. Use your equation to find exactly how long it will take for the concentration to reach the minimal effective concentration of 4 mg/L for this medication. Does your answer match your graph?

The population growth of an emerging African nation is defined by the function $P(t) = A_0(1.4)^t$, where A_0 is the present population and t is the time in decades. The present population is 4,200,000.

6. Write a specific function that describes the population of that nation in terms of decades from now.

7. Find the population in 5 years. Pay close attention to what t represents!

8. If the population continues growing at the same rate, how long will it take to reach 6 million?

5-6 Applications

Name ______________________________

Questions 9 and 10 use the compound interest formula, $A = P\left(1 + \frac{r}{n}\right)^{nt}$.

9. Facing some stiff gambling debts, Lucky Louie turns to a quick loan from a local loan shark. Louie borrows \$2,500 at 70% interest compounded daily. The friendly neighborhood loan shark informs Louie that his thumbs will be broken unless he pays the loan off in full in exactly 1 year. How much will he owe at that point?

10. Louie decides that in the interest of avoiding broken thumbs, he'd just as soon stay well away from the 1-year deadline. Instead, after hitting a winning streak, he decides to pay the loan back when the balance reaches \$3,500. When should he pay?

Questions 11 and 12 use the carbon dating formula, $f(x) = A_0\, 2^{-0.0175x}$. Recall that A_0 is the original amount of carbon-14 in a specimen, and x is the number of centuries since the specimen died.

11. In 2003, Japanese scientists announced the beginning of an effort to bring the long-extinct woolly mammoth back to life using modern cloning techniques, a goal that I think we can all agree is pretty cool. Their efforts were focused on an especially well-preserved specimen discovered frozen in the Siberian ice. Nearby samples of plant material were found to have 28.9% of the amount of carbon-14 in a living sample. If A_0 is the amount of carbon-14 originally in the specimen, write an expression that describes how much was left in 2003.

Leonello Calvetti/Stocktrek Images/Getty Images

5-6 Applications

Name ______________________________

12. What was the approximate age of these samples?

You probably know that when an earthquake is reported, how powerful it was (known as the magnitude) is given as a decimal number somewhere between 0 and 10. The scale used to measure the magnitude is known as the Richter scale, named in honor of a seismologist in California who devised the first version of the scale in 1935. Magnitude is given by the function

$$M(E) = \frac{2}{3}\log\frac{E}{10^{4.4}}$$

where E is the amount of energy released by the quake. Use this information for Questions 13 and 14.

13. On March 11, 2011, a massive earthquake struck off the coast of Japan. The quake and the resulting tsunami killed over 15,000 people and resulted in a well-publicized nuclear emergency when three power plants were damaged. The quake released 7.94×10^{17} joules of energy. What did it measure on the Richter scale?

14. The largest and most powerful nuclear weapon ever detonated was tested by the Soviet Union on October 30, 1961, on an island in the Arctic Sea. The blast was so powerful that there were reports of windows breaking in Finland over 700 miles away. The detonation released about 2.1×10^{17} joules of energy. What would be the magnitude of an earthquake that released that much energy?

Math in Drug Administration REVISITED

SofikoS/Shutterstock

1. Football player: 647 mg; girl: 153 mg
2. The potentially lethal dose for the girl is 574 mg, so his dose could kill her.
3. The minimum effective dose for the player is 404 mg, so her dose will do him no good.
4. The BSA ratio is $\frac{2.60}{1.02} \approx 2.54$. The weight ratio is $\frac{275}{65} \approx 4.23$. The ratio is much bigger for weight than for BSA.
5. 1.95
6. The recommended dosages are now 533 mg for the player and 209 mg for the girl. The lethal dose for the girl is 785 mg, so she'll be okay with his dose. The minimum effective dose for the player is now 333 mg, so her dosage will still do him no good.
7. Conclusions can vary, but it certainly seems like if the dosage is based on weight rather than BSA, it exaggerates the differences in dosage.

Unit 6
The Joy of Sets

John Lund/Blend Images

Outline

Math in Diversity

Demography is the statistical study of the changing characteristics of people in a population. If there's one thing we can guarantee for certain in this field, it's that the racial makeup of the United States population has undergone a radical shift in recent years and that shift is only expected to speed up.

The bar graph below shows the percentage of the total population that fell into four distinct categories in 1960, 2005, and 2020, and what those percentages are projected to be in 2050 by the Pew Research Center, an organization that ranks among the world leaders in population statistics.

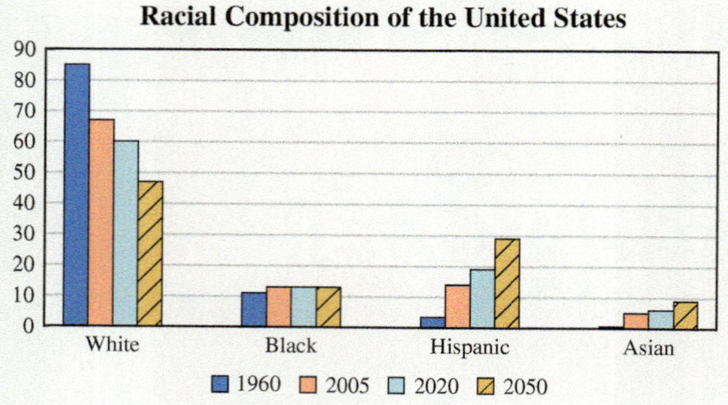

There's a lot going on here: The percentage of whites decreases dramatically, with Asians and Hispanics both showing significant gains, and Blacks staying more or less the same.

No matter how you slice it, our society is becoming more and more diverse every year, to the point where even defining what we mean by "race" isn't so easy anymore. The most common racial groups referred to in population statistics are white, Black, Asian, and Hispanic, but of course there are many people who fall into more than one category (including the 44th president of the United States, as well as the Vice President from 2020 to 2024). Actually, it's even more confusing than that: "Hispanic" isn't really a race, but rather an ethnic group, and many Hispanics also report themselves as either white or Black. (For the data in the bar graph, "white" means "non-Hispanic white.")

Sorting it all out to get any sort of meaningful picture of what we as Americans look like isn't easy, and the techniques of set theory are very useful tools in trying to do so. In this chapter, we'll define what we mean by sets, and we'll study sets and how they can be used to organize information in an increasingly complex world. The concept of sets has been used extensively since people began studying mathematics, but it wasn't until the late 1800s that the theory of sets was studied as a specific branch of math. One of the major tools that we will use to study sets, the Venn diagram, was introduced in an 1880 paper by a man named John Venn. These diagrams enable us to picture complicated relationships between sets of objects, like people of certain races.

Because race and ethnicity are self-reported in a variety of different ways, it's very difficult to find detailed data on the breakdown of races, but there are some reasonable estimates out there. The following estimates were cobbled together from a number of different sources. In a group of 1,000 randomly selected Americans, 727 would self-report as white, 134 as Black, and 185 as Hispanic. In addition, 18 would self-report as Black and white, 16 as Black and Hispanic, and 119 as white and Hispanic. Finally, 11 would self-report as all three. Based on these estimates, after completing this chapter, you should be able to answer the following questions:

1. How many of the original 1,000 report as white only, Black only, and Hispanic only?
2. How many report as Hispanic and Black, but not white?
3. How many report as either Hispanic or Black?
4. How many report as neither white, Black, nor Hispanic?
5. Based on the bar graph to the left, which set will have more elements in 2050: the set of Americans reporting as white, or those reporting as nonwhite?

For answers, see Math in Diversity Revisited on Page 730.

Lesson 6-1 Prep Skills

SKILL 1: CLASSIFY NUMBERS INTO CATEGORIES

The real numbers can be divided into two categories: rational numbers and irrational numbers. Within the rational numbers, we have integers, whole numbers, and natural numbers. The following figure shows examples of each classification and how the different classifications are related.

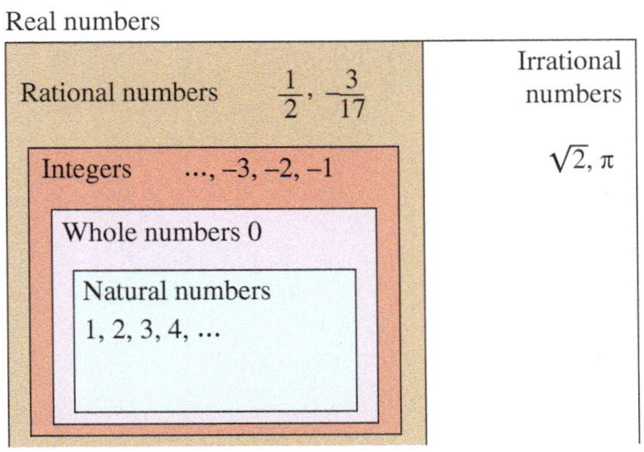

Note that this figure shows that:

The natural numbers consist of 1, 2, 3, . . .
The whole numbers consist of 0, 1, 2, 3, . . .
The integers consist of . . . , −3, −2, −1, 0, 1, 2, 3, . . .
All natural numbers are whole numbers.
All whole numbers are integers.
All integers are rational numbers.
No real number can be both rational and irrational.

SKILL 2: DECIDE IF A GIVEN NUMBER APPEARS ON A LIST

Sometimes lists of numbers are described verbally or are illustrated by enough numbers that you can find a pattern. In that case, to decide if a given number is or is not on the list, the key is to clearly identify the characteristics in the description, or the pattern.

For example, if I describe the group of all even numbers that are less than 30, you could decide that 18 is on that list because 18 is in fact even, and of course it's less than 30. If I ask you if 30 is on the list 4, 8, 12, 16, 20, . . . , you could recognize that each number on the list is four more than the previous one. In that case, the next three numbers would be 24, 28, and 32. This shows that 30 is not on the list.

PREP SKILLS QUESTIONS

For each number, decide if it is an integer, a whole number, an even number, or an odd number. Most will fall into more than one category.

1. 53

2. −8

3. 16

4 0

5. Would 23 be on the list described by 1, 4, 7, 10, 13, . . .?

6. Is 56 on a list of even integers between 12 and 56?

Lesson 6-1 Setting Up

LEARNING OBJECTIVES

- ☐ 1. Define sets and use different methods to represent them.
- ☐ 2. Identify when sets are equivalent.
- ☐ 3. Study cardinality for finite and infinite sets.

"Individual commitment to a group effort—that is what makes a team work, a company work, a society work, a civilization work."
—Vince Lombardi

Robert Churchill/Vetta/Getty Images

Have you ever thought about the role that grouping things plays in everyday life? Think of all the subgroups just within the people you know: You have a group of close friends, a group of Instagram followers, a group of family members, a group of casual acquaintances, a group of classmates, a group of professors, a group of coworkers. . . . You also have a group of keys, groups of clothes, electronics, foods, TV shows, and many others. Our entire world is divided into groups of things, or what we call *sets*. So studying sets from a mathematical standpoint is a good opportunity to study how math is used in our world. Our entire study of sets will be set up by the topics in this lesson. Set up—get it? You don't get that quality of bad humor in most textbooks.

0. What do you think a good definition for the word *set* would be?

6-1 Class

Election laws specify who is allowed to vote in an election. They have to be very specific in order to make an election fair and legal. Consider this group of citizens from a town, let's just call it Springfield: Homer, Marge, Edna, Bart, Ned, Apu, Moe, and Mr. Burns. Homer, Marge, Edna, Ned, and Moe are all registered to vote. Mr. Burns isn't registered, Apu is a foreign national in the States on a work visa, and Bart is 10 years old. This group of people is referred to in Questions 1, 2, 3, and 4.

PictureLux/The Hollywood Archive/Alamy Stock Photo

1. List the citizens from the above group that are eligible to vote in the next election.

In general, a **set** is a collection of objects. For our purposes, we will focus on *well-defined sets.* A set is **well-defined** if for any given object, we can objectively decide whether it is or is not in the set. In short, to be well-defined, the definition of what is or is not in a set has to be based on facts, not opinions.

Solved Example

Recognizing a Well-Defined Set

Decide if each set is well-defined, and describe your reasoning.

a. The set of people in the paragraph at the beginning of the Class portion who are allowed to vote in the next election

b. The set of funny characters in that paragraph

SOLUTION

The question we have to ask ourselves is this: Based on the description of the set, could you decide if any potential element belongs in the set without having to use opinion?

a. This set is well-defined: There are very specific laws governing who can and cannot vote in elections. You have to be a legal citizen, you have to be at least 18 years of age, and you have to be registered. Some states have laws restricting voting rights for convicted criminals. But in any case, these laws explicitly state who can vote; nobody's opinion matters. That makes the set well-defined.

b. This is a classic example of a set that is not well-defined. The problem is the "funny" part. I happen to think Homer Simpson is hilarious, but you might disagree. Nobody can objectively decide whether an individual belongs in a set based on a description like "funny," so the set is not well-defined.

2. There are many other laws that might affect certain sets, like the set of businesses in a certain industry. These laws need to clearly define the set the law applies to. Why is it important that the set of businesses that a certain law applies to is well-defined?

Did You Get It

Try this problem to see if you understand the concepts we just studied. The answer can be found at the end of the portfolio section.

1. Write an example of a set that IS well-defined, and an example of one that is not.

Each object in a set is called an **element** or a **member** of the set.

One method of designating a set is called the **roster method,** in which elements are listed between braces, with commas between the elements. The order in which we list elements isn't important: {2, 5, 7} and {5, 2, 7} are the same set. Often, we will name sets by using a capital letter.

3. The set of citizens from the paragraph at the beginning of the Class portion who are allowed to vote in that district of Springfield in the next election could be written using the roster method as:

V = {__________, __________, __________, __________, __________}

4. Write the set of Simpsons characters in the paragraph at the beginning of the Class portion whose names begin with a letter that comes before M in the alphabet. Is this set well-defined? Why or why not?

> **Math Note**
>
> The commas between elements make it clear that the elements of the set are the names, not the individual letters.

In math, the set of counting numbers or natural numbers is defined as **N** = {1, 2, 3, 4, . . .}. (When we are designating sets, the three dots, or ellipsis, mean that the list of elements continues indefinitely in the same pattern.) The set **E** = {2, 4, 6, 8, . . .} is the set of even natural numbers, and the set **O** = {1, 3, 5, 7, . . .} is the set of odd natural numbers.

5. Use the roster method to write the set of natural numbers less than 6.

> **Math Note**
>
> You can list an element of a set more than once if it means a lot to you, but it's common to choose not to list repeats. For example, the set of letters in the word letters is written as {l, e, t, r, s}.

6. Write the set of odd natural numbers greater than 4.

Students often wonder how many elements of a set to write before ending with an ellipsis. The correct answer is "seven." Just kidding—there is no set rule. Just make sure to include enough initial numbers so that the pattern is clear. For Question 6, just writing {5, 7, . . .} would leave any number of possible interpretations: {5, 7, 8, 10, 11, 13, 14, 16, . . .}, {5, 7, 10, 14, . . .} are two that come to mind.

Did You Get It

2. Write the set of months that end with the letter y.
3. Write each set, using the roster method.
   a. The set of even natural numbers from 80 to 90.
   b. The set of odd natural numbers greater than 10.

The symbol $\in$ is used to show that an object is a member or element of a set.

7. For example, if *A* is the set of days of the week, we could write Monday $\in A$ and read this as "__________ is an element of set ______."

8. Likewise, we could write __________ $\in$ ______ to indicate that "Friday is an element of set *A*."

When an object is not a member of a set, we use the symbol $\notin$.

9. Since "Icecreamday" is not a day of the week (although it probably should be), write a statement using symbols that means "Icecreamday is not an element of *A*."

Math Note

Be sure to use correct symbols when you show membership in a set. For example, the notation $\{6\} \in \{2, 4, 6\}$ is incorrect since the set $\{6\}$ is not a member of this set: only the number 6 is. So it WOULD be correct to write $6 \in \{2, 4, 6\}$.

For Questions 10–12, decide whether each statement is true or false.

10. Oregon $\notin A$, where *A* is the set of states west of the Mississippi River.

11. $29 \in \{1, 5, 9, 13, 17, \ldots\}$

12. $z \notin \{v, w, x, y, z\}$

The **descriptive method** is a second way to describe a set. It uses a short statement to describe the set. To see a couple of examples, look back at Questions 5 and 6: We used the descriptive method to define a set that you were then asked to write using the roster method.

13. Use the descriptive method to describe the set B containing 2, 4, 6, 8, 10, and 12.

Did You Get It

4. Decide whether each statement is true or false.
   a. July $\in A$, where A is the set of months between Memorial Day and Labor Day.
   b. $21 \in \{2, 5, 8, 11, \ldots\}$
   c. map $\notin \{\text{m, a, p}\}$
5. Use the descriptive method to describe the set A containing $-3, -2, -1, 0, 1, 2, 3$.

The third method of designating a set is **set-builder notation,** which uses variables. Recall that a variable is a symbol that can represent different elements of a set. We usually use letters for variables, but any symbol that's not a numeral will do.

Set-builder notation uses a variable, braces, and a vertical bar | that is read as "such that." For example, the set $\{1, 2, 3, 4, 5, 6\}$ can be written in set-builder notation as $\{x \mid x \in \mathbf{N} \text{ and } x < 7\}$. Let's break this down and see how we would read it aloud.

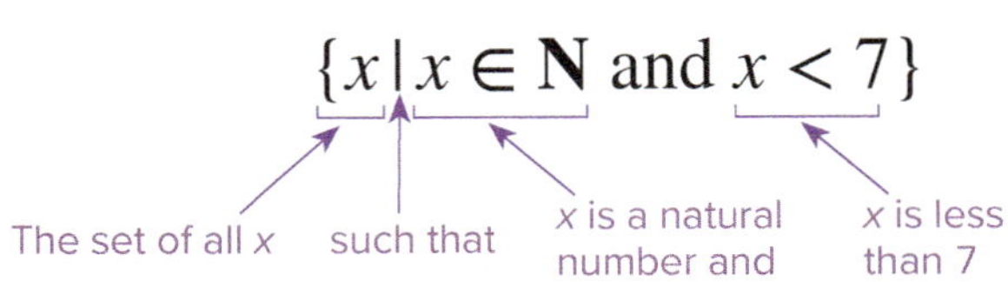

Math Note

Actually, a variable is something that can vary, not the symbol that we use to represent something that can vary. But we often use the term variable to describe the symbol used as well as the variable quantity or object itself.

If you think about it, set-builder notation is very much like the descriptive method. But instead of writing the description out in words, we use symbols to replace certain words to make the description shorter.

14. Write how this set would be described aloud: $K = \{x \mid x \in \mathbf{O} \text{ and } 9 \leq x \leq 39\}$

A note on using an ellipsis: If a set contains enough elements that it's cumbersome to list them all, we can again use an ellipsis to represent the missing elements as long as we illustrate a clear pattern. For example, the set in Question 14 could be written as $\{9, 11, 13, 15, \ldots, 39\}$. Likewise, the set $\{\text{a, b, c}, \ldots, \text{x, y, z}\}$ includes all the letters of the alphabet.

Solved Example 2

Writing a Set in Set-Builder Notation

Write the set $\{-7, -5, -3, -1\}$ using set-builder notation.

SOLUTION

The first thing we need to do is find the connection between the given elements. In this case, all four numbers are negative and all four are odd. Looking a little deeper, these are the four negative odd numbers that are between −8 and 0. We could also say between −9 and 1, or if we wanted to be wise guys, between −7.234 and 0.13. But we'll stick with saying they're all between −8 and 0. So we'd write the set as

$$\{x \mid x \in \mathbf{O} \text{ and } -8 < x < 0\}$$

This can be read aloud as "The set of all x such that x is an odd integer and x is between −8 and 0."

Use set-builder notation to write the set in Questions 15 and 16, then write how your answer would be read aloud.

15. The set R contains the elements 2, 4, and 6.

16. The set W contains the elements red, yellow, and blue.

In Questions 17–19, designate the set S with elements 32, 33, 34, 35, . . . using the given method.

17. The roster method.

18. The descriptive method.

19. Set-builder notation.

20. Discuss which you think is the most efficient way to write this particular set.

Did You Get It ?

6. Use set-builder notation to designate each set, then write how your answer would be read aloud.
   a. The set K contains the elements 10, 12, 14, 16, 18.
   b. The set W contains the elements Democrat and Republican.

21. As of 2025, the set of Asian-American presidents of the U.S. would contain how many elements?

A set with no elements is called an empty set or null set. The symbols used to represent the empty set are { } or ∅. So if we were asked to write the set of all Asian-American U.S. presidents, our answer would look like this: ∅.

22. Which of the following sets are empty?
a. The set of woolly mammoth fossils in museums
b. $\{x \mid x$ is a living woolly mammoth$\}$
c. { }
d. $\{x \mid x$ is a natural number between 1 and 2$\}$

Math Note

Make sure you don't write the empty set as {∅}. This set isn't empty: It's the set containing the empty set, so it has one element.

6-1 Group

Now that we're into Unit 6, we're hoping that you're familiar with the idea of exchanging information with your new group mates. Exchange away!

Cardinal Numbers

1. Earlier in the lesson, we looked at the set $K = \{x \mid x \in \mathbf{O} \text{ and } 9 \leq x \leq 39\}$. How many elements are in the set?

So what is a cardinal number? It's 11, because the St. Louis Cardinals have won 11 World Series titles. Note: That's completely untrue. (Not the 11 World Series part—that's true.) The **cardinal number** (or **cardinality**) of a set is the number of elements in that set. For a set A the symbol for the cardinality is $n(A)$, which is read as "n of A."

2. Use this notation for cardinal number, and your answer from Question 1, to write the cardinality of the set K.

In Questions 3–6, find the cardinal number of each set and write it using $n(A)$ notation.

3. $A = \{5, 10, 15, 20, \ldots, 50\}$

4. $B = \{x \mid x \in \mathbf{N} \text{ and } x < 16\}$

5. $C = \{\varnothing\}$

6. $D = \{\}$

Finite and Infinite Sets

Sets come in two delightful flavors: *finite* and *infinite.* A set is called **finite** if it has no elements, or has cardinality that is a natural number. A set that is not finite is called an **infinite** set.

7. Why do you think we didn't define an infinite set to be a set with cardinality ∞?

8. The set $K = \{x \mid x \in \mathbf{O} \text{ and } 9 \le x \le 39\}$ is ________ because it contains ________ elements. The set $J = \{x \mid x \in \mathbf{O}$ and $x \ge 39\}$ is ________ because it has an unlimited number of elements.

In Questions 9–12, classify each set as finite or infinite.

9. $Z = \{x \mid x \in \mathbf{N} \text{ and } x > 100\}$.

10. U is the set of all stars in the universe.

11. $D = \{100, 102, 104, 106\}$

12. S is the set of songs that can be written.

Did You Get It

7. Find the cardinality of each set, or classify it as infinite.
a. $\{1, 3, 5, 7, 9\}$ b. $\{x \mid x \in \mathbf{E}$ and x is between 1 and $19\}$ c. $\{x \mid x \in \mathbf{N} \text{ and } x > 1{,}000{,}000\}$

Equal and Equivalent Sets

As we study sets in this unit, it's important to understand the concepts of *equal sets* and *equivalent sets.* Two sets A and B are **equal** (written $A = B$) if they have exactly the same elements. Two finite sets are said to be **equivalent** (written $A \cong B$) if they have the same number of elements ($n(A) = n(B)$).

13. The sets A = {Mars, Saturn, Venus, Uranus} and B = {Uranus, Mars, Saturn, Venus} are ________ because ____________________.

14. The sets A = {Mars, Saturn, Venus, Uranus} and B = {Mercury, Neptune, Jupiter, Pluto} are ________ because ____________________.

In Questions 15–20, decide if the two given sets are equal, equivalent, both, or neither.

15. {p, q, r, s}; {a, b, c, d}

16. {8, 10, 12}; {12, 8, 10}

17. {213}; {2, 1, 3}

18. {1, 2, 10, 20}; {2, 1, 20, 11}

19. {even natural numbers less than 10}; {2, 4, 6, 8}

20. {4, 5, 6}; {4, 4, 5, 6}

21. Based on your observations, are equal sets also equivalent, or are equivalent sets also equal?

It can be very interesting to study trends in the top college majors in terms of degrees awarded over some time period. In this case, we have the number of bachelor's degrees awarded in 2000, 2010, and 2022 for the 10 majors that were most popular in 2022. There's a lot of information in this table and using sets to analyze some of it can be useful.

Major	2000	2010	2022
Business	263,515	365,133	375,418
Social sciences and history	128,036	177,169	151,109
Education	105,458	104,008	89,410
Health professions	75,933	143,463	263,765
Psychology	73,645	100,906	129,609
Visual and performing arts	61,148	93,939	90,241
Biological and biomedical sciences	60,576	89,984	131,462
Engineering	58,209	76,356	123,017
Communication and journalism	58,013	83,231	86,043
Computer and information sciences	44,142	43,066	108,503

Source: National Center for Education Statistics

22. List the set of majors that increased in popularity every year listed.

23. List the set of majors that didn't increase in popularity from 2000 to 2022.

24. List the set of majors that increased in popularity by at least 30% from 2010 to 2022. (Recall that percent increase was covered in Lesson 2-1.)

6-1 Portfolio

Name ______________________________

Check each box when you've completed the task. Remember that your instructor will want you to turn in the portfolio pages you create.

ALEKS Practice

1. ☐ Include any written work from the ALEKS assignment along with any notes or questions about this lesson's content.

Applications

1. ☐ Complete the applications problems.

Reflections

Type a short answer to each question.

1. ☐ What's the difference between equal and equivalent sets?
2. ☐ Describe what it means for a set to be well-defined, as well as a situation in which something not being well-defined would lead to negative consequences.
3. ☐ Write an example of an infinite set and describe how you know it's infinite.
4. ☐ Name one thing you learned or discovered in this lesson that you found particularly interesting.
5. ☐ What questions do you have about this lesson?

Looking Ahead

1. ☐ Complete the prep skills for Lesson 6-2.
2. ☐ Read the opening paragraph in Lesson 6-2 carefully and answer Question 0 in preparation for that lesson.

Answers to "Did You Get It?"

1. Answers vary widely. **2.** {January, February, May, July}
3. a. {80, 82, 84, 86, 88, 90} **b.** {11, 13, 15, 17, 19, . . .}
4. a. True **b.** False **c.** True
5. One answer: The set of integers that are between −4 and 4.
6. a. $\{x \mid x$ E and $9 < x < 19\}$; the set of all x such that x is an even number between 9 and 19.
b. $\{x \mid x$ is one of the major American political parties$\}$; the set of all x such that x is one of the major American political parties **7. a.** 5 **b.** 9 **c.** Infinite

Answers to Prep Skills Questions

1. Integer, whole number, odd number **2.** Integer, even number
3. Integer, whole number, even number **4.** Integer, whole number
5. No **6.** No

6-1 Applications

Name ______________________________

For Questions 1–4, write each set using the roster method. You may have to do a little Internet research, which is a good thing.

1. R is the set of odd natural numbers between 2 and 16.

2. B is the set of integers that are less than -4.

3. C is the set of the seven mainland countries in Central America.

4. O is the set of colors in the flags of the states whose names begin with the letter O.

For Questions 5–8, decide if the set is well-defined or not, and explain your reasoning.

5. The set of all Americans

6. The set of all lawyers in Alaska

7. The set of luxury cars in the 2024 model year

6-1 Applications

Name ______________________________

8. The set of all mothers

Identity theft is one of the most costly types of crime committed in the United States, and younger adults are often the targets. In 2023 alone, almost half a million people under the age of 30 were victims of financial fraud. The following charts show types of identity theft fraud reported in 2023 and the percentage of victims by age. Use this information for Questions 9–16.

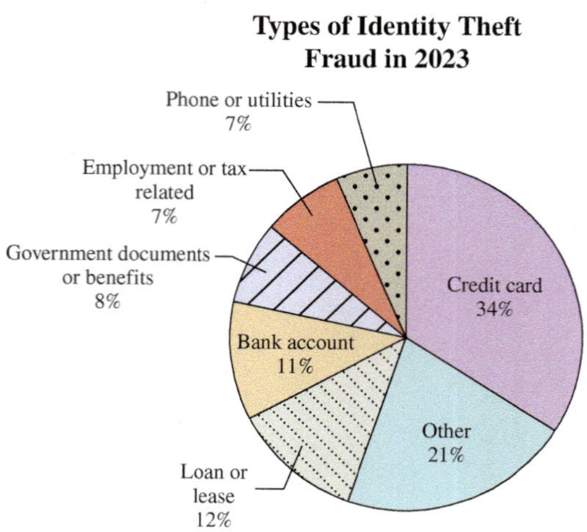

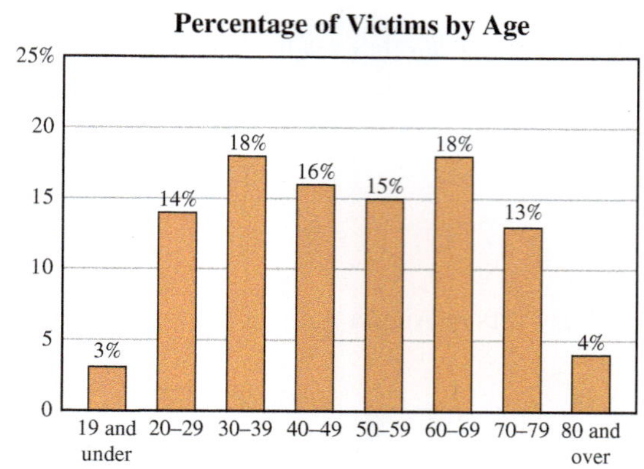

9. List the set of types of identity fraud that made up less than 20% of all reported cases.

6-1 Applications

Name ______________________________

10. List the set of age groups that made up less than 14% of I.D. theft victims.

11. List the set $\{x \mid x$ is a percentage for an age group of those 40 and over who are victims of I.D. theft$\}$.

12. Write the set {Bank Account, Loan or Lease, Other, Credit Card} using set-builder notation.

13. Find the cardinal number for the set $F = \{x \mid x$ is a type of I.D. theft with a percentage less than 25$\}$. Use the notation for cardinality that we learned in this lesson.

14. Write two sets that are equivalent but not equal based on the data in the charts.

15. Write the set of identity theft types that made up more than 50% of all reported claims.

16. Are the two sets listed below equal, equivalent, or neither?

$A = \{x \mid x$ is an I.D. theft victim less than 70 years old$\}$

$B = \{y \mid y$ is an age group that did not have an I.D. theft percentage between 10 and 20%$\}$

Lesson 6-2 Prep Skills

SKILL 1: USE THE ROSTER METHOD

To designate a set using the roster method, list the elements between braces with commas between the elements.

- $\{A, B, C, D\}$ — This is the roster method for the set consisting of the first four letters in the alphabet.

SKILL 2: USE SET-BUILDER NOTATION

To designate a set using set-builder notation, use a variable followed by a vertical line, then a description of what qualifies a given item to be in the set. The description should use symbols to make it as concise as possible. The variable, the vertical line, and the description should all be enclosed in a set of braces.

- $\{x \mid x \in \mathbf{N} \text{ and } x \leq 4\}$ — This is the set-builder notation for the numbers 1, 2, 3, and 4.

SKILL 3: USE INDUCTIVE REASONING

In Lesson 1-1, we learned that inductive reasoning is the process of reasoning that arrives at a general conclusion based on the observation of specific examples.

- Use inductive reasoning to predict the next three numbers on the list.

  2, 4, 6, 8, ____, ____, ____

  2, 4, 6, 8, 10, 12, 14 — This list appears to be a list of consecutive even integers. If this pattern continues, the next three numbers would be 10, 12, and 14.

SKILL 4: CALCULATE POWERS OF 2

The powers of 2 are useful to have memorized up to a certain point. After the point where you have them memorized, you can continue the list by just doubling the previous number repeatedly.

$2^0 = 1$

$2^1 = 2$

$2^2 = 2 \cdot 2 = 4$

$2^3 = 2 \cdot 2 \cdot 2 = 8$

$2^4 = 2 \cdot 2 \cdot 2 \cdot 2 = 16$

$2^5 = 2 \cdot 2 \cdot 2 \cdot 2 \cdot 2 = 32$

$2^6 = 2 \cdot 2 \cdot 2 \cdot 2 \cdot 2 \cdot 2 = 64$

$2^7 = 2 \cdot 2 \cdot 2 \cdot 2 \cdot 2 \cdot 2 \cdot 2 = 128$

- Calculate 2^9.

  Start with $2^7 = 128$.

  $2^8 = 128 \cdot 2 = 256$

  $2^9 = 256 \cdot 2 = 512$

  Start with a power of 2 you remember. In this case, $2^7 = 128$. Double this value 2 more times to get to $2^9 = 512$.

PREP SKILLS QUESTIONS

1. Write the set of days of the week using the roster method.

2. Write the set of even integers greater than 5 using the roster method.

3. Write the set {4, 5, 6, 7, 8, 9,10} using set-builder notation.

4. Write the set {e, f, g, h, i, . . . , y, z} using set-builder notation.

5. Use inductive reasoning to predict the next three numbers on each list.

   4, 8, 16, 32, 64, ______, ______, ______

   10, 30, 50, 70, ______, ______, ______

6. Compute each power of 2.

   2^4

   2^7

Lesson 6-2 Busy Intersections, More Perfect Unions

LEARNING OBJECTIVES

- ☐ 1. Find the complement and all subsets for a given set.
- ☐ 2. Evaluate set statements involving subset notation.
- ☐ 3. Perform and apply set operations: union, intersection, subtraction.

David Planchet/McGraw Hill

What is the intersection between technology, art and science? Curiosity and wonder, because it drives us to explore, because we're surrounded by things we can't see.

—Louie Schwartzberg

We've seen that set theory is about identifying relationships between things that are grouped together for some reason. Taking that idea a little further, sets often have relationships with other sets, and that's when things get a bit complicated. In that case, a system for displaying and studying those relationships will come in handy, which ultimately is kind of the main point of studying set theory. So far, we've just scratched the surface. For example, you are a member of both the set of college students and the set of students taking a college math course. You could be in the set of sophomores or the set of juniors, but not in both. You might be in the set of students living off campus and the set of students who walk to class. Maybe you're in the set of students that eats lunch in the cafeteria and the set of students that think the french fries are too soggy, but not in the set of people who load on the ketchup and eat the lousy things anyhow. Let's see what we can do about organizing all of these complicated connections between sets.

0. Describe a set that you and your closest friend are both in, and one that you are in that they are not.

6-2 Class

To begin we'll need to introduce a new concept called a *universal set.*

The **universal set** is the set of all objects in the universe.

Just kidding. (No harm in trying to have a little fun in a math class.) We now return to our regularly scheduled definition:

The **universal set** for a given situation, symbolized by U, is the set of all objects that are reasonable to consider in that situation.

For example, all of the sets described in the opening paragraph contain people, so theoretically we could use the set of all human beings as U. But it would be much more reasonable to assign $U = \{\text{College students}\}$.

Once we define a universal set in a given setting, we are restricted to considering only elements from that set. If $U = \{1, 2, 3, 4, 5, 6, 7, 8\}$, then the only elements we can use to define other sets in this setting are the integers from 1 to 8. In the remainder of this chapter, we'll use a clever method for visualizing sets and their relationships called a Venn diagram (so named because it was developed by a man named John Venn in the 1800s).

Here is an example of a Venn diagram.

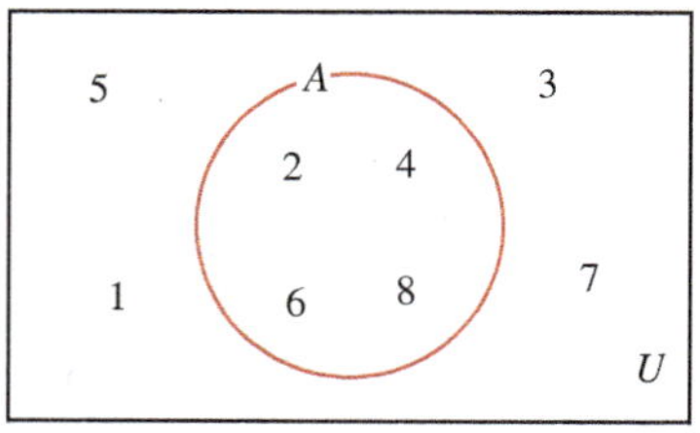

You can get a lot of information from this simple diagram, in which a set called A is being defined.

1. The universal set from which elements of A can be chosen is

$U = \{$_____, _____, _____, _____, _____, _____, _____, _____$\}$.

2. The set $A = \{$_____, _____, _____, _____$\}$, and the elements not in A are $\{$_____, _____, _____, _____$\}$.

We call the elements in U that are not in A the **complement of A,** and denote it A'. This is a good opportunity to review set-builder notation.

3. For a given set A,

$A' = \{x \mid x \in$ _____ and $x \notin$ _____ $\}$

In a Venn diagram, the complement of a set A is all the things inside the rectangle that are not inside the circle representing set A.

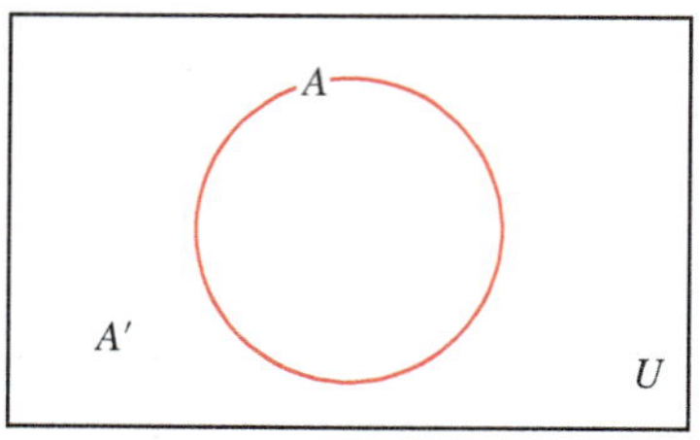

Math Note

The complement of the empty set is the universal set, and the complement of the universal set is the empty set.

2023-2024 Final AP Top Ten

RANK	TEAM	RECORD
1	South Carolina	38–0
2	Iowa	34–5
3	Connecticut	33–6
4	NC State	31–7
5	Southern California	29–6
6	Louisiana State	31–6
7	Texas	33–5
8	Oregon State	27–8
9	Stanford	30–6
10	UCLA	27–7

Here we have listed the top 10 teams from the final AP Women's College Basketball poll for the 2023–2024 season. Let U be the set of teams in the top 10. Note that a record of 34–5 means a team won 34 games and lost 5.

$U = \{$SCar, Iowa, Conn, NCSt, SCal, LSU, Tex, OrSt, Stan, UCLA$\}$. Now we'll define a new set: $D = \{$SCar, Iowa, Tex$\}$.

Solved Example 1

Working with Complements

For the sets U and D defined above:

a. Find the set D' using proper set notation.

b. Provide a verbal description of the sets D and D'.

c. Draw a Venn diagram that illustrates all of the sets in this example.

SOLUTION

a. D' is the complement of set D, making it all of the teams in the universal set that are NOT in D.

$$D' = \{\text{Conn, NCSt, SCal, LSU, OrSt, Stan, UCLA}\}.$$

Solved Example 1 *Continued*

b. This is kind of the key idea in set theory in my view: There's not much point in grouping things into sets unless you think about how they're connected. Superficially, you could say that all three are college basketball teams, all three have an animal in their logo, or all three are schools in the Eastern or Central time zones. And you wouldn't necessarily be wrong. But many of the other schools in U fit those criteria as well, and there has to be SOME reason that these particular three were grouped together.

Looking at the team's records, something jumps out at me: These three teams have the smallest number of losses among the teams in the top 10. In fact, all lost five games or less. So there's the key connection, which leads to our verbal description. D is the set of teams in the top 10 that lost less than six games during the 2023–2024 season.

That makes D' the set of teams that lost six games or more. A more complete description, though, would be teams in the top 10 that lost six or more games.

c. Our Venn diagram will consist of a rectangle representing the universal set U and a circle inside representing set D. The three schools in set D will be placed inside the circle, and the seven schools in D' will be placed outside the circle, but still inside the rectangle.

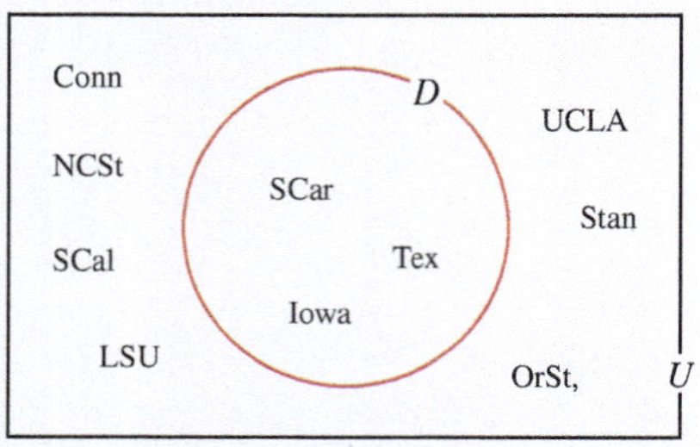

> **Math Note**
>
> You have to be a little careful here since the elements contain letters: Notice that we put the letters D and U on the borders to indicate that they're set names rather than elements.

We'll use these sets to illustrate some of the concepts we just learned.

4. Using the roster method, write the set W of all teams that won 30 or more games.

5. Write the set W'.

6. Draw a Venn diagram illustrating the sets W and W'.

Sidelight **Now and Venn**

Venn diagrams are generally credited to the British mathematician John Venn, who introduced them in an 1880 paper as they are used today. That makes it sound like a pretty old concept, but the general idea can be traced back much further. The great mathematician Leonhard Euler used similar diagrams in the 1700s, and other figures like them can be traced as far back as the 1200s!

While it's absolutely true that John Venn was a classic academic—at one time writing or lecturing in each of morality, mathematics, logic, probability theory, philosophy, metaphysics, and history—he had a somewhat surprising hobby: building machines. In particular, he was best known for building a machine for bowling cricket balls (which is roughly analogous to pitching in baseball). His machine was so good that in 1909 it "clean bowled" one of the top cricket players of the time on four occasions, which is kind of like a turn-of-the-century pitching machine striking out Babe Ruth.

Subsets

At the beginning of the lesson, we pointed out that you're in both the set of college students and the set of students taking a college math course. Notice that everyone in the second set is automatically in the first one. (Obviously, if you're taking a college math course, then you have to be a college student.) We could say that the set of students taking a college math course is contained in the set of all college students. When one set is contained in a second set, we call the smaller set a *subset* of the larger one.

If every element of a set A is also an element of a set B, then A is called a **subset** of B. The symbol $\subseteq$ is used to designate a subset; in this case, we write $A \subseteq B$.

Barry Barker/McGraw Hill

There are many subsets of this set of spring breakers: the subset of female students, the subset of guys who yell "WOO!" every time someone points a camera at them, the subset of locals who absolutely cannot wait until spring break season is over, and so on.

Speaking of college, back to basketball. Look back at the verbal description of set D in Solved Example 1.

7. If we define set B as the list of teams in the top seven, list set B.

8. Does every team in set D appear in set B?

9. What does this mean in terms of subsets? Write an answer in words and symbols.

10. Does every team in set D appear in set D? No, this is not a typo. Just think and answer yes or no.

11. What does this mean in terms of subsets? Write an answer in words and symbols.

Having run out of lemon (no big deal) and blue raspberry (now THAT'S a problem), a snow cone stand is offering only grape, cherry, and lime today. So the set of flavors is $S = \{\text{Grape, Cherry, Lime}\}$. The goal of Questions 12–18 will be to find all subsets of this set, which is all the choices of flavors we could make if we're undecided on how many snow cones we feel like choking down.

12. Set S contains _____ elements.

13. We could choose all three flavors, which would almost certainly lead to brain freeze. That would be the set:

Jill Fromer/Getty Images

14. We could choose two of the three flavors. List each subset containing exactly two flavors.

15. We could choose only one flavor. List each subset containing one flavor.

16. We could decide we don't want a snow cone. That could be represented by the set:

17. The set S, which has three elements, has a total of _____ subsets.

18. How many subsets are not equal to the original set S?

These subsets that aren't equal to the original set are called *proper subsets* of S. The Venn diagram for an example of a proper subset is shown below. In this case, $U = \{1, 2, 3, 4, 5\}$, $A = \{1, 3, 5\}$, and $B = \{1, 3\}$.

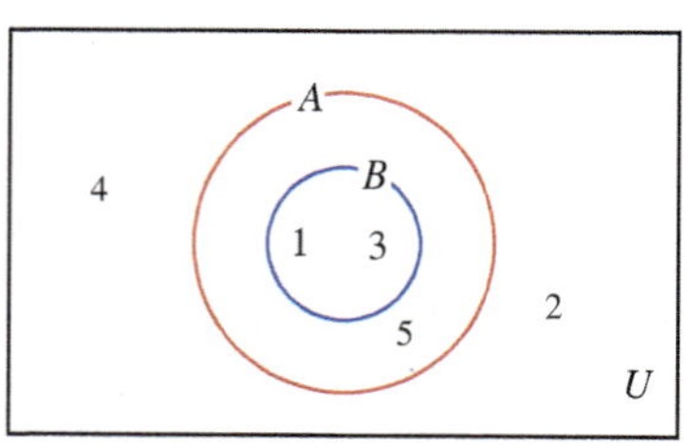

If a set A is a subset of a set B and is not equal to B, then we call A a **proper subset** of B, and write $A \subset B$. That is, $A \subseteq B$ and $A \neq B$.

19. More blanks to fill in: the set {Grape, Lime} is a proper subset of {Grape, Cherry, Lime} because {Grape, Lime} _____ {Grape, Cherry, Lime} and _______ $\notin$ {Grape, Lime}.

20. The two longest-running "reality" shows on American television, at least as of the end of 2024, are included in the set T = {*The Real World*, *Survivor*}. Write all of the subsets of this set. You may want to look back at Questions 12–18 to help make sure you don't miss any.

CBS Photo Archive/Getty Images

21. What does the group of subsets you found in Question 20 describe about these shows? Write in plain English, not using set terminology.

22. How many of the subsets you found are proper subsets?

Did You Get It

Try these problems to see if you understand the concepts we just studied. The answers can be found at the end of the portfolio section.

1. List all subsets of the set D from Solved Example 1.
2. How many proper subsets does the set {♦, ♥, ♠, ♣} have? List them.

To indicate that a set is not a subset of another set, the symbol $\not\subseteq$ is used. The symbol $\not\subset$ is used to indicate that the set is not a proper subset.

Solved Example 2

Understanding Subset Notation

Decide if each statement is true or false and briefly explain your choice.

a. $\{1, 3, 5\} \subset \{1, 3, 5, 7\}$

b. $\{a, b\} \subset \{a, b\}$

c. $\{x \mid x \in \mathbf{E} \text{ and } x > 10\} \subset \mathbf{N}$

SOLUTION

a. This statement reads "The set $\{1, 3, 5\}$ is a proper subset of the set $\{1, 3, 5, 7\}$." That's really a combination of two statements: that the first set is a subset of the second, and that the two sets are not equal. Each of those is true in this case, so the overall statement is true.

b. This statement is false. It looks true at first glance, because $\{a, b\}$ is in fact a subset of itself. But remember, the symbol $\subset$ means "is a proper subset of." That's not the case here, because obviously the two sets are equal. So $\{a, b\}$ is not a proper subset of itself.

c. This starts out with practice in reading set-builder notation. The first set is the set of even numbers that are more than 10. If you're bigger than 10 you are, of course, positive. And all even numbers are integers. That makes every number in the first set a positive integer, which in turn makes each a natural number. So the first set is definitely a subset of **N.** It's clearly not the entire set of natural numbers since it's missing all odd numbers, so the subset is proper, making the statement true.

23. Decide if each statement is true or false and briefly explain your choice.

a. $\{r, s, t\} \not\subset \{t, s, r\}$

b. {Lake Erie, Lake Huron} $\not\subset$ The set of Great Lakes

c. $\varnothing \subset \{5, 10, 15\}$

d. $\{u, v, w, x\} \subseteq \{x, w, u\}$

e. $\{0\} \subseteq \varnothing$

Math Note

Make sure you don't confuse subsets and elements. The statement $6 \in \{2, 4, 6\}$ is true, since 6 is an element of that set. And the statement $\{6\} \subseteq \{2, 4, 6\}$ is true for the same reason. But $\{6\} \in \{2, 4, 6\}$ isn't. The number 6 is an element of that set, but the set containing 6 is not.

24. Since every set is a subset of itself, a set with no elements still has ______ subset(s).

25. A set with one element has how many subsets? Describe them.

26. We've already seen that if a set has two elements, there are ________ subsets, and if a set has three elements, there are ________ subsets. This is an excellent opportunity to use the inductive reasoning that we practiced in Unit 1!

27. Complete the table as much as possible using the information from the previous questions.

Number of Elements	0	1	2	3	4	5
Number of Subsets						

28. Based on this pattern, make a conjecture about the number of subsets of a set with four elements. What about five elements? Add your answers to the table.

29. Write a formula for the number of subsets of a set with n elements.

30. Use that formula to verify the results in the table above for each value of n.

31. Describe the relationship between the number of proper subsets of a set and the total number of subsets.

32. Write a formula for the number of proper subsets of a set with n elements.

33. Use that formula to predict the number of proper subsets of a set with nine elements.

Did You Get It

3. Find the number of subsets and number of proper subsets for the set D' from Solved Example 1.

6-2 Group

The Intersection and Union of Sets

At the beginning of this lesson, we pointed out that you might be in both the set of students living off campus and the set of students who walk to class. We will identify objects that are common to two or more sets by using the term *intersection.*

The **intersection** of two sets A and B, symbolized by $A \cap B$, is the set of all elements that are in both sets. In set-builder notation, $A \cap B = \{x \mid x \in A \text{ and } x \in B\}$.

For example, if $A = \{10, 12, 14, 15\}$ and $B = \{13, 14, 15, 16, 17\}$, then the intersection $A \cap B = \{14, 15\}$, since 14 and 15 are the only elements that are common to both sets. The Venn diagram for $A \cap B$ is shown below. Notice that the elements of A are placed inside the circle for set A, and the elements of B are inside the circle for set B. The elements in the intersection are placed into the portion where the circles overlap: $A \cap B$ is the shaded portion.

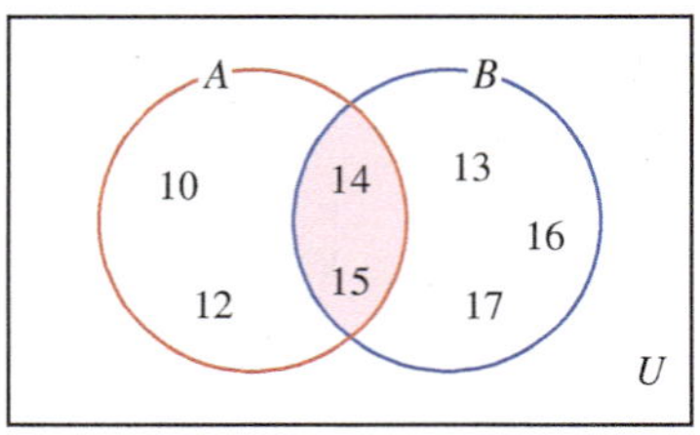

1. Look back at the set D from Solved Example 1, and also at set B from Question 7 following that example. Write a verbal description of the set $D \cap B$, then list out the set using the roster method.

Intersection is an example of a **set operation**—a rule for combining two or more sets to form a new set. (If you think about it, that's what familiar operations on numbers do: For example, when you add 10 and 35, you're combining them to get a new number, 45.) The intersection of three or more sets consists of the set of elements that are in every single set. Note that the word *and* is sometimes used to indicate intersection; $A \cap B$ is the set of elements in A and B.

If A = {Cleveland, Indianapolis, Chicago, Des Moines, Detroit}, B = {New York, Los Angeles, Chicago, Detroit}, and C = {Seattle, Los Angeles, San Diego}, find each requested set in Questions 2–4.

2. $A \cap B$

3. $B \cap C$

4. $A \cap B \cap C$

Did You Get It

4. If $A = \{5, 10, 15, 20, 25\}$, $B = \{0, 10, 20, 30, 40\}$, and $C = \{30, 50, 70, 90\}$, find $A \cap B$, $B \cap C$, and $A \cap B \cap C$.

When the intersection of two sets is the empty set, the sets are said to be **disjoint.** For example, the set of students who stop attending class midway through a term and the set of students earning A's are disjoint, because you can't be a member of both sets. The Venn diagram for a pair of disjoint sets *A* and *B* is shown below. It's pretty easy to recognize: If the sets have no elements in common, the circles representing them don't overlap at all.

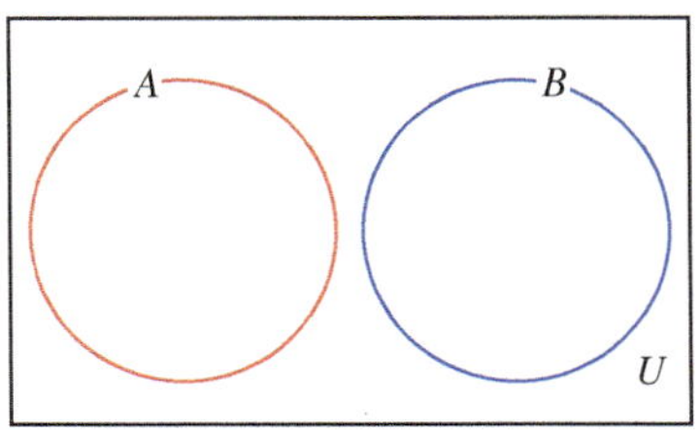

Another way of combining sets to form a new set is called *union.*

The **union** of two sets *A* and *B,* symbolized by $A \cup B$, is the set of all elements that are in set *A* or set *B* (or both). In set-builder notation, $A \cup B = \{x \mid x \in A \text{ or } x \in B\}$.

For example, if $A = \{5, 10, 15, 20\}$ and $B = \{5, 20, 30, 45\}$, then the union $A \cup B = \{5, 10, 15, 20, 30, 45\}$. Even though 5 and 20 are in both sets, we list them only once in the union. The Venn diagram for $A \cup B$ is shown below. The set $A \cup B$ is the shaded area consisting of all elements in either set.

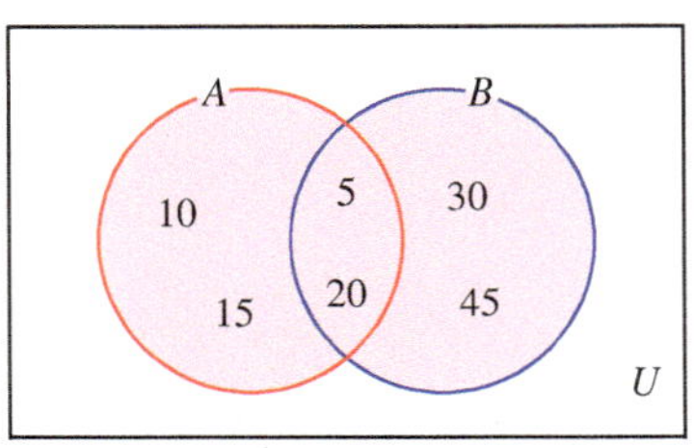

5. Why do you think the words "intersection" and "union" were chosen for these operations? An answer based on Venn diagrams would be especially appreciated.

Three experimental medications are being evaluated for safety. Each has a list of side effects that has been reported by at least 1% of the people trying the medication. This a blind trial, so the medications are simply labeled A, B, and C. The side effects for each are listed below.

A = {nausea, night sweats, nervousness, dry mouth, swollen feet}
B = {weight gain, nausea, nervousness, blurry vision, fever, trouble sleeping}
C = {dry mouth, nausea, blurry vision, fever, weight loss, eczema}

For Questions 6–8, find each requested set. Then provide a verbal description of what the set represents.

6. $A \cup B$

7. $A \cup C$

8. $A \cup B \cup C$

Did You Get It

5. If $A = \{a, b, c, d, e\}$, $B = \{a, c, e, g, i\}$, and $C = \{b, d, f, h, j\}$, find $A \cup B$, $A \cup C$, and $A \cup B \cup C$.

What about operations involving more than two sets and more than one operation? Just like with operations involving numbers, we use parentheses to indicate an order of operations.

For Questions 9–11, use the sets of side effects from Questions 6–8. Find the requested set, and again describe verbally what it represents.

9. $(A \cup B) \cap C$

10. $A \cap (B \cup C)$

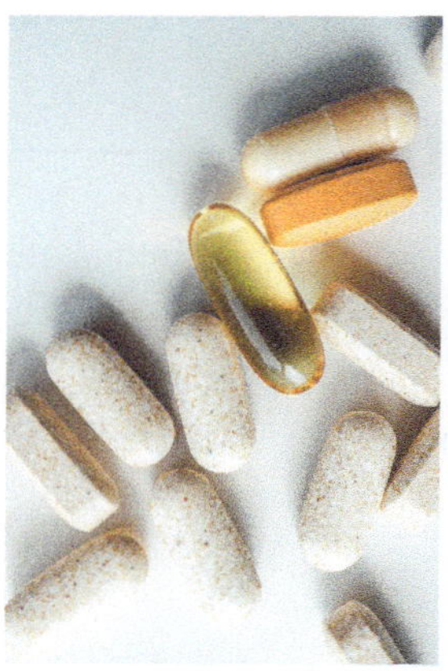

shine.graphics/Shutterstock

11. $(A \cap B) \cup C$

When combining union and intersection with complements as we will do next, we'll have to be extra careful. Pay particular attention to the parentheses and to whether the complement symbol is inside or outside the parentheses. And remember to do any operation in parentheses first! For example, to find $(A \cap B)'$, you would first find the intersection of A and B, then find all elements of the universal set not in that intersection.

Going back to our sets of side effects, recall that these were reported by at least 1% of users. The universal set below is the set of all side effects reported by ANY users. Use this to find the sets requested in Questions 12–14.

U = {nausea, night sweats, nervousness, dry mouth, swollen feet, weight gain, blurry vision, fever, trouble sleeping, weight loss, eczema, motor mouth, darting eyes, uncontrollable falling down}

12. $A' \cap C'$

13. $(A \cap B)' \cap C$

Math Note

Don't forget: The complement of a set *A* is all of the elements in the universal set that are not in *A*, not all of the objects in the universe that are not in *A*.

14. $B' \cup (A \cap C')$

Did You Get It

6. Let $U = \{1, 2, 3, 4, 5, 6, 7, 8\}$, $A = \{1, 3, 5, 7\}$, $B = \{2, 4, 6, 8\}$, and $C = \{2, 3, 5, 7\}$. Find each set.
   a. $A \cup (B \cap C)$ b. $(A \cup B)'$

Union and intersection are very applicable concepts, even when you don't necessarily realize that you're using them. In Questions 15 and 16, describe each set verbally.

15. The intersection of U.S. citizens older than 17 and the set of U.S. citizens that have not been convicted of a felony.

Pressmaster/Shutterstock

16. The union of my mother's parents and my father's parents.

Set Subtraction

Another set operation we can study is the *difference* of sets. If you're in the set of students that live off campus but not in the set of students that walk to school, then you're in the difference of those two sets.

The **difference** of two sets A and B, symbolized by $A - B$, is the set of all elements that are in A but not in B. In set-builder notation, $A - B = \{x \mid x \in A \text{ and } x \notin B\}$.

17. While it's often most natural to think of the difference of two sets in terms of set subtraction, you can actually describe the difference of sets A and B using operations that we studied earlier in the lesson. How? (Hint: The description of difference in set-builder notation is VERY helpful.)

18. For the side effect sets A, B, and C from earlier in the lesson, find the differences $A - B$ and $B - C$, and describe what each represents.

6-2 Portfolio

Name ______________________________

Check each box when you've completed the task. Remember that your instructor will want you to turn in the portfolio pages you create.

ALEKS Practice

1. ☐ Include any written work from the ALEKS assignment along with any notes or questions about this lesson's content.

Applications

1. ☐ Complete the applications problems.

Reflections

Type a short answer to each question.

1. ☐ Describe each of these important terms in your own words: subset, union, intersection, complement.
2. ☐ Write your own example of a union problem, and an intersection problem. No abstract sets! Explain the context.
3. ☐ Name one thing you learned or discovered in this lesson that you found particularly interesting.
4. ☐ What questions do you have about this lesson?

Looking Ahead

1. ☐ Complete the prep skills for Lesson 6-3.
2. ☐ Read the opening paragraph in Lesson 6-3 carefully and answer Question 0 in preparation for that lesson.

Answers to "Did You Get It?"

1. {SCar, Iowa, Tex}, {SCar, Iowa}, {SCar, Tex}, {Iowa, Tex}, {Scar}, {Iowa}, {Tex}, ∅ **2.** 15: {♦, ♥, ♠}, {♦, ♥, ♣}, {♦, ♠, ♣}, {♥, ♠, ♣}, {♦, ♥}, {♦, ♠}, {♦, ♣}, {♥, ♠}, {♥, ♣}, {♠, ♣}, {♦}, {♥}, {♠}, {♣}, ∅ **3.** 128 and 127 **4.** $A \cap B = \{10, 20\}$; $B \cap C = \{30\}$; $A \cap B \cap C = \varnothing$ **5.** $A \cup B = \{a, b, c, d, e, g, i\}$; $A \cup C = \{a, b, c, d, e, f, h, j\}$; $A \cup B \cup C = \{a, b, c, d, e, f, g, h, i, j\}$ **6. a.** $A \cup (B \cap C) = \{1, 2, 3, 5, 7\}$ **b.** $(A \cup B)' = \varnothing$

Answers to Prep Skills Questions

1. {Sunday, Monday, Tuesday, Wednesday, Thursday, Friday, Saturday}
2. $\{6, 8, 10, 12, 14, \ldots\}$
3. $\{x \mid x \in \mathbf{N} \text{ and } 4 \leq x \leq 10\}$ is one potential answer.
4. $\{x \mid x$ is a letter of the alphabet that comes after d$\}$
5. 128, 256, 512; 90, 110, 130
6. 16, 128

6-2 Applications

Name ______________________________

1. The U.S. Department of Health and Human Services recommends that everyone healthy enough to exercise should get either 150 minutes of moderate physical activity (like walking or stretching) per week or 75 minutes of vigorous activity (like running or biking) per week. Let's say that the gym at your campus has treadmills, stationary bikes, and elliptical machines. You can choose none, some, or all of those as part of a workout plan. Write all the possible choices you could make, then explain what in the world that has to do with our study of sets in this lesson.

JGI/Jamie Grill/Blend Images LLC

2. When playing five-card poker, players are originally dealt five cards (because six would just be silly) and can choose to discard any number of them from none to all five. How many different choices do you have of discards after the original draw? Your explanation in Question 1 and a formula you developed in this lesson will help.

3. This interesting ad was posted on Craigslist: "Assistant wanted: must be able to walk long distances over rough terrain or own an ATV. Must have shovel. I supply duct tape and tarpaulins." Doesn't sound suspicious at all. Write a statement involving three sets, intersection, and union that describes the requirements for this highly sought-after position.

In Questions 4–6, write a description of the complement of each set. You'll need to decide on an appropriate universal set, so make sure you describe that as well.

4. Your Facebook friends (If you're not on Facebook, play along and pretend that you are.)

6-2 Applications

Name ________________________________

5. The courses you'll need to take to fulfill your degree requirements.

6. NASA missions that resulted in a successful launch.

Questions 7–13 use the following sets. For each question, write a verbal description of the set.

U = the set of all people who have been charged with a felony
A = the set of people who are on trial or awaiting trial on felony charges
B = the set of people who have been convicted of a felony
C = the set of people who have been convicted of a felony and have been released from prison
D = the set of people who were charged with a felony and found not guilty
E = the set of people who were charged with a felony and had charges dropped before standing trial

7. $B \cup D$

8. C'

9. $B \cap C$

6-2 Applications

Name ______________________________

10. $C \cap B'$

11. $(A \cup B)'$

12. $B \cap D$

13. $A - (B \cap C)$

In Questions 14–17, decide if the statement is true or false. You may have to do some poking around on the Internet for information.

14. $\varnothing \subseteq A$, where A is the set of four-leaf clovers.

15. $\varnothing \subset B$, where B is the set of living unicorns.

6-2 Applications

Name ____

16. {Elvis Presley, Elvis Costello} $\subset$ {x | x is a member of the Rock and Roll Hall of Fame}

17. {Neil Armstrong, John Glenn, Alan Shepard, Buzz Aldrin} $\not\subset$ {y | y is a person that has landed on the moon}

Lesson 6-3 Prep Skills

SKILL 1: INTERPRET A VENN DIAGRAM

A Venn diagram is a useful tool for visualizing sets.

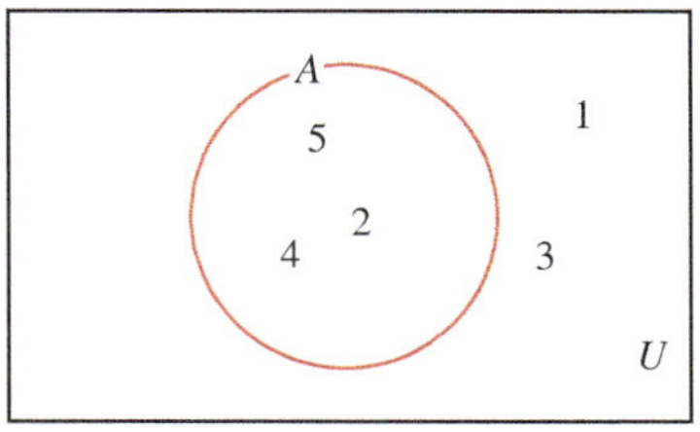

The Venn diagram above tells us the following:

- $U = \{1, 2, 3, 4, 5\}$ The universal set consists of all elements inside the rectangle.
- $A = \{2, 4, 5\}$ Set *A* contains all elements inside the circle representing set *A*.
- $A' = \{1, 3\}$ The set A', representing the complement of set *A*, contains all elements inside the rectangle that are not inside the circle representing set *A*.

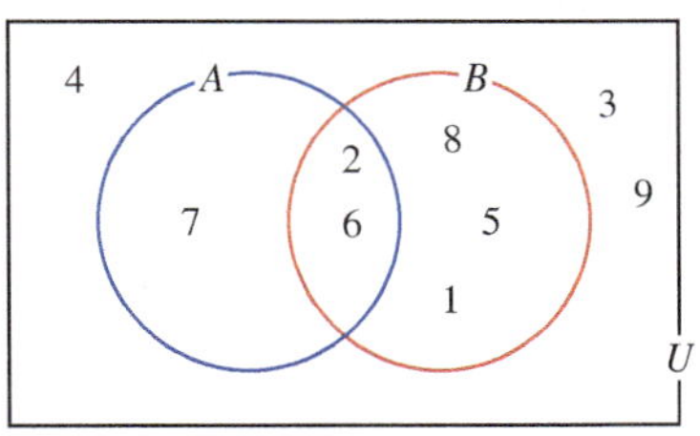

The Venn diagram above tells us the following:

- $U = \{1, 2, 3, 4, 5, 6, 7, 8, 9\}$ The universal set consists of all elements inside the rectangle.
- $A = \{2, 6, 7\}$ Set *A* contains all elements inside the circle representing set *A*.
- $B = \{1, 2, 5, 6, 8\}$ Set *B* contains all elements inside the circle representing set *B*.
- $A' = \{1, 3, 4, 5, 8, 9\}$ The set A', representing the complement of set *A*, contains all elements inside the rectangle that are not inside the circle representing set *A*.
- $B' = \{3, 4, 7, 9\}$ The set B', representing the complement of set *B*, contains all elements inside the rectangle that are not inside the circle representing set *B*.

SKILL 2: USE SET OPERATIONS

Remember that the intersection of two sets A and B, symbolized by $A \cap B$, is the set of all elements that are in both sets. In set-builder notation, $A \cap B = \{x \mid x \in A \text{ and } x \in B\}$.

Also, remember that the union of two sets A and B, symbolized by $A \cup B$, is the set of all elements that are in set A or set B (or both). In set-builder notation, $A \cup B = \{x \mid x \in A \text{ or } x \in B\}$.

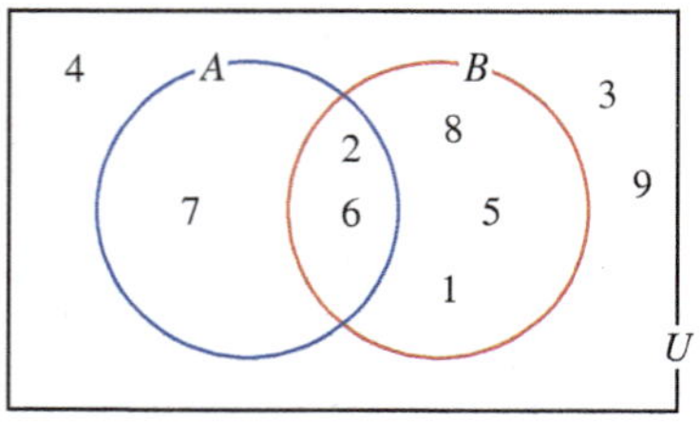

The Venn diagram above tells us the following:

- $A \cap B = \{2, 6\}$

  The intersection of sets A and B is the set of elements that are in both sets. To find the intersection of two sets, it helps to list (or at least think about) the elements in the two sets separately.
  $A = \{2, 6, 7\}$ and $B = \{1, 2, 5, 6, 8\}$
  The elements common to both sets are 2 and 6.

- $A \cup B = \{1, 2, 5, 6, 7, 8\}$

  The union of sets A and B is the set of elements that are in set A or set B (or both). To find the union of two sets, it helps to list (or at least think about) the elements in the two sets separately.
  $A = \{2, 6, 7\}$ and $B = \{1, 2, 5, 6, 8\}$
  The union consists of the elements that are in either set (or both). This means we merge the two sets (but don't list repeats).
  The elements from combining both sets are 1, 2, 5, 6, 7, and 8.

SKILL 3: FIND THE CARDINALITY OF A SET

Remember from Lesson 6-1 that $n(A)$, which is read as "n of A," indicates the number of elements in set A.

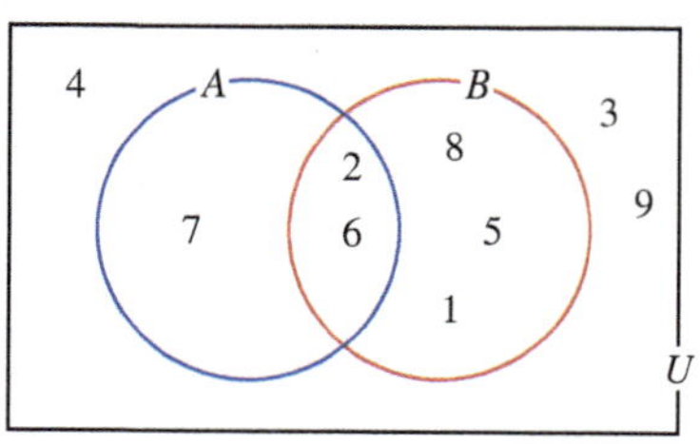

- $n(A) = 3$

  $A = \{2, 6, 7\}$
  Set A contains three elements, so $n(A) = 3$.

- $n(B) = 5$

  $B = \{1, 2, 5, 6, 8\}$
  Set B contains five elements, so $n(B) = 5$.

- $n(U) = 9$

  $U = \{1, 2, 3, 4, 5, 6, 7, 8, 9\}$
  Set U contains nine elements, so $n(U) = 9$.

SKILL 4: COMPUTE THEORETICAL PROBABILITIES

As we saw in Lesson 3-1, the probability of an event *E* in a sample space *S* occurring is given by

$$P(E) = \frac{\text{The number of outcomes in } E}{\text{The number of outcomes in } S} = \frac{n(E)}{n(S)}$$

You'll notice that this uses a similar notation to that used for cardinality.

Using *U* as the sample space, we can compute theoretical probabilities based on set notation.

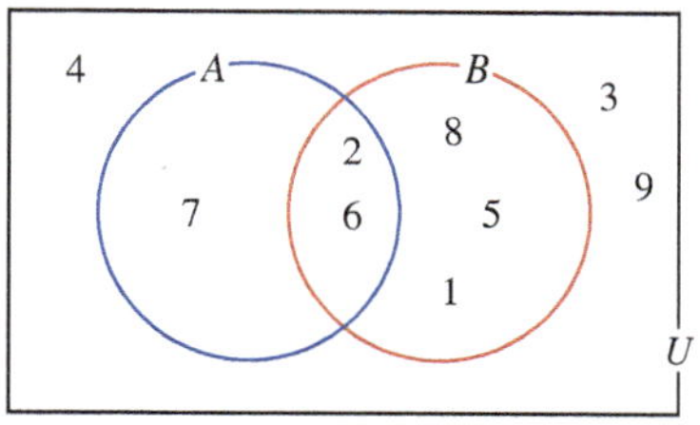

- If you choose a random element from the universal set *U*, what is the probability that it would be in set *A*?

$$P(A) = \frac{n(A)}{n(U)} = \frac{3}{9} = \frac{1}{3}$$

There are three elements in set *A*, and there are nine elements in the sample space *U*.

- If you choose a random element from the universal set *U*, what is the probability that it would be in $A \cap B$?

$$P(A \cap B) = \frac{n(A \cap B)}{n(U)} = \frac{2}{9}$$

There are two elements in $A \cap B$ and nine elements in the sample space *U*.

PREP SKILLS QUESTIONS

1. Based on the Venn diagram below, write sets *A*, *B*, and *A′* using the roster method.

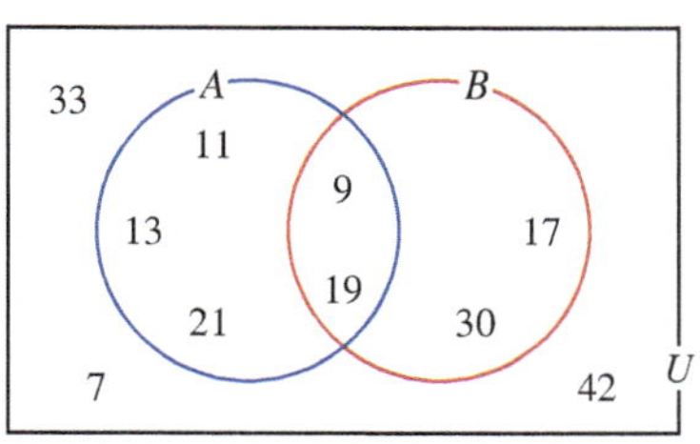

2. For the sets in Question 1, find $A \cup B$, $A \cap B$, and $B' \cup A$.

3. For the sets in Question 1, find $n(B)$ and $n(U)$.

4. If you were to choose a random element of the universal set in Question 1, what's the probability that it would be in set *A*?

5. What's the probability that it would be in set *U*?

Lesson 6-3 Worlds Collide

LEARNING OBJECTIVES

- ☐ 1. Illustrate sets with two-circle Venn diagrams.
- ☐ 2. Develop and use De Morgan's laws.
- ☐ 3. Use Venn diagrams to decide if two sets are equal.
- ☐ 4. Review how Venn diagrams can be used in probability.

David Sobecki

The interesting thing about the Internet is that it has created a kind of alternative circle of friends for people.
—Joanne Harris

Have you ever wanted to post something on Facebook or Instagram, then decided not to because it may not be something you want everyone you know to see? Some people are social media friends only with their closest buddies: More people are friends with hundreds of people, from their Mom to casual work acquaintances. My closest friends and I have a secret private group on Facebook for exactly this reason: The set of things I feel comfortable saying to my college buddies is most definitely not equal, or even equivalent, to the set of things I feel comfortable saying to my Mom and my golf pro. Most of us have many distinct circles of friends and contacts, and when those worlds collide the results can be unpredictable, and maybe unintentionally hilarious.

One good way to get a handle on complicated interplay between varying groups is with diagrams. And can you guess what kind of diagrams we'll choose to accomplish this? If you didn't say "Venn," please go back and rework the previous lesson. We'll wait here for you.

0. Think of two groups of people you know that intersect, and two that do not.

6-3 Class

Notice that there are four distinct regions in a Venn diagram illustrating two sets A and B. We'll want to number the regions for reference; we use Roman numerals so that we don't confuse the number of the region with elements in the set or the cardinality of the set.

1. Write the appropriate description next to each region. Pick your answer from the following list.

the elements in set B that are not in set A
the elements in both sets A and B
the elements in the universal set that are in neither set A nor set B
the elements in set A that are not in set B

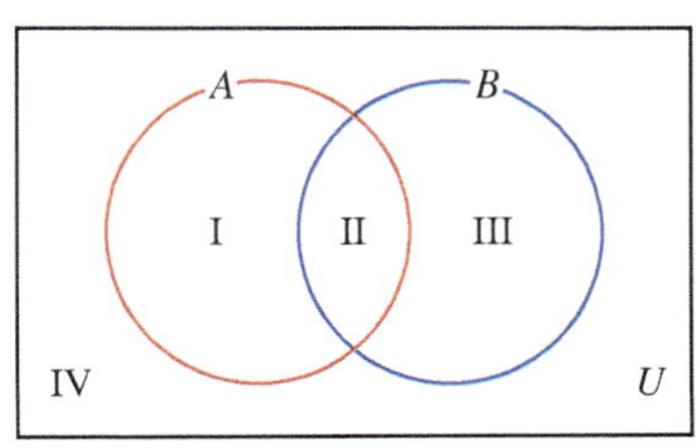

Region I represents

Region II represents

Region III represents

Region IV represents

The procedure that we use to illustrate set statements is found in the following box.

Illustrating a Set Operation with a Venn Diagram

Step 1. Draw a diagram for the sets, with Roman numerals in each region.

Step 2. Using those Roman numerals, list the regions described by each set.

Step 3. Find the set of numerals that correspond to the set given in the set operation.

Step 4. Shade the area corresponding to the set of numerals found in Step 3.

Solved Example

Illustrating Set Operations with a Venn Diagram

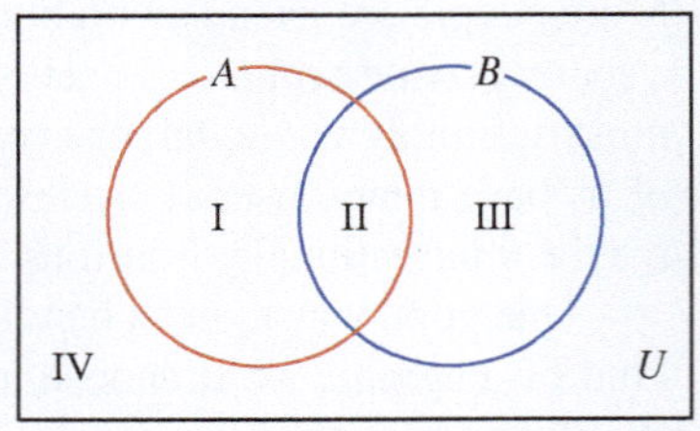

Draw a Venn diagram to illustrate the set $(A \cup B')'$.

SOLUTION

We'll begin with a generic Venn diagram (Step 1), which is always a good start.

Now we'll work our way through Steps 2–4.

Step 2: There are two sets in this operation: A and B'. Set A consists of the two regions inside the red circle, I and II. Set B', as the complement of B, is all of the stuff outside of the blue circle, which is regions I and IV. In summary:

$A = \{\text{I, II}\}$ $B' = \{\text{I, IV}\}$

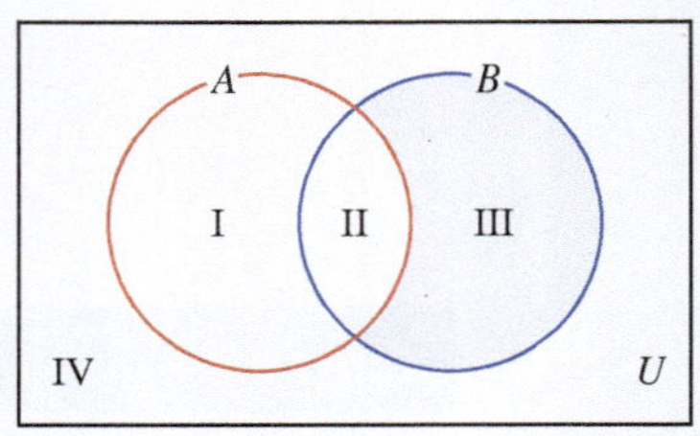

Step 3: The set described by the set operations here is $(A \cup B')'$. First we need to perform the union because it's in parentheses. Looking at the sets A and B' from Step 2, we can see that the union consists of regions I, II, and IV. That is,

$(A \cup B') = \{\text{I, II, IV}\}$.

The set we want is the complement of this set, which is simply region III.

$A \cup B')' = \{\text{III}\}$

Step 4: To illustrate the set $(A \cup B')'$ we shade the region found in Step 3, shown in the completed diagram.

2. Draw a Venn diagram to illustrate the set $A \cap B'$. (We thought about providing a blank diagram, but decided it wouldn't hurt to practice drawing them on your own. Sorry, man.)

> **Math Note**
>
> In any problem where we're asked to illustrate a set operation involving two sets, Steps 1 and 2 will be exactly the same.

3. Based on your Venn diagram, think of another way to write $A \cap B'$ using an operation from Lesson 6-2. (This is one of the big things we'll use Venn diagrams for: to identify when two sets that are written differently are actually the same.)

Did You Get It

Try these problems to see if you understand the concepts we just studied. The answers can be found at the end of the portfolio section.

1. Draw a Venn diagram to illustrate the set $A' \cap B$.
2. Draw a Venn diagram to illustrate the set $A' \cup B$.

De Morgan's Laws

There are two very well-known formulas that are useful in simplifying some set operations. They're named in honor of a 19th-century mathematician named Augustus De Morgan. First, we'll write the first of the formulas and illustrate it with an example. Then we'll see how Venn diagrams can be used to prove the formula.

De Morgan's First Law

For any two sets A and B,

$$(A \cup B)' = A' \cap B'$$

4. The first law states that the complement of the ________________ of two sets will always be equal to the ________________ of the complements of each set.

5. If $U = \{a, b, c, d, e, f, g, h\}$, $A = \{a, c, e, g\}$, and $B = \{b, c, d, e\}$, find $(A \cup B)'$ and $A' \cap B'$.

6. If we use the result of Question 5 to conclude that De Morgan's first law is likely to be true, we're using what type of reasoning, inductive or deductive?

7. Explain why this doesn't prove that the law is true.

Now we'll prove that the law is true. The plan is to use Venn diagrams to show that $(A \cup B)'$ and $A' \cap B'$ are the same set.

8. In the Venn diagrams below, set U contains which regions?

9. Set A contains which regions?

10. Set B contains which regions?

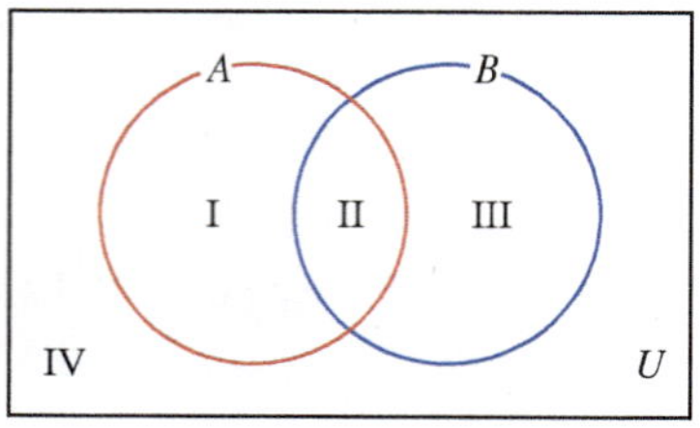

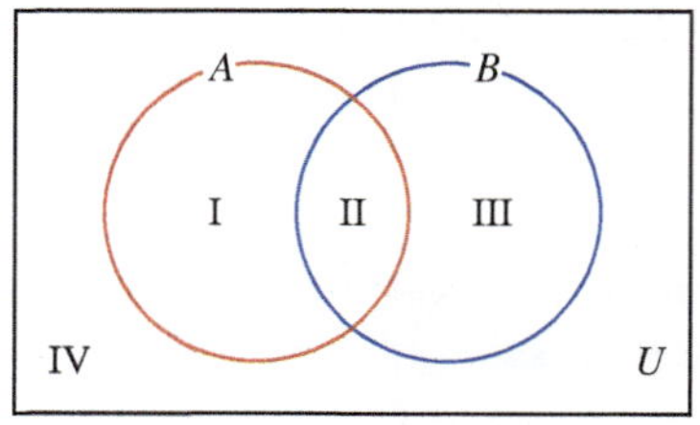

Fill in numbers of the regions in the preceding Venn diagram to answer Questions 11, 12, 14, and 15. Write as sets using the roster method.

11. $A \cup B =$

12. So $(A \cup B)' =$

13. Shade the region on the first diagram to illustrate $(A \cup B)'$.

14. $A' =$

15. $B' =$

16. So $A' \cap B' =$

17. Shade the region on the second diagram to illustrate $A' \cap B'$.

18. Since the diagrams for each side of the equation are identical, we use deductive reasoning to conclude that

______________________.

19. The second of De Morgan's laws involves an alternate form for the set $(A \cap B)'$. Based on the first law, make a conjecture as to what you think the alternate form for that set is.

20. Draw two Venn diagrams: one for $(A \cap B)'$, and one for the set that you believe it's equal to (from the previous question). Does this confirm your conjecture? If yes, great job. If not, revise your conjecture and try again.

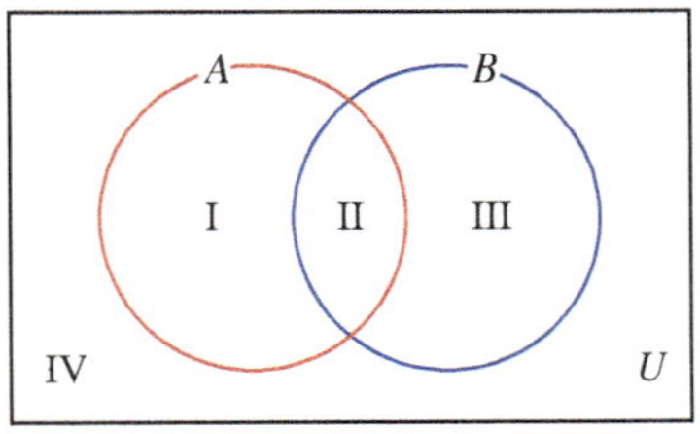

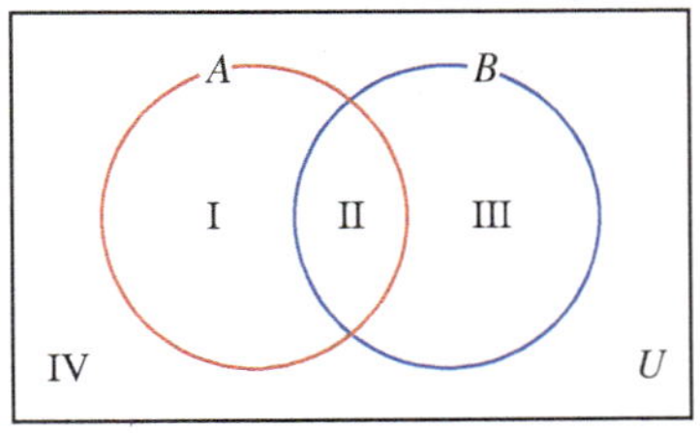

21. Fill in your result in the box below, then congratulate yourselves on being real mathematicians, because you just discovered AND proved a result.

De Morgan's Second Law

For any two sets *A* and *B*,

$$(A \cap B)' = \_\_\_\_\_\_$$

22. Even after proving a result, it's a good idea to use it in a specific example to make sure that it appears to be working correctly. For the sets $U = \{10, 11, 12, 13, 14, 15, 16\}$, $A = \{10, 11, 12, 13\}$, and $B = \{12, 13, 14, 15\}$, find $(A \cap B)'$ and $A' \cup B'$. Does the result confirm your version of De Morgan's second law? Does your result prove it?

Did You Get It

3. If $U = \{\text{ABC, NBC, CBS, Fox, USA, TBS, TNT, MTV}\}$, $A = \{\text{NBC, Fox, USA, TBS}\}$, and $B = \{\text{ABC, NBC, CBS, Fox}\}$, find $(A \cap B)'$ and $A' \cup B'$.

4. Use Venn diagrams to show that $B' \cap A = A - B$.

6-3 Group

The Cardinal Number of a Union

If 10 of your friends belong to the set of students taking a math class, and 14 belong to the set of students taking an English class, how many are in the union of those two sets? If your first instinct is 24, you're not alone—that's sort of the standard guess. And it might actually be right, but only if none of your friends are taking both a math and an English class. If any of them are in both classes, you'd be counting them twice by just adding the number of friends in each set. Venn diagrams can be used to analyze this situation.

As you can see in the figure below, if we start with the number of elements in *A*, we're counting all the members in regions I and II. When we add the number of elements in *B*, we're counting all the members in regions II and III. Do you see the issue? The elements in region II get counted twice.

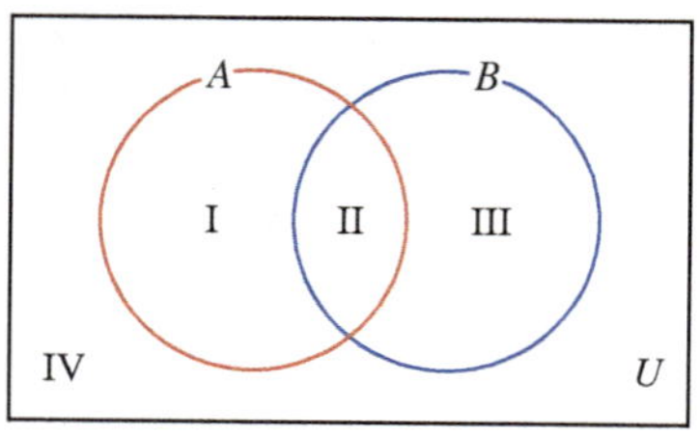

1. After adding the number of elements in sets *A* and *B*, what can we do to turn the result into the correct cardinality of the set $A \cup B$?

2. Use your answer to Question 1 to fill in the blank in the following box. Then congratulate yourselves again, because you've developed another formula! (Check with your instructor to make sure you have the right formula before moving on.)

The Cardinality of a Union

If $n(A)$ represents the cardinality of set *A*, then for any two finite sets *A* and *B*,

$$n(A \cup B) = n(A) + n(B) \_\_\_\_\_\_\_\_\_\_$$

Next, we'll use your shiny new formula to solve a problem.

3. In a survey of 100 randomly selected freshmen walking across campus, it turns out that 42 are taking a math class, 51 are taking an English class, and 12 are taking both. How many students are taking either a math class or an English class?

Digital Vision/Getty Images

4. What's the probability that one of those freshmen is taking either a math class or an English class?

5. Explain how the procedure you developed for finding the cardinality of a union is used in developing the second addition rule for probability: $P(A \text{ or } B) = P(A) + P(B) - P(A \text{ and } B)$.

Did You Get It

5. A poll of 200 doctors across the nation found that 112 were assisted in their office by registered nurses, 83 were assisted by licensed practical nurses, and 21 were assisted by both. How many were assisted by at least one type of nurse?

Digital Vision Ltd./SuperStock

We live in the information age—every time you turn around, somebody somewhere is trying to gather information about you, your opinions, and (most commonly) your spending habits. Surveys are conducted by the thousands every day, and every person, pet, pastime, and product are classified. The things we've learned about sets can be very helpful in interpreting information from surveys and classifications, and that will be the focus of the remainder of this lesson.

The plan is to give you some guidance on the first problem, providing steps that you can use for subsequent problems.

The problem: In 2025, there were 42 states that had some form of casino gambling in the state, 45 states that sold lottery tickets of some kind, and 39 states that had both casinos and lotteries. Draw a Venn diagram to represent these data, and find how many states have only casino gambling, how many states have only lotteries, and how many states have neither.

6. Draw a Venn diagram with two circles, and label the circles *L* and *C* for lottery and casino, respectively.

7. How many states have both casinos and a lottery? Put this value in the intersection of *C* and *L*.

8. Since 42 states have casino gambling and 39 have both, how many have casino gambling but not a lottery? Put that number in the portion of the circle labeled *C* that is outside of the intersection of *C* and *L*.

9. Since 45 states have a lottery and 39 have both, how many have a lottery but no casinos? Put that number in the portion of the circle labeled *L* that is outside of the intersection of *C* and *L*.

10. Add up the numbers in your diagram so far. There are 50 states (duh), so you can subtract from 50 to find how many states go inside the universal set but outside both circles. Do that.

11. Now that you have the diagram completed, answer all the questions in the original problem. Make sure you answer ALL of the questions.

The key starting point in that problem is knowing the number that goes into the intersection region of the Venn diagram. Once you know that, if you're given the total number that goes inside one of the circles, you can subtract the number in the intersection from that number to find the number that goes in that circle, but outside the intersection. Now you should be able to use the procedure outlined above to solve the next problem.

12. A dietitian studying the nutritional value of food at eight popular fast-food restaurants examines the amount of protein and the amount of fat in all of the breakfast meals available. Out of 93 meals available, 43 have more than 20 grams of protein, 65 have more than 25 grams of fat, and 31 have both more than 20 grams of protein and more than 25 grams of fat. How many meals have more than 20 grams of protein and less than 25 grams of fat? How many have less than 25 grams of fat and less than 20 grams of protein?

One of the most useful applications of Venn diagrams is for studying the results of surveys. Whether they are for business-related research or just to satisfy curiosity, surveys seem to be everywhere these days, especially online. Many news websites include daily surveys, like this one from cnn.com.

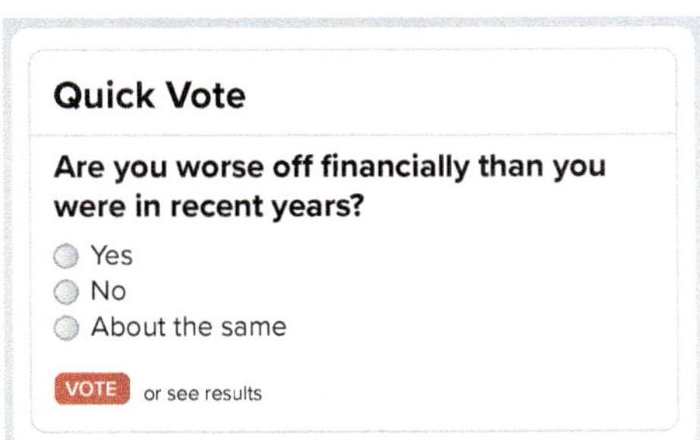

Our last problem, based on survey results, also uses a two-circle Venn diagram, but it is just a bit different: We're not given the number that goes in the intersection.

In a survey published in the *Journal of the American Academy of Dermatologists,* 500 people were asked by random telephone dialing whether they have a tattoo and/or a body piercing. Of these, 79 reported having a tattoo only, 31 reported having a piercing only, and 151 reported having at least one of the two. Draw a Venn diagram to represent these results and use your diagram to find the percentage of respondents that have a tattoo, that have a piercing, that have both, and that have neither.

13. Draw a two-circle Venn diagram, with one circle labeled to represent people with tattoos and the other labeled to represent people with piercings.

Neil Marriott/Digital Vision/ Getty Images

14. The key difference in this problem is that we're given the number of people with tattoos only (79). Does this number get entered into your Venn diagram? Where? Why?

15. Use what you discovered in Question 14 to finish the diagram and answer all questions in the problem.

Did You Get It ?

6. According to an online survey on Howstuffworks.com, 12,595 people gave their thoughts on Coke vs. Pepsi. Of these, 5,786 drink only Coke, 3,763 drink only Pepsi, and 11,405 drink at least one. Draw a Venn diagram to represent these results and use your diagram to find the percentage of respondents that drink Coke, that drink Pepsi, that drink both, and that drink neither.

6-3 Portfolio

Name ______________________________

Check each box when you've completed the task. Remember that your instructor will want you to turn in the portfolio pages you create.

ALEKS Practice

1. ☐ Include any written work from the ALEKS assignment along with any notes or questions about this lesson's content.

Applications

1. ☐ Complete the applications problems.

Reflections

Type a short answer to each question.

1. ☐ Describe the process of using a Venn diagram to decide if two sets that are written differently are actually the same.
2. ☐ Explain what types of surveys lend themselves to being analyzed with Venn diagrams.
3. ☐ Name one thing you learned or discovered in this lesson that you found particularly interesting.
4. ☐ What questions do you have about this lesson?

Looking Ahead

1. ☐ Complete the prep skills for Lesson 6-4.
2. ☐ Read the opening paragraph in Lesson 6-4 carefully and answer Question 0 in preparation for that lesson.

Answers to "Did You Get It?"

1.

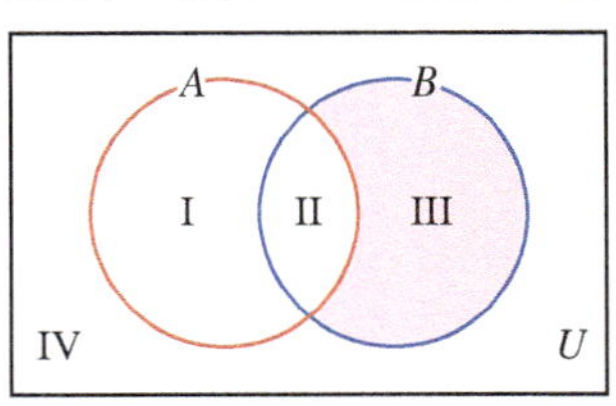

2.

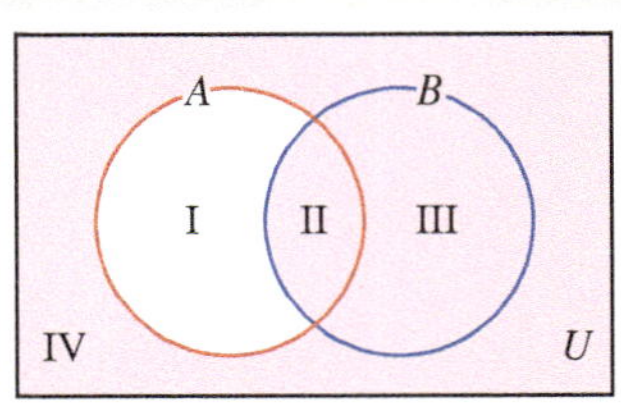

4. Both diagrams are

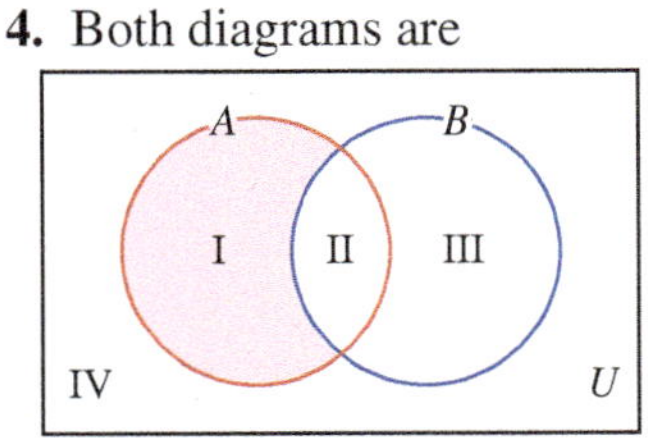

3. Both are {ABC, CBS, USA, TBS, TNT, MTV}

5. 174 **6.** Coke: 60.7%; Pepsi: 44.6%; both: 14.7%; neither: 9.4%

Answers to Prep Skills Questions

1. $A = \{9, 11, 13, 19, 21\}$, $B = \{9, 17, 19, 30\}$, $A' = \{7, 17, 30, 33, 42\}$
2. $A \cup B = \{9, 11, 13, 17, 19, 21, 30\}$, $A \cap B = \{9, 19\}$, $B' \cup A = \{7, 9, 11, 13, 19, 21, 33, 42\}$
3. $n(B) = 4$, $n(U) = 10$
4. $\frac{1}{2}$
5. 1

6-3 Applications

Name ______________________________

1. Draw a Venn diagram for each set described and use your diagrams to decide if the two sets are equal or not. $A \cup B'$ and $A' \cap B$

In Questions 2 and 3, A = {People majoring in a science} and B = {People majoring in Botany, Zoology, or Biochemistry}. Draw a Venn diagram that represents each set, and write a sentence describing what the set represents. Think carefully about the relationship between these sets before you draw your diagram.

2. $A \cup B$

Math Note

Contrary to popular belief, "zoology" is pronounced, "ZO-ology", not "ZOO-ology." It is NOT the study of zoos.

3. $A \cap B$

6-3 Applications

Name ______________________________

The table and Venn diagram below are for Questions 4–7. The table shows the 10 states that had the highest percentage of people voting Republican in the 2012 and 2024 presidential elections. For each question, write which region of the Venn diagram would include the state listed. Set A represents states in the top 10 in 2012, and set B the top 10 in 2024.

Year	2012	2024
States	Utah Wyoming Oklahoma Idaho West Virginia Alabama Arkansas Kentucky Nebraska Kansas	Wyoming West Virginia North Dakota Idaho Oklahoma Alabama Kentucky Tennessee Arkansas South Dakota

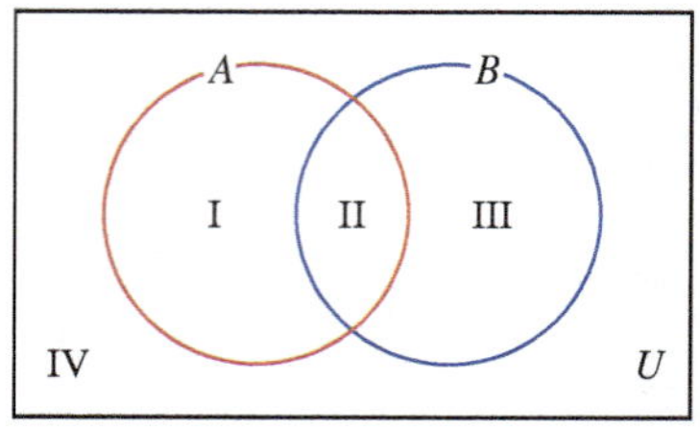

4. Arkansas

5. South Dakota

6. Utah

7. Florida

8. According to a survey conducted in 2024, out of 400 cars that were given inspections while getting an oil change, 138 were in need of brake repair, 94 needed new tires, and 78 needed both. Use a Venn diagram to find the number of cars that didn't need either repair, and the number with good tires that needed new brakes.

6-3 Applications

Name ______________________________

Use this information for Questions 9–14. Five hundred people are involved in a drug trial, testing two new acne medications. The two drugs involved are called Felafin and Gantex, and participants could have been given either, both, or neither medication. Let's call the set of people who were given Felafin F, and the set who were given Gantex G. (Pretty original, right?)

9. Write a description of what the set $F \cup G$ represents.

10. Using your answer to Question 9, what does set $(F \cup G)'$ represent?

11. Draw a Venn diagram illustrating the set $(F \cup G)'$.

12. Write descriptions of what the sets F' and G' represent.

13. Using your answer to Question 12, what does the set $F' \cap G'$ represent?

6-3 Applications

Name ______________________________

14. Compare your responses to Questions 10 and 13. Are they the same? Explain why they should be the same.

15. Based on an online poll of 70,000 people last February, 49% had eaten at McDonald's at least once in the previous month, 26% had eaten at Burger King in the last month, and 41% had eaten at neither restaurant. Draw a Venn diagram with the NUMBER of people who ate at these restaurants and use it to answer Questions 16–18.

16. How many of the people polled ate at both restaurants at least once in the last month?

17. How many had eaten at McDonald's but not Burger King?

18. What percentage of those polled had visited at least one of the two restaurants?

Lesson 6-4 Prep Skills

SKILL 1: FILL IN A VENN DIAGRAM

When filling in a Venn diagram from a verbal description, pay careful attention to whether we are being given the total for a given set or the number of elements that are only in that set and no others.

- Suppose a group of students conducted a survey of 200 individuals about soft drink preferences. There were 90 who reported drinking regular soda, 75 who drink diet soda, and 13 who drink both. Fill in the Venn diagram below.

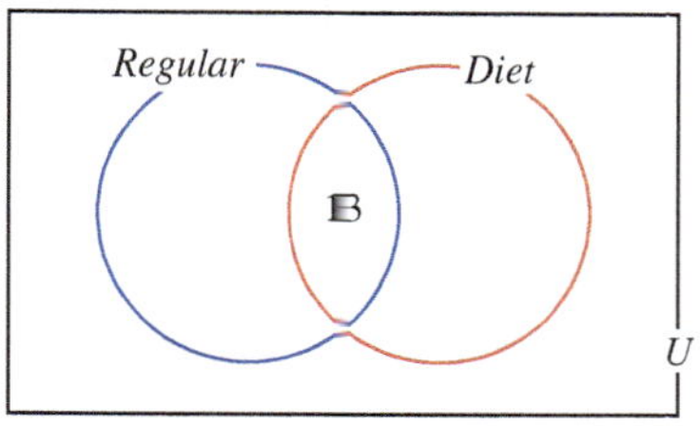

The first label we fill in is 13. Write 13 in the intersection of the two sets since that represents the number of people who reported drinking both types of soda.

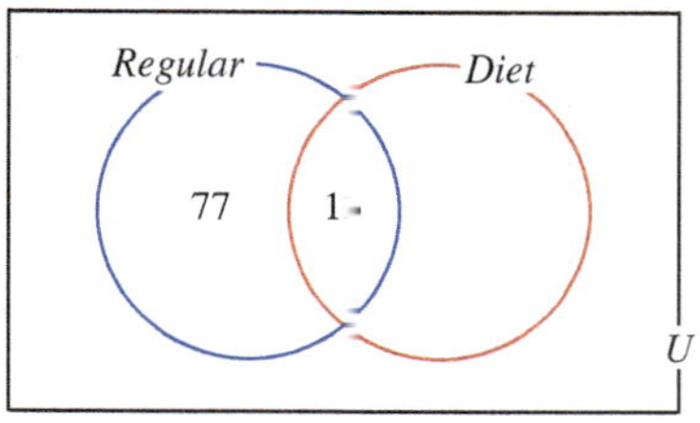

A common error is to misinterpret the 90 as the number who drink only regular soda. The **total** number of people who reported drinking regular soda is 90. That means $90 - 13 = 77$ people reported drinking only regular soda. We write 77 in the space inside the Regular circle but not in the Diet circle.

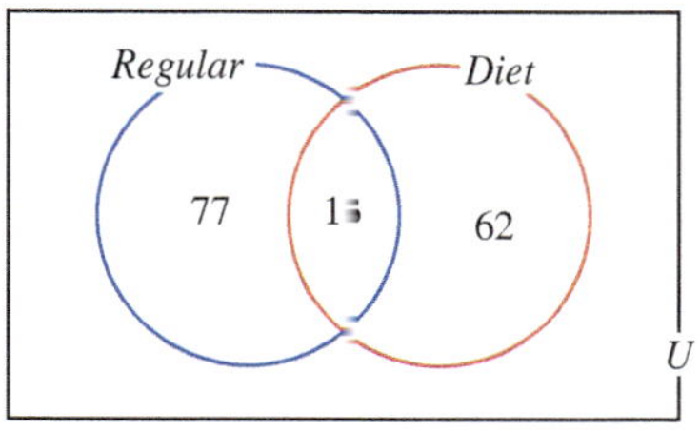

Another common error is to misinterpret 75 as the number who drink only diet soda. In fact, 75 is the total number of people who indicated drinking diet soda. That means $75 - 13 = 62$ people reported drinking only diet soda. We write 62 in the space inside the Diet circle but not in the Regular circle.

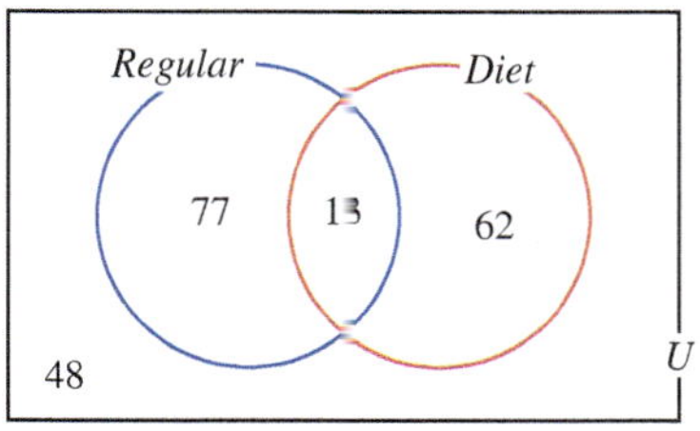

The last label to fill in is the space in the rectangle but not inside either circle. This represents the number of people who didn't report drinking either type of soda. For that, we'll need to do some arithmetic. $77 + 13 + 62 = 152$. That means 152 of the 200 people reported drinking soda of any type. $200 - 152 = 48$ people didn't report drinking any soda.

Note that the total of all of the labels in the diagram ($77 + 13 + 62 + 48 = 200$) checks with the information given in the question.

SKILL 2: INTERPRET THE SHADED PORTION OF A VENN DIAGRAM

There are four distinct regions in a Venn diagram illustrating two sets *A* and *B*.

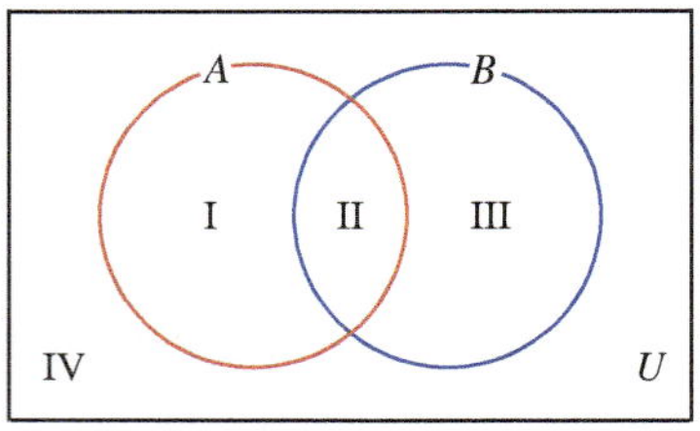

Region I represents the elements in set *A* that are not in set *B*.

Region II represents the elements in both sets *A* and *B*.

Region III represents the elements in set *B* that are not in set *A*.

Region IV represents the elements in the universal set that are in neither set *A* nor set *B*.

PREP SKILLS QUESTIONS

1. From a universal set with 90 elements, there are 8 elements that are in both sets *A* and *B*, 24 in set *A*, and 50 in set *B*. Fill in the number of elements in each region of the Venn diagram provided.

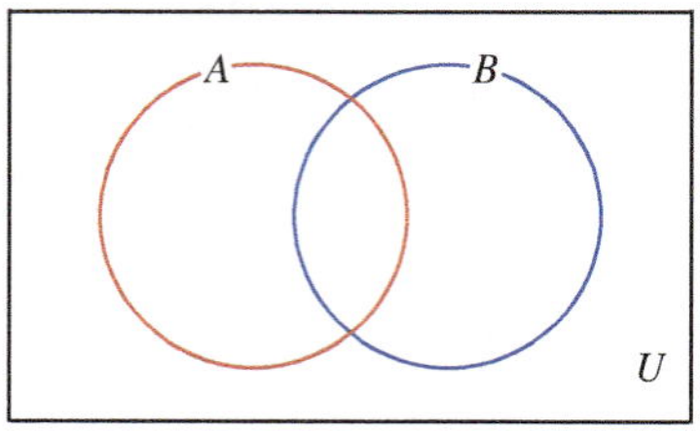

2. Write the set that is represented by the shaded portion of the Venn diagram.

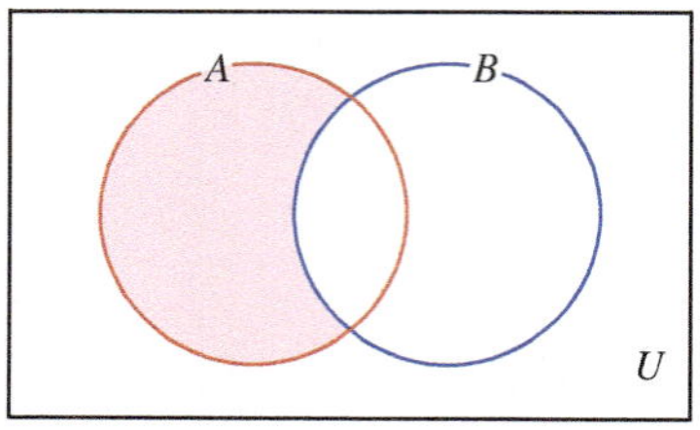

3. Write the set that is represented by the shaded portion of the Venn diagram.

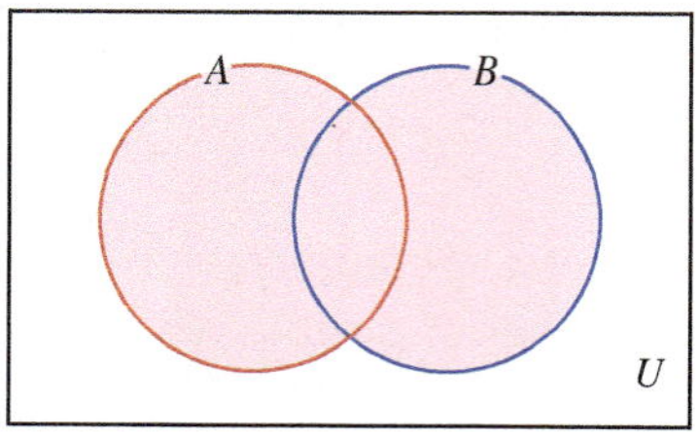

Lesson 6-4 A Dollar for Your Thoughts

LEARNING OBJECTIVES

- ☐ 1. Illustrate sets with three-circle Venn diagrams.
- ☐ 2. Decide if two sets are equal using three-circle Venn diagrams.
- ☐ 3. Solve a variety of applied problems using Venn diagrams.

Terry Vine/Blend Images LLC

Prejudice is a great time saver. You can form opinions without having to get the facts.
—E.B. White

Communication in our society is becoming cheaper, easier, and more effective all the time. In the age of smart phones and online communication, businesses are finding it simpler than ever to contact people for their opinions, and more and more people are finding out that companies are willing to pay to hear what they have to say. There are literally hundreds of companies in the United States today whose main function is to gather opinions on everything from political candidates to potato chips. In fact, almost \$50 *billion* is spent on market research in the U.S. each year. Maybe you'll think twice the next time somebody asks you for your opinion for free.

With all the money at stake, not surprisingly it's important to organize all of the information that gets gathered. That sure sounds a lot like what we've been using set theory for! When the interplay between sets becomes more complicated, a more flexible tool is needed, and so we bring you the three-circle Venn diagram. In my opinion, it's a great way to organize information gathered from surveys and other sources. (Oh, and no charge—you can have that opinion for free.)

0. Think of some situations where you might be offered money for your opinions.

6-4 Class

Here's some good news: The procedures we used for organizing and evaluating data in the last lesson work just fine for situations involving three sets: You just get a more complicated diagram.

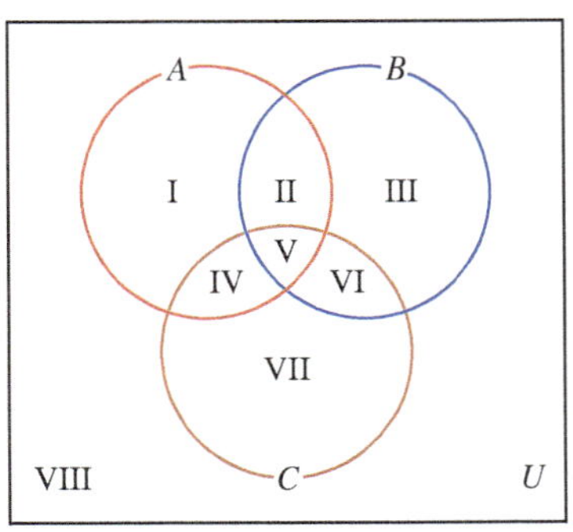

For Questions 1–8, fill in the blanks to describe each region in the diagram.

1. Region I represents the elements in set ____ but not in set ____ or set ____.

2. Region II represents the elements in set ____ and set ____ but not in set ____.

3. Region III represents the elements in set ____ but not in set ____ or set ____.

4. Region IV represents the elements in sets _____ and _____ but not in set _____.

5. Region V represents the elements in sets _____, _____, and _____.

6. Region VI represents the elements in sets _____ and _____ but not in set _____.

7. Region VII represents the elements in set _____ but not in set _____ or set _____.

8. Region VIII represents the elements in the universal set _____ but not in set _____, _____, or _____.

A survey is conducted about the radio listening habits of adults in the 18–45 age bracket. Specifically, those surveyed were asked whether they listen to terrestrial radio over the airwaves, Internet radio, or satellite radio. To organize the results, the marketing group conducting the survey defines sets as follows:

A = people who listen to terrestrial radio
B = people who listen to Internet radio
C = people who listen to satellite radio

Our goal in organizing this information will be to study the set $A \cap (B \cap C)'$.

9. What do you think this set represents in the context of this problem? (Not so easy, is it? If you can't decide, that's fine. The point is to try quickly.)

Maybe you're a super genius and it was clear to you what that set represents in this context. The rest of us will use Venn diagrams to figure it out in the Solved Example.

Solved Example

Organizing Information with a Venn Diagram

For the survey responses described above, what does the set $A \cap (B \cap C)'$ represent?

SOLUTION

We'll use the technique we learned in Lesson 6-3 to illustrate this set on a Venn diagram. If we're fortunate, that will make it a lot easier to figure out what set is being described. First, we'll want a three-circle Venn diagram with the regions numbered.

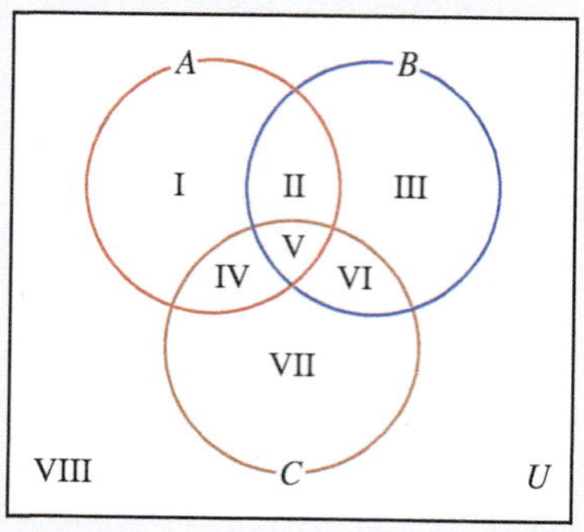

Solved Example *Continued*

As always in math, we'll note that the parentheses are there for a reason and start by working with what's inside. In this case, that's $B \cap C$. There are two regions that make up the stuff in both set B and set C: regions V and VI.

Next, notice that we're interested in the COMPLEMENT of $B \cap C$. That's all of the regions except V and VI. So

$$(B \cap C)' = \{\text{I, II, III, IV, VII, VIII}\}$$

The diagram shows us that

$$A = \{\text{I, II, IV, V}\}$$

The regions that are common to these two sets are I, II, and IV. So

$$A \cap (B \cap C)' = \{\text{I, II, IV}\}$$

This is shaded in the next diagram.

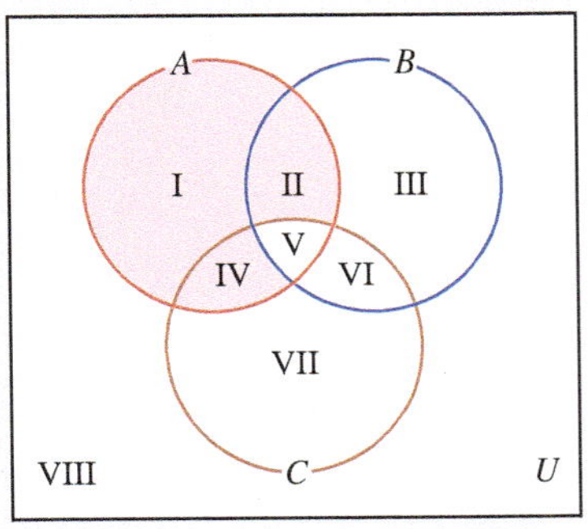

Now for the interpretation. All of the shaded stuff is definitely inside the circle for set A, which means all those people listen to terrestrial radio. But one part of A is missing: region V. This is the region inside all three circles, which represents the folks who listen to all three types of radio. So the set $A \cap (B \cap C)'$ represents everyone who listens to terrestrial radio, but not all three types. And that was a whole lot easier to figure out than it was from the original description of the set in terms of operations.

10. Using the Solved Example as a guide, draw a Venn diagram that represents the set $(A' \cup B) \cap C$.

11. Describe what the set in Question 10 represents in terms of the survey on radio listening habits.

Did You Get It

Try this problem to see if you understand the concepts we just studied. The answer can be found at the end of the portfolio section.

1. Draw a Venn diagram to illustrate the set $(A \cap B') \cup C$.

To get even better at working with Venn diagrams, it's helpful to turn the process around, starting with a shaded diagram and figuring out what set it represents.

12. Write the set illustrated by the Venn diagram. Your answer should use set operation symbols, not words. There may be more than one way to do so.

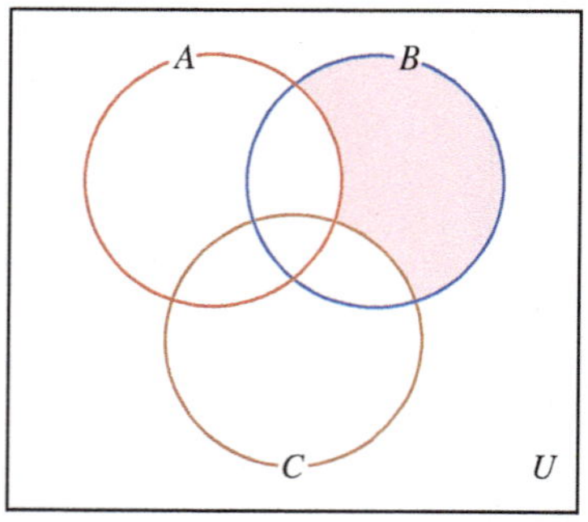

13. Give a verbal description of your answer to Question 12. Then explain where in this lesson you've already answered this question.

Did You Get It

2. Write the set illustrated by the Venn diagram to the right. Your answer should use set operation symbols, not words. There may be more than one way to do so.

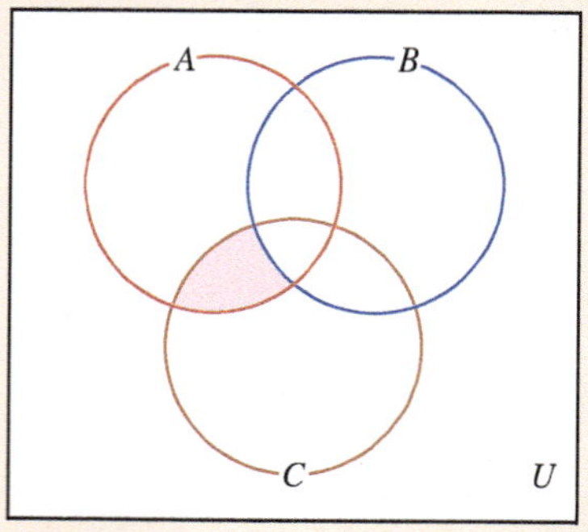

In Questions 14–19, the goal is to decide if the two sets are equal using Venn diagrams:

$(A \cup B) \cap C$ and $(A \cap C) \cup (B \cap C)$.

14. The set $A \cup B$ consists of which regions?

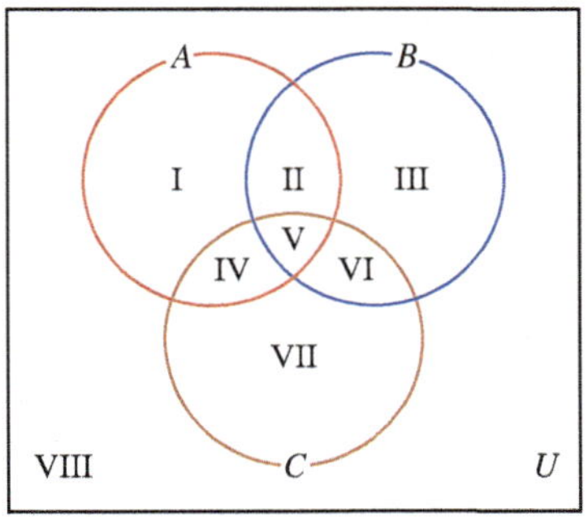

15. So $(A \cup B) \cap C$ consists of which regions?

16. The set $A \cap C$ consists of which regions?

17. The set $B \cap C$ consists of which regions?

18. So $(A \cap C) \cup (B \cap C)$ consists of which regions?

19. True or false: $(A \cup B) \cap C = (A \cap C) \cup (B \cap C)$. Describe how you decided on your answer.

Did You Get It

3. Decide if the two sets are equal using Venn diagrams: $B \cup (A \cap C)$ and $(A \cup B) \cap (B \cup C)$.

Sidelight The Lighter Side of Polls

Most of us have gotten pretty accustomed to public opinion polls that deal with weighty issues—politics, climate change, the economy—but not every survey is quite that serious. Let's have a look at some odd facts that came out of surveys.

- Just about 50% of lost remote controls are found in the cushions of furniture. About 4% end up in the fridge or freezer, and about 2% are found outside or in a car.
- 29% of Americans think that "cloud computing" involves actual clouds in the sky.
- Almost 10% of American college graduates believe that Judge Judy is a member of the United States Supreme Court.
- Over 60% of self-proclaimed vegetarians admitted to eating meat in the previous 24 hours.
- 47% of Americans don't put a single penny of their paycheck into long-term savings.
- 52% of Americans sing in the shower, 47% do something in the shower that is more traditionally done in a different bathroom facility, and 7% claim to never bathe at all.
- (This is my favorite.) 2% of voters polled during the 2012 election season thought that Republican presidential nominee Mitt Romney's full first name was "Mittens".

6-4 Group

A criminal justice major is studying the frequency of certain types of crimes in a nearby county. He studies the arrest records of 300 inmates at the county jail, specifically asking about drug-related offenses, domestic violence, and theft of some sort. He finds that 194 had been arrested for theft, 210 for drug offenses, and 170 for domestic violence. In addition, 142 had arrests for both theft and drugs, 111 for both drugs and domestic violence, 91 for both theft and domestic violence, and 45 had been arrested for all three. Our goal will be to draw a Venn diagram to represent these results. Once we've done so, we'll be able to use it to get a wealth of information about this study.

The first diagram is for reference, with the regions numbered as usual. The second diagram is the one we'll fill in with information specific to this study.

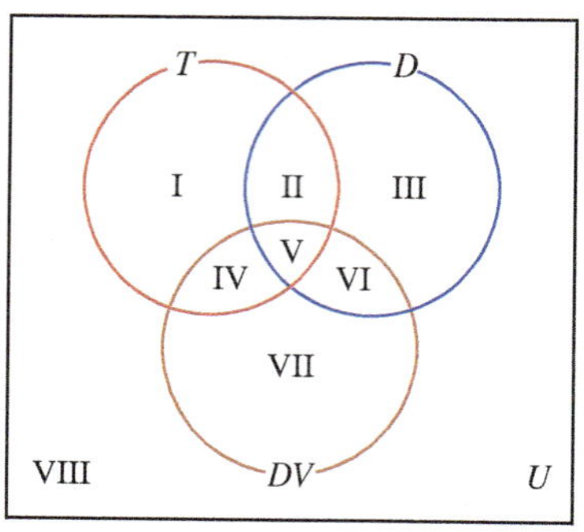

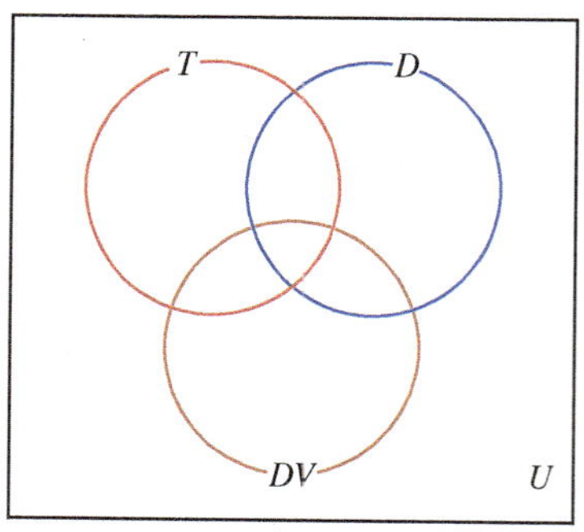

1. The only region we know for sure from the given information is region V—the number of inmates arrested for all three offenses. So we begin by putting _____ in region V.

2. There are ______ inmates with arrests for both theft and drugs, but we have to subtract the number arrested for all three offenses to find the number in region II:

   ______ – _____ = _____. Make sure you write this number into your diagram.

3. In the same way, find (and write in) the values in regions IV and VI.

   The number in region IV =

   The number in region VI =

Math Note

An important observation: We started with the number of elements in the innermost region and worked our way outward.

4. Now we're almost ready to find the number of elements in regions I, III, and VII. There were ______ inmates arrested for theft, but the values in regions II, IV, and V are already accounted for in the diagram.

5. Add the values for II, IV, and V here:

6. Subtract the value from Question 5 from the total number of inmates arrested for theft. That leaves ____ in region I. Write it into your diagram.

7. Using the steps in Questions 5 and 6 as a model, find and write in the values in regions III and VII.

   The number in region III =

   The number in region VII =

8. Adding up all the numbers in the diagram so far, we get ______. That leaves _____ in region VIII. Write it in.

Now that we have the diagram completed, we're ready to answer some questions that would have been very difficult to answer (to say the least) without first organizing/understanding the data in a Venn diagram.

9. Find the number of inmates who had been arrested for only drug-related offenses.

10. What percentage of inmates arrested for theft were also arrested for drug-related offenses?

11. Find the number of inmates who had been arrested for theft and domestic violence but not drugs.

12. Find the number of inmates who had been arrested for theft or drugs.

13. Find the number of inmates who had been arrested for none of these offenses.

When solving problems with Venn diagrams, the key is to write in the given information that applies exactly to certain regions, then use subtraction to find other regions one by one.

Did You Get It

4. Follow the steps laid out in Questions 1–8 to draw a Venn diagram for the following survey, then use it to answer the questions.

   An online music service surveyed 500 customers and found that 270 listen to hip-hop music, 320 listen to rock, and 160 listen to country. In addition, 140 listen to both rock and hip-hop, 120 listen to rock and country, and 80 listen to hip-hop and country. Finally, 50 listen to all three. Draw a Venn diagram to represent the results of the survey and find the number of customers who

   a. Listen to only hip-hop.
   b. Listen to rock and country but not hip-hop.
   c. Don't listen to any of these three types of music.
   d. Don't listen to country music.

In this example, rather than walking you through the steps, we're going to just provide the problem and a blank diagram and let you and your group mates work it out. The key difference here is that you won't be provided with the number of elements in the intersection of all three sets, so you'll have to read the information very carefully, and work from "out to in," rather than from the innermost region outward, like in the last example. Subtraction is still your friend.

Three of the most dangerous risk factors for heart attack are high blood pressure, high cholesterol, and smoking. In a survey of 690 heart attack survivors, 62 had only high cholesterol among those three risk factors; 36 had only smoking; and 93 had only high blood pressure. There were 370 total with high cholesterol, 159 with high blood pressure and cholesterol that didn't smoke, and 23 that smoked and had high cholesterol but not high blood pressure. Finally, 585 had at least one risk factor.

14. Draw a Venn diagram representing this information and use it to answer the following questions.

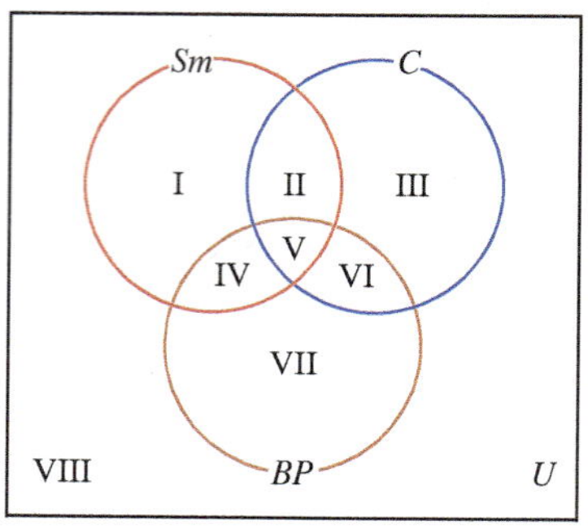

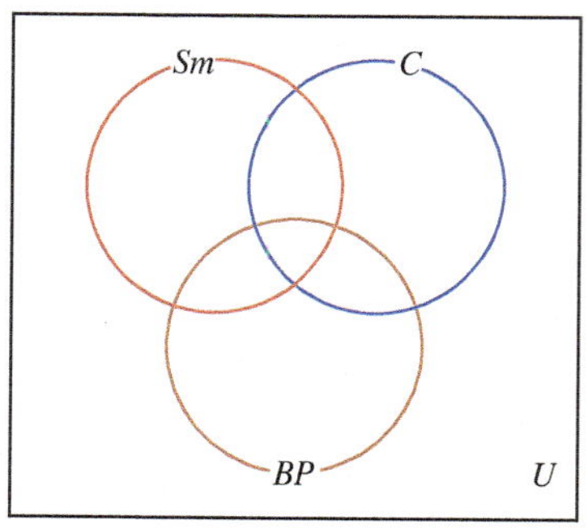

15. How many survivors had all three risk factors?

16. How many had exactly two of the three risk factors?

17. How many had none?

18. What percentage were smokers?

Did You Get It

5. Three other risk factors are obesity, family history of heart disease, and stress. Of the group of heart attack patients in Questions 14–18, 213 had a family history of heart disease; 47 of those with family history also suffered from stress but not obesity, 60 were obese but had no stress issues, and 12 were neither stressed nor obese. Stress was a factor for 170 total, 8 of whom had no family history of heart disease and weren't obese. There were 396 patients with none of these three risk factors.

   a. How many of the patients were obese?
   b. What percentage had all three of these risk factors?
   c. How many were not obese and did not have a family history of heart disease?

6-4 Portfolio

Name ______________________________

Check each box when you've completed the task. Remember that your instructor will want you to turn in the portfolio pages you create.

ALEKS Practice

1. ☐ Include any written work from the ALEKS assignment along with any notes or questions about this lesson's content.

Applications

1. ☐ Complete the applications problems.

Reflections

Type a short answer to each question.

1. ☐ Explain why Venn diagrams are so useful in organizing survey information with three responses.
2. ☐ If you see a Venn diagram online, on a news broadcast, or in print, do you feel like you'll be better able to process the information it provides now? Explain.
3. ☐ Name one thing you learned or discovered in this lesson that you found particularly interesting.
4. ☐ What questions do you have about this lesson?

Looking Ahead

1. ☐ Complete the prep skills for Lesson 7-1.
2. ☐ Read the opening paragraph in Lesson 7-1 carefully and answer Question 0 in preparation for that lesson.

Answers to "Did You Get It?"

1.

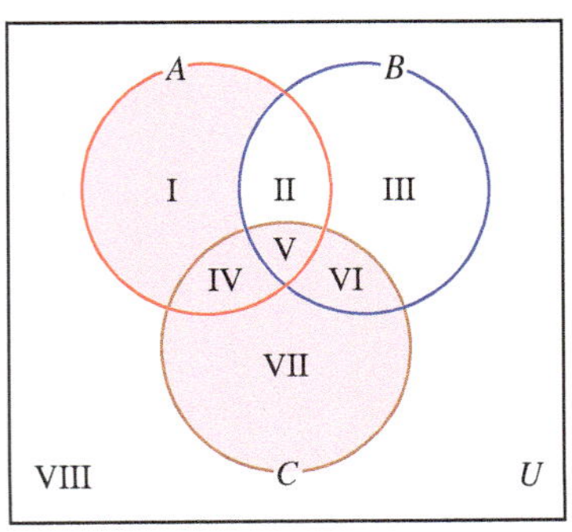

3. Both diagrams are:

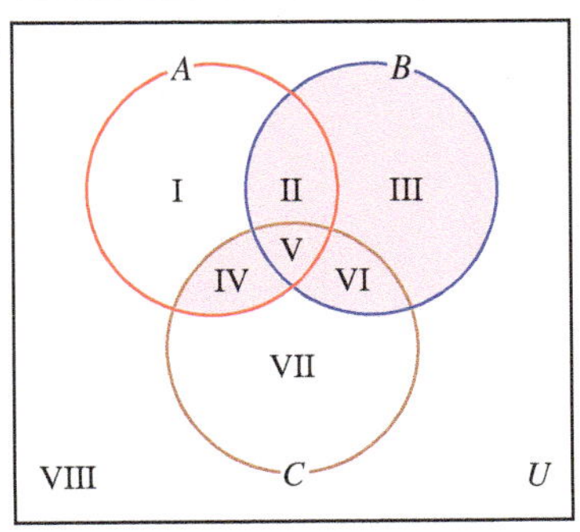

2. $(A \cap C) - B$ or $(A \cap C) \cap B'$

4. a. 100 **b.** 70 **c.** 40 **d.** 340

5. a. 227 **b.** 13.6% **c.** 404

Answers to Prep Skills Questions

1.

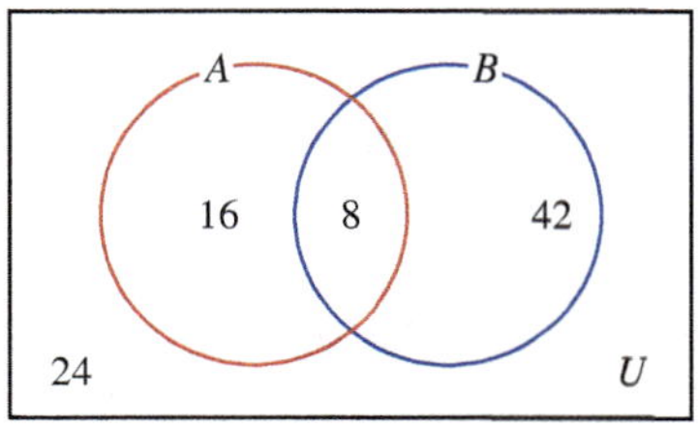

2. $A \cap B'$ or $A - B$

3. $A \cup B$

6-4 Applications

Name ____________________

1. Draw a Venn diagram for each set described and use your diagrams to decide if the two sets are equal or not.

$(A \cup B) \cup C'$ and $A \cup (B \cap C)$

The table and Venn diagram below are for Questions 2–7. The table shows the five states that had the highest percentage of people living in poverty for the years 2014, 2018, and 2022. For each question, write which region of the Venn diagram would include the state listed. Set *A* represents states in the top five in 2014, set *B* in 2018, and set *C* in 2022.

Year	2014	2018	2022
States	Mississippi Louisiana New Mexico Arizona Texas	Mississippi New Mexico Louisiana Alabama Kentucky	New Mexico Mississippi Louisiana Arkansas Kentucky

Source: U.S. Census Bureau

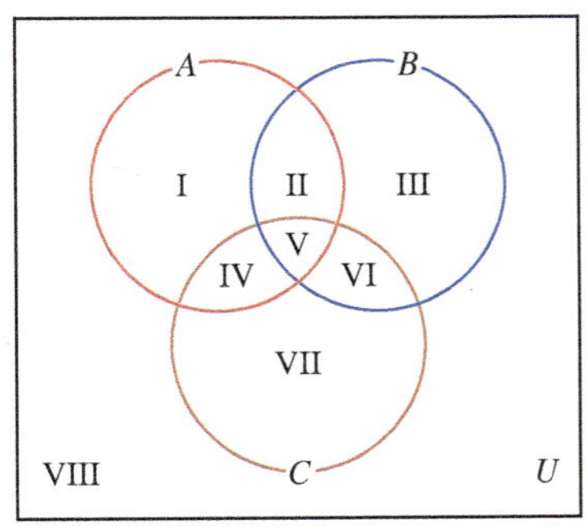

2. Arizona

3. New Mexico

4. Ohio

6-4 Applications

Name ______________________________

5. Kentucky

6. Alabama

7. Arkansas

In Questions 8 and 9, S = {U.S. senators}, D = {Elected officials who are Democrats}, and R = {Elected officials who are Republicans}. Draw a Venn diagram that represents each set, and write a sentence describing what the set represents.

8. $D \cup (R \cap S)$

9. $S - (D \cup R)$

6-4 Applications

Name ______________________________

A large inpatient addiction treatment facility currently houses 80 patients. Thirty-two of them have an addiction to alcohol, 25 have an addiction to narcotics, and 4 are addicted to both. Fill in the Venn diagram and use it to answer Questions 10–12.

10. How many patients are addicted only to narcotics?

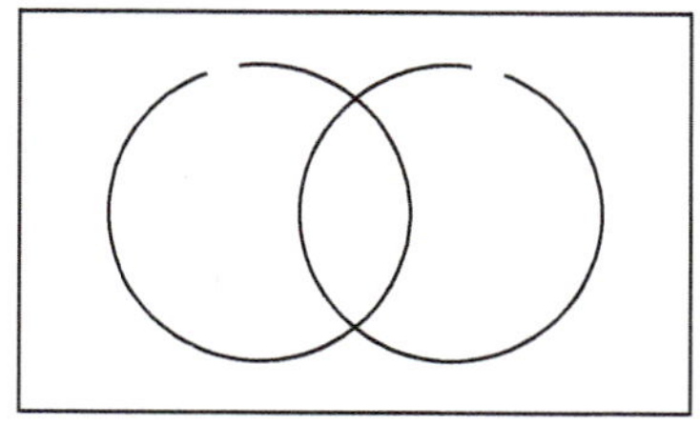

11. How many aren't addicted to narcotics?

12. How many are addicted to something other than alcohol or narcotics?

There's no official definition of a mountain, but one generally accepted standard is an elevation of at least 1,000 feet over the surrounding terrain. By that definition, there are 18 states that have at least one mountain, but no coastline (if we define coastline to mean areas where the land meets an ocean). There are 7 states that have coastline, but no mountains. Nine states have neither a mountain nor coastline. Draw a Venn diagram and use it to answer Questions 13–15.

13. How many states have at least one mountain?

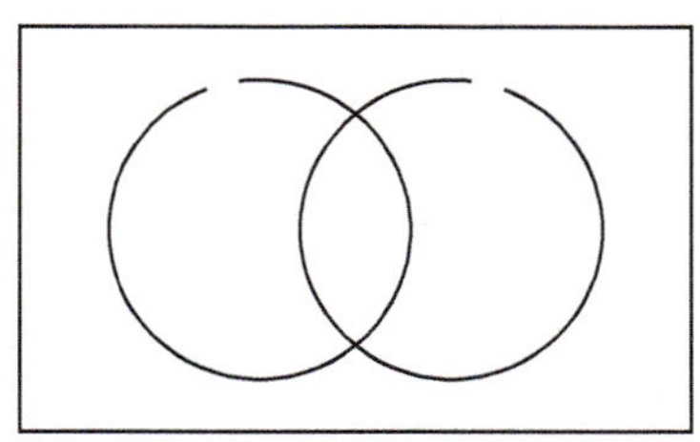

14. How many have coastline?

15. How many states have both?

6-4 Applications

Name ______________________________

The arts communities in 230 cities across the country were rated according to whether or not they have an art museum, a symphony orchestra, and a ballet company. There are 119 cities with an art museum; 20 of those also have ballet but no orchestra, and 41 have an orchestra but no ballet; 30 have neither. Of the 75 cities with a ballet company, 10 have an orchestra as well, but they lack an art museum. Twenty-two cities have only an orchestra. Fill in the Venn diagram and use it to answer Questions 16–18.

16. What percentage of the cities have an art museum?

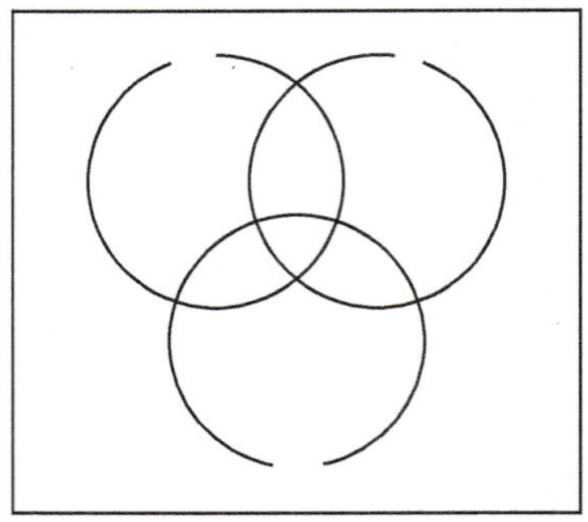

17. How many more cities are there with none of these three than with all of them?

18. If you pick a city at random from this list to travel to and really want to go either to an art museum or an orchestral concert, what is the percent chance that you'll end up disappointed?

Math in Diversity REVISITED

John Lund/Blend Images

The Venn diagram is based on the given demographic estimates:

1. White only: 601; Black only: 111; Hispanic only: 61
2. Hispanic and Black but not white: 5
3. Hispanic or Black: 303
4. Neither white, Black, nor Hispanic: 96
5. This is simple if you think about it: the percentage for whites is less than 50, so there will be more nonwhites

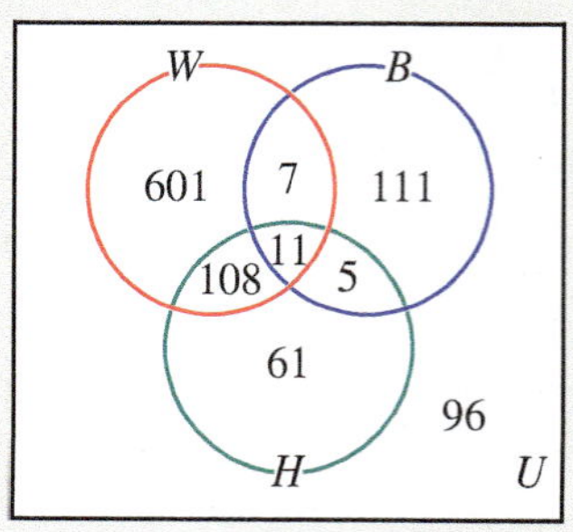

Unit 7
Uncommon Sense

Colin Anderson/Blend Images

Outline

Math in Mind Control

The term *mind control* is an absolute favorite of conspiracy theorists and science fiction writers, but is it real? That largely depends on what you mean by the term, but I can guarantee you one thing: There are a LOT of people and organizations out there that are trying to control the way you think each and every day. Everywhere you turn in modern society, someone is trying to convince you of something. "Buy my product!" "Vote for the guy I like!" "You can't afford NOT to lease a car from our dealership!" "Climate change is a hoax!" "You should go out with me, and soon!" "Pledge our fraternity—all the other ones stink!" Should I go on?

Logic is sometimes defined as correct thinking or correct reasoning, but many folks refer to logic by a more casual name: common sense. Regardless of what you call it, the ability to think logically is crucial for all of us because our lives are inundated daily with advertisements, contracts, product and service warranties, political debates, and news commentaries, to name just a few. People often have problems processing these things because of misinterpretation, misunderstanding, and faulty logic.

You can look up the truth or falseness of a fact on the Internet, but that won't help you in analyzing whether a certain claim is logically valid. The term *common sense* is really misleading, because evaluating logical arguments involves skills that are anything but common—but we're here to help! Forget what you've seen about mind control in the movies: The best way to control someone's mind is to bias their thinking by using their emotions and opinions against them. And that's where this chapter comes in. To decide whether or not an argument makes sense, our goal will be to study the structure of the argument without thinking about the particular topic: To make that possible, we'll use variables to stand in for sentences so that we won't know what the exact topic is. This enables us to focus simply on whether you can logically draw a conclusion based on certain types of statements.

To that end, we've written some claims below. To be honest, some of them are a little controversial, touching on hot-button issues like religion, politics, and, the most contentious of all, college football. We sincerely apologize if any of these statements offends you, but our aim is to illustrate how difficult it can be to decide if an argument is logically sound when you're distracted by emotion and strongly held beliefs. Your job is to evaluate each claim as a logical argument, and decide if the argument is valid: that is, if the conclusion can be logically drawn from a set of statements. The skills you learn in this chapter will help you to do so, but making your best guess now might prove to be very enlightening later on.

1. Where there's smoke, there's fire.
2. People with lots of money are always happy. My neighbor is so happy you just want to smack him sometimes, so I guess he has a lot of money.
3. Every team in the SEC is good enough to play in a bowl game. Florida State isn't in the SEC, so it doesn't belong in a bowl game.
4. Scripture is the word of God. I know this because it says so in the Bible.
5. If Iraq has weapons of mass destruction, we should go to war to make sure they never use them. It turned out they didn't have WMDs, so we shouldn't have started the war.
6. The sun will go out before the Texas Rangers win the World Series. Wait, what? They won in 2023? I better buy some heat lamps.

For answers, see Math in Mind Control Revisited on page 812.

Lesson 7-1 Prep Skills

SKILL 1: RECOGNIZE SENTENCES THAT CAN BE TRUE OR FALSE

In determining whether a statement can be (objectively) classified as true or false, start by asking yourself whether it is something that can be identified as a fact. Pay attention to the difference between facts and opinions (a HUGELY important skill in the modern political climate), and think about the fact that things like questions, requests, and commands can't be classified as true or false.

- *I never lose when I play Fortnite.* This statement can easily be verified. If I play and I lose, then this statement is false. If I play Fortnite and actually never lose, then this statement is true.
- *The Cubs are terrible.* This statement is opinion. By many measures, the Cubs have been terrible in the past and they might be terrible right now, but the statement that they are terrible is overly general and can't be known as fact.
- *Why is the sky blue?* This is a question. Questions can't be classified as true or false.
- *Make sure not to miss class!* This is a command. Commands can't be classified as true or false.

PREP SKILLS QUESTIONS

For each sentence below, decide if it's possible to objectively classify the statement as either true or false.

1. Seventy percent of voters believe that Congress is ineffective.
2. Congress is ineffective.
3. Where did you get those shoes?
4. I'm going to work or to sleep after class.
5. Bring me my brown pants.
6. If we don't get this patient to surgery stat, she won't survive.
7. Nobody can beat me at *Call of Duty*.

Lesson 7-1 Opening Statements

LEARNING OBJECTIVES

- ☐ 1. Define and identify statements.
- ☐ 2. Define the logical connectives and identify their use.
- ☐ 3. Recognize and write negations of statements.
- ☐ 4. Write statements symbolically, and translate symbolic statements back to verbal.

Jill Braaten/McGraw Hill

You can get along very well in this world by simply coming up with a quantity of reasonably valid statements.

—B. F. Skinner

Traditional gender stereotypes in America are becoming more outdated every year. In fact, a recent survey showed that slightly more than half of American women consider themselves the primary breadwinner in their household. But you might not know that if all you watch is television commercials. There are many household products that are still marketed primarily to women, often products associated with taking care of a family. The implication in these ads, sometimes overt and sometimes subconscious, is that if you're a good mom that cares about your family, you'll buy this particular product. It's easy to say that people can't possibly fall for stunts that blatant, but here's all the evidence you need that it works: Advertisers keep doing it.

This is exactly the sort of mind control we talked about in the unit opener: Advertisers don't want you to think about the fact that their actual argument is silly, so they try to tug at your heartstrings. But today, you begin to fight back by beginning your study of symbolic logic, which uses letters to represent statements and special symbols to represent words like *and, or,* and *not.* This will enable us to remove our emotional bias from an argument so that we can analytically evaluate the logic behind it. It begins with studying what exactly can be evaluated for logical correctness.

0. What does the word *logic* mean to you?

7-1 Class

Statements

In the English language there are many types of sentences; a few of the types are

- Factual statements (You have to pass the bar exam to practice law.)
- Commands (Get out of my face!)
- Opinions (Chocolate cake with cream cheese icing is the best dessert EVER.)
- Questions ('Sup with you?)
- Exclamations (Holy cow!)

In the objective study of logic, we will use only factual statements—it's pretty hard to decide if "Get out of my face" is true or false. And by "pretty hard" I mean "completely impossible."

A **statement** is a declarative sentence that can be objectively determined to be either true or false, but not both.

1. Decide whether or not each sentence is a statement. If it is a statement, decide whether it's true or false. If not, describe what type of sentence it is.

Sentence	Statement? Yes or No	If statement, true or false?	If not, what type of sentence?
Where in the world did you find gold suede sneakers?			
Moose are the largest members of the deer family.			
Give me a bottle of anything and a glazed donut—to go.			
Radioactive spiders can make you a superhero.			
$2 + 10 = 12$			
Dude, that is awesome!			
$10 - 6 = 143$			
The guy sitting next to me smells funny.			

Decide which of the following are statements and which are not. If a sentence is not a statement, explain why not.

2. Most scientists agree that climate change is a threat to the environment.

3. Is that your laptop?

4. Man, that hurts!

5. $432 + 8 \div 1.3 = \sqrt{115{,}000}$

6. This book is about database management.

7. Watching reality shows turns your brain to mush.

Sidelight **When True or False Fails**

Did it strike you as odd that the definition of statement requires a sentence to be "true or false, but not both"? Isn't something either true or false? How can it be both? Glad you asked.

Consider this fun example: "This sentence isn't true." If the sentence is in fact true, it means that it isn't true. But if it's false, it means that it is in fact true. Isn't that bizarre? This is a classic example of what's known as a **paradox:** something that appears to be a statement, but contradicts itself. It's literally impossible to decide if a paradox is true or false, because it's either both or neither depending on whether you're a "glass half full" type of person. An Internet search for the string "paradox example" is hours of fun (but not while you're supposed to be doing your homework). Here are some examples I liked. See if you can figure out why they're paradoxical.

- There's an exception to every rule.
- I know nothing at all.
- Moderation in all things, including moderation.
- What would happen if Pinocchio said, "My nose is about to grow"?
- Does the set of all sets that do not contain themselves contain itself?

Did You Get It

Try this problem to see if you understand the concepts we just studied. The answer can be found at the end of the portfolio section.

1. Decide which of the following are statements and which are not.
   a. Dude, those pants rock!
   b. $12 - 8 = 5$
   c. Jimmy Kimmel is the host of the *Daily Show.*
   d. Cat can shoot slow-motion video with her phone.
   e. When does the party start?
   f. History is interesting.

Simple and Compound Statements

Statements can be classified as simple or compound. A simple statement contains only one idea. Each of these statements is an example of a simple statement.

- Your jeans are torn.
- My dorm room has a pile of dirty socks in it.
- Daytona Beach is in Florida.

A statement like "I'm taking chemistry this semester and I'm going to get an A" is called a *compound statement* because it's formed from more than one simple statement.

A **compound statement** is a statement formed by joining two or more simple statements with a **connective.** There are four basic connectives used in logic: and (the **conjunction**), or (**disjunction**), if . . . then (**conditional**), and if and only if (**biconditional**).

Solved Example 1

Identifying Connectives in Statements

For each compound statement, identify the connective used, as well as the two simple statements that it connects.

a. Immigration reform and gun control are two controversial political issues.
b. If a politician makes a strong statement on immigration, then she risks alienating many voters.
c. Luisa said she'll vote to reelect her senator if and only if he advocates for restrictions on assault weapons.
d. In the upcoming presidential debate, I expect my favorite candidate to talk about the Second Amendment or border security.

SOLUTION

a. The key word here is "and." This is a conjunction connecting the statements "Immigration reform is a controversial political issue" and "Gun control is a controversial political issue." Notice that the conjunction isn't written as "Immigration is a controversial political issue and gun control is a controversial political issue." There's nothing wrong with that statement, and in fact it makes it a lot easier to recognize the conjunction. But people don't typically talk like that, and one of the issues we'll deal with in this chapter is recognizing formal logic statements in plain spoken English.

b. The "if … then" structure here makes it pretty easy to see that this is a conditional statement. The two simple statements are "A politician makes a strong statement on immigration" and "She risks alienating many voters."

c. The biconditional might be the easiest to recognize because "if and only if" isn't something you hear every day—it should kind of jump out at you. This particular biconditional connects "Luisa said she'll vote to reelect her senator" and "Her senator advocates for restrictions on assault weapons."

d. Most of the time when you see "or" in a statement, it's going to indicate a disjunction, and that's the case here. This is another example where you have to think a little bit and write more formally to identify the two simple statements connected: "In the upcoming presidential debate, I expect my favorite candidate to talk about the Second Amendment," or "In the upcoming presidential debate, I expect my favorite candidate to talk about border security."

Here are some more examples of compound statements using connectives. For each compound statement, state the connective used and then identify the simple statements that are connected.

8. We will win the game if and only if we score more points than the other team.

9. John studied for 5 hours, and he got an A.

Math Note

In standard usage, the word "then" is often omitted from a conditional statement; instead of "If it snows, then I will go skiing," you'd probably just say, "If it snows, I'll go skiing."

10. If I get 80% of the questions on the LSAT right, then I will get into law school.

11. Min Jee will run in a mini triathlon or play in the campus tennis tournament.

Now let's compare the two main types of statements we've learned about. Classify each statement as simple or compound. If it is compound, state the name of the connective used.

12. Our school mascot is a moose.

13. If you register for WiFi service, you will get 3 days of free access.

14. Tomorrow is the last day to register for classes.

15. In the interest of saving the planet, I plan to buy either a hybrid or a motorcycle.

> **Math Note**
>
> Technically, we've given the names conjunction, disjunction, conditional, and biconditional to the connectives themselves, but from now on, we'll refer to whole statements using these connectives by those names. For example, we would call the compound statement in Question 15 a disjunction.

We'll worry later about determining whether statements involving quantifiers and connectives are true or false. For now, focus on learning and understanding the terms.

Did You Get It

2. Classify each statement as simple or compound. If it is compound, state the name of the connective used.
   a. My jacket is both warm and light.
   b. This is an informative website on STDs.
   c. If it doesn't rain tomorrow, I'm going windsurfing.
   d. I'm going to eat at Taco Bell or Wendy's today.
   e. Yesterday was the deadline to withdraw from a class.

Quantified Statements

Quantified statements involve terms like *all, each, every, no, none, some, there exists,* and *at least one.* Notice that the first five *(all, each, every, no, none)* refer to including or excluding all members of some set. These are known as **universal quantifiers.** (If you're familiar with set theory, this refers to the universal set.) The other three *(some, there exists, at least one)* describe that some things have a certain characteristic but don't claim that everything does. These are called **existential quantifiers.** Here are some examples of quantified statements:

- Every student that comes to class and studies will pass.
- No nursing student is also majoring in criminal justice.
- Some people who are Miami Hurricane fans are also Miami Dolphin fans.
- There is at least one professor in this school who does not have brown eyes.

The first two statements use universal quantifiers, while the third and fourth use existential quantifiers. Note that the statements using existential quantifiers are not "all inclusive" (or all exclusive) as the other two are.

16. Write an example of a statement using a universal quantifier about college courses.

17. Write an example of a statement using an existential quantifier about pizza. (Note: Mmmmm, pizza! is not a statement.)

Math Note

The words "each," "every," and "all" mean the same thing, so what we say about all in this lesson applies to the others as well. Likewise, "some," "there exists," and "at least one" are considered to be the same and are treated that way as well.

7-1 Group

Negations

The negation of a statement is a corresponding statement with the opposite truth value. This means that if a statement is true its negation is false, and if a statement is false its negation is true.

Consider the statement "My dorm room is blue."

1. Can the statement "My dorm room is white" represent the negation of the first statement?

Fuse/Getty Images

2. Could both statements be true? Explain how.

3. Could both statements be false? Explain how.

4. Can the statement "My dorm room is not blue" represent the negation of the statement "My dorm room is blue?"

5. Could both statements be true? Explain.

6. Could both statements be false? Explain.

7. Explain how the previous two questions agree with the definition of negation.

To make sure that you have a correct negation, check that if one of the statements is true, the other must be false, and vice versa. The simplest way to negate a simple statement is by adding the word *not.* Write the negation of each statement.

8. Auburn will win their game this Saturday.

9. I took a shower today.

10. My car is clean.

11. We said that adding "not" is the *simplest* way to negate a statement. That doesn't mean it's the *only* way, or even necessarily the best way. Rewrite each of the negations in Questions 8–10 in a more natural, conversational way.

Negations for quantified statements take a little more thought, as we'll see in Solved Example 2.

Solved Example 2

Writing the Negation of a Quantified Statement

Write the negation of the statement "Everyone is on social media."

SOLUTION

The most common answer to this question turns out to be wrong, and understanding why is the key to negating quantified statements. That common answer is "No one is on social media." There's a pretty good chance that's exactly what you were thinking, and that's okay. It sounds reasonable, doesn't it? But here's why it's wrong.

The definition of negation requires that two statements have the opposite truth value. Now suppose it is the case that some people are on social media and others are not. Let's look at the two statements:

Everyone is on social media. This statement is false.
No one is on social media. This statements is also false.

Do you see the problem? These two statements are both false in this case, which means they do NOT have opposite truth values, and so are not negations of each other.

Now let's talk about the correct negation. The statement claims that all people are on social media. If there are some people who aren't, the claim is false. So the opposite of "Everyone is on social media" is "Some people are not on social media." Give it some thought, and you'll see that both statements can't be simultaneously true.

12. Write the negation of "All dogs are fuzzy."

The negations of quantified statements are summarized in the following table.

Statement contains . . .	Example	Negation	Example
All do	All of my meals are low in fat.	Some do not, or not all do	Some of my meals are not low in fat.
Some do	Some majors require 5 years of study.	None do, or all do not	There are no majors that require 5 years of study.
Some do not	Some people don't go to football games.	All do	Everyone goes to football games.
None do	No airlines include checked bags for free.	Some do	At least one airline allows a checked bag for free.

One final reminder: be careful when negating statements. It's always tempting to say that the negation of a statement like "Every student will pass" is "Every student will fail," but it isn't. Not only is it possible that both statements are false, in most classes it's *likely* that both are false.

In Questions 13–16, write the negation of each of the following quantified statements.

13. Every student taking Quantitative Reasoning this semester will pass.

14. Some people who are Miami Hurricane fans are also Miami Dolphin fans.

15. There is at least one professor in this school who does not have brown eyes.

16. No nursing student is also majoring in criminal justice.

Did You Get It ?

3. Write the negation of each of the following quantified statements.
   a. All cell phones have cameras.
   b. No woman can win the lottery.
   c. Some professors have PhDs.
   d. Someone in this class will get a B.

Symbolic Notation

Recall that one of our goals in this lesson is to write statements in symbolic form to help us evaluate logical arguments objectively. Now we'll introduce the symbols and methods that will be used. The symbols for the connectives and, or, if … then, and if and only if are shown in the following table:

Connective	Symbol	Name
and	$\wedge$	Conjunction
or	$\vee$	Disjunction
if . . . then	$\rightarrow$	Conditional
if and only if	$\leftrightarrow$	Biconditional

Simple statements in logic are usually denoted with lowercase letters like p, q, and r.

17. Suppose we use p to represent the statement "I get paid Friday" and q to represent the statement "I will go out this weekend." Then the conditional statement "If I get paid Friday, then I will go out this weekend" can be written in symbols as ____________.

We have symbols for the connectives, but what about negations? The symbol ~ (tilde) represents a negation.

18. If p still represents "I get paid Friday," then $\sim p$ represents the statement

"________________________________."

When writing more complicated statements in symbols, we'll make use of parentheses often to specify an order. (We'll deal with this issue more in the next lesson.) To begin studying the use of parentheses, we'll study these two statements: $\sim p \wedge q$ and $\sim(p \wedge q)$. Both could be read as "Not p and q" or "The negation of p and q." But are they the same?

19. The statement $\sim p \land q$ means "First negate the statement p, then use that negation in a conjunction with statement q." If p is the statement "Large Coney is a dog" and q is the statement "Guinness is a cat," describe what the statement $\sim p \land q$ means.

20. The statement $\sim(p \land q)$ means "Negate the conjunction of the statement p and the statement q." If p is once again "Large Coney is a dog" and q is once again "Guinness is a cat," what does the statement $\sim(p \land q)$ mean? Is this the same meaning as the statement in Question 19? (By the way, Large Coney is in fact a dog. See page 5.)

The same line of reasoning applies when the negation is used with other connectives. For example, $\sim p \rightarrow q$ means $(\sim p) \rightarrow q$, which is not the same thing as $\sim(p \rightarrow q)$.

For Questions 21–24, let p represent the statement "It is cloudy" and q represent the statement "I will go to the beach." Write each statement in symbols.

21. I will not go to the beach.

22. It is cloudy, and I will go to the beach.

23. If it is cloudy, then I will not go to the beach.

24. I will go to the beach if and only if it is not cloudy.

John Dakers/Life File/Getty Images

Did You Get It

4. Let p represent the statement "I will buy a Coke" and q represent the statement "I will buy some popcorn." Write each statement in symbols.
   a. I will buy a Coke, and I will buy some popcorn.
   b. I will not buy a Coke.
   c. If I buy some popcorn, then I will buy a Coke.
   d. I will not buy a Coke, or I will buy some popcorn.

You probably noticed that some of the compound statements we've written sound a little awkward. It isn't always necessary to repeat the subject and verb in a compound statement using *and* or *or*. Here are some examples of statements written formally, then rewritten in a more natural way.

Formal: "It is cold, and it is snowing"
More natural: "It's cold and snowing."

Formal: "I will go to a movie, or I will go to a play"
More natural: "I'll go to a movie or a play."

Also the words *but* and *although* can be used in place of *and*.

Formal: "I will not buy a DVD player, and I will buy a video streaming device."
More natural: "I won't buy a DVD player, but I will buy a video streaming device."

Now that we've practiced writing verbal statements in symbols, it's a good time to practice the opposite skill: translating from symbolic to verbal form. In Questions 25–29, write each statement in words. Let p = "My dog is a golden retriever" and q = "My dog is fuzzy." First, write in formal language. Then try to write in a more natural, conversational way.

25. $\sim p$

Formal:

More natural:

26. $p \vee q$

Formal:

More natural:

27. $\sim p \rightarrow q$

Formal:

More natural:

28. $q \wedge p$

Formal:

More natural:

If this is your dog (which it's not, because it's mine), the statement in Question 28 describes it pretty well.
David Sobecki

29. $q \leftrightarrow p$

Formal:

More natural:

Did You Get It

5. Write each statement in words. Let p = "My friend is a football player" and q = "My friend is smart."
   a. $\sim p$ b. $p \vee q$ c. $\sim p \rightarrow q$
   d. $p \leftrightarrow q$ e. $p \wedge q$

Using Technology: Logical Tests in Excel

Excel has logical functions built into it that enable you to test the truth value of a statement. We'll make use of these throughout this unit. For now, we'll just learn how to do a basic logical test based on the size of the output for a given function.

In this case, the question is to find whether or not the outputs for the function $y = -2x^2 + 30x + 100$ are more or less than 150 for whole number inputs. In the first column of the sheet below, whole numbers are entered, starting with 1. In the second column, we have the outputs for the above function. To get these, we entered into cell B2: "= −2*A2^2+30*A2+100", then copied down to the cells underneath. Then the key cell is D2: Look carefully at the formula entered in that cell. The IF command is used to test the truth value of the statement "The output of the function is less than 150." So we began with "=IF(B2<150". After that, we had to tell Excel what to enter in the cell depending on whether the statement B2<150 is true or false. You can enter numbers if you like (it's common in computer science to enter 1 for true and 0 for false), but we chose the truth values TRUE and FALSE. When entered in all capitals, Excel recognizes these are truth values: We'll need that later in the unit.

D2 fx =IF(B2<150,TRUE,FALSE)

	A	B	C	D	E	F
1	Input (x)	Output		Test		
2	1	128		TRUE		
3	2	152		FALSE		
4	3	172		FALSE		
5	4	188		FALSE		

See the Using Tech video on Logical Tests in Excel for further information.

7-1 Portfolio

Name ______________________________

Check each box when you've completed the task. Remember that your instructor will want you to turn in the portfolio pages you create.

Technology

1. ☐ In this Excel activity, we'll use a logical test to test whether the outputs of a quadratic function are positive.

ALEKS Practice

1. ☐ Include any written work from the ALEKS assignment along with any notes or questions about this lesson's content.

Applications

1. ☐ Complete the applications problems.

Reflections

Type a short answer to each question.

1. ☐ Describe the difference between a sentence and a statement.
2. ☐ List the logical connectives and give an example of at least one statement that uses each.
3. ☐ Name one thing you learned or discovered in this lesson that you found particularly interesting.
4. ☐ What questions do you have about this lesson?

Looking Ahead

1. ☐ Complete the prep skills for Lesson 7-2.
2. ☐ Read the opening paragraph in Lesson 7-2 carefully and answer Question 0 in preparation for that lesson.

Answers to "Did You Get It?"

1. b, c, and d are statements. **2. a.** Compound, conjunction **b.** Simple **c.** Compound, conditional **d.** Compound, disjunction **e.** Simple
3. a. Some cell phones do not have cameras. **b.** At least one woman can win the lottery. **c.** No professors have PhDs. **d.** No one in this class will get a B.
4. a. $p \wedge q$ **b.** $\sim p$ **c.** $q \rightarrow p$ **d.** $\sim p \vee q$
5. a. My friend is not a football player. **b.** My friend is a football player or my friend is smart. **c.** If my friend is not a football player, then my friend is smart. **d.** My friend is a football player if and only if my friend is smart. **e.** My friend is a football player and my friend is smart.

Answers to Prep Skills Questions

1. Yes **2.** No **3.** No **4.** Yes **5.** No **6.** Yes **7.** Yes

7-1 Applications

Name ______________________________

In Questions 1–9, a boatload of sentences are provided. (Well, nine actually.) First, decide if each sentence is or is not a statement. If it's not, explain why and move on. If it is, classify the statement as simple or compound. Finally, if the statement is compound, identify the type of compound statement that it is.

1. Porsche now makes coupes and sedans.

2. Please close the door on the way out.

3. Anyone that votes for that guy for president is either drunk or stupid.

4. If 1,200 more people sign this petition, a ballot initiative to legalize marijuana will be voted on in November.

5. Holy cow, a UFO!

Brand X/Superstock

6. The two square roots of 49 are 7 and 11.

7. I'm going to quit my job if and only if I get a huge inheritance when my aunt passes.

8. Who do you think I am, Bill Gates?

7-1 Applications

Name ______________________________

9. For the last 11 years, the champion of the American Football Conference West Division has been either Denver or Kansas City.

In Questions 10–13, write the negation of the statement.

10. I've been running miles each morning before work for the last 6 weeks.

11. Everyone I hang out with regularly has a smartphone.

12. Some of the new shows that premiered last week will be canceled before their fifth episode airs.

13. Nobody puts guacamole on a hot dog.

For Questions 14–17, let p = "Kal sees a physical therapist" and q = "Kal is prescribed a narcotic pain reliever." Write each statement in symbols.

14. Kal is prescribed a narcotic pain reliever or he's seeing a physical therapist.

15. Kal is seeing a physical therapist and he hasn't been prescribed a narcotic pain reliever.

16. If Kal sees a physical therapist, then he won't be prescribed a narcotic pain reliever.

7-1 Applications

Name ____

17. Kal isn't prescribed a narcotic pain reliever if and only if he isn't going to physical therapy.

In Questions 18–22, let r = "A suspect in the case has been arrested" and s = "Our investigation has been concluded." For each compound statement in symbols, first write a formal translation using the simple statements above. Then write the same statement in a more natural, conversational way.

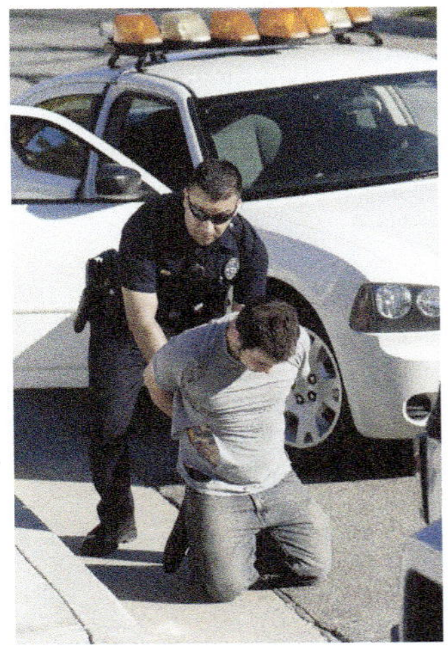

moodboard/123RF

18. $\sim s$

19. $r \wedge s$

20. $\sim r \vee s$

21. $s \rightarrow r$

22. $\sim s \leftrightarrow \sim r$

Lesson 7-2 Prep Skills

SKILL 1: RECOGNIZE TYPES OF COMPOUND STATEMENTS

Remember from Lesson 7-1 that a compound statement is a statement formed by joining two or more simple statements with a connective. There are four basic connectives used in logic: and (the conjunction), or (disjunction), if . . . then (conditional), and if and only if (biconditional).

- I went to a movie, and I got some popcorn.

This statement is an example of a conjunction. The word "and" connects the two simple statements "I went to a movie" and "I got some popcorn."

- Oscar will go to the party or stay home to study for his test.

This statement is an example of a disjunction. The word "or" connects the two simple statements "Oscar will go to the party"and "Oscar will stay home to study for his test."

- If the Cardinals go to the World Series, Jake is going to one of the games.

This is an example of a conditional statement. The "if . . . then" structure connects the two simple statements "The Cardinals go to the World Series" and "Jake is going to one of the games."

- You'll get an A in this course if and only if your final percentage is 90% or higher.

This is an example of a biconditional statement. The words "if and only if" connect the two simple statements "You'll get an A in this course" and "Your final percentage is 90% or higher."

SKILL 2: TRANSLATE SYMBOLIC STATEMENTS INTO WORDS

When translating symbolic statements into words, you'll need to remember the meaning of each symbol. The following table should help.

Connective	Symbol	Name
and	$\wedge$	Conjunction
or	$\vee$	Disjunction
if . . . then	$\rightarrow$	Conditional
if and only if	$\leftrightarrow$	Biconditional

Also, remember that the symbol $\sim$ represents a negation.

Let p = "You eat pizza" and q = "You eat a salad"

- $\sim q$ would translate as "You don't eat a salad."

$\sim q$ is the negation of q.

- $p \wedge q$ would translate as "You eat pizza, and you eat a salad."

The symbol $\wedge$ indicates a conjunction, so we use "and" to connect the statements.

- $p \rightarrow q$ would translate as "If you eat pizza, then you eat a salad."

The symbol $\rightarrow$ is used to indicate a conditional statement, so we use the "if . . . then" structure to connect the statements.

PREP SKILLS QUESTIONS

Identify the type of each compound statement.

1. This is my birthday, and I'm planning to bury my face in cake.

2. If my party is canceled I'm going to hit the roof.

3. I'll go to bed tonight if and only if I run out of cash.

4. I won't open all my gifts at dinner.

5. I plan to have pasta or steak for dinner.

Let p = "The patient is in critical condition" and q = "A surgery room is being prepped." Write a verbal translation of each statement.

6. $p \rightarrow q$

7. $\sim p \vee q$

8. $\sim(q \wedge p)$

9. $q \leftrightarrow p$

Lesson 7-2 Finding the Truth

LEARNING OBJECTIVES

- ☐ 1. Build truth tables for negations, disjunctions, and conjunctions.
- ☐ 2. Build truth tables for conditional and biconditional statements.
- ☐ 3. Build truth tables for compound statements.
- ☐ 4. Use the hierarchy of connectives, and compare it to order of operations.

RGR Collection/Alamy Stock Photo

Three things cannot be long hidden: the sun, the moon, and the truth.
—Buddha

"You can't believe everything you hear." Chances are you were taught this when you were younger, and it's pretty good advice. In an ideal world, everyone would tell the truth all the time, but in this world, it is extremely important to be able to separate fact from fiction. When someone is trying to convince you of some point of view, the ability to logically evaluate the validity of an argument can be the difference between being informed and being deceived—and maybe between keeping and being separated from your hard-earned money! (Perhaps by someone like the dude in the picture.)

This lesson is all about deciding when a compound statement is or is not true, based not on the topic of that statement, but simply on the structure of the statement and the truth of the underlying components. We learned about logical connectives in Lesson 7-1. In this lesson, we'll analyze these connectives using *truth tables.* A **truth table** is a diagram in table form that is used to show when a compound statement is true or false based on the truth values of the simple statements that make up that compound statement. This will allow us to analyze arguments objectively.

0. Write about a way that someone tried to deceive you in the past week and how you knew they weren't being truthful.

7-2 Group

Negation

According to our definition of statement, a statement is either true or false, but never both. Consider the simple statement p = "Today is Tuesday."

1. If it is in fact Tuesday, then p is ________, and its negation ($\sim p$), which is "Today is not Tuesday" is ________.

2. If it's not Tuesday, then p is ________ and ($\sim p$) is ________.

3. A truth table is a table used to summarize the possible truth values for statements. When a statement is true, we put a T in the table to represent that. When false, we put an F in the table. Use your answers from Questions 1 and 2 to build a truth table for the negation. Notice that we've filled in a T and an F in the first column under p: This is to indicate that statement p has two possible truth values, true or false.

p	$\sim p$
T	
F	

4. For each condition for the statement p—true or false—the table tells us that the negation $\sim p$ has the

_______________ truth value.

Conjunction

If we have a compound statement with two component statements p and q, there are four possible combinations of truth values for these two statements.

5. Complete the symbolic values of p and q for each possibility in the following table.

Possibilities	**Truth Value for Each** p	q
Both p and q are true.		
p is true and q is false.		
p is false and q is true.		
Both p and q are false.		

So when setting up a truth table for a compound statement with two component statements, we'll need a row for each of the four possibilities.

Now we're ready to analyze conjunctions. Recall that a conjunction is a compound statement involving the word "and." Suppose a friend who's prone to exaggeration tells you, "I bought a new laptop and a new iPad."

6. This compound statement can be symbolically represented by ________, where p = "I bought a new laptop" and q = "I bought a new iPad."

Mark Dierker/McGraw Hill

7. Under what circumstances would that conjunctive statement be true? When would it be false? Explain in as much detail as you can.

The best way to describe the purpose of a truth table is to put some organization to that stuff you just wrote. Let's do that here.

8. If your friend actually had made both purchases, then of course the statement "I bought a new laptop and a new iPad" would be ________. In terms of a truth table, that tells us that if p and q are both true, then the conjunction $p \wedge q$ is ________ as well. Complete the first row of the truth table.

9. On the other hand, suppose your friend bought only a new laptop or only a new iPad, or maybe neither of those things. Then the statement "I bought a new laptop and a new iPad" would be ________. In other words, if either p or q or both are false, then the compound statement $p \wedge q$ is ________ as well. With this information, complete the second, third, and fourth rows of the truth table for a basic conjunction.

p	q	$p \wedge q$
T	T	
T	F	
F	T	
F	F	

Truth Values for a Conjunction

The conjunction $p \wedge q$ is true only when both p and q are true.

Did You Get It

Try this problem to see if you understand the concepts we just studied. The answer can be found at the end of the portfolio section.

1. Let p = "Those pants were designed by Michael Kors" and q = "Those pants cost more than an average citizen of Jamaica makes in a week." Under what circumstances would the conjunction $p \wedge q$ be true?

Disjunction

Next, we'll look at truth tables for or statements. Suppose your friend from the previous example made the statement, "I bought a new laptop or a new iPad" (as opposed to "and").

10. If your friend actually did buy both, then this statement would be ________. Complete the first row of the truth table.

11. If your friend actually did buy one or the other, then this statement would be ________. Complete rows 2 and 3 of the truth table.

12. And if he or she bought neither, then the statement would be ________. Complete the last row of the truth table.

p	q	$p \vee q$
T	T	
T	F	
F	T	
F	F	

Important note: In the study of logic, we use what's known as the *inclusive or.* That means that we'll consider a disjunction, like "I bought a new laptop or a new iPad" to be true if both of the component statements are true. In short, we're interpreting "or" to mean "one or the other, or both."

Truth Values for a Disjunction

The disjunction $p \vee q$ is true unless both p and q are false.

Did You Get It

2. Let p = "Global temperatures have been rising for the last 50 years" and q = "Climate change poses a threat to human survival." Under what circumstances would the disjunction $p \vee q$ be true?

Conditional Statement

Quick review: A conditional statement, which is sometimes called an *implication,* consists of two simple statements using the connective if . . . then. For example, the statement "If I bought a ticket, then I can go to the concert" is a conditional statement. The first component, in this case "I bought a ticket," is called the **antecedent.** The second component, in this case "I can go to the concert," is called the **consequent.**

Conditional statements are used all the time in math, not just in logic. "If an element is in both set A and set B, then it's in the intersection of A and B" is one of MANY examples from earlier in the book. To build the truth table for the conditional statement, we'll think about the following example:

If the Cubs win tomorrow, they make the playoffs.

13. We'll use p = "the Cubs win tomorrow" and q = "they make the playoffs." This makes our conditional statement ________.

14. We'll be considering four cases to fill out the truth table. Describe the four cases verbally.

For Questions 15–18, fill in the blanks for each case. Then use that information to complete the truth table.

15. Case 1: The Cubs win tomorrow, and they make the playoffs (both p and q are true). The statement was that if the Cubs won, they would make the playoffs, so if they win and make the playoffs, the statement was definitely ________.

16. Case 2: The Cubs win tomorrow but don't make the playoffs (p is true, but q is false). I told you that if the Cubs won, they'd make the playoffs; if they won and didn't make the playoffs, I'm a liar liar pants on fire, and the conditional statement is ________.

17. Case 3: The Cubs lose tomorrow and still make the playoffs (p is false and q is true). This requires some serious thought. My claim was that if the Cubs won, they'd make the playoffs. In order for that claim to be false, the Cubs would have to win and not make the playoffs. That's not the case if they didn't win, so the statement is not false. And we know that if a statement isn't false, it's ________! That makes the conditional statement ________.

p	q	$p \to q$
T	T	
T	F	
F	T	
F	F	

18. Case 4: The Cubs lose tomorrow and don't make the playoffs (p and q are both false). This is pretty much the same as Case 3: The statement is only false if the Cubs win and don't make the playoffs, so again if the Cubs lose, the statement isn't false, making it ________.

For Cases 3 and 4, it might help to think of it this way: We'll be optimists and consider a statement to be true unless we have absolute proof that it's false. If you're totally unconvinced by the discussions in Cases 3 and 4, hang in there. . . . In the applications for this lesson, we'll develop a way to rewrite a conditional statement as a disjunction, in which case we can use what we already know about disjunctions to study truth values.

Truth Values for a Conditional

The conditional statement $p \to q$ is false only when the antecedent p is true and the consequent q is false.

Did You Get It

3. Let p = "Ebola outbreaks are becoming more frequent" and q = "The CDC infectious disease team is traveling more than in previous years." Under what circumstances would the conditional $p \to q$ be true?

Biconditional Statement

A biconditional statement is really two statements; it's the conjunction of two conditional statements. For example, the statement "I will stay in and study Friday if and only if I don't have any money" is the same as "If I don't have any money, then I will stay in and study Friday and if I stay in and study Friday, then I don't have any money." In symbols, we can write either $p \leftrightarrow q$ or $(p \to q) \wedge (q \to p)$. Since the biconditional is a conjunction, for it to be true, both of the statements $p \to q$ and $q \to p$ must be true. We will once again look at cases to build the truth table. Don't forget to fill in the truth table.

Case 1: Both p and q are true.

19. What are the truth values for $p \to q$ and for $q \to p$? (Two questions here.)

20. What is the truth value of $(p \to q) \wedge (q \to p)$ (which is also $p \leftrightarrow q$)?

p	q	$p \leftrightarrow q$
T	T	
T	F	
F	T	
F	F	

Case 2: p is true and q is false.

21. What are the truth values for $p \rightarrow q$ and for $q \rightarrow p$?

22. What is the truth value of $(p \rightarrow q) \wedge (q \rightarrow p)$, which is also $p \leftrightarrow q$?

Case 3: p is false and q is true.

23. What are the truth values for $p \rightarrow q$ and for $q \rightarrow p$?

24. What is the truth value of $(p \rightarrow q) \wedge (q \rightarrow p)$, which is also $p \leftrightarrow q$?

Case 4: p is false and q is false.

25. What are the truth values for $p \rightarrow q$ and $q \rightarrow p$?

26. What is the truth value of $(p \rightarrow q) \wedge (q \rightarrow p)$, which is also $p \leftrightarrow q$?

Truth Values for a Biconditional

The biconditional statement $p \leftrightarrow q$ is true when p and q have the same truth value and false when they have opposite truth values.

27. A technician who designs an automated irrigation system needs to decide whether the system should turn on if the water in the soil falls below a certain level or if and only if the water in the soil falls below a certain level. Describe what the difference would be. Don't think in terms of logic vocabulary: Describe in terms of how this would affect the operation of the system.

Here's a quick summary of all the truth tables you've built so far in this lesson. DO NOT TRY TO MEMORIZE THESE TABLES! There's just too much there, and the Ts and Fs start to all look the same. The reason we built the tables from scratch is to help you *understand* them so that you don't need to rely on memorization.

Summary of Truth Values

Negation (not)

p	$\sim p$
T	F
F	T

Conjunction (and)

p	q	$p \wedge q$
T	T	T
T	F	F
F	T	F
F	F	F

Disjunction (or)

p	q	$p \vee q$
T	T	T
T	F	T
F	T	T
F	F	F

Conditional (if . . . then)

p	q	$p \rightarrow q$
T	T	T
T	F	F
F	T	T
F	F	T

Biconditional (if and only if)

p	q	$p \leftrightarrow q$
T	T	T
T	F	F
F	T	F
F	F	T

Using Technology: Logical Functions in Excel

The negation and the four connectives (conjunction, disjunction, conditional, and biconditional) can be built into Excel.

The **NOT** function performs a negation and uses the syntax =NOT(logical). The NOT function returns a truth value opposite the truth value of the logical argument.

The **AND** function is used for a conjunction. The syntax is =AND(logical1, logical2). The AND function returns a value of TRUE when both logical values are true. Otherwise, the AND function returns a value of FALSE.

The **OR** function is used for a disjunction. The syntax is =OR(logical1, logical2). The OR function returns a value of TRUE when one or both logical values are true. Otherwise, the OR function returns a value of FALSE.

The **IF** function is used for the conditional and biconditional. The syntax is =IF(logical_test, [value_if_true], [value_if_false]).

Since the conditional statement $p \rightarrow q$ is false only when p is true and q is false, we enter the following in cell C3: =IF(AND(A3=TRUE,B3=FALSE),FALSE,TRUE). The IF function in column C looks for where both conditions from the AND function are met. If that happens, then the value in column C is FALSE. Otherwise, the value in column C is TRUE.

C3 *fx* =IF(AND(A3=TRUE,B3=FALSE),FALSE,TRUE)

	A	B	C	D	E	F	G	H
1		Conditional						
2	p	q	p→q					
3	TRUE	TRUE	TRUE					
4	TRUE	FALSE	FALSE					
5	FALSE	TRUE	TRUE					
6	FALSE	FALSE	TRUE					

McGraw Hill

Since the biconditional statement $p \leftrightarrow q$ is true only when p and q have the same truth value, we enter the following in cell C3: =IF(A3=B3,TRUE,FALSE). The IF function in column C looks for the case the values in columns A and B match. If that happens, then the value in column C is TRUE. Otherwise, the value in column C is FALSE.

C3 *fx* =IF(A3=B3,TRUE,FALSE)

	A	B	C	D	E	F
1		Biconditional				
2	p	q	p↔q			
3	TRUE	TRUE	TRUE			
4	TRUE	FALSE	FALSE			
5	FALSE	TRUE	FALSE			
6	FALSE	FALSE	TRUE			

McGraw Hill

See the Using Tech video on Logical Functions in Excel for further information.

7-2 Class

Truth Tables for Compound Statements

Once we know truth values for the basic connectives, we can use truth tables to find the truth values for any logical statement. The key to the procedure is to take it step by step, so that in every case, you're deciding on truth values based on one of the truth tables in the colored box at the end of the Class portion of this lesson.

"My leg isn't better, or I'm taking a break" is an example of a statement that can be written as $\sim p \vee q$.

1. Think about the situations that would make this statement true. Can you describe them all?

Photodisc Inc./Getty Images

Completing a truth table can help organize your thoughts, and this is the real value of truth tables. Your instructor should have given you thin strips of paper or card stock: You'll use those in the remainder of the lesson.

Solved Example 1

Building a Truth Table

Build a truth table for the statement $\sim p \vee q$ and use it to decide when that statement is true.

SOLUTION

Step 1: Set up a table as shown. We start with two columns for the simple statements p and q. The final statement involves $\sim p$, so we'll want a column for $\sim p$ and then one for the overall statement we're studying. By the way, the order in which you list the Ts and Fs doesn't matter as long as you cover all the possible combinations. For consistency in this book, we'll always use the order TTFF for p and TFTF for q when these are the only two letters in the logical statement.

p	q	$\sim p$	$\sim p \vee q$
T	T		
T	F		
F	T		
F	F		

Step 2: Now we'll fill in the $\sim p$ column. This is the negation of p, which means it has the exact opposite truth values of those in the first column.

p	q	$\sim p$	$\sim p \vee q$
T	T	F	
T	F	F	
F	T	T	
F	F	T	

Solved Example 1 *Continued*

Step 3: Now we're ready to find the truth values for the disjunction $\sim p \vee q$. We'll want to focus only on the columns for $\sim p$ and q, so here's a great idea: Use one of the strips your instructor provided to cover up the p column, which can't do anything other than distract you:

	q	$\sim p$	$\sim p \vee q$
	T	F	
	F	F	
	T	T	
	F	T	

Now we're simply using the truth values for the disjunction that we developed in the Group portion of this lesson. A disjunction is true unless both statements are false. That occurs in the second row of the table, so that row gets an F in the last column, while the other rows get T.

p	q	$\sim p$	$\sim p \vee q$
T	T	F	T
T	F	F	F
F	T	T	T
F	F	T	T

Step 4: Interpret the result. Looking at the last column, the only F is in the second row. Looking back at the columns for p and q, we see that in the second row p is true and q is false. So we can conclude that the compound statement $\sim p \vee q$ is true unless p is true and q is false.

2. Use the results of Solved Example 1 to describe situations when the statement "My leg isn't better, or I'm taking a break" is true. Then explain why using a truth table here was useful.

When a compound statement has multiple connectives, it will sometimes have parentheses to indicate the order in which we should work with those connectives. It is important to follow the order of operations—specifically, find the truth value of statements in parentheses first. (If this doesn't remind you of the order of operations for arithmetic and algebra, you're just not paying attention.)

"It is not true that if it rains, then we can't go out" is an example of a statement that can be written as $\sim(p \rightarrow \sim q)$. In order to build a truth table for this statement, we'll need to have a column for the conditional statement $p \rightarrow \sim q$, and then find the negation of that statement.

Creatas/Getty Images

3. This means we'll need columns for p and q (as usual), and also columns for ________, ____________, and ______________.

4. Fill in the column headers in the truth table below using your answers to Question 3.

p	q			
T	T			
T	F			
F	T			
F	F			

5. The only information needed to fill in the $\sim q$ column is the q column, so cover up the p column, then use what you learned about truth values for negations to fill in the $\sim q$ column.
6. Which columns will you cover up to help decide on truth values for the next column in the table?
7. Cover up the column or columns listed in Question 6, then fill in the second-last column in the truth table.
8. Finish the truth table by finding the negations of the values in the second-last column. Then describe under what circumstances the statement "It is not true that if it rains, then we can't go out" is true.

Did You Get It

4. Using the technique from Solved Example 1, build a truth table for the statement $p \leftrightarrow (p \wedge q)$.

We can also construct truth tables for compound statements that involve three or more components. For a compound statement with three simple statements p, q, and r, there are eight possible combinations of Ts and Fs to consider. These are listed in the truth table we'll use next. The order of the Ts and Fs doesn't matter as long as all the possible combinations are listed. Whenever there are three letters in the statement, we suggest using the order shown below for consistency.

p	q	r	$q \rightarrow r$	$p \vee (q \rightarrow r)$
T	T	T		
T	T	F		
T	F	T		
T	F	F		
F	T	T		
F	T	F		
F	F	T		
F	F	F		

"I'll do my math assignment, or if I think of a good topic, then I'll start my English essay" is an example of a statement that can be written as $p \vee (q \rightarrow r)$.

Rubberball/Getty Images

9. Cover up the p column so that you can focus strictly on q and r, then use what you learned about truth values for conditionals to fill in the $q \rightarrow r$ column.

10. Now cover up every column except the p column and the $q \rightarrow r$ column, then use those and what you learned about truth values for disjunctions to finish the truth table.

11. Describe situations under which the statement "I'll do my math assignment, or if I think of a good topic, then I'll start my English essay" is true.

In the statement $p \vee (q \rightarrow r)$, the parentheses are significant: They tell us to first find truth values for $q \rightarrow r$, then form a disjunction with p. This doesn't necessarily have the same meaning, or truth values, as $(p \vee q) \rightarrow r$. (If you feel like extra practice on truth tables would be a good idea for you, build a truth table for $(p \vee q) \rightarrow r$ and compare to the one you built in Questions 9 and 10. You'll find out that the last column of the table works out to be different.)

One way to make sure the intended groupings are always clear would be to include parentheses 100% of the time. Once again, this sounds an awful lot like the order of operations for arithmetic: In order to avoid *always* having to use parentheses, we agree on a standard order for multiple operations, using parentheses only when the intent is to violate that order.

To accomplish the same thing in dealing with statements in logic, a **hierarchy of connectives** has been agreed upon somewhere along the line. This hierarchy tells us which connectives should be done first when there are no parentheses to guide us:

1. Negation ($\sim$)
2. Conjunction ($\wedge$) or disjunction ($\vee$)
3. Conditional ($\rightarrow$)
4. Biconditional ($\leftrightarrow$)

The connectives higher on the list are done first.

Solved Example 2

Using the Hierarchy of Connectives

What type of statement is $p \wedge q \rightarrow r$?

SOLUTION

First, let's notice that parentheses would make this a really easy question. If the statement were written as $(p \wedge q) \rightarrow r$, then we'd find the conjunction first, making the overall statement a conditional statement. That's worth noting: Whatever connective we'd use LAST tells us the type of overall statement we have. So if the statement were written as $p \wedge (q \rightarrow r)$, we'd use the conditional connective first and the conjunction last, making this a conjunction.

Now to the question at hand. Lacking parentheses, we'll need to look at the hierarchy of connectives to decide what comes first. In this case, the conjunction comes before the conditional. In other words, the conditional would be last, meaning the statement $p \wedge q \rightarrow r$ is a conditional.

When a compound statement has both a conjunction and a disjunction, we'll need to use parentheses to indicate which should be considered first.

12. A statement like $p \wedge q \vee r$ is ambiguous without parentheses: $(p \wedge q) \vee r$ is a ________________ and $p \wedge (q \vee r)$ is a ________________.

If we decide to use parentheses to make a statement clearer, the parentheses go around the connectives that would be done first. So we'd write $p \leftrightarrow q \rightarrow r$ as $p \leftrightarrow (q \rightarrow r)$ since the conditional comes before the biconditional in the hierarchy of connectives.

For each statement in Questions 13–16, identify the type of statement using the hierarchy of connectives, and rewrite using parentheses to indicate order.

13. $\sim p \vee \sim q$

14. $p \rightarrow \sim q \wedge r$

15. $p \vee q \leftrightarrow q \vee r$

16. $p \rightarrow q \leftrightarrow r$

Did You Get It

5. For each, identify the type of statement using the hierarchy of connectives, and rewrite using parentheses to indicate order.
   a. $\sim p \vee q$
   b. $p \vee \sim q \rightarrow r$
   c. $p \vee \sim q \leftrightarrow \sim p \vee \sim q$
   d. $p \wedge \sim q$
   e. $p \leftrightarrow q \rightarrow r$

17. Use the truth value of each simple statement to determine the truth value of the compound statement. If you need help finding the truth value for the simple statements, either use the Internet as a resource or ask your professor very, very nicely for help. Don't overthink this one: In essence, you're doing one row of a truth table.

p: Kate Middleton married Prince William in 2011.
q: Prince William's mother was the Queen of England.
r: Barring death or divorce, Kate Middleton will become queen one day.

Statement: $p \vee q \rightarrow r$

Did You Get It

6. Using the simple statements in Question 17, find the truth value of the statement $(\sim p \wedge \sim q) \rightarrow r$.

7-2 Portfolio

Name ______________________________

Check each box when you've completed the task. Remember that your instructor will want you to turn in the portfolio pages you create.

Technology

1. ☐ In this Excel activity, we'll build some truth tables for a few compound statements.

ALEKS Practice

1. ☐ Include any written work from the ALEKS assignment along with any notes or questions about this lesson's content.

Applications

1. ☐ Complete the applications problems.

Reflections

Type a short answer to each question.

1. ☐ Explain when a conjunction is true, and when a disjunction is true, and why that makes sense.
2. ☐ If you had to describe to someone not in this class what we used truth tables for in this lesson, what would you say?
3. ☐ Name one thing you learned or discovered in this lesson that you found particularly interesting.
4. ☐ What questions do you have about this lesson?

Looking Ahead

1. ☐ Complete the prep skills for Lesson 7-3.
2. ☐ Read the opening paragraph in Lesson 7-3 carefully and answer Question 0 in preparation for that lesson.

Answers to "Did You Get It?"

1. If the pants were both designed by Michael Kors and cost more than the average Jamaican citizen makes in a week. **2.** If global temperatures actually have been rising for the last 50 years or climate change poses a threat to human survival, or both.

3. It's always true unless Ebola outbreaks are becoming more frequent and the CDC infectious disease team is not traveling more than in previous years.

4.

p	q	$p \wedge q$	$p \leftrightarrow p \wedge q$
T	T	T	T
T	F	F	F
F	T	F	T
F	F	F	T

5. **a.** Disjunction; $(\sim p) \vee q$ **b.** Conditional: $(p \vee \sim q) \rightarrow r$ **c.** Biconditional: $(p \vee \sim q) \leftrightarrow (\sim p \vee \sim q)$ **d.** Conjunction: $p \wedge (\sim q)$ **e.** Biconditional: $p \leftrightarrow (q \rightarrow r)$

6. The statement is true.

Answers to Prep Skills Questions

1. Conjunction
2. Conditional
3. Biconditional
4. Negation
5. Disjunction
6. If the patient is in critical condition, then a surgery room is being prepped.
7. The patient is not in critical condition or a surgery room is being prepped.
8. It is not the case that a surgery room is being prepped, and the patient is in critical condition.
9. A surgery room is being prepped if and only if the patient is in critical condition.

7-2 Applications

Name ______________________________

For each compound statement in Questions 1–4, identify the truth value for each simple statement, then use that and the basic truth tables we built in this lesson to decide if the compound statement is true or false. You may need to do some poking around to find the truth value of the simple statements.

1. p = Ebola is highly contagious, q = There is a vaccine that protects against Ebola infection; $p \vee q$

2. Same p and q: $p \rightarrow q$

3. p = Recreational marijuana is legal in every state, q = Millions of people have been arrested for possessing marijuana; $p \wedge q$

4. Same p and q: $p \leftrightarrow q$

The manager of a resort tells the staff "You will be receiving holiday bonuses and an extra day off next year."

5. This compound statement is made up of two simple statements. Identify the statements and assign a letter to each.

6. Write the compound statement in symbols using a connective.

7. Build a truth table for the compound statement, then describe under what conditions the manager's claim is not true.

7-2 Applications

Name ____________________

An advertiser is claiming that if you buy their foreign language system and use it for at least an hour every single day, you'll be able to speak a new language conversationally in 6 weeks.

8. This compound statement is made up of three simple statements. Identify them and assign a letter to each.

9. Write the compound statement in symbols using connectives.

10. Build a truth table for the compound statement.

11. If you buy their Spanish language product and use it only for a half hour a day, then can't speak conversational Spanish in 6 weeks, is the claim made by the advertiser true or false?

12. If you buy their French language product and use it for an hour 5 days a week, and can speak conversational French in 6 weeks, is their claim true or false?

13. If you buy and use their Chinese language product for 75 minutes every day and can speak just a little bit of Chinese 6 weeks later, is their claim true or false?

7-2 Applications

Name ____

14. If you can speak conversational German in 6 weeks, are there any circumstances under which their claim can be considered false? Explain.

In a nationally televised debate, a politician claimed that if elected, he would build a wall between the U.S. and Mexico and force the Mexican government to pay for it.

15. This compound statement is made up of three simple statements. Identify them and assign a letter to each.

16. Write the compound statement in symbols using a connective.

17. Build a truth table for the compound statement, then describe under what conditions the politician's claim is true.

7-2 Applications

Name ____________________

18. Let p = "My favorite band is dropping a new album next Tuesday" and q = "I'm going to order the album from Amazon." Build a truth table for the statement $p \rightarrow q$ and describe the circumstances under which the statement is true.

19. For the same two statements p and q, build a truth table for the statement $\sim p \vee q$, then describe the circumstances under which the statement is true. What do you notice?

Lesson 7-3 Prep Skills

SKILL 1: IDENTIFY TRUTH VALUES FOR CONNECTIVES

To identify the circumstances when a symbolic statement is true or false, make sure you're familiar with the summary of truth values below. But remember, the goal is NOT to memorize these tables! It's to understand when each type of statement is true and false.

Summary of Truth Values

Negation (not)

p	$\sim p$
T	F
F	T

Conjunction (and)

p	q	$p \wedge q$
T	T	T
T	F	F
F	T	F
F	F	F

Disjunction (or)

p	q	$p \vee q$
T	T	T
T	F	T
F	T	T
F	F	F

Conditional (if . . . then)

p	q	$p \rightarrow q$
T	T	T
T	F	F
F	T	T
F	F	T

Biconditional (if and only if)

p	q	$p \leftrightarrow q$
T	T	T
T	F	F
F	T	F
F	F	T

SKILL 2: COMPLETE A TRUTH TABLE FOR A COMPOUND STATEMENT

Now that we know the truth values for the basic connectives, we can use those truth tables to find truth values for any logical statement.

Don't forget about what amounts to the order of operations for connectives. Called the hierarchy of connectives, the following list tells us which connectives should be done first when there are no parentheses to guide us:

1. Negation ($\sim$)
2. Conjunction ($\wedge$) or disjunction ($\vee$)
3. Conditional ($\rightarrow$)
4. Biconditional ($\leftrightarrow$)

The connectives higher on the list are done first.

The statement "I'm going to the movies, and I'm not getting popcorn" could be written in symbols by $p \wedge \sim q$, where $p =$ "I'm going to the movies" and $q =$ "I'm getting popcorn."

To complete a truth table for this statement, start with two columns for the simple statements p and q.

p	q
T	T
T	F
F	T
F	F

The starting point for all truth tables based on two simple statements will look like this table. Nothing says the Ts and Fs in each column have to appear in the order they're in, but starting them consistently will help.

We'll need to add a column for $\sim q$.

p	q	$\sim q$
T	T	F
T	F	T
F	T	F
F	F	T

The truth values for the $\sim q$ column are determined by the opposite of those in the q column.

Now we're ready to find $p \wedge \sim q$. Complete the truth values for the last column using only the p and $\sim q$ columns.

p	q	$\sim q$	$p \wedge \sim q$
T	T	F	F
T	F	T	T
F	T	F	F
F	F	T	F

The truth values for the $p \wedge \sim q$ column are determined by the truth values of the conjunction based on the p column and the $\sim q$ columns. When completing this last column, it may be helpful to cover up the q column with a strip of paper.

PREP SKILLS QUESTIONS

1. Under what circumstances is the negation $\sim p$ true?
2. Under what circumstances is the conjunction $p \wedge q$ false?
3. Under what circumstances is the disjunction $r \vee s$ true?
4. Under what circumstances is the conditional $r \rightarrow p$ true?
5. Under what circumstances is the biconditional $q \leftrightarrow p$ false?
6. Complete a truth table for the statement $q \vee \sim p \rightarrow r$. Draw in columns as necessary.

p	q	r	
T	T	T	
T	T	F	
T	F	T	
T	F	F	
F	T	T	
F	T	F	
F	F	T	
F	F	F	

Lesson 7-3 To Be and Not to Be

LEARNING OBJECTIVES

- ☐ 1. Classify a statement as a tautology, a self-contradiction, or neither.
- ☐ 2. Identify statements that are logically equivalent.
- ☐ 3. Write and recognize negations of compound statements.
- ☐ 4. Write and recognize the converse, inverse, and contrapositive of a statement.
- ☐ 5. Evaluate logical connections between a statement and its converse, inverse, and contrapositive.

Jill Braaten/McGraw Hill

"Good morning" is a contradiction of terms.
—Jim Davis

It's no secret that weight loss has become big business in the U.S. It seems like almost every week, a new company pops into existence with the latest miracle pill to turn you into a supermodel. A typical advertisement will say something like "Use of our product may result in significant weight loss." That sounds great, but think about what that statement really means. If use of the product "may" result in significant weight loss, it also may not result in any weight loss at all, it may result in weight gain, or it may result in you turning into a pumpkin. In fact, the statement could be translated into "You will lose weight or you will not lose weight." Of course, this statement is always true. In this lesson, we'll study statements of this type (and others).

0. Write a statement similar to the one in the paragraph above that is always true when you analyze it.

7-3 Class

Tautologies and Self-Contradictions

In our study of truth tables in Lesson 7-2, we saw that most compound statements are true in some cases and false in others. What we haven't done is think about whether that's how it works for every compound statement. Some simple examples should be enough to convince you that this is most definitely not the case.

Solved Example 1

Identifying a Statement That Is Always True

Think about the simple statements "I'm going to Cancun for spring break this year" and "I'm not going to Cancun for spring break this year."

a. What is the relationship between these statements?

b. If one is true, what can we say about the other?

c. When is the statement "I'm going to Cancun for spring break this year or I'm not going to Cancun for spring break this year" true?

Solved Example 1 *Continued*

SOLUTION

a. As we learned in Lesson 7-1, these statements are negations of each other.

b. Since they're negations, we know for sure that if one is true, the other one is false.

c. There are two ways to answer this question, and I'll discuss both. First, let's look at it from a formal logic standpoint, seeing as this is a unit on, you know, formal logic. That means making a truth table. I'll use p to represent "I'm going to Cancun for spring break this year" and q to represent "I'm not going to Cancun for spring break this year."

Of course, we usually have four rows in a truth table with two simple statements involved. But in this case, we only need two of the rows. Why? Since p and q are negations of each other, they can't both be true or both be false. So there are only two cases of interest to us: p is true and q is false, and p is false and q is true.

p	q	$p \vee q$
T	F	
F	T	

Now we use what we know about disjunctions. If either of the simple statements is true, then the disjunction is. So we can see in this case that the disjunction is always true.

Now let's forget about formal logic and use common sense. Either you're going to Cancun or you're not, so just by thinking about it for a moment I can decide that the statement "I'm going to Cancun for spring break this year or I'm not going to Cancun for spring break this year" is always true. Fortunately, this matches what we found using formal logic.

If you're thinking that the formal logic approach in Solved Example 1 was silly because it's reasonably clear that the statement is always true, that's GREAT! It's exactly how you should be thinking about this unit. But we've seen more complicated statements where it can be really hard to decide on truth values, and that's where the formal logic approach is so useful. It doesn't care how simple or complicated a statement is: The truth table pretty much always works the same.

Next, think about this compound statement: "I'm going to Cancun for spring break this year, and I'm not going to Cancun for spring break this year."

1. Describe what would make this statement true.

2. Describe what would make this statement false.

We've now seen an example of a statement that's always true, and one that's always false. These types of statements are given special names in logic.

A **tautology** is a compound statement that's always true, regardless of the truth values of the simple statements that make it up. A **self-contradiction** is a compound statement that is always false.

3. Is every compound statement either a tautology or a self-contradiction? Discuss.

Did You Get It

Try this problem to see if you understand the concepts we just studied. The answer can be found at the end of the portfolio section.

1. Describe each statement as tautology, self-contradiction, or neither.
   a. Mornings are great and mornings aren't great.
   b. Mornings are great and waffles aren't great.
   c. Mornings are great or mornings aren't great.

Now we'll study some more complicated statements, where formal logic will be more useful. There are many ways to state things in a confusing manner that make it hard to decide whether they actually make sense or not.

For what it's worth, this photo was in fact taken when we were going to a concert.

Consider the statements p = "I am going to a concert" and q = "I will wear black." Translate each statement in Questions 4–6 into a word statement using this choice of p and q.

4. $(p \wedge q) \rightarrow p$

5. $(p \vee q) \rightarrow q$

6. $(p \wedge q) \wedge (\sim p \wedge \sim q)$

7. Can you predict which statements are tautologies, self-contradictions, or neither? Give it a shot.

Solved Example 2

Using a Truth Table to Classify a Statement

Use a truth table to decide if the statement "If I go to a concert and wear black, then I go to a concert" is a tautology, a self-contradiction, or neither.

SOLUTION

We already know that this statement can be written in symbols as $(p \wedge q) \rightarrow p$, so we'll need four columns in our truth table: p, q, $p \wedge q$, and $(p \wedge q) \rightarrow p$.

p	q	$p \wedge q$	$(p \wedge q) \rightarrow p$
T	T		
T	F		
F	T		
F	F		

The conjunction $p \wedge q$ is true only when both p and q are true:

p	q	$p \wedge q$	$(p \wedge q) \rightarrow p$
T	T	T	
T	F	F	
F	T	F	
F	F	F	

Now we'll cover up the q column and use the truth values for the conditional, which is true unless $p \wedge q$ is true and p is false:

p		$p \wedge q$	$(p \wedge q) \rightarrow p$
T		T	T
T		F	T
F		F	T
F		F	T

This shows that the statement is always true, making it a tautology.

In Questions 8 and 9, make a truth table to decide if each of the statements in Questions 5 and 6 is a tautology, a self-contradiction, or neither. Make sure you answer the question after completing each table!

8. $(p \vee q) \rightarrow q$

p	q	$p \vee q$	$(p \vee q) \rightarrow q$
T	T		
T	F		
F	T		
F	F		

9. $(p \wedge q) \wedge (\sim p \wedge \sim q)$

p	q	$\sim p$	$\sim q$	$p \wedge q$	$\sim p \wedge \sim q$	$(p \wedge q) \wedge (\sim p \wedge \sim q)$
T	T					
T	F					
F	T					
F	F					

10. How do you feel about your answers to Question 7 now? Did the truth tables make it easier to recognize tautologies or self-contradictions?

Did You Get It

2. Decide if each statement is a tautology, a self-contradiction, or neither.
   a. $(p \vee q) \wedge (\sim p \rightarrow q)$
   b. $(p \wedge \sim q) \wedge \sim p$
   c. $(p \rightarrow q) \vee \sim q$

Logically Equivalent Statements

In Applications 18 and 19 of Lesson 7-2, we studied the two logical statements $p \rightarrow q$ and $\sim p \vee q$. The truth tables for the two statements are combined into one here:

p	q	$\sim p$	$p \rightarrow q$	$\sim p \vee q$
T	T	F	T	T
T	F	F	F	F
F	T	T	T	T
F	F	T	T	T

Notice that the truth values for both statements are *identical:* TFTT. When this occurs, the statements are said to be *logically equivalent;* this means that if we start with the same simple statements, then $p \rightarrow q$ and $\sim p \vee q$ will always have the same truth value.

Two compound statements are **logically equivalent** if and only if they have the same truth values for all possible combinations of truth values for the simple statements that compose them. The symbol for logically equivalent statements is ≡, but the symbol ⇔ is often used as well.

Let p = "Erin earns at least an 80% on her final exam" and q = "She passes her math class."

11. Write the word statement for $p \rightarrow q$.

12. Write the word statement for $\sim q \rightarrow \sim p$.

13. Complete the truth table to decide if the two statements $p \rightarrow q$ and $\sim q \rightarrow \sim p$ are logically equivalent.

p	q	$\sim p$	$\sim q$	$p \rightarrow q$	$\sim q \rightarrow \sim p$
T	T				
T	F				
F	T				
F	F				

Did You Get It

3. Decide which two statements are logically equivalent. Then write examples of simple statements p and q and write each compound statement verbally.
   a. $\sim(p \wedge \sim q)$
   b. $\sim p \wedge q$
   c. $\sim p \vee q$

7-3 Group

We've seen that being able to recognize when two statements are logically equivalent can be helpful in getting to the real meaning of statements that may be confusing, either intentionally or not. De Morgan's laws for logic are a commonly used tool for recognizing logically equivalent statements.

De Morgan's Laws for Logic

For any two statements p and q:

- The statement $\sim(p \vee q)$ is logically equivalent to $\sim p \wedge \sim q$.
- The statement $\sim(p \wedge q)$ is logically equivalent to $\sim p \vee \sim q$.

If you don't remember De Morgan's laws for sets, now would be a good time to look back at Lesson 6-3 for a review. We'll wait here for you. If you haven't covered Unit 6 (The Joy of Sets), you can safely skip the next paragraph.

Do you see the similarities between De Morgan's laws for logic and the ones for sets? We're exchanging sets like A and B for statements p and q, and exchanging intersection and union for "and" and "or." (It's not a coincidence that the symbols for "and" and "or" look suspiciously like the symbols for intersection and union.) We can prove De Morgan's laws using truth tables, but it's more fun to make you do it in the applications, so that's what we did.

De Morgan's laws are most often used to write the negation of conjunctions and disjunctions. For example, the negation of the statement "I will go to work or I will go to the beach" is "I will not go to work and I will not go to the beach." Notice that when you negate a conjunction, it becomes a disjunction, and when you negate a disjunction, it becomes a conjunction—that is, the *and* becomes an *or,* and the *or* becomes an *and.* That's the way you should remember them, by the way. Don't think in terms of the symbols in the purple box: Think about what it says in words.

In Questions 1–4, write the negations of the following statements, using De Morgan's laws.

1. Studying is necessary and I am a hard worker.

2. Shoplifting is a felony or a misdemeanor.

3. I will pass this test or I will drop this class.

4. The patient needs an RN or an LPN, and they're very sick. (This one requires two applications of De Morgan.)

Did You Get It

4. Write the negations of the following statements, using De Morgan's laws.
   a. I will study for this class or I will fail.
   b. I will go to the dance club and the restaurant.
   c. The movie is a comedy or a thriller, and it is awesome.

Earlier in this lesson, we saw that the two statements $p \to q$ and $\sim p \lor q$ are logically equivalent. Now that we know De Morgan's laws, we can use this fact to find the negation of the conditional statement $p \to q$.

5. The first step in finding the negation is to replace $p \to q$ with its equivalent form from the paragraph above. Do this to fill in the blank:

$\sim(p \to q) \equiv \sim($__________$)$

6. Now the right side is the negation of a disjunction, so you can use De Morgan's law for the disjunction to rewrite it. Then you can simplify, removing any parentheses. (Hint: What is the negation of a negation?)

7. Complete the truth table below to verify the formula you developed for the negation of an implication. First you'll need to fill in your formula at the top of the last column.

p	q	$\sim q$	$p \to q$	$\sim(p \to q)$	
T	T				
T	F				
F	T				
F	F				

8. Do the last two columns confirm your formula for the negation of a conditional? If not, go back and try Questions 5–7 again. If yes, write your formula below:

The negation of $p \to q$ is

Let's try our shiny new formula in an example. If you say "It is not the case that if I get a biology degree then I'll be guaranteed a good job," that's the same thing as saying "I got a biology degree and wasn't guaranteed a good job."

In Questions 9 and 10, we'll use our formula to write the negation of the statement "If I get that promotion, I'll buy a sports car."

9. If we want to write that statement as $p \to q$, identify the statements p and q.

$p =$ "______________________________

______________________________"

and $q =$ "______________________________

______________________________"

Dave Sobecki

10. Now use your formula to write the negation of the statement "If I get that promotion, I'll buy a sports car."

"____"

Did You Get It

5. Write the negation of the statement "If the video I made goes viral, it will be on the Today show."

Variations of a Conditional Statement

In addition to their obvious importance in logic, conditional statements play a huge role in math in general. Many of the most important results in the history of math are if . . . then statements. Consider the famous Pythagorean theorem, which almost everyone remembers as simply $a^2 + b^2 = c^2$. But with no context, and especially without describing what quantities a, b, and c represent, that equation is completely meaningless.

So here's the context. First, we need to know that those variables represent the lengths of sides in a right triangle. Second, we get more specific: a and b are the lengths of the legs (the two shorter sides), and c is the length of the hypotenuse, which sounds a lot like a Muppet, but is actually the side across from the right angle. If all of those conditions are met, we can conclude that the famous equation is a true statement.

11. Write the Pythagorean theorem as a conditional statement. All the information you need is in the preceding paragraph.

One of the ways we can learn more about conditional statements is to study three related statements, the *converse,* the *inverse,* and the *contrapositive.*

Variations of a Conditional Statement

Name	In Symbols	In Words
Conditional	$p \rightarrow q$	If p, then q.
Converse	$q \rightarrow p$	If q, then p.
Inverse	$\sim p \rightarrow \sim q$	If not p, then not q.
Contrapositive	$\sim q \rightarrow \sim p$	If not q, then not p.

12. Discuss this among your group and write any conclusions you agree on: Which, if any, of the three variations on a conditional do you think have the same meaning as $p \rightarrow q$?

Dave Sobecki

Now let's see how you did, with the help of a nice doggie named Bear. (Her real name is Tessa, but everyone calls her Bear.)

We'll use the statement "If Bear is a chocolate Lab, then Bear is brown" to illustrate the variations of a conditional, and to decide if each is logically equivalent to the original conditional statement. Note that the original conditional statement is true—the thing that makes a Labrador retriever a chocolate Lab is brown fur.

13. The converse is "If Bear is brown, then Bear is a chocolate Lab." Is this true? Discuss.

14. The inverse is "If Bear isn't a chocolate Lab, then Bear isn't brown." True or false? Discuss.

15. The contrapositive is "If Bear isn't brown, then Bear isn't a chocolate Lab." True or false? Discuss.

16. Based on the above examples, only one of the three variations on $p \rightarrow q$ can possibly be logically equivalent to the original conditional. Which one, and how do we know that? (Hint: Questions 13–15 deal with a *specific example.*)

17. If only we had a way to check your conclusion from Question 16 . . . oh wait, there is: a truth table! This time, you're on your own: Set up and complete a truth table comparing truth values for a conditional $p \rightarrow q$ and the variation that you chose in Question 16. If they're logically equivalent, then your conjecture was correct and you win! If not, reevaluate your answer to Question 16 and try again.

Next, consider the conditional "If you earned a bachelor's degree, then you got a high-paying job."

18. Write the converse, and discuss if it does or does not have the same meaning as the original statement.

19. Write the inverse, and discuss if it does or does not have the same meaning as the original statement.

20. Write the contrapositive, and discuss if it does or does not have the same meaning as the original statement.

You now know that the converse of a conditional is NOT logically equivalent to the original conditional. That's a good thing, because maybe the most common flaw in logic is thinking that they *are* equivalent. If you watch out for it, you can see people making this mistake all over the place, from daily conversation all the way up to presidential debates.

Did You Get It

6. Write the three variations of the statement "If the video I made goes viral, it will be on the Today show." and classify each one as equivalent or not equivalent to the original statement.

Earlier in this unit, we pointed out that in the study of formal logic, we tend to use very formal wording for statements, which often sounds kind of stiff and unnatural. There are other wordings that can be used for conditional statements aside from the standard "if . . . then" connective. To close the lesson, we're going to ask you to think about some alternative words that are used, and decide how to use them in statements.

Our base statement will be "If you are convicted of a felony, you will serve time in jail." Listed below are alternative wordings for a conditional.

p implies q
q if p
p only if q
p is sufficient for q
q is necessary for p
All p are q

21. Your job is to figure out how to write a statement using each of those wordings that has the same meaning as the original statement. Go.

7-3 Portfolio

Name ______________________________

Check each box when you've completed the task. Remember that your instructor will want you to turn in the portfolio pages you create.

Technology

1. ☐ In this Excel activity, we'll use the logically equivalent form for conjunctions in a truth table.

ALEKS Practice

1. ☐ Include any written work from the ALEKS assignment along with any notes or questions about this lesson's content.

Applications

1. ☐ Complete the applications problems.

Reflections

Type a short answer to each question.

1. ☐ Explain what it means for two statements to be logically equivalent.

2. ☐ What is a tautology? What is a self-contradiction?

3. ☐ Describe the process of using truth tables to decide if two statements are logically equivalent.

4. ☐ Name one thing you learned or discovered in this lesson that you found particularly interesting.

5. ☐ What questions do you have about this lesson?

Looking Ahead

1. ☐ Complete the prep skills for Lesson 7-4.

2. ☐ Read the opening paragraph in Lesson 7-4 carefully and answer Question 0 in preparation for that lesson.

Answers to "Did You Get It?"

1. a. Self-contradiction **b.** Neither **c.** Tautology **2. a.** Neither **b.** Self-contradiction **c.** Tautology **3.** a and c are equivalent; answers vary on second part. **4. a.** I will not study for this class and I will not fail.
b. I will not go to the dance club or the restaurant.
c. The movie is not a comedy and it is not a thriller, or it is not awesome.
5. The video I made went viral and it was not on the Today show.
6. Converse: "If my video is on the Today show, then it went viral." Inverse: "If my video doesn't go viral, then it won't be on the Today show." Contrapositive: "If my video isn't on the Today show, then it didn't go viral." Only the contrapositive is equivalent to the original statement.

Answers to Prep Skills Questions

1. When p is not true
2. If either p or q or both are false
3. If either r or s or both are true
4. It's true unless r is true and p is false.
5. It's false if q and p have different truth values.
6.

p	q	r	$\sim p$	$q \vee \sim p$	$q \vee \sim p \rightarrow r$
T	T	T	F	T	T
T	T	F	F	T	F
T	F	T	F	F	T
T	F	F	F	F	T
F	T	T	T	T	T
F	T	F	T	T	F
F	F	T	T	T	T
F	F	F	T	T	F

7-3 Applications

Name ______________________________

In Questions 1–4, use De Morgan's laws to write the negation of each statement.

1. I'm flying to Houston tomorrow and staying at the Westin.

2. I'll pick up a sandwich on the way or eat dinner at the airport on the way out.

3. I'm going through the TSA precheck lane and I won't remove my shoes while going through security.

4. I won't fall asleep during the flight or I won't have a complementary beverage.

In Questions 5–7, decide if the two statements are negations. If you can, use De Morgan's laws. If not, write each statement in symbols and use truth tables.

5. 1: This session of Congress has a record number of women serving, and a record number of people of color.
2: This session of Congress doesn't have a record number of women, or it doesn't have a record number of people of color.

7-3 Applications

Name ____

6. 1: If social workers get a pay raise, then children in our community will be at less risk.
2: It's not the case that if children in our community will be at less risk, then social workers get a pay raise.

7. 1: You won't get a good job with a major in medieval history, and you will regret that major.
2: You won't regret majoring in medieval history or you will get a good job.

Let p = "Lobbyists play a key role in shaping public policy" and q = "Campaign donations influence politicians' positions on issues." In Questions 8–10, translate each symbolic statement into words, then use truth tables to decide if each is a tautology, a self-contradiction, or neither.

8. $(p \vee q) \vee (\sim p \wedge \sim q)$

7-3 Applications

Name ______________________________

9. $(p \wedge q) \wedge (\sim p \vee \sim q)$

10. $(p \leftrightarrow q) \wedge (\sim p \leftrightarrow \sim q)$

For each statement in Questions 11–13, write the negation, converse, inverse, and contrapositive. Then explain why the contrapositive says the same thing as the original statement.

11. If work calls, you should tell them I have Legionnaire's disease.

7-3 Applications

Name ______________________________

12. Getting no raise this year implies that my boss thinks I'm doing a rotten job.

13. Signing a waiver is necessary for participating in the drug trial.

Lesson 7-4 Prep Skills

SKILL 1: WRITE STATEMENTS IN SYMBOLS

When writing statements in symbols, you'll need to remember the meaning of each symbol. The following table should help.

Connective	Symbol	Name
and	$\wedge$	Conjunction
or	$\vee$	Disjunction
if . . . then	$\rightarrow$	Conditional
if and only if	$\leftrightarrow$	Biconditional

Also, remember that the symbol $\sim$ represents a negation.

- Statement: "I did all my homework for this unit, and I earned a B on the test."

  Identify letters:
  Let p = I did all my homework for this unit.
  Let q = I earned a B on the test.

  These are the two simple statements from the original statement.

  Symbols: $p \wedge q$

  Since "and" was used to connect the statements, we use the symbol $\wedge$ between p and q.

- Statement: "If I hold the rope tightly, I won't fall."

  Identify letters:
  Let p = I hold the rope tightly.
  Let q = I will fall.

  These are the two simple statements from the original statement. Note that "I won't fall" is the negation of "I will fall."

  Symbols: $p \rightarrow \sim q$

  Since the statement has the "if . . . then" structure, we use the symbol $\rightarrow$ between p and $\sim q$.

SKILL 2: BUILD TRUTH TABLES

We have been building truth tables throughout this unit. Here's a summary of the process along with some hints along the way. Follow these steps to build a truth table for any logical statement:

1. Start with columns for the individual statements. If there are two individual statements, for consistency we use TTFF and TFTF to complete the first two columns. If there are three individual statements, for consistency we use TTTTFFFF, TTFFTTFF, and TFTFTFTF to complete the first three columns.
2. Add columns for any negations of individual statements.
3. Use the hierarchy of connectives to add columns for each statement involving a connective that may be used to build the overall statement. The last column should be for the truth value of the entire statement.
4. Use the truth values for the basic connectives to complete each new column. For a negation, focus on only the one column being negated. For the other connectives, focus only on the two columns corresponding to the statements being connected. Using a strip of paper to cover columns not needed can be helpful.

PREP SKILLS QUESTIONS

Write each statement in symbols. Identify what the letters in your symbolic statement represent.

1. I ate an entire side of beef for lunch and my stomach does NOT feel good.
2. If I don't take some medicine soon, I might pass out.
3. I'm going to work tomorrow if and only if I feel better.
4. Next time I go to that restaurant I won't order the same thing or I'll call an ambulance in advance.
5. Build a truth table for the statement $p \wedge (\sim q \vee p)$.
6. Build a truth table for the statement $r \rightarrow (p \wedge \sim q)$.

Lesson 7-4 Being Argumentative

LEARNING OBJECTIVES

- ☐ 1. Identify the difference between a valid argument and a fallacy.
- ☐ 2. Use truth tables to evaluate the validity of arguments.
- ☐ 3. Determine the validity of common argument forms.
- ☐ 4. Use common argument forms to decide if arguments are valid.

SAHACHAT/iStock/Getty Images Plus

Be able to defend your arguments in a rational way. Otherwise, all you have is an opinion.

—Marilyn vos Savant

Common sense is a funny thing in our society: We all think we have it, and we also think that most other people don't. This thing that we call common sense is really the ability to think logically, to evaluate an argument or situation and decide what is and is not reasonable. It doesn't take a lot of imagination to picture how valuable it is to be able to think logically. We're pretty well protected by parents for our first few years of life, but after that the main tool we have to guide us through the perils of life is our brain. The more effectively that brain can analyze and evaluate the mass of information we're all exposed to every day, the more successful we're likely to be.

The work we've done in building the basics of symbolic logic in the first three lessons in this unit has prepared us for the real point: analyzing logical arguments objectively. That's the topic of this important lesson. Remember why we're using letters to represent statements: Your goal is to ignore the TOPIC of an argument, and simply focus on whether or not a conclusion can be reasonably drawn from a preliminary set of statements. Check your emotions and opinions at the door, and get ready to be argumentative.

0. Write at least three situations where thinking logically is really important in your life.

7-4 Class

Valid Arguments and Fallacies

A logical argument consists of two parts: a set of premises and a conclusion based on those premises. Premises are statements that are offered as supporting evidence for the conclusion. Our goal is to decide whether an argument is *valid* or *invalid.*

An argument is **valid** if the conclusion necessarily follows from the premises, and **invalid** if it's not valid. An error in reasoning that leads to an invalid argument is known as a **fallacy.**

There's one word conspicuously missing from our definition of valid: "True." The reason why is so important that it deserves special emphasis.

The Key to Understanding Valid Arguments in Logic

Logic is not about deciding if a claim is true: It's about deciding if the claim can be deduced from the given premises.

If you don't keep this idea at the front of your thinking, you'll really struggle with this lesson.

Let's look at an example of a logical argument.

Premise 1: All students in this class will pass.

Premise 2: Shaquille is a student in this class.

Conclusion: Shaquille will pass this class.

1. What are we told about who will pass this class in premise 1?

2. Is Shaquille a student in this class?

3. What can you say about the validity of the conclusion? Why?

It's very important at this point to understand the difference between a true statement and a conclusion to a valid argument. A statement that is known to be false can still be a valid conclusion if it follows logically from the given premises.

Random question: Is Los Angeles in Mexico?

4. The statement "Los Angeles is in Mexico" is definitely ________. (Hint: A good answer will end in "alse.")

Consider this argument:

Premise 1: Los Angeles is in California or Mexico.

Premise 2: Los Angeles is not in California.

Conclusion: Therefore, Los Angeles is in Mexico.

5. What does the statement "Los Angeles is in California or Mexico" mean?

> **Math Note**
>
> Interesting but not particularly relevant note: There are no words in common spoken English that exactly rhyme with "false."

6. What does the statement "Los Angeles is not in California" mean? This one is kind of obvious.

7. If we accept both of the two previous statements (premises), then the statement "Therefore, Los Angeles is in Mexico" must be ________. (Hint: Your answer should rhyme with "shrew.")

This is a valid argument: If we accept the two premises, then Los Angeles would in fact be in Mexico. We know, however, that Los Angeles is NOT in Mexico, and there's the tricky part. *To be valid, the conclusion of an argument has to follow from the premises whether they're true or not.* In this case, we're accepting the premise "Los Angeles is not in California" even though we know that it's actually false. We can then deduce that the conclusion (Los Angeles is in Mexico) follows logically from that premise. Again, this emphasizes that the validity of an argument is not about whether or not the conclusion is a true statement.

Truth Table Method

One way to decide if an argument is valid is based on truth tables. We'll use the following procedure.

Testing the Validity of an Argument

Step 1. Write the argument in symbols.

Step 2. Write the argument as a conditional statement; use a conjunction between all premises and the implication ($\Rightarrow$) for the conclusion. (Note: The symbol $\Rightarrow$ means the same as $\rightarrow$ but will be used to designate an argument.)

Step 3. Set up and construct a truth table as follows:

Symbols	Premise $\wedge$ Premise $\Rightarrow$ Conclusion

Step 4. If all truth values under $\Rightarrow$ are Ts (that is, the statement in the last column is a tautology), then the argument is valid; otherwise, it is invalid.

Notice that we're not looking for all Ts under the conclusion, because we don't care if the conclusion is true or not. We're looking for all Ts under the *implication of the conclusion from the premises* because we're interested in the conclusion following logically from the premises. Think about that carefully, because it's the key to understanding the process.

Now let's use our procedure to decide if an argument is valid.

Solved Example

Deciding if an Argument Is Valid

Decide if the following argument is valid.

If the ad campaign is a success, the marketing manager won't lose her job.

The marketing manager lost her job.

Therefore, the ad campaign wasn't a success.

Math Note

There can be more than two premises in an argument; in that case, put the conjunction sign between all premises.

SOLUTION

Step 1 *Write the argument in symbols.* Let p = "The ad campaign is a success," and let q = "The marketing manager loses her job." Let's look at each statement in the argument individually.

If the ad campaign is a success, the marketing manager won't lose her job.

Solved Example *Continued*

This translates as "If p then not q," which in symbols is $p \to {\sim}q$.

The marketing manager lost her job.

This is exactly statement q, so that's a freebie.

The ad campaign wasn't a success.

This is the negation of p, which we write as ${\sim}p$. Now we'll write the argument in symbols:

$p \to {\sim}q$	(Premise)
q	(Premise)
$\therefore {\sim}p$	(Conclusion)

A line is used to separate the premises from the conclusion, and the three triangular dots ∴ mean "therefore."

Math Note

Notice that we're putting our Ts and Fs under the connective symbol in the later columns, not centered in the column. I've found that this really helps me to focus on the truth values for the given connective without getting distracted by all the other stuff.

Step 2 *Write the argument as an implication* by connecting the premises with a conjunction and implying the conclusion as shown.

Premise 1		**Premise 2**		**Conclusion**
$(p \to {\sim}q)$	$\wedge$	q	$\Rightarrow$	${\sim}p$

Step 3 *Construct a truth table* as shown.

p	q	${\sim}p$	${\sim}q$	$p \to {\sim}q$	$(p \to {\sim}q) \wedge q$	$[(p \to {\sim}q) \wedge q] \Rightarrow {\sim}p$
T	T	F	F	F	F	T
T	F	F	T	T	F	T
F	T	T	F	T	T	T
F	F	T	T	T	F	T

Step 4 *Determine the validity of the argument.* Since all the values under the ⇒ are true, the argument is valid.

In Questions 8–11, we will decide if this argument is valid: Emerging research shows that if cancer patients undergo aromatherapy, depression is lessened. One of my cancer patients is reporting a decrease in depression, so they must have undergone aromatherapy.

8. **Step 1:** *Write the argument in symbols.*

Let p = "____"

and let q = "____."

9. Write the premise "If cancer patients undergo aromatherapy, depression is lessened" in symbols using your statements p and q.

10. Write the premise "One of my cancer patients is reporting a decrease in depression" in symbols.

11. Write the conclusion "They must have undergone aromatherapy" in symbols, then write the entire argument in symbols below.

(Premise)

(Premise)

______

(Conclusion)

TRBfoto/Getty Images

12. Follow the remaining steps in the procedure we've outlined for using a truth table to determine the validity of an argument. A truth table has been provided to help.

p	q			
T	T			
T	F			
F	T			
F	F			

Did You Get It

Try these problems to see if you understand the concepts we just studied. The answers can be found at the end of the portfolio section.

Use the procedure you just practiced to decide if each argument is valid or not.

1. I will run for student government or I will join the athletic boosters.

   I'm not joining the athletic boosters.
   ______

   Therefore, I will run for student government.

2. John's boss warned him that if he blew off work to go to the playoff game, he'd get fired. I heard John got fired, so I guess he must have gone to that playoff game. Cool!

Sidelight Logic and The Art of Forensics

Many students find it troubling that an argument can be considered valid even if the conclusion is clearly false. But arguing in favor of something that you don't necessarily believe to be true isn't a new idea by any means—lawyers do it all the time, and it's commonly practiced in the area of formal debate, a style of intellectual competition that has its roots in ancient times.

Junial Enterprises/Shutterstock

In formal debate (also known as forensics), speakers are given a topic and asked to argue one side of a related issue. Judges determine which speakers make the most effective arguments and declare the winners accordingly. One of the most interesting aspects is that in many cases, the contestants don't know which side of the issue they will be arguing until right before the competition begins. While that aspect is intended to test the debater's flexibility and preparation, a major consequence is that opinion, and sometimes truth, is taken out of the mix, and contestants and judges must focus on the validity of arguments.

A variety of organizations sponsor national competitions in formal debate for colleges. The largest is an annual championship organized by the National Forensics Association. Students from well over 100 schools participate in a wide variety of categories. In 2024, Western Kentucky University won its fourth consecutive overall team championship.

We haven't bugged you for a while about the connection (or lack of connection) between the truth of a conclusion and the validity of an argument, so think about each of these arguments:

Either $2 + 2 \neq 4$ or $2 + 2 = 5$.

$2 + 2 = 4$.

So, $2 + 2 = 5$.

13. Is this argument valid?

14. Is this conclusion true?

If $2 + 2 \neq 5$, then my feet hurt.

My feet don't hurt.

So, $2 + 2 \neq 5$.

15. Is this argument valid?

16. Is this conclusion true?

Did we mention that the validity of arguments is not about the conclusion being true? In the Applications for this lesson, you'll build truth tables to check the validity of these arguments.

You might be thinking that you could just figure out whether the arguments we've studied are valid or not, without using truth tables. You might even be right. But when arguments get more complicated, the value of having a systematic approach becomes more apparent. When an argument has three premises, that's usually the case. In this case, the last column will contain a conjunction of three premises.

In Questions 17–19, we'll decide if this argument is valid:

$p \to r$

$q \wedge r$

p

$\therefore \sim q \to p$

17. *Write the argument as an implication.* Make a conjunction of all three premises and imply the conclusion:

18. *Construct a truth table.* When there are three premises, we will begin by finding the truth values for each premise and then work the conjunction from left to right as shown.

p	q	r	$\sim q$	$p \to r$	$q \wedge r$	$\sim q \to p$	$(p \to r) \wedge (q \wedge r) \wedge p$	$[(p \to r) \wedge (q \wedge r) \wedge p] \Rightarrow (\sim q \to p)$
T	T	T						
T	T	F						
T	F	T						
T	F	F						
F	T	T						
F	T	F						
F	F	T						
F	F	F						

19. Is the argument valid? Why or why not?

Did You Get It

3. Decide if the argument is valid.

$p \vee q$

$q \vee \sim r$

$\therefore q$

7-4 Group

Common Valid Argument Forms

We have seen that truth tables can be used to test an argument for validity. But some argument forms are common enough that they are recognized by special names. When an argument fits one of these forms, we can decide if it is valid or not just by knowing the general form, rather than constructing a truth table. We'll start with a description of some commonly used valid arguments.

Law of Detachment	**Example**
$p \to q$	If our team wins Saturday, then they go to a bowl game.
p	Our team won Saturday.
$\therefore q$	Therefore, our team goes to a bowl game.

Law of Contraposition	**Example**
$p \to q$	If I try hard, I'll get an A.
$\sim q$	I didn't get an A.
$\therefore \sim p$	Therefore, I didn't try hard.

Law of Syllogism	**Example**
$p \to q$	If I make an illegal U-turn, I'll get a ticket.
$q \to r$	If I get a ticket, I'll get points on my driving record.
$\therefore p \to r$	Therefore, if I make an illegal U-turn, I'll get points on my driving record.

Law of Disjunctive Syllogism	**Example**
$p \lor q$	My top client demands the penthouse or an executive suite.
$\sim p$	He couldn't stay in the penthouse.
$\therefore q$	Therefore, he stayed in an executive suite.

1. Make up your own example that fits the law of detachment. Compare with others and make sure your examples fit.

2. Make up your own example that fits the law of contraposition. Compare with others and make sure your examples fit.

3. Make up your own example that fits the law of syllogism. Compare with others and make sure your examples fit.

> **Math Note**
>
> The law of syllogism is also known as the law of transitivity. This is because it looks just like the transitive property of equality: if $a = b$ and $b = c$, then $a = c$.

4. Make up your own example that fits the law of disjunctive syllogism. Compare with others and make sure your examples fit.

Common Fallacies

Next, we'll study some commonly used arguments that are invalid because they result from a fallacy—that is, an incorrect way of reasoning. Sometimes people use these without realizing they're being illogical. Other times, people use them in a deliberate attempt to confuse or deceive. Beware!

Fallacy of the Converse	**Example**
$p \to q$	If it's Friday, then I'll go to happy hour.
q	I'm at happy hour.
$\therefore p$	Therefore, it must be Friday.

5. Explain why this argument isn't valid.

Comstock Images/Comstock Images/Getty Images

Fallacy of the Inverse	**Example**
$p \rightarrow q$	If I exercise every day, then I will lose weight.
$\sim p$	I don't exercise every day.
$\therefore \sim q$	Therefore, I won't lose weight.

6. Explain why this argument isn't valid.

Fallacy of the Inclusive Or	**Example**
$p \vee q$	I'm going to take chemistry or physics.
p	I signed up for chemistry.
$\therefore \sim q$	Therefore, I'm not taking physics.

7. Explain why this argument isn't valid. You might find it helpful to look back at our discussion of truth values for disjunctions back in Lesson 7-2.

In Questions 8–12, use the known valid arguments and fallacies to decide if each argument is valid. Make sure you list the name of the argument form you used.

8. If you eat a healthy diet, you'll live past 70.

You've really made an effort to eat healthy foods.

Therefore, you'll live past 70.

9. You can access the Web at this hotel if you pay for WiFi.

You're too broke to pay for WiFi.

So you won't be able to access the Web.

10. If you watch *Big Brother*, you watch reality shows.

If you watch reality shows, you have time to kill.

Therefore, if you watch *Big Brother*, you have time to kill.

11. You planned to major in criminal justice or pre-law. Since you're majoring in criminal justice, I guess that means you're not majoring in pre-law.

12. Botany grads work in parks or at research labs. Linh got her degree in botany last year and doesn't work at a park, so she must work at a research lab.

13. Explain why the law of contraposition, the fallacy of the converse, and the fallacy of the inverse were given the names that they have.

To close the lesson, we'll talk about one final common fallacy that doesn't necessarily fit a particular form in terms of letters, like the others. It's known as **circular reasoning.** Think about this little story.

A suspect in a criminal investigation tells the police detective questioning him that his statements can be trusted because his friend Sue can vouch for him. The detective asks the suspect how he knows that Sue can be trusted, and he says "I can assure you of her honesty." Ultimately, the suspect becomes even more suspect because of circular reasoning: His argument boils down to "I am honest because I am honest."

This is the essence of circular reasoning: By making the conclusion one of the premises, you can pretty much guarantee that if the premises are true, so is the conclusion. You might think this example seems blatantly silly, but you'd be surprised at how often people try to get away with this fallacy. A Google search for the string "circular reasoning" brings up hundreds of arguments that are thought to be circular.

14. Write an example of an argument that uses circular reasoning.

Did You Get It

4. Use the known valid arguments and fallacies to decide if each argument is valid. Make sure you list the name of the argument form you used.
   a. If you do well in hospitality management, you'll get a job at a great resort.

      You graduated with a 3.8 GPA in hospitality management.
      ___
      Therefore, you should get a job at a great resort.

   b. If you work hard, you will be a success.

      You are not a success.
      ___
      I guess you don't work hard.

   c. Jon is either really cheap, or flat broke. I got a look at his checking account statement and he's not broke, so he must just be cheap.
   d. If my lab rats don't die, I'll get a passing lab grade. Since I passed bio lab, that means the rats didn't die.

7-4 Portfolio

Name ______________________________

Check each box when you've completed the task. Remember that your instructor will want you to turn in the portfolio pages you create.

Technology

1. ☐ In this Excel activity, we'll evaluate common arguments and fallacies.

ALEKS Practice

1. ☐ Include any written work from the ALEKS assignment along with any notes or questions about this lesson's content.

Applications

1. ☐ Complete the applications problems.

Reflections

Type a short answer to each question.

1. ☐ Looking back over this unit, what is the point of studying logical arguments using symbols rather than the original words they're stated in?
2. ☐ Describe the process of using truth tables to evaluate a logical argument.
3. ☐ Why is the last column of a truth table used to evaluate the validity of an argument always an implication?
4. ☐ Name one thing you learned or discovered in this lesson that you found particularly interesting.
5. ☐ What questions do you have about this lesson?

Looking Ahead

1. ☐ Complete the prep skills for Lesson 8-1.
2. ☐ Read the opening paragraph in Lesson 8-1 carefully and answer Question 0 in preparation for that lesson.

Answers to "Did You Get It?"

1. Valid **2.** Invalid **3.** Invalid **4. a.** Valid: law of detachment
b. Valid: law of contraposition **c.** Valid: law of disjunctive syllogism
d. Invalid: fallacy of the converse

Answers to Prep Skills Questions

1. $p \wedge q$; p = I ate an entire side of beef for lunch, q = my stomach does NOT feel good
2. $\sim p \rightarrow q$; p = I take some medicine soon, q = I might pass out
3. $p \leftrightarrow q$; p = I'm going to work tomorrow, q = I feel better
4. $p \vee q$; p = Next time I go to that restaurant I'm ordering something different, q = I'll call an ambulance in advance
5.

p	q	$\sim q$	$\sim q \vee p$	$p \wedge (\sim q \vee p)$
T	T	F	T	T
T	F	T	T	T
F	T	F	F	F
F	F	T	T	F

6.

p	q	r	$\sim q$	$p \wedge \sim q$	$r \rightarrow (p \wedge \sim q)$
T	T	T	F	F	F
T	T	F	F	F	T
T	F	T	T	T	T
T	F	F	T	T	T
F	T	T	F	F	F
F	T	F	F	F	T
F	F	T	T	F	F
F	F	F	T	F	T

7-4 Applications

Name ______________________________

1. For the two arguments in Questions 13–16 of the Class portion of this lesson, write each argument in symbols, then use truth tables to decide which (if either) is valid. Then talk about the relationship between an argument being valid and the conclusion being true.

2. A logical argument is written below. Just based on reading it, do you think this argument is valid?

   The play opens tonight and I won't be in attendance.

   If I didn't have to work, I'd attend.

   ∴ I'll be attending.

3. Assign letters to each of the simple statements involved in this argument.

4. Write the argument in symbols.

   ____________ Premise 1

   ____________ Premise 2

   ∴ ____________ Conclusion

5. Write the argument as a conditional statement by connecting the premises with a conjunction.

   $\wedge$ $\Rightarrow$

7-4 Applications

Name ______________________________

6. Build a truth table for this argument, then state whether the argument is valid or invalid.

In Questions 7–9, an argument is given. Using Questions 2–6 as a model, assign letters to each of the simple statements in the argument, write the argument in symbols, then use a truth table to decide if it's valid.

7. I'll take the job if it meets my salary requirements or offers rapid advancement.

The salary they're offering is way too low.

So, I'm not taking the job.

7-4 Applications

Name ______________________________

8. If you text while driving, you're endangering other drivers. Either you endanger other drivers, or you think you're invincible. That means that if you think you're invincible, you won't text and drive.

9. I'm changing my major if and only if I get an A in biology this semester. I'll pass that class, or I won't change my major. Therefore, if I changed my major, then I passed the class.

7-4 Applications

Name ______________________________

In Questions 10–13, an argument is given. Write the argument in symbols, then decide whether the argument is valid by using the common forms of valid arguments and fallacies.

10. If background checks for buying guns are expanded, it will be harder for felons to obtain guns. It won't be harder for felons to obtain guns. Therefore, background checks won't be expanded.

11. If the interest rate goes any lower, I'll pull my money out of savings and bury it in my backyard. The interest rate didn't go lower, so I won't pull my money out of savings and bury it in the backyard.

12. I majored in culinary arts or I got a job working in an art gallery after graduation. I didn't get a job working in an art gallery, so I majored in culinary arts.

13. If the climate continues to warm, the polar ice caps will melt. If the polar ice caps melt, New Orleans will be under water. So if the climate continues to warm, New Orleans will be under water.

Math in Mind Control REVISITED

Colin Anderson/Blend Images

All of the arguments listed are invalid except the last one. If you were to write that argument using connectives, it would be "Texas wins World Series → sun goes out; Texas won the World Series; therefore, sun will go out." Of course this conclusion is false (at least as of this writing), but this is because the first premise is false, not because the argument isn't valid.

The first argument is basically "Fire → smoke, so smoke → fire." This is the fallacy of the converse. So is the second: money → happiness, so happiness → money. The third argument looks like this: being in SEC → good enough for bowl, so not being in SEC → not good enough for bowl. This is the fallacy of the inverse. The fourth argument is classical circular reasoning. The fifth is another fallacy of the inverse: weapons of mass destruction → war, so no weapons of mass destruction → not war.

Unit 8
How Do You Measure Up?

clintscholz/iStock/Getty Images

Outline

Math in Seeing the World

I love my house, I love my yard, I love my neighborhood, and I REALLY love my dogs. But I still get the urge to travel—a lot. Traveling to new cities, basking in new surroundings, experiences, cultures—there's just something magical about travel that makes you want to do it more and more once you catch the bug. When you start to travel outside of the States, you realize that a lot of things are different, among them the way that many things are measured.

To measure something means to assign a number that represents its size. In fact, measurement might be the most common use of numbers in everyday life. Numbers are used to measure heights, weights, distances, grades, weather, sizes of homes, capacities of bottles and cans, and much more. Even in our monetary system, we're using numbers to measure sizes: little ones, like the cost of a candy bar, and big ones, like an annual salary.

The one thing that every measurement has in common is that it's good for absolutely nothing unless there are units attached to the number. If a friend asks you how far you live from campus and you answer "three," they're going to think you're a goof, right? An answer like that is meaningless because without units, there's no context to describe what the number represents. And that's entirely because there are many different units that can be used to describe similar measurements. That number 3 could mean 3 miles, kilometers, blocks, houses, even parsecs if you come from a galaxy far, far away.

In this unit, we'll study the different ways that things are measured in our world, with a special focus on the different units that are used. You probably have at least a passing familiarity with the metric system, since measurements in units like liters and kilometers are becoming more common. When you gather information of interest to you, you don't always get to choose the units that the information is provided in, so the ability to convert measurements from one unit to another is a particularly useful skill in the information age.

One obvious application of this skill comes into play when traveling outside the U.S. Most countries use the metric system almost exclusively, meaning that to understand various bits of information, you need to be able to interpret measurements in an unfamiliar system. Suppose that you and your sweetie are planning a dream vacation to Australia. Who wouldn't want to go to a place where you might see a kangaroo hop across the road in front of you? Seriously, it happens all the time. The little guys are like squirrels over there. The following questions might be of interest to you on the trip. By the time you finish this unit, you'll be able to answer these and other questions, making you a more well-informed traveler.

1. A check of a tourist website informs you that the temperature is expected to range from 24° to 32° Celsius. How should you pack?
2. The airline that will fly you down under allows a maximum of 30 kilograms for each piece of luggage. If your empty suitcase weighs 4 pounds, how many pounds can you pack into it?
3. The resort you will be staying at is 4.2 kilometers from the airport and 0.6 kilometers from downtown. Can you walk downtown? And how long will it take to drive your rental car from the airport at 35 miles per hour?
4. The gas station closest to the airport caters to many tourists, so it has prices listed in both Australian and American dollars. Gas costs \$1.43 per liter (in U.S. dollars). If you use the car's entire 16-gallon tank, how much will it cost to refill?
5. At the time of this writing, one Australian dollar is worth 0.63 U.S. dollars. If the resort quotes you a price of 329 dollars per night Australian, how much are you paying in American dollars?

For answers, see Math in Seeing the World Revisited on page 82.

Lesson 8-1 Prep Skills

SKILL 1: PERFORM BASIC UNIT CONVERSIONS

Basic unit conversions can be performed using a little common sense and multiplication or division. In order to convert a distance of 10 feet to inches, we use the fact that there are 12 inches in 1 foot. But do we multiply by 12 or divide by 12? Notice that $10 \cdot 12 = 120$ and $10 \div 12 = \frac{5}{6} \approx 0.83$. Since there are 12 inches in a single foot, there will be more inches than feet in a given length. So we choose the operation (multiplication) that produces the larger number. (That's the common sense part—think about whether your answer makes sense!) Ten feet is equivalent in length to 120 inches.

- Convert 2 days to hours.
  $2 \cdot 24 = 48$
  2 days is equivalent to 48 hours.

  Since there are 24 hours in a single day, there will be more hours than days in a length of time. So to convert from days to hours, we multiply by 24.

- Convert 300 seconds to minutes.
  $300 \div 60 = 5$
  300 seconds is equivalent to 5 minutes.

  Since there are 60 seconds in a single minute, there will be more seconds than minutes in a length of time. So to convert from seconds to minutes, we divide by 60.

SKILL 2: DIVIDE OUT COMMON FACTORS

Dividing out common factors involves the same skill as reducing fractions. When you reduce $\frac{10}{15}$ to $\frac{2}{3}$, you divide the numerator and the denominator by 5. It can be helpful to think of $\frac{10}{15}$ as $\frac{2 \cdot 5}{3 \cdot 5}$, and then divide both the numerator and denominator by 5. Sometimes this is shown by crossing out the common factors $\frac{2 \cdot \cancel{5}}{3 \cdot \cancel{5}} = \frac{2}{3}$.

But DON'T THINK OF THIS AS JUST CROSSING OUT STUFF THAT LOOKS ALIKE. Think of it as what it is: division.

Simplify each fraction by dividing out the common factors.

- $\frac{z}{2z} = \frac{\cancel{z}}{2 \cdot \cancel{z}}$
  $= \frac{1}{2}$

  Divide numerator and denominator by the common factor of *z*.

- $\frac{20xy}{5y} = \frac{4 \cdot \cancel{5} \cdot x \cdot \cancel{y}}{\cancel{5} \cdot \cancel{y}}$
  $= \frac{4x}{1} = 4x$

  Divide numerator and denominator by the common factors of 5 and *y*.

PREP SKILLS QUESTIONS

Convert each measurement below to the new unit requested.

1. 27 hours to minutes
2. 410 minutes to hours
3. 97 inches to feet
4. 12.3 feet to inches
5. 87 dollars to cents
6. 1,493 cents to dollars

Simplify each fraction by dividing out common factors.

7. $\frac{4x}{x}$
8. $\frac{2y}{12y}$
9. $\frac{40xyz}{8xy}$

Lesson 8-1 Going to Great Lengths

LEARNING OBJECTIVES

- ☐ 1. Understand the importance of units in measurement.
- ☐ 2. Understand how dimensional analysis makes converting units easy.
- ☐ 3. Identify the key components of the metric system.
- ☐ 4. Convert between U.S. and metric units of length, and describe perspective on the size of these measurements.

Jamie Grill/JGI/Blend Images

I want to run the beach's length, because it never ends.
—Deborah Ager

As I reflect back on the history of numeracy in my family (the human family, that is), counting up objects certainly came first. But I bet that second place goes to measuring the lengths of objects. In studying length, we're looking back into our distant past—the earliest units of measure. The fact that lengths can vary so widely describes why different units of measurement are necessary: while inches are perfectly good for measuring your waistline, the Earth is about 5,892,000,000,000 inches from the Sun. Clearly, a larger unit of measure would be a better idea for such a huge distance. In learning how to convert distances from one unit to another, you'll actually learn something much more valuable: a basic approach that can be used to convert any units of measure. Of course, it wouldn't be quantitative reasoning in our world if we didn't also study a variety of situations where this skill comes in handy.

0. Write as many units as you can think of for measuring length.

8-1 Class

The units of length that are commonly used in the U.S. are the inch, foot, yard, and mile. These are part of the **English system of measurement,** which has been in use for a very, very long time. A list of basic conversion factors between these units of measure is provided in the table below.

Units of Length in the English System

12 inches (in.) = 1 foot (ft)

3 ft = 1 yard (yd)

5,280 ft = 1 mile (mi)

One way to use this information to convert between different units is multiplication or division.

1. To convert 8 yards to feet, we could use the fact that there are 3 feet in 1 yard and ________ 8 by 3 to get ________ feet.

2. To convert 564 inches to feet, we could use the fact that there are 12 inches in a foot and __________ by 12 to get __________ feet.

This works okay, but has one major drawback: It requires some careful thought to make sure you're not multiplying when you should divide or vice versa. You might not think that's a big deal when converting feet to yards, but how about inches to miles? When conversions get more complicated, a systematic approach really comes in handy.

That approach is provided by **dimensional analysis,** which is based on a fiendishly simple idea: If measurements in two different units represent the same actual length, like 1 yard and 3 feet, then a fraction formed from dividing those units is just a fancy way to say 1.

3. Start with the fact that 1 yd = 3 ft. Divide both sides of this equation by 3 ft and write the equivalent equation you have remaining. Think about what happens to the units.

4. Start with the fact that 1 yd = 3 ft. Divide both sides of this equation by 1 yd and write the equivalent equation you have remaining.

If all went well, you've proved that two different fractions involving feet and yards are both equal to 1. We call a fraction of this type a **conversion factor.** This next point is hugely important in understanding unit conversions, so it deserves special emphasis.

The Key Idea in Converting Units

Multiplying any measurement by a conversion factor won't change the size of the measurement because we're just multiplying by 1. But it will change the units for the measurement.

Our clever procedure for converting units will be to multiply by one or more conversion factors in a way that the units we don't want will divide out and leave behind the units we do want. Dimensional analysis is emphasized in the next several questions, but it will be useful in most of the problems in this entire unit. In case you didn't catch the significance: *Understanding dimensional analysis is essential for success in this unit.*

The plan is to start with a solved example of how dimensional analysis works, then have you try some examples on your own.

Solved Example 1

Converting from Inches to Feet

a. Write two conversion factors that compare inches and feet.

b. Use one of those conversion factors to convert 204 inches to feet.

Solved Example 1 *Continued*

SOLUTION

a. Remember, a conversion factor is just a comparison between two measurements that are actually the same. In this case, we know that there are 12 inches in 1 foot, so 12 inches and 1 foot are the exact same length. That means if we divide one measurement by the other, the result is just 1. So that's how we get two conversion factors:

$$\frac{12\text{ in.}}{1\text{ ft}} \text{ and } \frac{1\text{ ft}}{12\text{ in.}}$$

b. Now we can use trial and error, or we can think. We'll have you try the trial-and-error approach in a bit. For now, we'll make a clever observation. Make sure you really think about this—once this makes sense to you, the rest of this unit gets a whole lot easier.

If we multiply the measurement we were given (204 inches) by the second conversion factor we wrote, something interesting happens:

$$204\text{ in.} \times \frac{1\text{ ft}}{12\text{ in.}}$$

Ignore the numbers and look at the units for a moment. In particular, notice that we have inches in the numerator of 204 in. $\left(\text{you can think of it as } \frac{204\text{ in.}}{1} \text{ if you like}\right)$, and inches in the denominator of $\frac{1\text{ ft.}}{12\text{ in.}}$. That's great—it means that the units of inches can be divided out. That's great because we were trying to get rid of inches as the units. Better still, look at what units get left behind: feet, which is exactly what we wanted.

$$204\,\cancel{\text{in.}} \times \frac{1\text{ ft}}{12\,\cancel{\text{in.}}} = \frac{204\text{ ft}}{12}$$

Now we can focus on the numbers and perform the arithmetic. The result of 204 divided by 12 is 17. Putting the pieces together, the entire calculation looks like this:

$$204\,\cancel{\text{in.}} \times \frac{1\text{ ft}}{12\,\cancel{\text{in.}}} = \frac{204\text{ ft}}{12} = 17\text{ ft}$$

5. The height of Mount Everest is 29,035 feet. Write two conversion factors that compare feet and miles. Once you have one, the other will be its reciprocal.

Pixtal/age fotostock

6. Multiply the measurement 29,035 feet by EACH of the conversion factors from Question 5, and look carefully at the units. Which of the multiplications will divide out feet and leave behind miles?

29,035 ft × ―――   29,035 ft × ―――

7. Complete the multiplication that converts 29,035 feet into miles, then congratulate yourself for successfully using dimensional analysis.

Did You Get It

Try this problem to see if you understand the concepts we just studied. The answer can be found at the end of the portfolio section.

1. Absolutely true story: I'm writing this question on an airplane on the way to Charlotte, NC, and according to the flight tracker on my American Airlines app, right now we're about 5.1 miles above the ground. Use dimensional analysis to find how many feet above ground we are. (I said it was a true story, not a good one.)

Sometimes we'll need to multiply by more than one conversion factor to complete a conversion. The distance by air from JFK airport to LaGuardia airport in New York is about 11 miles. How many inches is that?

8. Write a conversion factor that will convert from miles to feet. Think carefully about where to put the units you're trying to get rid of.

9. Write a conversion factor that will convert from feet to inches.

10. Use the conversion factors from Questions 8 and 9, and show the work of dimensional analysis (including dividing away units) to find the distance from JFK airport to LaGuardia airport in inches.

Math Note

Notice that when converting a length to different units, we put the units we want to eliminate in the denominator and the units we want to keep in the numerator. If you understand this simple idea, dimensional analysis will be a snap for you no matter how complicated the conversions might look.

Sidelight **A Brief History of Measuring Lengths**

Nobody knows how and when human beings started measuring physical objects, but historians seem to be in general agreement that the earliest recorded measurements were lengths. In particular, they were based on parts of the human body. If you search the Internet for early units of measurement, the most common one you will find is the **cubit,** first used by the Egyptians around 5,000 years ago. The cubit was equal to the distance from a person's elbow to the outstretched middle finger. (If you try to measure something that way, make sure to stretch out your other fingers as well. That's funny if you think about it.) Other early units were the **palm** (the distance across the base of a person's four fingers) and the **digit** (the thickness of a person's middle finger).

The fact that these measurements were based on body parts leads us to believe that one of two things was true: Either everyone was exactly the same size in ancient Egypt, or their units of measurement were pretty darn imprecise. Things got a little more specific sometime around the year 1100, when King Henry I of England declared that the distance from the tip of his nose to the end of his thumb would be known as a **yard.** (At least this distance was based on one specific person's physical size.) The oldest known yardstick still in existence, believed to have been made in 1445, is accurate to a modern yard within about one three-hundredths of an inch.

A couple hundred years later, King Edward I declared that one-third of a yard should be called a **foot.** In 1595, Queen Elizabeth changed the length of a **mile** from the Roman tradition of 5,000 feet to the current 5,280 feet so that it was exactly 8 **furlongs.** (Perhaps she was a big fan of horse racing, which still uses furlongs.) Eventually, the English system of measurement became standard throughout the world. Of course, these units are still used in the United States today. The real irony? England has been officially using the metric system since 1965, which would be sort of like England deciding not to speak English anymore. Go figure. In fact, only three countries in the world have not officially adopted the metric system, which we'll study later in this lesson: the United States, Liberia, and Myanmar.

Dimensional analysis can also be used to convert more complicated units, like those for speed. This one you'll do with a little less guidance. Be sure to show the multiplication of conversion factors and dividing away of units.

PhotoLink/Getty Images

11. A decent major league pitcher can throw a baseball 90 miles per hour. How fast is that in feet per second? (Hint: First use a conversion factor to eliminate miles from the numerator, then use another conversion factor or two to eliminate hours from the denominator.)

Did You Get It

2. Still on the plane, and it's NOT going the speed of sound. I'm a professor, not a test pilot. However, the speed of sound at sea level and room temperature is about 1,126 feet per second. How fast is that in miles per hour?

8-1 Group

The Metric System

If you grew up in the U.S., you're probably perfectly happy with the English system of measurement. You've been using it your whole life, after all. But the truth is that it's not that great of a system because there's no rhyme or reason to it. Why is 12 inches a foot? Why is 5,280 feet a mile? They just are. Somebody decided, and that was that. The metric system, on the other hand, makes a bit more sense because the different units of measure are all related by powers of 10. We'll see exactly what that means after introducing the base units. The metric system uses three basic units of measure: the *meter,* the *liter,* and the *gram.*

In this lesson, we'll study the **meter,** since it's the standard metric unit of length. One meter is a bit more than a yard, about 39.4 inches. The symbol for meter is "m." The basic unit for capacity (or volume) in the metric system is the **liter,** which is a little bit more than a quart. (You're probably familiar with 2-liter bottles of soda—that's close to a half gallon.) The symbol for liter is "L." The basic unit for measuring weight in the metric system is the **gram.** This is a very small unit—a nickel weighs about 5 grams. The symbol for gram is "g."

Now back to other units in the metric system. All units other than the three basic units are based on multiples of the basic units, with those multiples all being powers of 10. For example, a standard unit of weight for people in the metric system is the kilogram: This is equal to 1,000 grams, and 1,000 is 10 raised to the third power. A commonly used unit of length for smaller measurements is the centimeter, which is one-hundredth of a meter. This is the key idea in understanding the metric system: *It's the prefix in front of the basic unit that determines the size.*

The most common metric prefixes are listed in the table below.

Prefix	Symbol	Meaning	
kilo	k	1,000 units	Bigger than the basic unit
hecto	h	100 units	Bigger than the basic unit
deka	da	10 units	Bigger than the basic unit
	m, L, g	1 unit	
deci	d	$\frac{1}{10}$ of a unit	Smaller than the basic unit
centi	c	$\frac{1}{100}$ of a unit	Smaller than the basic unit
milli	m	$\frac{1}{1,000}$ of a unit	Smaller than the basic unit

Math Note

Notice that the prefixes for each of the three most common metric units that are smaller than the basic unit all end in the letter i: deci-, centi-, and milli-.

The relationship between units in the metric system is a lot like the units used in our monetary system.

1. Complete the table to demonstrate this comparison for the base unit of meter.

Metric Unit	kilometer	hectometer	dekameter	meter (base unit)	decimeter	centimeter	millimeter
Meaning	1,000 meters	100 meters	10 meters	1 meter	$\frac{1}{10}$ meter	$\frac{1}{100}$ meter	$\frac{1}{1,000}$ meter
Monetary Comparison			\$10	\$1			

Now that you've learned how to do dimensional analysis, you can use it to convert units within the metric system. We're not saying that you *should,* mind you—a handful of examples might lead us to a simpler approach.

Solved Example 2

Converting Units in the Metric System

a. Use dimensional analysis to find the length of a book in mm that is 25.4 cm long, and the diameter of a bolt in cm that is 3.25 mm across.

b. In each case, describe what calculation you performed numerically, and what it did to the decimal point in the measurement.

SOLUTION

a. In each case, we'll use an appropriate conversion factor based on the fact that 1 cm is 10 mm. For the 25.4 cm book, we want to eliminate centimeters, so we'll put 1 cm in the denominator.

$$25.4 \text{ cm} \times \frac{10 \text{ mm}}{1 \text{ cm}} = 254 \text{ mm}$$

For the 3.25 mm diameter, we want to eliminate millimeters, so this time 10 mm goes in the denominator.

$$3.25 \text{ mm} \times \frac{1 \text{ cm}}{10 \text{ mm}} = 0.325 \text{ cm}$$

b. The numerical result of converting cm to mm was multiplying by 10, which moved the decimal point one place to the right. The result of converting in the other direction, from mm to cm, was dividing by 10, which moved the decimal point one place to the left. There might be a pattern here. . . .

2. Find the height of a child in m who is 135 cm tall. Remember, one meter is 100 centimeters.

135 cm × ——— = ________ m

3. Numerically, what calculation did you perform? What did that do to the decimal point?

4. Find wheelbase of a car in cm that is 2.761 m.

2.761 m × ——— = __________ cm

5. Numerically, what calculation did you perform? What did that do to the decimal point?

6. To recap, each time you multiply by 10, the decimal point moves one place to the __________, and each time you divide by 10, the decimal point moves one place to the __________.

Rather than remembering or looking up a bunch of different conversion factors for all of the different metric prefixes, we simply move the decimal point based on the number of steps you move in the metric unit conversions table. Let's see how that works.

Conversions Within the Metric System

- To change a larger unit to a smaller unit in the metric system, multiply by 10^n, where n is the number of steps that you move down in the diagram below.
- To change a smaller unit to a larger unit in the metric system, divide by 10^n, where n is the number of steps that you move up in the diagram below.

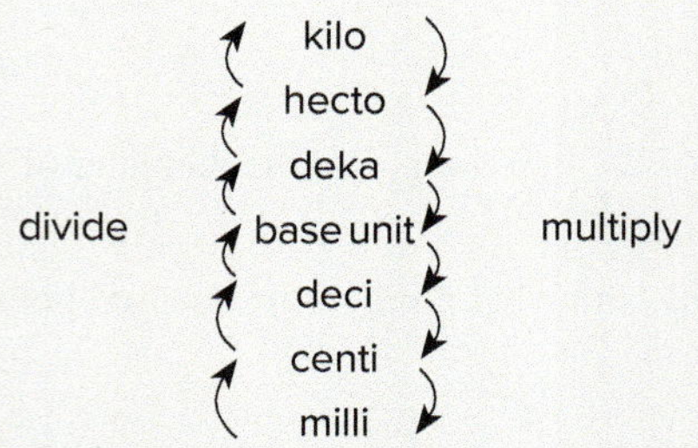

Now let's put the box above into use. Two rival colleges have campuses that are 16.9 km apart. How far is that in centimeters?

7. How many steps apart are km and cm in the diagram?

8. Based on their relative positions, should we multiply or divide to convert km into cm?

9. What power of 10 should we multiply or divide by?

10. How many places should we move the decimal place, and in which direction?

11. So how many cm apart are the campuses?

To keep the direction you move the decimal place straight, think of it this way: When changing to a smaller unit, you'll need a larger measurement, which will happen if you move the decimal place to the right. In a pinch, think about feet and inches. If something is a foot long, it's 12 inches long. Smaller measurement, bigger number.

Did You Get It

3. On a different plane now. Not the one I was supposed to be on, mind you, but that's a long story. Anyhow, the width of my seat is 28", which is 711.2 mm. How many dekameters is that?

12. A French dip sandwich from a certain fast-food chain contains a whopping 3,610 mg of sodium. If you ate that sandwich, you'd be ingesting way more than the recommended daily amount of sodium (2,300 mg) in one sandwich. How many grams over the recommended amount would you ingest if you ate that sandwich? Read the units in that question carefully, please.

Because most of the world uses the metric system while the U.S. uses the English system, it's often necessary to convert between the two systems. (This is especially true in the sciences, where the metric system is used almost exclusively.) We'll use dimensional analysis, along with the conversion factors for units of length listed below. Because we know dimensional analysis and how to convert metric units, we don't need an extensive list of conversion factors.

English-Metric Equivalents for Length

1 in. = 2.54 cm

1 mi $\approx$ 1.61 km

13. Many road signs now list mileages in both miles and kilometers. This sign can be found near my brother-in-law's house, just east of Louisville. (Hi, Kevin.) How accurate is the conversion?

David Sobecki

14. On a stretch of highway, the posted speed limit is 70 miles per hour. Find the speed limit in kilometers per minute.

McGraw-Hill Education

Looking at a ruler that has both inches and centimeters on it is a good way to get a feel for the connection between them.

15. Notice that 2 inches is just about 5 centimeters. Look carefully. Would you say that 2 inches is a little more or a little less than 5 cm?

16. Use dimensional analysis to find more precisely how many centimeters are equal to 2 inches.

17. At any given time, I can walk out my back door and find a half-dozen little lizards. That's life in Florida. The most common type around here is the brown anole, which has a typical length of about 96.3 millimeters. Convert this length to inches.

Cat Sobecki.

18. I've never driven a Tesla model S, but I sat in one a few minutes ago. And if you haven't, you really should because it's super-cool. It's also 4.98 m long. Convert this length to feet.

Did You Get It

4. The Airbus A320 that I'm in right now is 123 feet, 3 inches long. How many centimeters is that?
5. The wingspan is 35.8 meters, which is how many inches?

We began this lesson by explaining that lots of different units of length exist because there are so many different-sized objects in our world. Let's use this concept to close the lesson and see if you have some perspective on lengths in the metric system. In Questions 19–23, decide if millimeters, centimeters, meters, or kilometers would be the most appropriate unit for describing each length.

19. The height of an elephant.

20. The height of the letters in the Did You Get It header above.

21. The height reached by a typical hot air balloon flight.

22. The height reached by a typical rocket launch.

23. The thickness of the heaviest book you own.

8-1 Portfolio

Name ____

Check each box when you've completed the task. Remember that your instructor will want you to turn in the portfolio pages you create.

Technology

1. ☐ In this Excel activity, we'll use formulas to convert units of length.

ALEKS Practice

1. ☐ Include any written work from the ALEKS assignment along with any notes or questions about this lesson's content.

Applications

1. ☐ Complete the applications problems.

Reflections

Type a short answer to each question.

1. ☐ What is a conversion factor in measurement?
2. ☐ Explain how we use conversion factors in dimensional analysis to convert units.
3. ☐ What role do prefixes play in the metric system?
4. ☐ Name one thing you learned or discovered in this lesson that you found particularly interesting.
5. ☐ What questions do you have about this lesson?

Looking Ahead

1. ☐ Complete the prep skills for Lesson 8-2.
2. ☐ Read the opening paragraph in Lesson 8-2 carefully and answer Question 0 in preparation for that lesson.

Answers to "Did You Get It?"

1. 26,928 feet **2.** 767.7 mi/hr **3.** 0.07112 dam **4.** 3,756.66 cm
5. 1,409.45 in.

Answers to Prep Skills Questions

1. 1,620 minutes
2. About 6.83 hours
3. About 8.08 feet
4. 147.6 inches **5.** 8,700 cents
6. $14.93
7. 4
8. $\frac{1}{6}$
9. $5z$

8-1 Applications

Name ______________________________

1. A professional basketball court is 94 feet long. Find the length in both inches and meters.

2. An official Olympic-sized swimming pool is 50 meters long. How many yards long is it? How many feet would you swim in a 20-lap race?

The nutrition information label shown is for a generic brand of peanut butter. Use it to answer Questions 3–6.

Serving Size 32g

Amount Per Serving			
Calories 188		Calories from Fat 135	
		% Daily Value*	
Total Fat 16g			25%
Saturated Fat 3g			17%
Trans Fat 0g			
Cholesterol 0mg			0%
Sodium 147mg			6%
Total Carbohydrate 6g			2%
Dietary Fiber 2g			8%
Sugars 3g			
Protein 8g			
Vitamin A	0%	• Vitamin C	0%
Calcium	1%	• Iron	3%

3. How many milligrams of peanut butter make up one serving?

4. How many dekagrams of fat are there in one serving?

5. Based on the % daily value listed, how many centigrams of saturated fat are you recommended to ingest per day?

6. How many kilograms of carbohydrates are you recommended to take in per year?

8-1 Applications

Name ______________________________

7. The screen dimensions of a 70-inch HDTV are 34.3 in. high and 61.0 inches wide. Find the dimensions in centimeters and feet.

8. In April of 2015, a magnetic levitation train in Japan broke the speed record for a train, reaching a top speed of 603 kilometers per hour. How many miles can that baby go in an hour? How many feet in a minute?

In this lesson, we studied metric prefixes from milli- (one-thousandth) up to kilo- (thousands). There are, however, many other prefixes. For example, mega- refers to millions (10^6), giga- is billions (10^9), and micro- is one-millionths (10^{-6}).

9. In the movie *Back to the Future,* 1.21 gigawatts of power were needed to send the time machine, well, back to the future. How many watts is this?

10. The average cell in a human body is about 10 micrometers wide. How many millimeters is that?

11. The current largest cruise ship in the world weighs 248,663 megagrams. How heavy is that in pounds? (One kilogram is 2.2 pounds.)

Math Note

Interesting fact: Anyone familiar with modern computer technology is comfortable with the term gigabytes. But when *Back to the Future* was made (1985), giga- was so uncommon as a prefix that it's consistently mispronounced in the movie (jiga, rather than giga) and NOBODY involved in production of the movie knew to correct the cast.

Lesson 8-2 Prep Skills

SKILL 1: USE DIMENSIONAL ANALYSIS

As we saw in Lesson 8-1, multiplying any measurement by a conversion factor won't change the size of the measurement because we're just multiplying by one. But it will change the units for the measurement. Using dimensional analysis involves converting a quantity from one unit to another of equal size but with different units.

Use dimensional analysis to convert each measurement.

- 5 days to minutes

$$5\ \cancel{\text{days}} \cdot \frac{24\ \cancel{\text{hrs}}}{1\ \cancel{\text{day}}} \cdot \frac{60\ \text{min}}{1\ \cancel{\text{hr}}} = 7{,}200\ \text{min}$$

1 day = 24 hr and 1 hr = 60 min
Using these equivalences, form the correct conversion factors to divide away the unwanted units.

5 days is equal to 7,200 minutes.

- \$12 per hour to cents per minute

$$\frac{12\ \cancel{\text{dollars}}}{1\ \cancel{\text{hr}}} \cdot \frac{100\ \text{cents}}{1\ \cancel{\text{dollar}}} \cdot \frac{1\ \cancel{\text{hr}}}{60\ \text{min}} = \frac{20\ \text{cents}}{1\ \text{min}}$$

1 dollar = 100 cents and 1 hr = 60 min
Using these equivalences, form the correct conversion factors to divide away the unwanted units.

12 dollars per hour is equivalent to 20 cents per minute.

SKILL 2: USE THE METRIC SYSTEM

Use the following summary as a reminder of how we convert units within the metric system.

- To change a larger unit to a smaller unit in the metric system, multiply by 10^n, where n is the number of steps that you move down in the diagram below.
- To change a smaller unit to a larger unit in the metric system, divide by 10^n, where n is the number of steps that you move up in the diagram below.

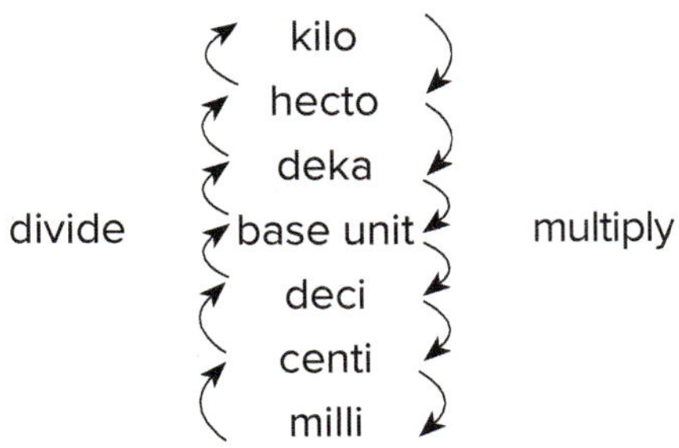

- Convert 55 meters to millimeters.

  55 m = 55,000 mm

Meters and millimeters are 3 steps apart in the diagram above. To make this conversion, we multiply by 10^3, which moves the decimal point 3 places to the right.

- Convert 55 centimeters to kilometers.

  55 cm = 0.00055 km

Centimeters and kilometers are 5 steps apart in the diagram above. To make this conversion, we divide by 10^5, which moves the decimal point 5 places to the left.

SKILL 3: USE AREA AND VOLUME FORMULAS

The area of a square is calculated using the formula $A = s^2$, where s is the length of a side of the square.

The volume of a cube is calculated using the formula $V = s^3$, where s is the length of a side of the cube.

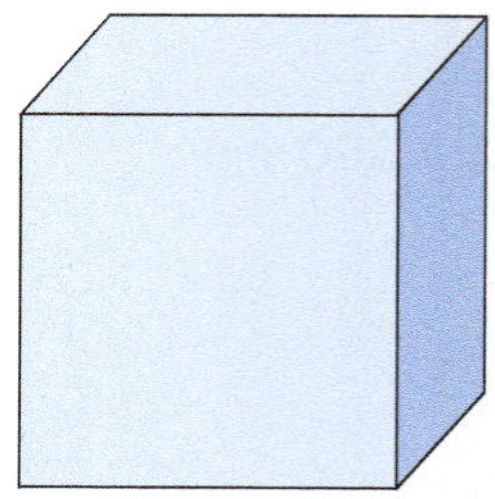

- Find the area of a square with sides of length 4 in.

  $A = (4\text{ in.})^2 = (4\text{ in.})(4\text{ in.})$
  $A = 16\text{ in.}^2$

  Use the formula $A = s^2$ with $s = 4$ in. Notice that the units work out to be square inches, which confirms that this is an area.

- Find the volume of a cube with side of length 4 in.

  $V = (4\text{ in.})^3 = (4\text{ in.})(4\text{ in.})(4\text{ in.})$
  $A = 64\text{ in.}^3$

  Use the formula $V = s^3$ with $s = 4$ in. Notice that the units work out to be cubic inches, which confirms that this is a volume.

PREP SKILLS QUESTIONS

1. Use dimensional analysis to convert 4 miles to yards.
2. Use dimensional analysis to convert 25 inches to centimeters.
3. Use dimensional analysis to convert 230 feet per minute into miles per hour.
4. How many milliliters are in 27 liters?
5. How many kilometers are in 347 dekameters?
6. How many centimeters are in 9,000 meters?
7. What is the area of a square with sides that are 4 ft long?
8. What is the volume of a cube with sides that are 1.3 in. long?

Lesson 8-2 New Dimensions

Trinette Reed/Blend Images

LEARNING OBJECTIVES

- ☐ 1. Understand the difference between unit conversions for length, and unit conversions for area and volume.
- ☐ 2. Convert area and volume measurements within the U.S. system, and describe perspective on sizes of these measurements.
- ☐ 3. Convert area and volume measurements between the U.S. and metric systems, and describe perspective on sizes of these measurements.

Like gods, we have created a new universe called cyberspace that contains great good and ominous evil. We do not know yet if this new dimension will produce more monsters than marvels, but it is too late to go back.

—David Horsey

It made perfect sense to begin our study of measurement with length, since that's probably the earliest and most basic form of measuring stuff. But units of length have a serious limitation: They can only be used to measure things in one dimension, and this isn't a cartoon: We all live in a three-dimensional world. If you're looking to buy a house, you might be somewhat interested to know that it's 52 feet long, but you'd surely be a lot more interested to know that it has 2,100 square feet of floor space. If you're looking to buy paint for your bedroom, knowing that the walls are 8 feet high won't help a whole lot if you don't also know the length, which would then allow you to find area. How much water does a 6-foot-deep swimming pool hold? There's no way to tell—to find the capacity, you need a lot more than just the depth. In order to measure things in our three-dimensional world, we'll become experts at working with area (two dimensions), volume, and capacity (both three dimensions). And not surprisingly, given the name, we'll find dimensional analysis to be particularly helpful. Don't you just love it when things make sense?

0. Write a couple of examples of things that are typically measured in more than one dimension. (Other than the ones provided in the opening paragraph, of course.)

8-2 Class

Conversions of Area

Units of length are also called linear units, because they are used to measure along a line. The units used to measure the size of an object in two dimensions are called square units. They are based on a very simple idea: One square unit is defined to be the area of a square that is one unit long on each side. For example, a square inch is the area of a square that is 1 inch on each side, as shown in the really lame figure below.

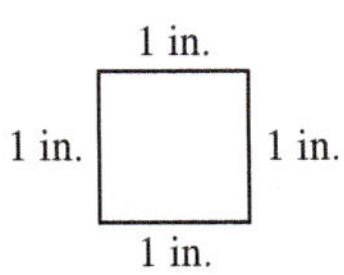

1. Write the formula for the area of a square.

2. Use this formula to write and simplify the calculation (including units) to find the area of the square on the previous page.

3. From algebra, how would we simplify the expression $(1x)(1x)$?

4. Describe the connection between your answers to Questions 2 and 3.

Math Note

You can write in.^{2} or sq. in., but in either case we pronounce it "square inches."

Continuing our course-long philosophy that understanding is ALWAYS better than just knowing, the point of Questions 1–4 was giving you a fundamental understanding of why it makes perfect sense to measure area in square units. Once we're comfortable with that idea, we define the area of other figures to be the number of square units that will fit inside the figure.

For example, the rectangle in the next figure is 3 inches by 5 inches, and we can see that __________ one-inch by one-inch squares will fit inside. So its area is __________ square inches, which we abbreviate as __________ in.^{2}

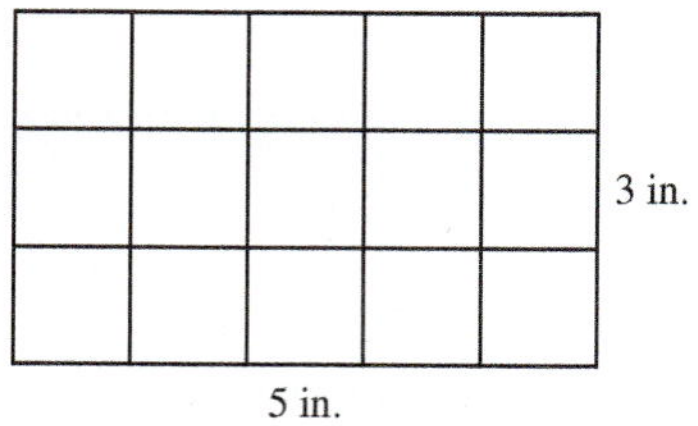

5. Write the formula for the area of a rectangle.

6. Use this formula to write and simplify the calculation (including units) to find the area of this rectangle.

7. From algebra, how would we simplify the expression $(3x)(5x)$? What does this have to do with anything?

8. At this point you should be pretty darn good at converting units of length. Next, we'll turn our attention to units of area. Rather than learning new conversion factors, we'll use ones that we already know and just be clever. For example, we know of course that 1 ft = 12 in. Square both sides of that equation and simplify the result, paying special attention to the units.

9. The figure below illustrates graphically why 1 square foot is ________ square inches. 1 square foot is made up of ________ one-square-inch boxes.

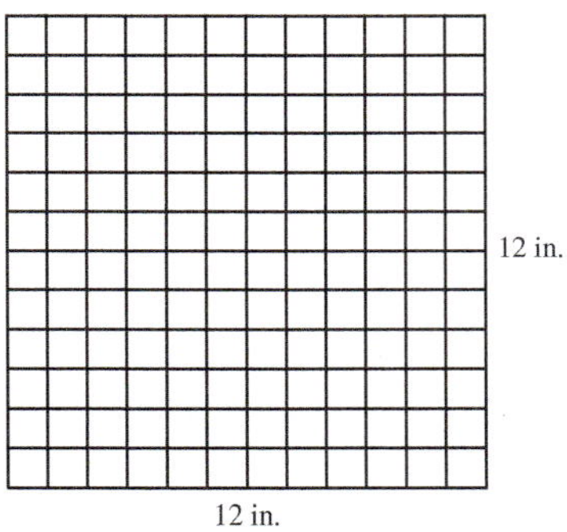

Solved Example 1

Converting an Area

A 52-in. wide-screen TV has a screen area of 1,064 in.^{2} Multiply that measurement by a conversion factor that will eliminate the inch units. Your conversion factor should be based on the fact that 1 square foot is the same as 144 square inches. Does the result of the multiplication have appropriate units?

SOLUTION

Since the measurement we're starting with has square inches in the numerator, we'll want a conversion factor that has square inches in the denominator:

$$1{,}064 \text{ in.}^2 \times \frac{1 \text{ ft}^2}{144 \text{ in.}^2} \approx 7.39 \text{ ft}^2$$

This worked out just like we would hope: The square inches divided out, and the resulting units are square feet, which are also perfectly good units for an area.

10. A 60-in. wide-screen TV has a screen area of 1,546 in.^{2} How many square feet is that? Use the fact that 1 ft = 12 in. and write a conversion factor with inches in the denominator.

$1{,}546\,\text{in.}^2 \times \underline{\qquad}$

11. The units don't work out quite right, do they? What if we multiply by that same conversion factor again? Give it a try, and if the units work out right, complete the conversion.

$1{,}546\,\text{in.}^2 \times \underline{\qquad} \times \underline{\qquad} \approx$

Did You Get It

Try this problem to see if you understand the concepts we just studied. The answer can be found at the end of the portfolio section.

1. A 32-in. wide-screen TV has screen area 3.05 ft^2. How many square inches is that?

The point of the last few questions is that there are two ways that we can convert units of area: we can develop new conversion factors, like $\frac{1\text{ ft}^2}{144\text{ in.}^2}$, or we can use the ones for length that we already know and just multiply by them twice. The advantage of the second approach is that it requires remembering less information.

8-2 Group

There's one very old measure of area that's still in common use that we'll need conversion factors for because it's not based on squared units of length: the acre. The following conversions will come in handy for some upcoming problems.

Conversion Factors for Area

1 acre = 43,560 ft^2	1 mi^2 ≈ 2.59 km^2
1 mi^2 = 640 acres	1 km^2 ≈ 247.1 acres

1. Find the area of Yellowstone National Park in square miles given that it has an area of 2,219,789 acres.

Math Note

The acre was originally defined to be the size of a field that one man could plow in one day behind one ox. That seems a bit subjective, no?

Solved Example 2

Converting Square Yards to Acres

An average of 2,191,781 square yards of carpet is discarded into landfills every day in the U.S. How many acres would that carpet cover?

SOLUTION

This question brings up a really important point: When using dimensional analysis, you won't always have a direct conversion from the given units to the ones you want. In that case, you'll need to come up with an intermediate step or two, converting a given measurement into something else that's easier to work with. In this case, we have three units we can convert into acres, using the conversion factors from the purple box: square feet, square miles, and square kilometers. So if we're going to have any success here, we'll need to convert what we're given (square yards) into one of those three units.

The best choice is square feet because we know a simple conversion from yards to feet. But we'll need to be careful because we're working with SQUARE units. We want to eliminate square yards, so we'll put TWO factors of yards in the denominator:

$$2{,}191{,}781\ \text{yd}^2 \times \frac{3\ \text{ft}}{1\ \text{yd}} \times \frac{3\ \text{ft}}{1\ \text{yd}} = 19{,}726{,}029\ \text{ft}^2$$

Now we're back on familiar ground: The table in the purple box shows us that 1 acre is the same as 43,560 square feet. We want to eliminate square feet and end up with acres, so we put 1 acre in the numerator of our conversion factor, and 43,560 square feet in the denominator.

$$19{,}726{,}029\ \text{ft}^2 \times \frac{1\ \text{acre}}{43{,}560\ \text{ft}^2} \approx 452.8\ \text{acres}$$

2. It's common to estimate an acre as being about the size of a football field. But how accurate is that comparison? An American football field is 100 yards long (without the end zones) and 160 feet wide. How close is the area to one acre?

Did You Get It

2. The state of Delaware has an area of 2,490 square miles. How many acres is that?
3. Promoters are planning to unfurl a giant flag covering an entire football field at a playoff game, including the end zones. The flag will cover an area of 1.32 acres. How many square yards of fabric will be needed?

3. Recall that 1 in. = 2.54 cm. Use this to build two conversion factors between feet and centimeters.

4. The plans for building a deck call for 28 square meters of deck boards. If you already have 300 square feet of boards on hand, do you have enough?

Wests/iStockphoto.com

5. Prince Edward County in Ontario, Canada, has an area of 1,050.1 square kilometers. Is it larger or smaller than Butler County, Ohio, which is 470 square miles?

6. Find what percent larger the bigger county is.

7. Find what percent smaller the smaller county is. (If you think this is the same question as number 6, think about it more carefully.)

Did You Get It

4. The campus of the University of Guelph in Ontario has an area of 4.1 km^2. How many square miles is that?

The next example shows a very common and very useful application of area conversions for homeowners.

8. Jane plans to have new carpet installed in her living room. She measures and finds that the room is 17 ft wide and 20 ft long. She found a carpet option she likes equally well at two different stores and decides she'll go with the cheaper one. The carpet she wants sells for \$21 per square yard at one store and \$2.50 per square foot from another store. Assuming she goes the cheaper route, how much will it cost to have the room carpeted?

Did You Get It

5. Find the cost of carpeting Jane's basement, which is 520 square feet, with cheaper carpet costing $14.50 per square yard.

Conversions of Volume and Capacity

Just as the measurements of area are based on the two-dimensional square, measurements of volume are based on the three-dimensional analog, the cube. So volume is measured in cubic units.

In the English system, this is typically cubic inches (in.^3), cubic feet (ft^3), and cubic yards (yd^3). For example, 1 cubic foot consists of a cube whose measure is 12 inches on each side.

9. Write the formula for the volume of a cube.

10. Use this formula to write and simplify the calculation (including units) to find the volume of both cubes in the figure.

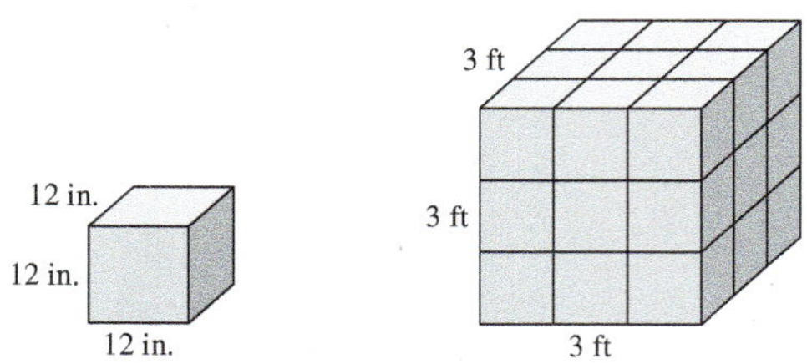

11. From algebra, how would we simplify the expression $(3x)(3x)(3x)$? What's the point?

Measures of volume also include measures of *capacity*. The **capacity** of a container is equal to the amount of fluid the container can hold. In the English system, capacity is measured in fluid ounces, pints, quarts, and gallons. The conversion factors for capacity are shown in the following box.

Units of Capacity in the English System

1 pint (pt) = 16 fluid ounces (oz)
1 quart (qt) = 2 pints (pt)
1 gallon (gal) = 4 quarts (qt)

Since volume and capacity are both measures of size in three dimensions, we can compare the units. A cubic foot (ft^3) of water is about 7.48 gallons, and a cubic yard of water is about 202 gallons. If a gallon of water is poured into a container, it would take up about 231 cubic inches. Finally, a cubic foot of freshwater weighs about 62.5 pounds, and a cubic foot of seawater weighs about 64 pounds. These measures are summarized in the next fancy colored box.

Conversion Factors for Capacity in the English System

$1 \text{ ft}^3 \approx 7.48 \text{ gal}$

$1 \text{ yd}^3 \approx 202 \text{ gal}$

$1 \text{ gal} \approx 231 \text{ in.}^3$

$1 \text{ ft}^3 \text{ freshwater} \approx 62.5 \text{ lb}$

$1 \text{ ft}^3 \text{ seawater} \approx 64 \text{ lb}$

12. A water tank at an aquarium has a volume of 3,000 cubic feet and weighs 3,200 pounds empty. How many gallons of water will the tank hold, and now much will it weigh if it's filled with seawater?

Erica Simone Leeds

Did You Get It

6. How many cubic feet of water are in a 100-gallon freshwater tank, and how much does that water weigh?

13. The team managers for a college football team carry quart bottles of water for the players to drink during breaks at practice. The bottles are stored in carrying racks that weigh 6 pounds and hold 16 bottles. How heavy (in pounds) is a rack when full? (Ignore the weight of the empty bottles.)

In the metric system, a cubic centimeter consists of a cube whose measure is 1 centimeter on each side. Let's build a conversion factor to relate cubic meters and cubic centimeters.

14. Begin with the fact that there are ________ centimeters in 1 meter.

15. Write this as an equation.

16. Raise both sides of that equation to the ________ power to obtain cubic centimeters and cubic meters as units. Simplify the result.

17. Does your answer match what's below the figure?

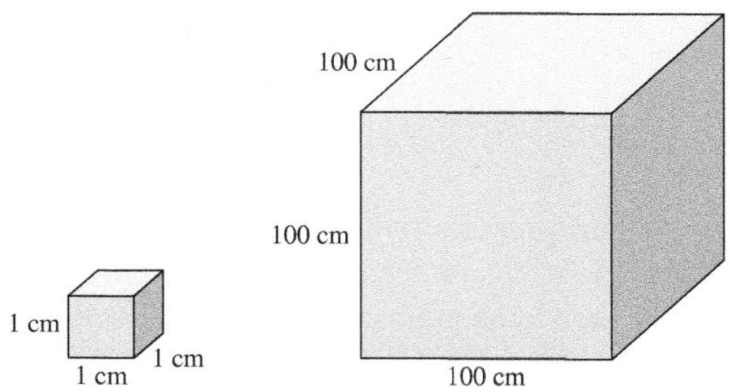

1 cubic centimeter

1 cubic meter = 1,000,000 cm^3

The base unit for volume in the metric system is the liter, defined to be 1,000 cubic centimeters. Cubic centimeters are abbreviated cm^3 or cc.

18. What are the dimensions of a cube with a volume of 1,000 cubic centimeters? Label those dimensions on the figure below.

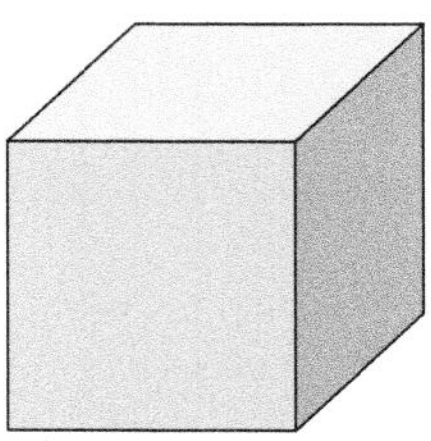

1 liter = 1,000 cm^3

19. This definition of liter provides a simple and interesting connection between two of the base units in the metric system, meters and liters: 1 cubic centimeter is the same as ___________ milliliter. (Remember what the prefix milli- means!)

20. A typical backyard inground pool has a volume of about 82.5 cubic meters. How many milliliters of water are needed to fill such a pool?

David Sobecki

We really only need one conversion factor for computing units of capacity between the English and metric systems, so that guy will get a colored box all to itself:

English-Metric Conversion of Capacity

1 liter ≈ 1.06 quarts

21. The owner of a pool the size of the one in Question 20 has an algae problem (which I can assure you is about half as much fun as a toothache). He buys a special algaecide, and the directions on the bottle instruct him to apply 16 oz for every 10,000 gallons of water in the pool. How much algaecide should he pour in?

Did You Get It

7. On average, 2,476,000 cubic centimeters of Coke are consumed worldwide every second. How many kiloliters is that? How many gallons?

8-2 Portfolio

Name ____________________

Check each box when you've completed the task. Remember that your instructor will want you to turn in the portfolio pages you create.

ALEKS Practice

1. ☐ Include any written work from the ALEKS assignment along with any notes or questions about this lesson's content.

Applications

1. ☐ Complete the applications problems.

Reflections

Type a short answer to each question.

1. ☐ Explain what "square units" mean. Use examples like "square inches" and "square miles."
2. ☐ We know that 1 yd = 3 ft. So how come 1 $yd^2 \neq 3\ ft^2$?
3. ☐ What's the connection between the base units of length and capacity in the metric system?
4. ☐ Describe three things in your house, dorm, or apartment that could be measured using cubic units, and explain why those units would be appropriate.
5. ☐ Name one thing you learned or discovered in this lesson that you found particularly interesting.
6. ☐ What questions do you have about this lesson?

Looking Ahead

1. ☐ Complete the prep skills for Lesson 8-3.
2. ☐ Read the opening paragraph in Lesson 8-3 carefully and answer Question 0 in preparation for that lesson.

Answers to "Did You Get It?"

1. 439.2 $in.^2$ **2.** 1,593,600 acres **3.** 6,388.8 yd^2 **4.** 1.58 mi^2 **5.** \$837.78
6. 13.37 ft^3, 835.625 lb **7.** 2.476 kL, 656.14 gal

Answers to Prep Skills Questions

1. 7,040 yd **2.** 63.5 cm **3.** About 2.61mi/hr **4.** 27,000 mL
5. 3.47 km **6.** 900,000 cm **7.** 16 ft^2 **8.** 2.197 $in.^3$

8-2 Applications

Name ____

In Questions 1–6, which unit would be most appropriate: acres, square miles, square feet, liters, milliliters, or kiloliters?

1. The amount of coffee served between noon and 1 P.M. at the nearest Starbucks

2. The average lot size in a middle-class community in Florida

3. Key West, FL

McGraw Hill

4. The daily dosage for an infant pain reliever

5. The nearest golf course

6. The amount of gas sold in a year at a fueling station in Toronto.

7. When planning to build a concrete patio in his backyard, KJ lays out the space and finds that the dimensions he wants are 146 inches by 183 inches. First find the area in square inches, then use dimensional analysis to convert that to square feet.

8. KJ (Question 7) does some online research and finds that 4 inches is the standard thickness for pouring concrete in nonindustrial applications. Convert this to feet and multiply by your answer to Question 7 to find the volume of concrete that he needs.

8-2 Applications

Name ______________________________

9. Feeling pretty darn proud of himself for doing all the calculations correctly, KJ calls a company that delivers premixed concrete and is asked how many cubic yards he needs. Rats! The volume he has is in cubic feet. Help this poor sap out, please. How many cubic yards does he need?

10. Most companies will only allow you to order concrete by the full cubic yard. Find the relative difference between what KJ has to order and what he needs.

11. Each spring, my yard has about 900 square feet of space to cover with mulch. One year, trying to save some cash, I bought cheapo overseas mulch from a gas station, and the packaging said that each bag covered 1.8 square meters. How many bags did I need?

12. A store has 122 1-quart containers of milk, but the manager wants to pour the milk into gallon containers to sell at a higher price. How many gallon containers can the manager fill, allowing 5% extra for spillage?

13. Measure the room that your class meets in and find the area. Then write that area in square feet, square inches, square meters, and square centimeters.

8-2 Applications

Name ______________________________

14. The Pentagon in suburban Washington, DC, has total floor area of just about 6,500,000 square feet. If a campus building consisted of 100 identical copies of your classroom, 8 restrooms, each with an area of 320 square feet, and 150 square meters of hallway, how would it compare in size to the Pentagon?

Hisham F. Ibrahim/Photodisc/Getty Images

15. The capacity of a dump truck is listed by the manufacturer as 10 yd^3. A construction site requires the removal of 200 m^3 of dirt, and the contractor has three identical trucks. How many trips will each need to make?

In 1970, the largest engine available on a Ford Mustang was a 429-cubic inch model that produced 375 horsepower. In 2024, you could top out at a 5-liter engine that produced 500 horsepower.

16. Which of the two cars had the bigger engine? How much bigger was that engine by percentage?

17. Rate each engine in terms of horsepower per cubic inch of volume.

Lesson 8-3 Prep Skills

SKILL 1: USE DIMENSIONAL ANALYSIS

As we have seen in Lessons 8-1 and 8-2, multiplying any measurement by a conversion factor won't change the size of the measurement because we're just multiplying by 1. But it will change the units for the measurement. Using dimensional analysis involves converting a quantity from one unit to another of equal size but with different units.

Use dimensional analysis to convert the measurement.

- 10 ft to centimeters

12 in. = 1 ft and 2.54 cm = 1 in.
Using these equivalences, multiply by correct conversion factors to divide away the unwanted units.

$$10\ \cancel{\text{ft}} \cdot \frac{12\ \cancel{\text{in.}}}{1\ \cancel{\text{ft}}} \cdot \frac{2.54\ \text{cm}}{1\ \cancel{\text{in.}}} = 304.8\ \text{cm}$$

10 ft is an equal length to 304.8 cm.

SKILL 2: USE THE METRIC SYSTEM

As we saw in Lessons 8-1 and 8-2, the following summary can be used as a reminder of how to convert units within the metric system.

- To change a larger unit to a smaller unit in the metric system, multiply by 10^n, where n is the number of steps that you move down in the diagram below.
- To change a smaller unit to a larger unit in the metric system, divide by 10^n, where n is the number of steps that you move up in the diagram below.

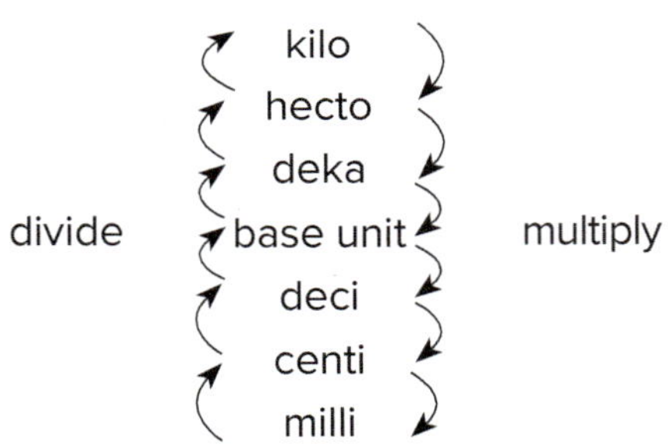

- Convert 2.3 kilometers to decimeters.

2.3 km = 23,000 dm

Kilometers and decimeters are 4 steps apart in the diagram above. To make this conversion, we multiply by 10^4, which moves the decimal point 4 places to the right. Note that decimeter is abbreviated as dm, while dekameter is abbreviated as dam.

SKILL 3: EVALUATE ALGEBRAIC EXPRESSIONS

To evaluate an expression (or a formula), replace all occurrences of the variable or variables with specific numbers. This allows us to find a numeric value.

- Evaluate $3x + 4$ for $x = -2$. This means to find the value of $3x + 4$ when $x = -2$.

  $3(-2) + 4 = -6 + 4 = -2$ Replace x with -2 and simplify the result.

PREP SKILLS QUESTIONS

1. Use dimensional analysis to convert 134 yards to inches.
2. Use dimensional analysis to convert 111 centimeters to inches.
3. Use dimensional analysis to convert 4 miles into kilometers.
4. How many decigrams are in 84.3 milligrams?
5. How many hectometers are in 94 meters?
6. How many dekaliters are in 431.2 liters?
7. Evaluate the following expression for $x = 48$.

   $\frac{7}{4}x + 51$

8. Evaluate the following expression for $t = -4.3$.

   $\frac{4}{7}(t - 40)$

Lesson 8-3 Weighty Matters

LEARNING OBJECTIVES

- ☐ 1. Convert weight and temperature measurements within the U.S. and metric systems.
- ☐ 2. Convert weight and temperature measurements between the U.S. and metric systems.
- ☐ 3. Demonstrate an understanding of the sizes of measurements in these systems.

Patience and tenacity are worth more than twice their weight of cleverness.
—Thomas Huxley

EMMANUEL DUNAND/AFP/
Getty Images

According to the Nintendo Corporation, the weight limit for the fitness board that can be used with its Wii game system is 150 kg. If you're like the average American, you have no idea exactly how heavy you can be and still safely use the board. This, of course, is because you are so accustomed to weights being measured in pounds. But like any other form of measurement, there are a variety of units that can be used for weight. We will study them in this lesson, then conclude with a quick look at measuring temperatures, like the 150 or so degrees the guy in this picture must be feeling while working out in that getup.

0. About how many pounds do you think 150 kg is? Once you learn conversions, check back and see how good your estimate was.

8-3 Group

Conversions for Measures of Weight

Before we begin our study of measuring weight, there's a technical issue we need to deal with. We defined the gram as the standard unit of weight back in Lesson 8-1, but that's not quite accurate: The gram is a measure of *mass,* not weight. The weight of an object is the measure of the gravitational pull on it, so your weight can vary depending on where you are. Mass, on the other hand, is a constant measure of the physical size of an object regardless of location. Fifty kilograms is fifty kilograms, whether it's at the bottom of the ocean, the top of Mount Everest, or orbiting Earth in the International Space Station. While the gravitational pull of Earth can vary slightly depending on altitude, that variance is so tiny as to be mostly negligible anywhere on the planet. Since we all live on Earth, in this book we'll consider mass and weight to be synonymous. If you don't live on Earth, my apologies, and take me to your leader.

In the English system, weight is most often measured in ounces, pounds, and tons. In the metric system, we know that the base unit for weight is the gram, but this is a very small unit—a dollar bill weighs about a gram, for example. So a more commonly used unit is the kilogram, which is 1,000 grams. There is also a metric equivalent of the ton: A metric ton, sometimes written "tonne," is 1,000 kg. The basic conversions for weight, along with the symbols used to represent units, are shown in the box at the top of the next page.

Units of Weight

English Units	Metric Units	English/Metric Conversions
16 ounces (oz) = 1 pound (lb)	1 kg = 1,000 g	1 kg ≈ 2.2 lb
2,000 lb = 1 ton (T)	1,000 kg = 1 metric ton (t)	1 oz ≈ 28 g

Actually, the first equivalence under Metric Units above isn't necessary, as we've already studied conversions within the metric system. We gave you that one for free. For converting metric units of weight, we'll go back to moving decimal places based on the prefix.

For reference, here's a reminder on the order of metric prefixes going from largest (1,000) down to smallest (one-thousandth): kilo, hecto, deka, base unit, deci, centi, milli

In Questions 1–3, convert each measurement.

1. 150 grams of protein powder to milligrams

2. 23 grams of salt to hectograms

3. 3.5 kilograms of chicken to pounds

Math Note

Recall that when converting metric units, we multiply or divide by powers of 10. When moving three steps, we multiply or divide by 10^3, or 1,000. When moving six steps, we multiply or divide by 10^6, or 1,000,000.

Did You Get It

Try these problems to see if you understand the concepts we just studied. The answers can be found at the end of the portfolio section.

1. One of the weight classes in Olympic boxing is 79 kg. How heavy is that in grams? Pounds?
2. An average golden retriever puppy weighs 12½ ounces at birth. How many grams does it weigh?

Conversions for Measures of Temperature

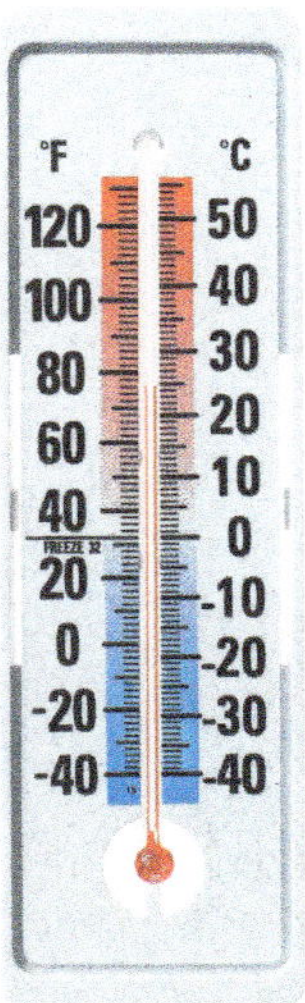

Ken Cavanagh/ McGraw Hill

Temperature is measured in the English system using degrees **Fahrenheit.** On this scale, water freezes at 32° and boils at 212°. The average temperature of the human body is 98.6°. In the metric system, temperatures are measured in degrees **Celsius** (also known as **centigrade**). Like all metric measurements, the Celsius scale is based on powers of 10 in some form: Water freezes at 0° and boils at 100°. The average temperature of the human body is 37°.

4. How can you tell from the information in the paragraph above that a change of one degree on the Celsius scale is NOT the same as a change of one degree on the Fahrenheit scale?

5. Compare the difference in temperature Fahrenheit between freezing and boiling water to the difference in temperature Celsius by forming a ratio (and reducing). This will tell you how many degrees Fahrenheit corresponds to one degree Celsius.

Using what you learned in Question 5, it's not too difficult to develop formulas that convert between the two measures of temperature.

Fahrenheit-Celsius Conversions

Converting from Celsius to Fahrenheit:

$$F = \frac{9}{5}C + 32$$

Converting from Fahrenheit to Celsius

$$C = \frac{5}{9}(F - 32)$$

6. In preparing for vacation, Randy and Catalina check the Internet and find that the average temperature over the next week at their destination is predicted to be 28° Celsius. Should they pack coats or bathing suits?

Math Note

It would be nice if there were two separate symbols for degrees in the two systems. It would also be nice if I had a Ferrari. Alas, neither of those things was in the cards. With temperatures, you just have to pay attention to context.

7. A frozen pizza needs a 450° Fahrenheit oven to cook. Suppose you bought a cheap oven on eBay, and found that it had been scavenged from Romania, so the temperatures on the dial are Celsius. At what temperature should you set the oven?

Did You Get It

3. The average high temperature for March in Moose Jaw, Saskatchewan, is −1° Celsius, which sounds dreadful at first if you're not a fan of the cold. Is it? Find the temperature on the Fahrenheit scale, then comment on how cool a name Moose Jaw, Saskatchewan, is.
4. The pizza box from Question 7 recommends 225° Fahrenheit for rewarming. Where should the dial be set for rewarming?

8. According to the Guinness Book of World Records, the largest healthy baby ever delivered weighed 360 ounces. How much did he weigh in pounds? How does this weight compare to the average weight of a newborn boy, which is 7.5 lb?

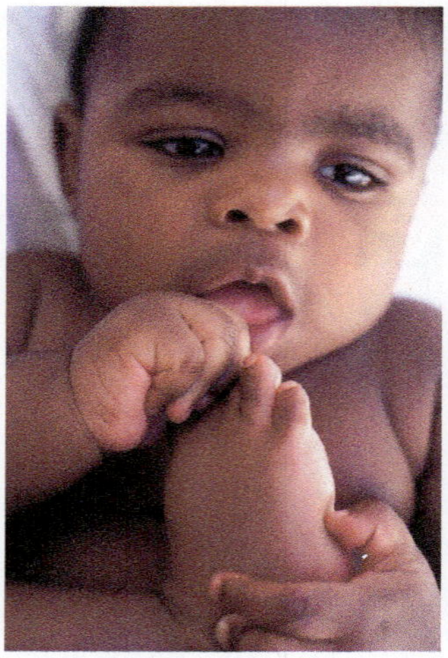
Arthur Tilley/Getty Images

9. Suppose the child was a little older and weighed 50 lb when his parents took him to the doctor, who read his temperature as 40°C. Does he have a fever? If so, how many degrees Fahrenheit above normal body temperature is the poor little guy? (Little being a relative term in this case.) Recall that 98.6°F is considered the normal body temperature for humans.

10. If he has a fever, the recommended dose is 15 mg of acetaminophen per kg of body weight. If there are 160 mg of acetaminophen in 5 mL of children's Tylenol, how many mL of children's Tylenol should the child receive?

11. If the entire bottle of liquid children's Tylenol contains 4 fluid ounces, how many mg of acetaminophen are in the bottle? How many grams? How many kilograms?

Did You Get It

5. How many ounces is a 12-pound bowling ball?
6. The temperature in a bowling alley is important for consistent lane conditions. At 74°F, what is the equivalent temperature in Celsius?

12. An empty Boeing 747 weighs about 175 tons. How much does it weigh in ounces?

David R. Frazier Photolibrary/Alamy

13. If this same 747 burns about 10 tons of fuel per hour when at cruising altitude, calculate the number of gallons per minute this jet consumes (assuming jet fuel weighs about 6.8 lb per gallon).

14. At $4.55 per gallon for jet fuel, how much will fuel cost for a 4-hour flight?

15. A 747 can carry up to 467 passengers. How much would each have to pay just to cover fuel costs for the flight?

16. Jayden decides that he absolutely will not go to Florida for spring break unless he loses 15 pounds in the next 3 months. He finds an advertisement online for a British weight loss system that guarantees a loss of 3 kg per month. If that claim is true, will he reach his goal?

Did You Get It

7. An average male African elephant weighs about 176,000 ounces. How many tons does he weigh?
8. Bart's football coach tells him that he'll start next year if he gains 10 pounds in the 14 weeks between spring practice and summer drills. He hires a personal trainer who promises him a gain of 1 kg every 3 weeks. Will he make his goal?

8-3 Portfolio

Name ______________________________

Check each box when you've completed the task. Remember that your instructor will want you to turn in the portfolio pages you create.

Technology

1. ☐ In this Excel activity, we'll convert a list of temperatures from Celsius to Fahrenheit and create a bar graph.

ALEKS Practice

1. ☐ Include any written work from the ALEKS assignment along with any notes or questions about this lesson's content.

Applications

1. ☐ Complete the applications problems.

Reflections

Type a short answer to each question.

1. ☐ Explain the difference between mass and weight. In most applications, weight and mass are treated as if they mean the same thing. Why is it reasonable to do that?
2. ☐ Even though the base unit of mass/weight in the metric system is the gram, kilograms are used far more commonly. Why is that?
3. ☐ Measuring temperatures in metric units is different from the other measurements we studied. How?
4. ☐ Name one thing you learned or discovered in this lesson that you found particularly interesting.
5. ☐ What questions do you have about this lesson?

Looking Ahead

1. ☐ Complete the prep skills for Lesson 8-4.
2. ☐ Read the opening paragraph in Lesson 8-4 carefully and answer Question 0 in preparation for that lesson.

Answers to "Did You Get It?"

1. 79,000 g, or about 173.8 lb **2.** About 350 g **3.** 30.2°F
4. About 110°C (107.2°C is exact.) **5.** 192 oz **6.** About 23.3°C
7. 5½ T **8.** Yes, by about 0.3 lb

Answers to Prep Skills Questions

1. 4,824 in. **2.** About 43.7 in. **3.** About 6.44 km
4. 0.843 dg **5.** 0.94 hm **6.** 43.12 daL
7. 135 **8.** About −25.3

8-3 Applications

Name ______________________________

In Questions 1–4, which would be the most appropriate unit of weight: gram, kilogram, or metric ton?

1. A cruise ship

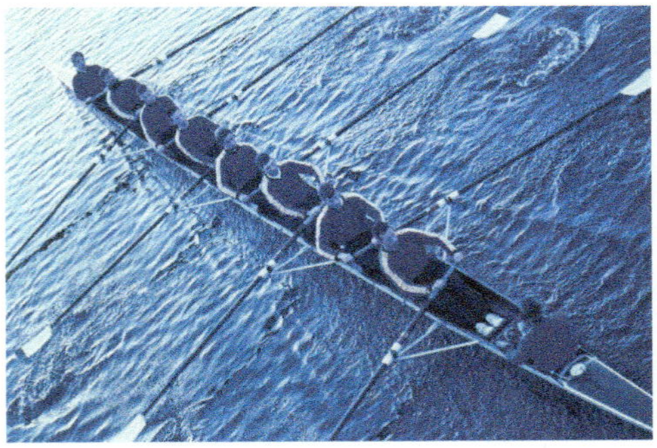

Purestock/SuperStock

2. The eight-man rowing crew at Harvard

3. A crew-neck sweater

4. Tom Cruise

5. If the temperature is 26°C, are you more likely to be in Seattle or San Antonio?

6. If the temperature is 9°C, are you more likely to be in Louisville or Anchorage?

7. If you step into a bath at 90° F, what will be the result?

8. If you step into a bath at 90°C, what will be the result?

9. Blaine weighs 112 kg. Is he more likely to be a football player or a golfer?

10. Daisuke weighs 50 kg. Is he more likely to be a school bus driver or one of the kids riding the bus?

8-3 Applications

Name ______________________________

11. My favorite recipe (spaghetti pie) calls for 12 ounces of ground sausage. Since it's awesome, everybody wants some, so I decided to make five pies and pass them out to the select few in my inner circle. The sausage comes in 1-pound tubes. How many tubes did I need, and how many grams of delicious sausage were left over for my omelet the next morning? (Feel free to email me for the recipe. Seriously dude, it's awesome.)

12. After giving birth to her first child, Tricia was informed that she weighed 111 ounces. Her first thought was "Holy smokes—no wonder it felt like I was carrying around a bowling ball." The average newborn girl in the U.S. weighs about 7 pounds, 2 ounces, and an average bowling ball weighs 12 pounds. Find the baby's weight in pounds and rate Tricia's first instinct for accuracy.

13. A large construction crane is rated to lift weights up to 16 metric tons. The design for a new building in downtown Houston calls for four large chillers to be placed on the roof, each of which weighs 8,200 pounds. Can the crane handle them all at once?

14. Before heading to Switzerland for a skiing vacation, Jamel checks a weather report, which predicts temperatures from $-3°C$ to $5°C$ while he and his family are there. Worried that he's going to almost freeze to death and have to get rescued by one of those St. Bernards with a barrel of wine around its neck, he checks a temperature conversion app. What does he find out about the Fahrenheit temperatures?

8-3 Applications

Name ______________________________

15. The temperature started out at 39°F at 7:30 this morning, and rose steadily at the rate of 3.2°F per hour until noon. What was the temperature in Celsius at noon, and what was the rate of increase in degrees Celsius per hour?

16. A pharmacist receives a shipment of two-thirds of a kilogram of a prescription powder used to treat early-stage flu. She plans to divide it into 90 bottles for distribution to customers, and the manufacturer recommends 6,000 mg per customer. Does she have enough?

Lesson 8-4 Prep Skills

SKILL 1: FIND THE AREA OF A CIRCLE

The area, A, of a circle is calculated using the formula $A = \pi r^2$, where r is the radius of the circle.

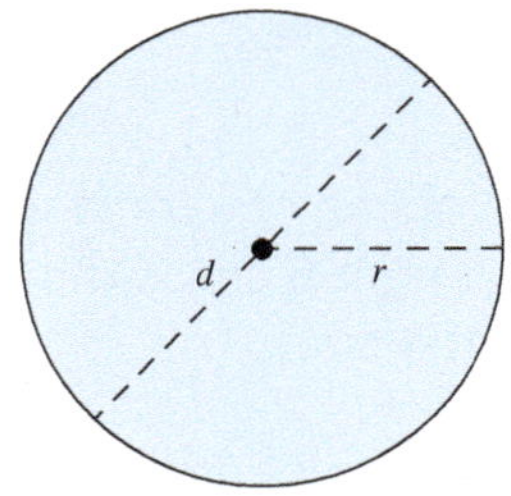

- Find the area of a circle with diameter 12 cm.

$A = \pi(6 \text{ cm})^2$

$A \approx 113.1 \text{ cm}^2$

If the diameter is 12 cm, the radius is 6 cm.

SKILL 2: FIND THE CIRCUMFERENCE OF A CIRCLE

The circumference, C, of a circle is the length around the outside of the circle. It's calculated using the formula $C = \pi d$, where d is the diameter of the circle or $C = 2\pi r$, where r is the radius of the circle. Note that these two formulas are really the same; the diameter of a circle is twice the radius, and if we replace d in the first formula with $2r$, we get the second formula.

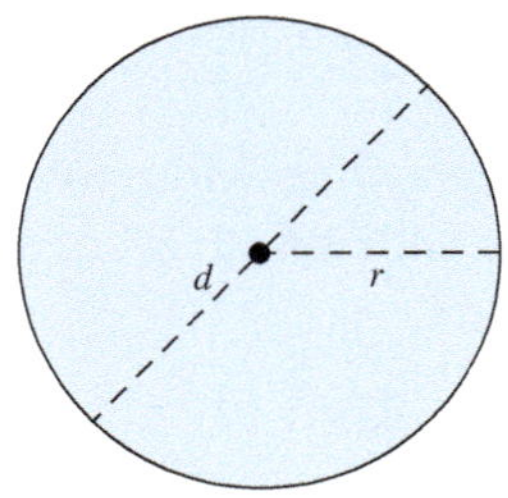

- Find the circumference of a circle with diameter 12 cm.

$C = \pi(12 \text{ cm})$

$C \approx 37.7 \text{ cm}$

Since we are given the diameter, we use the formula $C = \pi d$.

SKILL 3: FIND THE AREA OF A RECTANGLE

The area, A, of a rectangle is calculated using the formula $A = lw$, where l is the length of the rectangle and w is the width of the rectangle.

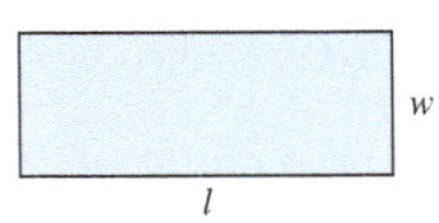

- Find the area of a rectangle with length 8 cm and width 5 cm.

$A = (8 \text{ cm})(5 \text{ cm})$

$A = 40 \text{ cm}^2$

Use the formula $A = lw$ with $l = 8$ cm and $w = 5$ cm.

SKILL 4: FIND THE VOLUME OF A RECTANGULAR BOX

The volume, V, of a rectangular box is calculated using the formula $V = lwh$, where l is the length of the rectangle and w is the width of the rectangle.

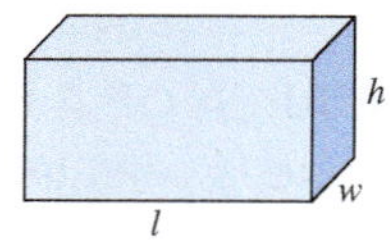

- Find the volume of a rectangular box with length 8 cm, width 5 cm, and height 4 cm.

$V = (8\text{ cm})(5\text{ cm})(4\text{ cm})$ — Use the formula $V = lwh$ with $l = 8$ cm, $w = 5$ cm, and $h = 4$ cm.

$A = 160\text{ cm}^3$

SKILL 5: WRITE AND INTERPRET A RATIO

Often written as a fraction, a ratio is a way of comparing the relative size of one quantity compared to another. If the wingspan on a model airplane is 2 ft and the wingspan on the full-sized version of that same plane is 28 ft, then a ratio formed by dividing the wingspan of the model by that of the full-sized plane tells us the scale used to make the model airplane. Writing this ratio as $\frac{2\text{ ft}}{28\text{ ft}}$, which can be reduced to $\frac{1}{14}$, tells us that the model plane is 1/14th the size of the full-sized airplane.

- Find the length-to-width ratio of a rectangle with length 8 cm and width 5 cm.

$\frac{8\text{ cm}}{5\text{ cm}} = 1.6$ — The length is 1.6 times as long as the width.

- Find the width-to-length ratio of a rectangle with length 8 cm and width 5 cm.

$\frac{5\text{ cm}}{8\text{ cm}} = 0.625$ — The width is 0.625 times as long as the length. Interpreting the decimal as a percent, we'd say that the width is 62.5% of the length.

PREP SKILLS QUESTIONS

1. Find the area of a circle with radius 4.5 cm.
2. Find the area of a circle with diameter 11 in.
3. What is the circumference of the circle in Question 1?
4. Find the area of a rectangle with length 4 in. and width 2 ft.
5. A rectangular box has the rectangle in Question 4 as its base, and its height is 16 in. What is its volume?
6. Write and simplify the ratio of 48 in. to 8 in. What does the result tell you about the relationship between those two measurements?

Lesson 8-4 Stocking the Shelves

LEARNING OBJECTIVES

- ☐ 1. Develop surface area and volume formulas for rectangular solids and cylinders.
- ☐ 2. Study the volume-to-surface-area ratio for different product packages and identify its significance.
- ☐ 3. Develop methods for deciding on optimal size and shape given certain goals.

Andrew Resek/McGraw Hill

Sometimes life gives us lessons sent in ridiculous packaging.
—Dar Williams

When you're at the grocery store looking at a stocked shelf, what makes you choose one product over another? In an ideal world that choice would be based on the quality of the actual product. But in this world, far too many people don't put that much thought into the quality of the product; buying choices are often made based on advertising or the appearance of the container that products are packaged in. Human psychology is a strange thing, and even the shape of packaging can have a subtle effect on how likely you are to notice a given item. When it comes to food packaging, volume has to be the first consideration: Packages have to be designed to hold a certain amount of product. But another important consideration is the cost of materials, so it's in the manufacturer's best interest to limit the amount of metal, plastic, or cardboard used in packaging. As we'll see in this lesson, however, efficiency is often sacrificed in the interest of aesthetics because keeping costs down is good, but if nobody buys your product it doesn't really matter what your costs look like—you're out of business, pal.

0. After reading the opening paragraph, what do you think the main topic of this lesson will be?

8-4 Class

Surface Area and Volume of a Rectangular Box

David Sobecki

We'll begin our mathematical study of food packaging with my all-time favorite breakfast cereal, Count Chocula, which is not as good as it was when I was a kid, but that's probably because it was like 93% sugar back then. It's still pretty awesome, though. Anyhow, the goal is to study the efficiency of this particular cereal's packaging.

1. How many sides does a cereal box have?

The dimensions of the family-size box of Count Chocula I bought today are 23.7 cm by 6 cm by 31.5 cm.

2. Assign variable names l, w, and h to represent these quantities. Think about it, and try to assign variable names to each dimension that make the most sense. Looking at the picture of the box will help.

3. Find the area of the bottom of this box.

4. Is there another surface with the same area? Which one?

5. Write a formula for the area of the bottom in terms of l, w, and/or h.

6. Find the area of the front of the box (the part with our friend the Count on it).

7. Is there another surface with the same area? Which one?

8. Write a formula for the area of the front in terms of l, w, and/or h.

9. Find the area of the side of the box (the part that usually has the nutrition label).

10. Is there another surface with the same area? Which one?

11. Write a formula for the area of one side in terms of l, w, and/or h.

12. At this point, you should be able to use all of the stuff you've thought about up until now to write a formula for the combined area of all of the sides of the box. This is known as the **surface area** of a figure. Write your formula below in the colored box.

The Surface Area of a Rectangular Box

For a rectangular box with base length l, base width w, and height h, the surface area is

$A =$ ____________________

13. Use your formula to find the surface area of the Count Chocula box.

14. The way we compute the volume of a box is pretty straightforward ($V = l \cdot w \cdot h$). You were asked about it in the prep assignment. What is the volume of this box of yummy treats?

15. Find the ratio of the volume to that of the surface area. Try to think about what this value represents. You'll discuss it more soon.

Did You Get It

Try this problem to see if you understand the concepts we just studied. The answer can be found at the end of the portfolio section.

1. Find the volume, surface area, and ratio of volume to surface area for a shipping container that measures 8 ft by 6 ft by 12 ft.

Surface Area and Volume of a Cylinder

Solved Example

Finding the Surface Area of Parts of a Cylinder

A cylinder, like the soup can shown in the photo, is made up of three pieces: a circular top and bottom, and a rolled-up rectangle for the sides. The can in the photo has a radius of 3.6 cm (which is half of the distance across the top) and a height of 11.3 cm.

a. Find the area of the top and bottom.

b. Find the area of the sides.

SOLUTION

a. The top and bottom are each circles with radius 3.6 cm. From geometry (and the prep skills for this lesson), we know that a circle with radius r has area πr^2. So the area of the circular top and bottom is $\pi(3.6 \text{ cm})^2 \approx 40.7 \text{ cm}^2$.

b. To find the area of the sides, we'll have to be clever. Even though it appears round, it's really just a rectangle: Picture removing the top and bottom and unrolling the sides. And, of course, the area of a rectangle is length times width. The tricky part is figuring out what those dimensions are, based on information we were provided about the cylinder.

One of the dimensions is definitely the height of the can, which is given as 11.3 cm. The other dimension takes some visualization. If you picture unrolling the sides into a rectangle, the length of the top part of the rectangle is exactly the full length around the circular top and bottom. This is the circumference of those circles, which is 2π times the radius. In this case, we get a length of $2\pi(3.6) \approx 22.6$ cm. Now we multiply that length by the height of 11.3 cm to get an area of approximately 255.4 cm^2.

16. Use the results of the Solved Example to find the surface area of the entire soup can.

David Sobecki

17. Write a formula for the area of the side of the can if we use r to describe the radius and h to describe the height.

18. Use what you've learned to write a formula for the surface area of a cylinder on the blank line in the purple box.

The Surface Area of a Cylinder

For a cylinder with radius r and height h, the surface area is

$A =$ ____________________

19. Use your formula to find the surface area of the soup can, and make sure it matches your answer from Question 16. There may be some minor difference due to rounding error.

20. The volume of a can is found by first calculating the area of the bottom of the can and multiplying by the height of the can. $V =$ ________ h

21. Can the volume of the box of cereal be calculated by multiplying the area of the base times the height? What does that tell you in general about volume?

22. Back to the soup can. Find the volume using the formula from Question 20, then the ratio of the volume to that of the surface area. Try to think about what this value represents. You'll discuss it more soon.

23. Before moving on to the experimental phase of this lesson, fill in one of the blanks below with "surface area," and one with "volume."

The ________________ of a container is a measure of how much food it can hold.

The ____________________ of a container is a measure of how much material is needed to build the container.

Did You Get It ?

2. A standard soda can is 4.83 in. high and has a radius of 1.3 in. Find the volume, surface area, and volume-to-surface-area ratio.

8-4 Group

LAB: Exploring Efficiency in Food Packaging

Supplies needed:

1. Have students and instructors bring in a wide assortment of rectangular boxes and metal cans of food. Better still, plan on donating the food to a local food pantry when you're done.
2. Rulers with centimeter measurements.

1. Directions:

a. Carefully measure each package, recording all dimensions shown on the table below.
b. Find the surface area and the volume of each.
c. Calculate the volume-to-surface-area ratio for each product according to type (cans or boxes).
d. Calculate the ratio of length to width for the bases of all boxes.

Rectangular Boxes						
Length	Width	Height	Volume	Surface Area	V to SA Ratio	L to W Ratio

Cylindrical Cans				
Radius	Height	Volume	Surface Area	V to SA Ratio

2. Look at your answers to Question 23 at the end of the Class portion of this lesson. Use them to explain why the ratio of volume to surface area is a measure of how efficient the containers are in terms of cost of production.

3. Which of the containers is most efficient in terms of construction cost? Which is the least?

4. Are cylinders more or less efficient than rectangular boxes on average? Explain.

5. Look at the boxes that ranked higher in terms of efficiency. What can you conclude about comparing their height to the size of the base?

6. Look at the cans that ranked higher in terms of efficiency. What can you conclude about comparing their height to the size of the base?

Now we're going to do some math modeling to study maximum efficiency. We'll begin with cylinders, as that's a bit more straightforward. Using the formulas you developed in this lesson and some pretty darn fancy algebra, I was able to develop a formula that describes the surface area for a cylinder with a given volume V and given height h:

$$A = \frac{2V}{h} + 2\sqrt{\pi V h}$$

For maximum efficiency in terms of material cost, we would like to find a height that makes the surface area as small as possible when we know the volume. Finding the height that does that is a classic problem in calculus, which this course is not. We can, however, do the next best thing: Graph the function on a graphing calculator and find the lowest point on the graph.

7. The volume of the soup can we studied was 460.1 cm^3. Plug this value into the surface area formula to find a function describing the surface area of a can with that volume. Your formula should have only one variable.

8. Graph your answer to Question 7 on a graphing calculator. Set the viewing window so that Xmin = 0 and Xmax is at least 20. Then find the lowest point on the graph, either by using trace or by using the built-in minimum command. If you're not familiar with that command, the Using Technology box below will help you out. Write both coordinates of the lowest point.

Using Technology: Using the Minimum Command on a Graphing Calculator

Step 1: With the function you want to find a minimum on graphed, press [2nd] [TRACE] to get to the Calculate menu. Then choose 3:minimum.

Step 2: At this point, the calculator will give you this prompt: Left Bound? It's asking you for an *x* value that is definitely to the left of the minimum you're looking for. You can either use the arrow keys to move the cursor to the left of the minimum, or key in a number that is on the screen but left of the *x* value at the minimum. Then hit [ENTER].

Step 3: Now the calculator will prompt you to enter a right bound. Either move the cursor to the right of the minimum, or key in an *x* value that is on the screen but right of the *x* value at the minimum. Then hit [ENTER].

Step 4: The next prompt the calculator provides is Guess?. Either move the cursor close to the minimum, or key in an *x* value that is close to the minimum and again hit [ENTER]. The calculator will then provide both coordinates of the minimum point.

The reason for all of the prompts is that if there's more than one low point on the visible part of the graph, the calculator needs to know which one you want to find. If the graph clearly shows only one minimum (which it should in Question 8) you can just key in Xmin and Xmax for the left and right bounds, and halfway between them as your guess.

See the Using Tech Video on Using the Minimum Command on a Graphing Calculator for further information.

9. One of those coordinates is the smallest surface area you could possibly have for a can with that volume. What is that minimal surface area?

10. Find the ratio of volume-to-surface area in that case. How does it compare to what we calculated for the actual can? What can you conclude?

11. The other coordinate of the minimum point you found can be used to find the dimensions of the most efficient can. (You'll need to use the volume as well.) Find those dimensions. How do they compare to the Chunky soup can?

12. Look back at your table on page 872. Did any of the cans have a higher ratio than the ideal can with volume 460.1 cm^3? If so, what do you think that says about finding the ideal can?

Now let's model efficiency for the rectangular boxes. This is a little more involved because there are three variable dimensions, as opposed to two for the cylinder. To construct a model, we have to choose a specific ratio of length to width for the bottom of the box, which is why you were asked to find those ratios. If we use R to represent the ratio of length to width (where we always take "length" to mean "the longer of those two dimensions"), then this function describes the surface area of a rectangular box in terms of the volume (V) and the width (w):

$$A = \frac{2V(1+R)}{Rw} + 2Rw^2 \qquad \text{(Trust me, I'm a doctor.)}$$

13. The Count Chocula box we studied earlier had a volume of 4,479.3 cm^3 and a length-to-width ratio of 3.95. Use that information to write a function describing its surface area in terms of the width of the base w.

14. Graph your answer to Question 13 on a graphing calculator. Set the viewing window so that Xmin = 0 and Xmax is at least 20. Then find the lowest point on the graph, either by using trace or by using the built-in minimum command. Write both coordinates of the lowest point.

15. One of those coordinates is the smallest surface area you could possibly have for a box with that volume. What is that minimal surface area?

16. Find the ratio of volume to surface area in that case. How does it compare to what we calculated for the actual box? What can you conclude?

17. What would the dimensions of the most efficient box be? How do they compare to the Count Chocula box?

18. Based on what you've learned, do you think the ratio for the most efficient box will change depending on the volume of the box? You'll be checking out some other boxes in the Applications; at this point, we're just asking for your best guess.

19. Given that in each case, we found that the packaging was efficient, but not perfectly efficient, what are some other factors that you think go into deciding on the dimensions for packaging? List as many different ideas as you can think of.

8-4 Portfolio

Name ______________________________

Check each box when you've completed the task. Remember that your instructor will want you to turn in the portfolio pages you create.

Technology

1. ☐ In this Excel activity, we'll calculate volume, surface area, and measurement ratios for various food packages.

ALEKS Practice

1. ☐ Include any written work from the ALEKS assignment along with any notes or questions about this lesson's content.

Applications

1. ☐ Complete the applications problems.

Reflections

Type a short answer to each question.

1. ☐ Describe what the surface area of an object is.
2. ☐ Why is the volume-to-surface-area ratio useful in studying the efficiency of packaging?
3. ☐ What did you learn about food packaging from this lesson?
4. ☐ Take another look at your answer to Question 0 at the beginning of this lesson. Would you revise your answer now that you've completed the lesson? How would you summarize the topic of this lesson now?
5. ☐ What questions do you have about this lesson?

Looking Ahead

1. ☐ Complete the prep skills for Lesson 9-1.
2. ☐ Read the opening paragraph in Lesson 9-1 carefully and answer Question 0 in preparation for that lesson.

Answers to "Did You Get It?"

1. Volume: 576 ft^3; surface area: 432 ft^2; ratio: 1.33
2. Volume: about 25.6 $in.^3$; surface area: about 50.1 $in.^2$; ratio: 0.51

Answers to Prep Skills Questions

1. About 63.6 cm^2
2. About 95.0 $in.^2$
3. About 28.3 cm
4. 96 $in.^2$
5. 1,536 $in.^3$
6. $\frac{48\text{ in.}}{8\text{ in.}} = 6$; The first measurement is six times as big.

8-4 Applications

Name ______________________________

Pick one of the rectangular boxes you measured in class, and write all of the relevant information about it below. You'll be copying this information from the table on page 872.

Length:

Width:

Volume:

Surface area:

Volume-to-surface-area ratio:

Length-to-width ratio:

1. Recall that the function describing the surface area of a rectangular box for a fixed volume in terms of the width of the base w is

$$A = \frac{2V(1 + R)}{Rw} + 2Rw^2$$

where V is the volume and R is the ratio of length to width. For the box above, use the values of V and R to write the surface area function.

2. Use a graphing calculator to draw the graph, then find the lowest point on the graph and record the coordinates here.

3. Describe what each of the coordinates of that point mean.

4. Find the volume-to-surface-area ratio for the most efficient box with that volume and discuss how efficient you think the box you measured is.

8-4 Applications

Name ______________________________

5. How does the volume-to-surface-area ratio for the ideal box of that volume compare to the ideal ratio for a box with a different volume we found in Question 16 of the Group portion? What do you think that comparison means?

Next, record below the information for one of the cans you measured. Again, you're just copying from the table at the beginning of the Group portion.

Radius:

Height:

Volume:

Surface area:

Volume-to-surface-area ratio:

Length-to-width ratio:

6. Recall that the function describing the surface area of a cylindrical can for a fixed volume in terms of the height h is

$$A = \frac{2V}{h} + 2\sqrt{\pi V h}$$

For the can above, use the values of V and h to write the surface area function.

7. Use a graphing calculator to draw the graph, then find the lowest point on the graph and record the coordinates here.

8. Describe what each of the coordinates of that point mean.

8-4 Applications

Name ______________________________

9. Find the volume-to-surface-area ratio for the most efficient can with that volume and discuss how efficient you think the can you measured is.

10. How does the volume-to-surface-area ratio for the ideal can of that volume compare to the ideal ratio for a can with a different volume we found in Question 10 of the Group portion? What do you think that comparison means?

11. Look at all of the data in the tables on page 872, with a focus on the volume-to-surface-area ratio column. Try to draw some general conclusions on some things that make containers more efficient in terms of cost of construction.

Math in Seeing the World REVISITED

clintscholz/iStock/Getty Images

1. The Fahrenheit equivalents are 75.2° and 89.6°, so beach wear is the way to go.
2. Thirty kilograms is about 66 pounds, so you can pack 62 pounds of clothes and personal items.
3. Walking to downtown shouldn't be a problem—0.6 kilometers is about 0.37 miles. The distance from the airport is about 2.6 miles, which will take just about 4½ minutes at 35 miles per hour.
4. You used 16 gallons, which is about 60.4 liters. At $1.43 per liter, the refill costs $86.37.
5. It will cost $207.27.

Index

D

E

Q

R

S

T

U

V

W

Y

Z

ALEKS®

ALEKS is a research-based, online learning program that offers course products for Math and Chemistry. Rooted in over 25 years of research and analytics, ALEKS helps educators understand each student's knowledge and learning progress in depth, and provides the individual support required for every student to achieve mastery.

ISBN 978-1-266-21451-6
MHID 1-266-21451-8

EAN

9 781266 214516 90000

mheducation.com/highered